Lecture Notes in Electrical Engineering

Volume 356

AF734357

Board of Series editors

Leopoldo Angrisani, Napoli, Italy
Marco Arteaga, Coyoacán, México
Samarjit Chakraborty, München, Germany
Jiming Chen, Hangzhou, P.R. China
Tan Kay Chen, Singapore, Singapore
Rüdiger Dillmann, Karlsruhe, Germany
Haibin Duan, Beijing, China
Gianluigi Ferrari, Parma, Italy
Manuel Ferre, Madrid, Spain
Sandra Hirche, München, Germany
Faryar Jabbari, Irvine, USA
Janusz Kacprzyk, Warsaw, Poland
Alaa Khamis, New Cairo City, Egypt
Torsten Kroeger, Stanford, USA
Tan Cher Ming, Singapore, Singapore
Wolfgang Minker, Ulm, Germany
Pradeep Misra, Dayton, USA
Sebastian Möller, Berlin, Germany
Subhas Mukhopadyay, Palmerston, New Zealand
Cun-Zheng Ning, Tempe, USA
Toyoaki Nishida, Sakyo-ku, Japan
Bijaya Ketan Panigrahi, New Delhi, India
Federica Pascucci, Roma, Italy
Tariq Samad, Minneapolis, USA
Gan Woon Seng, Nanyang Avenue, Singapore
Germano Veiga, Porto, Portugal
Haitao Wu, Beijing, China
Junjie James Zhang, Charlotte, USA

About this Series

"Lecture Notes in Electrical Engineering (LNEE)" is a book series which reports the latest research and developments in Electrical Engineering, namely:

- Communication, Networks, and Information Theory
- Computer Engineering
- Signal, Image, Speech and Information Processing
- Circuits and Systems
- Bioengineering

LNEE publishes authored monographs and contributed volumes which present cutting edge research information as well as new perspectives on classical fields, while maintaining Springer's high standards of academic excellence. Also considered for publication are lecture materials, proceedings, and other related materials of exceptionally high quality and interest. The subject matter should be original and timely, reporting the latest research and developments in all areas of electrical engineering.

The audience for the books in LNEE consists of advanced level students, researchers, and industry professionals working at the forefront of their fields. Much like Springer's other Lecture Notes series, LNEE will be distributed through Springer's print and electronic publishing channels.

More information about this series at http://www.springer.com/series/7818

Shengzhao Long · Balbir S. Dhillon
Editors

Proceedings of the 15th International Conference on Man–Machine–Environment System Engineering

Springer

Editors
Shengzhao Long
Astronaut Research and Training
Center of China
Beijing
China

Balbir S. Dhillon
University of Ottawa
Ottawa, ON
Canada

ISSN 1876-1100 ISSN 1876-1119 (electronic)
Lecture Notes in Electrical Engineering
ISBN 978-3-662-48223-0 ISBN 978-3-662-48224-7 (eBook)
DOI 10.1007/978-3-662-48224-7

Library of Congress Control Number: 2015947101

Springer Heidelberg New York Dordrecht London
© Springer-Verlag Berlin Heidelberg 2015
This work is subject to copyright. All rights are reserved by the Publisher, whether the whole or part of the material is concerned, specifically the rights of translation, reprinting, reuse of illustrations, recitation, broadcasting, reproduction on microfilms or in any other physical way, and transmission or information storage and retrieval, electronic adaptation, computer software, or by similar or dissimilar methodology now known or hereafter developed.
The use of general descriptive names, registered names, trademarks, service marks, etc. in this publication does not imply, even in the absence of a specific statement, that such names are exempt from the relevant protective laws and regulations and therefore free for general use.
The publisher, the authors and the editors are safe to assume that the advice and information in this book are believed to be true and accurate at the date of publication. Neither the publisher nor the authors or the editors give a warranty, express or implied, with respect to the material contained herein or for any errors or omissions that may have been made.

Printed on acid-free paper

Springer-Verlag GmbH Berlin Heidelberg is part of Springer Science+Business Media
(www.springer.com)

Xuesen Qian' Estimation

Grandness Scientist Xuesen Qian' Sky-high Estimation for the Man–Machine–Environment System Engineering

龙升照同志：

我收到您主编的《人机环境系统工程研究进展（第一卷）》，翻看了之后，感到非常高兴，1985年秋提出的一个想法，现在8年之后已赫然成书，500多页的巨卷！而且研究范围已大大超出原来航天，内容涉及航空、航天、航海、兵器、电子、能源、交通、电力、煤炭、冶金、体育、康复、管理……等领域！你们是在社会主义中国开创了这门重要现代科学技术！

此致

敬礼！

钱学森

1993.10.22

Xuesen Qian' Letter

Grandness Scientist Xuesen Qian' Congratulatory Letter to the 20th Anniversary Commemorative Conference of Man–Machine–Environment System Engineering Foundation

龙升照同志:

你的来信已收到。欣悉人-机-环境系统工程创立 20 周年纪念大会暨第五届全国人-机-环境系统工程学术会议即将召开，我向你们表示最热烈的祝贺!

20 年来，你们在人-机-环境系统工程这一新兴科学领域进行了积极的开拓和探索，并取得了非常可喜的成绩，我感到由衷的高兴。

希望你们今后再接再励，大力推动人-机-环境系统工程理论及应用的蓬勃发展，为中国乃至世界科学技术的进步作出积极贡献!

祝

工作顺利!

钱学森

2001年6月26日

Preface

In 1981, under the direction of the great scientist Xuesen Qian, an integrated frontier science—Man–Machine–Environment System Engineering (MMESE)—came into being in China. Xuesen Qian gave high praise to this emerging science. In a letter to Shengzhao Long, he pointed out, "**You are creating this very important modern science and technology in China!**" on October 22, 1993.

In the congratulation letter to the commemoration meeting of the twentieth anniversary of establishing the Man–Machine–Environment System Engineering, the great scientist Xuesen Qian stated, "You have made active development and exploration in this new emerging science of MMESE, and obtained encouraging achievements. I am sincerely pleased and hope you can do even more to make prosper development in the theory and application of MMESE, and **make positive contribution to the progress of science and technology in China, and even in the whole world**" on June 26, 2001.

October 22, which is the day the great scientist Xuesen Qian gave high praise to MMESE, was determined to be the Foundation Commemoration Day of MMESE by the second conference of the 5th MMESE Committee on October 22, 2010. On this very special day, the great scientist Xuesen Qian pointed out in a letter to Shengzhao Long, "**You are creating this very important modern science and technology in China!**" and the conference also determined that the annual conference on MMESE would be held from October 21–23 to cherish the memory of the great contributions that the great scientist Xuesen Qian had made to the MMESE!

The 15th International Conference on MMESE will be held in Hangzhou, China on October 21–23 of this year; hence, we will dedicate *Proceedings of the 15th International Conference on Man–Machine–Environment System Engineering* to our readers.

Proceedings of the 15th International Conference on Man–Machine–Environment System Engineering is the academic showcase of the 15th International Conference on MMESE joint held by MMESE Committee of China and Beijing KeCui Academe of MMESE in Guilin, China. The conference proceedings consisted of 85 excellent papers selected from more than 400 papers. Due to limitations of space, some excellent papers have been left out, and we feel deeply sorry for that. Crudeness in contents and possible incorrectness are inevitable due to the somewhat pressing editing time and we hope you kindly point them out promptly, and your valuable comments and suggestions are also welcome.

Proceedings of the 15th International Conference on Man–Machine–Environment System Engineering will be published by Springer-Verlag, German. Springer-Verlag is also responsible for the related matters on index of Index to EI, so that the world can know the research quality and development trend of MMESE theory and application. Therefore, the publication of *Proceedings of the 15th International Conference on Man–Machine–Environment System Engineering* will greatly promote the vigorous development of MMESE in the world, and realize the grand object of "**making positive contribution to the progress of science and technology in China, and even in the whole world**" proposed by Xuesen Qian.

We would like to express our sincere thanks to Springer-Verlag, German for their full support and help during the publishing process.

Beijing
July 2015

Prof. Shengzhao Long

Program and Technical Committee Information

General Chairman

Professor Shengzhao Long, Astronaut Research and Training Center of China

Program Committee Chairman

Professor Balbir S. Dhillon, University of Ottawa, Canada

Technical Committee Chairman

Professor Enrong Mao, College of Engineering, China Agricultural University, China

Program and Technical Committee Members

Professor Yanping Chen, University of Management and Technology, USA
Professor Hongfeng Gao, University of California, USA
Professor Birsen Donmez, University of Toronto, Canada
Professor Xiangshi Ren, Kochi University of Technology, Japan
Professor Kinhuat Low, Nanyang Technological University, Singapore
Senior Engineer Baoguo Luan, CH-AUTO Technology Co. Ltd, China
Professor Baoqing Xia, Weapon Industrial Hygiene Research Institute, China
Professor Chunhui Wang, Astronaut Research and Training Center of China, China
Professor Fang Jie, China North Vehicle Research Institute, China

Senior Engineer Fong Zhou, AVIC Aerospace Life-support Industries, Ltd, China
Professor Guangtao Ma, Shenyang Jianzhu University, China
Professor Guansheng Huang, Beijing Special Vehicle Institute, China
Professor Guohua Jiang, Astronaut Research and Training Center of China
Professor Guohua Shi, Engineering Academy of the Second Artillery, China
Professor Haiming Wu, Weapon Industrial Hygiene Research Institute, China
Professor Hao Li, Astronaut Research and Training Center of China, China
Professor Hongjun Xue, Northwestern Polytechnical University, China
Professor Honglei Li, Air Defense Forces Academy, China
Senior Engineer Jiajun Sun, Petrochina Changqing Oilfield Company, China
Professor Jian Zhang, Naval Medical Research Institute, China
Professor Junfeng Liu, Educational Technology Center of Dalian, China
Professor Lijing Wang, Beijing University of Aeronautics and Astronautics, China
Professor Long Ye, Beijing Jiaotong University, China
Professor Xiaochao Guo, Institute of Aviation Medicine, Air Force, China
Professor Xingwei Wang, Institute of Aviation Medicine, Air Force, China
Professor Ying Jin, Astronaut Research and Training Center of China, China
Professor Yonghua Tao, Naval Medical Research Institute, China
Professor Yongqing Hou, China Academy of Space Technology, China
Professor Yuhong Shen, Quartermaster Research Institute of the General Logistics Department of CPLA, China
Professor Zhibing Pang, Air Defense Forces Academy, China

Contents

Part II Research on the Machine Character

Biography of Editors in Chief

Prof. Shengzhao Long He is the founder of the Man–Machine–Environment System Engineering (MMESE), the Chairman of the Man–Machine–Environment System Engineering (MMESE) Committee of China, the Chairman of the Beijing KeCui Academy of Man–Machine–Environment System Engineering (MMESE), and the Former Director of Ergonomics Lab of Astronaut Research and Training Center of China. In October 1992, he is honored by the National Government Specific Allowance.

He graduated from the Shanghai Science and technology University in 1965, China. In 1981, directing under famous Scientist Xuesen Qian, founded MMESE theory. In 1982, he proposed and developed Human Fuzzy Control Model using fuzzy mathematics. From August 1986 to August 1987, he conducted research in Man–Machine System as a visiting scholar at the Tufts University, Massachusetts, U.S.A. In 1993, he organized Man–Machine–Environment System Engineering (MMESE) Committee of China. He published "Foundation of theory and application of Man–Machine–Environment System Engineering" (2004) and "Man–Machine–Environment System Engineering" (1987). He edited "Proceedings of the 1st–14th Conference on Man–Machine–Environment System Engineering" (1993–2014). e-mail: shzhlong@sina.com

Dr. Balbir S. Dhillon He is a professor of Engineering Management in the Department of Mechanical Engineering at the University of Ottawa, Canada. He has served as a Chairman/Director of Mechanical Engineering Department/Engineering Management Program for over 10 years at the same institution. He has published over 345 (i.e., 201 journal + 144 conference proceedings) articles on reliability, safety, engineering management, etc. He is or has been on the editorial boards of nine international scientific journals. In addition, Dr. Dhillon has written 34 books on various aspects of reliability, design, safety, quality, and engineering management published by Wiley (1981), Van Nostrand (1982), Butterworth (1983), Marcel Dekker (1984), Pergamon (1986), etc. His books are being used in over 85 countries and many of them are translated into languages such as German, Russian, and Chinese. He has served as General

Chairman of two international conferences on reliability and quality control held in Los Angeles and Paris in 1987.

Professor Dhillon has served as a consultant to various organizations and bodies and has many years of experience in the industrial sector. At the University of Ottawa, he has been teaching reliability, quality, engineering management, design, and related areas for over 29 years and he has also lectured in over 50 countries, including keynote addresses at various international scientific conferences held in North America, Europe, Asia, and Africa. In March 2004, Dr. Dhillon was a distinguished speaker at the Conference/Workshop on Surgical Errors (sponsored by White House Health and Safety Committee and Pentagon), held at the Capitol Hill (One Constitution Avenue, Washington, D.C.).

Professor Dhillon attended the University of Wales where he received a BS in electrical and electronic engineering and an MS in mechanical engineering. He received a Ph.D. in industrial engineering from the University of Windsor. e-mail: dhillon@genie.uottawa.ca

Part I
Research on the Man Character

Chapter 1
Experimental Study of Muscle Fatigue in a Standing Work Posture

Hao Yu, Duming Wang, Liezhong Ge and Chuan Wang

Abstract Fatigue can directly affect a worker's health. Using experimental analysis, we investigated whether a standing chair can reduce fatigue and how to optimize the design of work conditions for standing staff. Ten college students were selected to participate in fatigue comparative experiments. We scored the subjects' feelings and evaluated the subjects' muscle fatigue using a surface electromyogram (EMG) of the subject's leg muscles. There was significant difference in fatigue between standing posture and sitting/leaning postures. There was no significant difference between a sitting posture and leaning posture. Surface EMG showed that a leaning posture had the lowest fatigue for the three postures. The appropriate selection of an auxiliary chair can assist workers in effectively reducing muscle fatigue, even more than an ordinary office chair.

Keywords Standing work · Auxiliary chair · Fatigue

Many jobs require workers to use a standing posture, such as sailors who must steer and monitor conditions from the ship's cockpit. Workers who must stand for long periods of time often experience muscle fatigue. In addition, work productivity can be greatly affected due to a dramatically increased physical load [1]. To reduce the fatigue of personnel and ensure efficiency, a chair may be provided to reduce the fatigue associated with a standing posture. However, in a sitting posture, the angles of vision are different than in a standing posture, so the selection of an appropriate seat can be a problem. For workers requiring a standing posture, a working auxiliary chair can be conveniently moved, folded, and stored as needed. The chair also can be easily installed and used in the operation. However, the available studies

H. Yu · C. Wang (✉)
Naval Medical Research Institute, Shanghai 200433, China
e-mail: hg04381@163.com

D. Wang · L. Ge
Zhejiang Sci-Tech University, Hangzhou 310012, China

C. Wang
Naval University of Engineering, Wuhan 430033, China

© Springer-Verlag Berlin Heidelberg 2015

S. Long and B.S. Dhillon (eds.), *Proceedings of the 15th International Conference on Man–Machine–Environment System Engineering*, Lecture Notes in Electrical Engineering 356, DOI 10.1007/978-3-662-48224-7_1

have not indicated whether this work auxiliary chair can alleviate fatigue or improve efficiency.

In this study, we used experiment analysis to determine whether a standing-style auxiliary work chair can reduce fatigue for personnel in standing operations. Our study was designed to determine whether a ship could be equipped with a working chair for special standing work posts.

1.1 Object and Methods

1.1.1 Test Subjects

Ten volunteers (graduate students at our university) were selected as test subjects. All participants agreed to the research protocol, which was approved by the Institutional Review Board of Zhejiang Sci-Tech University. The average age of the subjects was 24.2 ± 1.5 years. No subject reported being diagnosed with a musculoskeletal disease.

1.1.2 Experimental Equipment and Materials

All fatigue tests were performed in the laboratory of the product usability testing center of Zhejiang Sci-tech University. The applied instruments included a Finnish Mega6000 with four-channel surface electromyography and a sampling frequency of 1000 Hz. In the evaluation of local muscle fatigue, the most commonly used indexes of surface electromyogram (SEMG) signals were the time-domain index and area indexes. Three kinds of standing posture work auxiliary chairs (manufactured in Germany; Fig. 1.1) and an ordinary office chair (Fig. 1.2) were tested.

The Borg rate of perceived exertion (RPE) (CR-20) Scale is most commonly used fatigue self-assessment scale in workload evaluations (Fig. 1.3). The RPE uses a 15-point scale, with a minimum score of 6 and a maximum score of 20. The reason for such a design is as follows: When G. Borg performed the research for hoisting heavy loads, RPE values were found to have a strong relationship with heart rate, such that heart rate = 10 × RPE. This scale has been widely verified for validity with regard to the correlation of physiological workload to heart rate.

1.1.3 Experiment Design and Tasks

Our experiment used a single-factor tested inner design. The variables were divided into three kinds of working conditions: work performed in a sitting posture in an

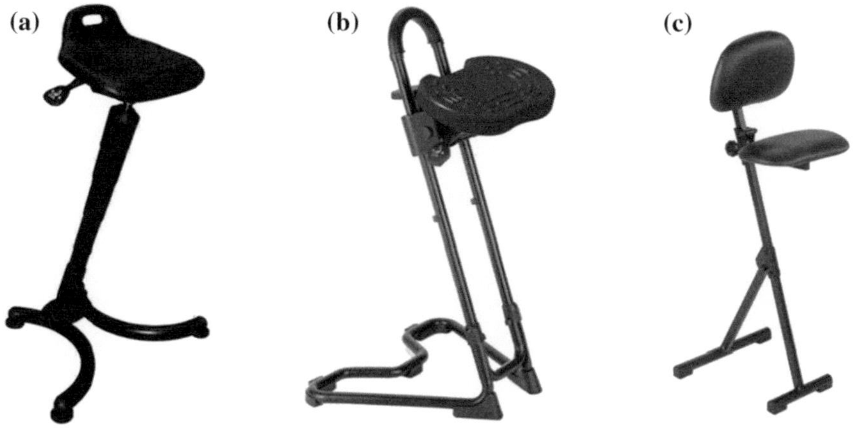

Fig. 1.1 Three working auxiliary chairs for a standing posture (Styles *A*, *B*, and *C*, respectively)

Fig. 1.2 Ordinary office chair

ordinary office chair, work performed in a standing posture work, and work performed in a leaning posture using a standing work auxiliary chair.

All participants underwent fatigue tests for the three working conditions, lasting 150 min for each condition.

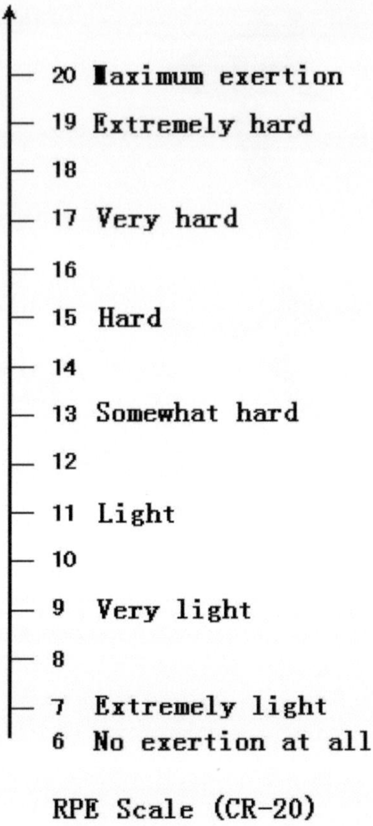

Fig. 1.3 Borg subjective rated scale

1.1.4 Statistical Analysis

The electromyogram (EMG) data was saved to the computer. The statistical analysis was performed on EMG data using SPSS11.0 statistical analysis software.

1.2 Results and Discussion

1.2.1 Evaluation of Seat

The fatigue experiment lasted 150 min for each condition. Therefore, it was necessary to select a suitable standing-system auxiliary working chair for the experiment. The evaluation flow is as follows: Users first dismount according to the requirements, then adjust the chair height and tilt angle according to their own needs. We then conducted the sitting experience, and finally performed the

Table 1.1 Seat evaluation

No.	Evaluation dimension	Description	Score
1	Structural simplicity	Is the seat structure design simple and clean?	
2	Ease of adjustment	Is it convenient to adjust the seat to sit face height, backrest tilt, etc.?	
3	Seating comfort	Is the seat comfortable when sitting (hips, lower back, etc.)?	
4	Easy storage capability	Is it easy to store the seat after use (easy storage, storage space)?	
5	Materials' skid resistance	What is the size of the coefficient of friction material between the seat and clothing (whether it is installed with anti-slip materials)?	
6	Overall evaluation		

evaluation and scoring according to Table 1.1. The evaluation used a seven-point scale, with a score of 3 being the best, –3 being the worst, and the middle value being score 0.

The subjective evaluation results for the three kinds of standing work auxiliary chairs are shown in Fig. 1.4.

Style A chair scores were ranked the lowest for each evaluation dimension, for which the average score was −0.8 in adjustment convenience (a negative

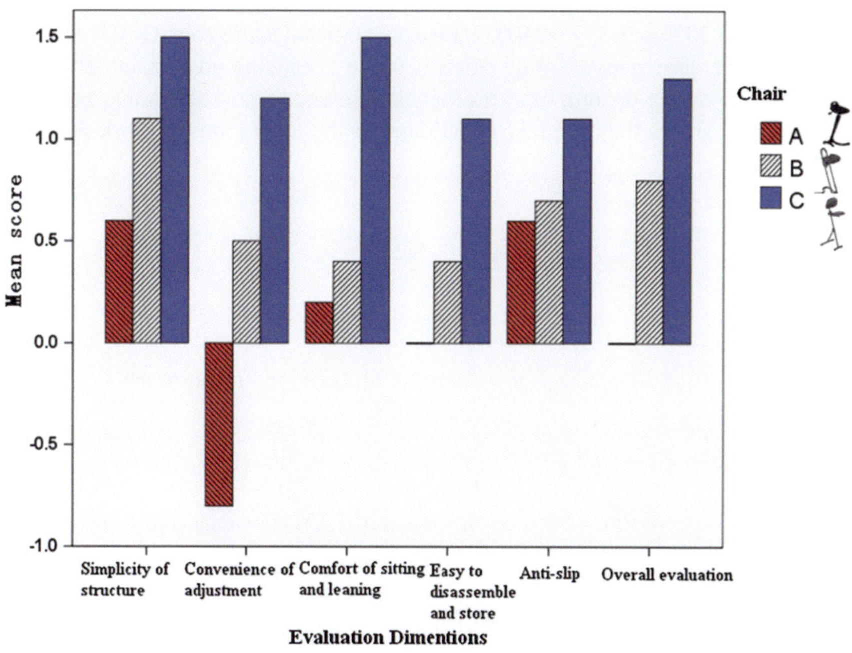

Fig. 1.4 Subjective results for the three standing working auxiliary chairs

evaluation). Style B chairs were ranked second for each dimension. Style C chair scores were the best for each subjective evaluation dimension. Therefore, the style C chair was chosen for the comparison between a standing-posture auxiliary work chair and a general office chair.

1.2.2 Contrast Experiment

The EMG tested the muscles that are directly related with the standing posture on one side of body. The electrode patches for measuring muscle were placed on the left erector spinae, the left tibialis anterior, lateral gastrocnemius, and the left side of the left femoral biceps. We aimed to eliminate the effects on experiment results that were caused by different data collection conditions. Every other 15 min, the subject took the standard standing posture for 1 min in order to collect EMG data. We used the BORG subjective scale to conduct subjective fatigue scoring at the same time. The self-assessed fatigue scores are shown in Fig. 1.5.

As shown in Fig. 1.5, there was no large difference in self-assessed fatigue between the three work postures during the first half of the experiment. However, as time progressed, a difference in the self-assessment for fatigue in the three kinds of work posture is shown in the second half of the experiment. These differences are mainly shown for fatigue in standing posture; the other two postures have little difference.

The results show an obvious discrepancy for fatigue in the three kinds of work postures ($F(2,287) = 3.8$, $P < 0.05$). Least-Significant Difference (LSD) analysis shows that the discrepancy mainly exists between a standing posture and other two postures. The sitting posture has no obvious difference from the leaning posture.

Figure 1.6 shows the original SEMG data for the three kinds of work postures.

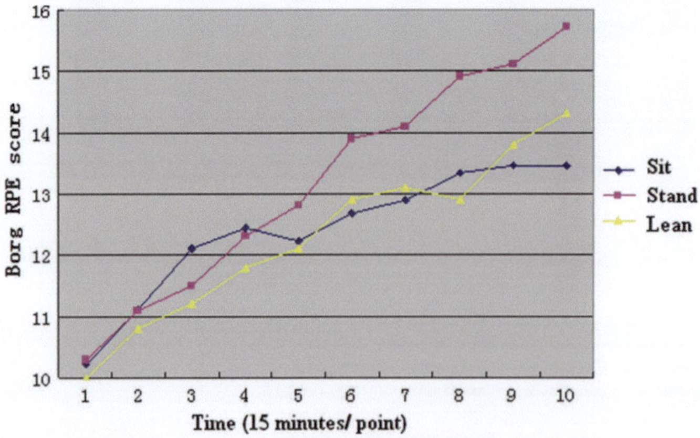

Fig. 1.5 Self-assessed working posture borg fatigue evaluation results

(a)

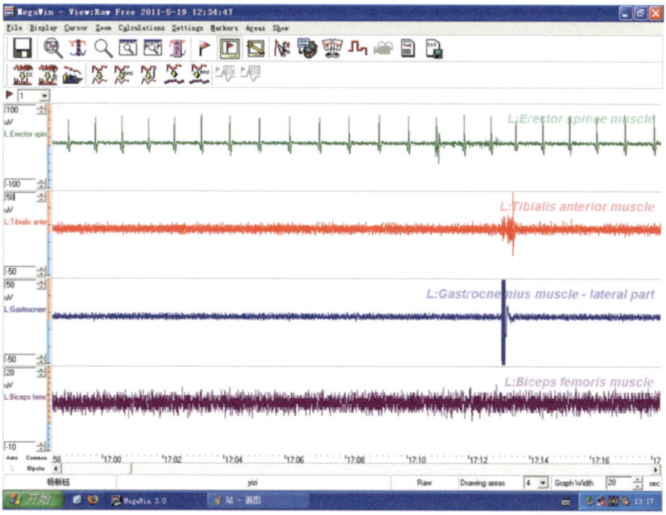

(b)

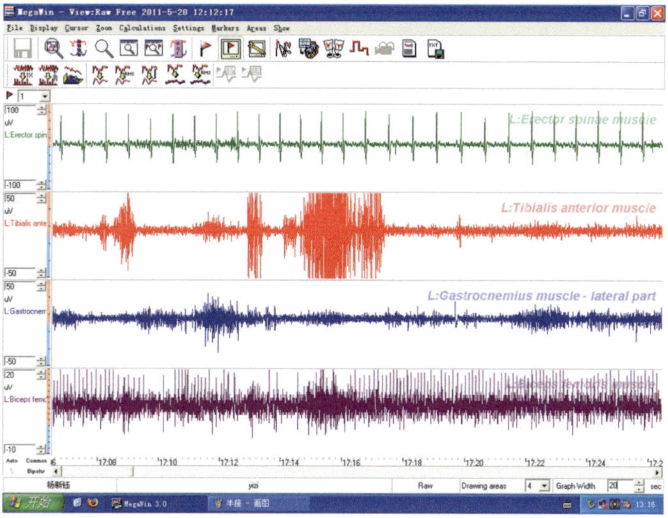

Fig. 1.6 EMG data for the three work postures. **a** Sitting posture. **b** Standing posture. **c** Leaning posture

Because it can be affected by the size of individual muscle, subcutaneous fat thickness, and other physiological and anatomical factors, SEMG signal amplitude has large differences between individuals. To ensure data comparability and accuracy and to eliminate the effect of individual differences, the SEMG data were standardized. The last-minute data index was taken as the standard, and data from each time point data were compared with the standard. This allowed us to obtain a

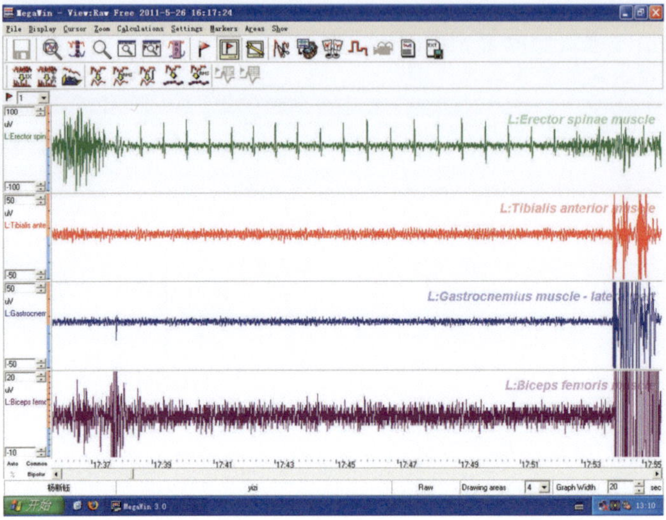

Fig. 1.6 (continued)

relative standard value (AEMG). The larger the AEMG value is, the higher is the degree of muscle fatigue (Fig. 1.7).

As it is seen from Fig. 1.7, the myoelectric signal AEMG index was lowest in the standing posture for three muscles (not the biceps femoris) when using the Style C standing-posture work auxiliary chair. The overall fatigue degree is also ranked lowest. These results indicate that the type C standing posture work auxiliary chair may be helpful in reducing the muscle fatigue of workers; its effect even exceeds the sitting posture work effect in the ordinary office chair.

The variance analyses of measurement were repeatedly performed on each muscle. The results show that the level of significance was not met for the difference between erector spinae and biceps femoris in the AEMG indicative posture ($F(2{,}198) = 0.88$, $P > 0.05$;$F(2{,}198) = 0.26$, $P > 0.05$); $F(2{,}198) = 0.26$, $P > 0.05$). The tibialis anterior muscle and gastrocnemius AEMG indicative posture differences were significant ($F(2{,}198) = 5.53$, $P < 0.05$; $F(2{,}198) = 10.38$, $P < 0.05$).

It is generally recognized that the integral electromyogram (IEMG) value is positively associated with the degree of fatigue in the process of moderate-intensity load-level muscle fatigue. The IEMG value would generally increase as the fatigue intensifies. The maturation-promoting factor (MPF) value would decrease as the fatigue intensifies (Fig. 1.8).

The variance analyses were performed with repeated measurements on each muscle. The results show that differences between the erector spinae, biceps femoris, and lateral gastrocnemius MPF indexes are not significant ($F(2{,}198) = 0.83$, $P > 0.05$; $F(2{,}198) = 0.98$, $P > 0.05$; $F(2{,}198) = 1.1$, $P > 0.05$).

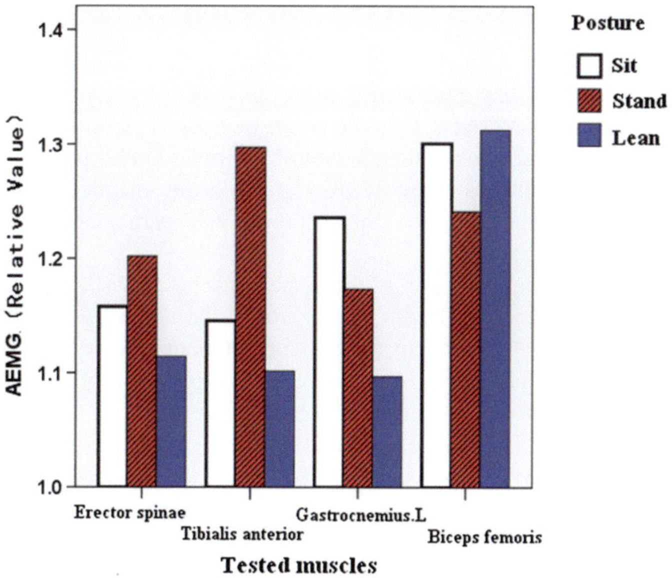

Fig. 1.7 AEMG normalized values throughout the four muscles

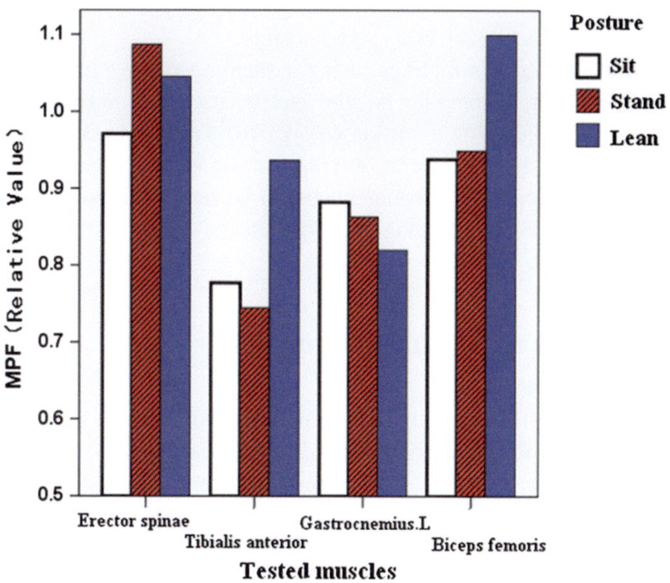

Fig. 1.8 MPF normalized values for the four muscles

1.3 Conclusion

The SEMG signal is a one-dimensional time signal of the neuromuscular system recorded from the skin surface. It is the result of various physical and chemical factors of the central nervous system's control information and peripheral muscle bioelectricity [2, 3]. Changes in the SEMG signal indicate chemical changes in the process of muscular fatigue; thus, this approach has particular importance for peripheral muscle fatigue [4].

A chair acts as an important part of the human–machine interface, and its design can directly affect a worker's health. For example, the incidence of tank driver's waist pain is up to 94.0 %, the main reason being an inappropriate human–machine interface design [5]. Maintaining one posture for a long period of time can lead to muscle and fascia tension for one side or both sides of the body, as well as fatigue or lumbago. A comfortable chair can be an effective measure to solve this problem. The existing studies are mostly focused on the comprehensive analysis and model evaluation of the human–machine interface, with little verification of design by experiment.

We evaluated the combined results of a subjective evaluation and objective SEMG signals. We found that a leaning work posture using the Style C standing work auxiliary chair resulted in higher surface muscle electric signals using an MPF index than other postures on average. The overall fatigue was the lowest for this condition [6, 7]. The experimental results show that the Style C chair is helpful in reducing a worker's muscle fatigue. Its effect is similar to or even partly exceeds that of an ordinary office chair in a sitting posture.

According to our study's findings, using a standing-posture auxiliary chair for job positions requiring a standing posture can be useful in reducing a worker's fatigue. It is a feasible solution to ensure an operator's effectiveness and continuous work capacity. The standing posture auxiliary chair is also required in a human engineering and application environment. It can be helpful to solve the problems associated with fatigue in a standing work posture.

1.4 Compliance with Ethical Standards

The study was approved by the Logistics Department for Civilian Ethics Committee of the Naval Medical Research Institute.

All subjects who participated in the experiment were provided with and signed an informed consent form.

All relevant ethical safeguards have been met with regard to subject protection.

References

1. Zhu Z (2003) Engineering psychology course, 1st ed. People's Education Press, Beijing, pp 130–171
2. Qi L, Mao E (2004) Comprehensive evaluation model tank driving seat static parameters. China Agric Univ 9(1):67–70
3. Wang Y (2007) Application of the sEMG technique on the muscular fatigue exercise. SiChuan Sports Sci 12(4):47–50
4. Chen S, Zhang L (2011) Emciency and sensitivity of assessment of muscle fatigue by utilizing sEMG parameters. J Wuhan Inst Phys Educ 45(12):71–77
5. Wang J (2004) Studies on the NOn-fatigue specificity of the fatigue-related sEMG signal parameters. Space Med Med Eng 17(1):39–43
6. Liu W, Yuan X, Zhang D (2003) Development of a comprehensive assessment of driving ergonomics of the cockpit bench. China Saf Sci 11(11):9–12
7. Song Z, Mao E, Zhou Y (2006) Study design optimization of mechanical systems model human-machine interface. Mech Des Manuf 2(2):35–37

Chapter 2
Customer Perceived Value as a Predictor of On-line Clothes Loyalty in a Chinese Sample

Xueyan Wei, Yahui Qi and Wenjie Liu

Abstract Research on the on-line shopping effects of positive on customer perceived value is limited, especially regarding its relation to the young generation's loyalty. This study examined the effect of the on-line customer perceived value (production perceived value, service perceived value, on-line shopping environment perceived value, and cost perceived value) on customer loyalty (attitude loyalty and behavior loyalty) in a Chinese sample of 280 young generation (e.g., post-80s and post-90s). Results showed that customer perceived value was associated with greater customer loyalty. Customer perceived value emerged as a significant predictor of customer loyalty. The service perceived value and product perceived value had the strongest relationship with loyal attitude. The product perceived value, service perceived value, and cost perceived value were the alertest predictors as customer behavior loyalty. Theoretical and practical implications for customer perceived value on on-line shopping are also discussed.

Keywords Customer perceived value · Customer loyalty · China · Young generation · On-line shopping

This research was founded by the National Science Foundation of China (71102094), foundation of Jiangsu educational plan program (D20131103) and by the fundamental research funds for the central universities (JUSRP1506XNC).

X. Wei (✉) · Y. Qi · W. Liu
College of Tianjiabing Education, Jiangnan University, No.1800#, LiHu Road, 214122 Wuxi, Jiangsu, People's Republic of China
e-mail: 8765330@qq.com

© Springer-Verlag Berlin Heidelberg 2015

S. Long and B.S. Dhillon (eds.), *Proceedings of the 15th International Conference on Man–Machine–Environment System Engineering*, Lecture Notes in Electrical Engineering 356, DOI 10.1007/978-3-662-48224-7_2

2.1 Introduction

In recent years, more and more customers are willing to use the Internet as one of main shopping channels, especially in developing country, such as China. Why do customers make full use of the Internet as a retailing platform? Research shows that, compared to traditional environments, e-shopping is considered to be both the more convenient access [1] as well as the more time-efficient mode to realize a purchase due to the minimal efforts. Gu [6] also found that e-product value, e-services value, e-cost value, and on-line environment value are the key perceive value in Chinese e-shopping process.

Value had been considered as important behavior predictors [4] and played an absolutely necessary role in persuading consumers to utilize the firm's offerings [10]. Customer perceived value (CPV) is defined as "the difference between all of the customer's expected benefits and offerings and the perceived choices from customer's evaluations" [7]. Previous empirical studies indicated that perceived value was an unidimensional and global structure of total customer value perceptions. Perceived value had been treated as multidimensional and could be measured by perceived value using various getting (benefits) and giving (sacrifices) dimension too [4]. "Getting and giving" could be considered as the attributes of conventional customer perceived value. According to Kotler [7], customer perceived value (CPV) is viewed as the difference between total customer value and total customer cost. Total customer value contains four factors: product value, services value, personnel value, and image value.

Cheng et al. [3] found that the value contents described above obtained considerable explanatory power. That explained the relevance of value is a decisive factor of the intention to utilize the Internet as an e-shopping environment. The study is endeavoring to identify e-customer perceived value, attempting to clarify the dimension' function and gain insight into the customer purchase process.

- Hypothesis 1: Product perceived value, service perceived value, cost perceived value, and on-line shopping environment perceived value would be significantly related to attitude loyalty and behavior loyalty.

  More recent TAM research indicates that "perception of usefulness" and "ease of use" of technology can be helpful to determine "intended usage" [9]. Chu and Cheng [5] found that customer value is the structure and customer loyalty is the implementation. The two coexist for a mutual goal of building a close relationship with customers. It is essential to identify culture-specific features or behavioral responses to customer loyalty so as to provide relevant training to employees in firms that operate in different cultures [2]. Chinese culture insists on the practical and utilitarianism such as fast, easy, and time-efficient [6].
- Hypothesis 2: Product perceived value, service product value, and on-line shopping environment value will positively predict attitude loyalty and behavior loyalty.
- Hypothesis 3: Cost perceived value will negatively predict attitude loyalty and behavior loyalty.

2.2 Method

2.2.1 Participants

Research participants were 339 college students (post-80 and post-90) from National University and Occupational University in mainland China. Data were collected from some key university in Shanghai, Beijing, Wuxi and Suzhou. Valid copies were recovered with an effective recovery rate of 82.6 %. For the sample, 36.8 % were male, with a mean age of 25.5 years of age. 70.8 % were college graduates, undergraduates, and Doctors; 29.2 % were Vocational and technical college students (3-year-course), and 81.8 % had 2–3 years of on-line shopping experience as on-line practitioners.

2.2.2 Measures

Subjects rated all questionnaires (CPVQ and CLQ). All measures were developed originally in English. Adapted Chinese version was examined and compared with the original one for equivalence and agreement [2, 6].

Customer Perceived Value—was derived from the Customer Perceived Value Questionnaire (CPVQ) and measured with the average of 25 items [7]. Each item was on a 5 point scale from 1(strongly disagree) to 5(strongly agree). High-internal consistency is supported by alpha coefficients of 0.73–0.83. The internal consistency reliability was 0.78 for all four subscales. Previous research has reported reliability of
$\alpha = 0.80$ [6].

Customer loyalty—Customer loyalty Questionnaire (CLQ)—derived from Parasuraman et al. [8] research, includes two dimensions with the average 6 items: attitude loyalty (3 items, $\alpha = 0.75$; "I would like to acclaim this shop as the one of the best on-line shops") and behavior loyalty (3 items, $\alpha = 0.70$; example item, "I will buy clothes on this on-line shop even though the price increases slightly"). The internal consistency reliability was 0.79 for all two subscales. Previous research has reported reliability of $\alpha = 0.85$ [2].

2.2.3 Procedure and Data Analysis

Data were analyzed using SPSS version 15.0. Descriptive statistics, Pearson correlation, and hierarchical multiple regression analysis were used in the study.

Table 2.1 Means, standard deviations, reliability estimates, and Pearson correlations (n = 280)

Variables	*M*	*SD*	Cronbach α	1	2	3	4	5	6
Sex									
Age education	25.5	5.9							
1. Product perceived value	18.59	4.13	0.73	1					
2. Service perceived value	26.63	4.75	0.78	0.41***	1				
3. Environment perceived value	17.04	3.84	0.76	0.34***	0.55***	1			
4. Cost perceived value	18.85	3.92	0.84	−0.17**	−0.11	0.01	1		
5. Attitude loyalty	9.19	2.32	0.75	0.38***	0.39***	0.30***	−0.15*	1	
6. Behavior loyalty	8.36	2.18	0.70	0.29***	0.28***	0.25***	−0.19**	0.52***	1

*$P < 0.05$, **$P < 0.01$, ***$P < 0.001$

2.3 Results

Correlations between scores on the four dimensions of customer perceived value and two dimensions of customer loyalty were significant (see Table 2.1).

Hierarchical regression results showed that the control variables explained a combined 6.9 % of variance for attitude loyalty and 7.3 % for behavior loyalty. Product perceive value and service perceive value combined 20.6 % of variance ($F = 36.05$; $p < 0.001$) for attitude loyalty. Product perceive value, service perceive value, and cost perceived value explained a combined 13.2 % of variance ($F = 26.29$; $p < 0.001$) for behavior loyalty (see Table 2.2).

2.4 Discussion

The results indicated three dimensions of customer perceived value were positively correlated with attitude loyalty and behavior loyalty. Consistent with previous research [2, 3, 6], Chinese young generation specially focused on the perceived cost value. If the customer found they do not need more energy and time, there will be extra satisfaction on the on-line shop. The attitude loyalty and behavior loyalty will be more obvious.

The robust linkage between product perceived value and behavior loyalty is well established. Product perceived value is the strongest predictor of behavior loyalty. Rational factors play an important role including product perceived value, service perceived value, and cost perceived value. On the other hand, fraud product is also unacceptable in China. Trust is the essential factor which builds reliable

Table 2.2 Hierarchical regression analysis for prediction of customer loyalty (n = 280)

Variables	Attitude loyalty				Behavior loyalty			
	β	p	R^2	ΔR^2	β	p	R^2	ΔR^2
Step1: Control variables								
Sex	0.06	<0.05			0.05	<0.05		
Age	0.33	0.004			0.32	0.004		
Education	0.00	0.46			0.01	94		
Control variables			6.9				7.3	
Step2: Product perceived value	0.279	<0.001	0.206	0.057	0.199	<0.001	0.086	0.086
Step3: Service perceived value	0.262	<0.001	0.149	149	0.156	<0.001	0.115	0.029
Step4: Cost perceived value	–	–	–	–	−0.133	<0.001	0.132	0.017
Step5: Environment value	–	–	–	–				
Full model								
F		36.05				26.29		
P		<0.001				<0.001		

*P < 0.05, **P < 0.01, ***P < 0.01

relationship between C2B. Because consumers can easily compare information and find other websites that provide similar products or services [11], once they found the purchase is reliable, they prefer to trust that on-line shop. That is also cost savings actually.

2.5 Compliance with Ethical Standards

The study had obtained approval from the Institute of psychology, Chinese Academy of Sciences Ethics Committee.

All subjects participated in the experiment had signed the informed consent.

All relevant ethical safeguards had been met in relation to subject protection.

References

1. Anderson RE, Srinivasan SC (2003) E-satisfaction and e-loyalty: a contingency framework. Psychol Mark 20(2):123–138
2. Cao L (2009) The key influencing factor on customer loyalty In Chinese C2C model. Chinese doctor theses, Tongji Universtiy
3. Cheng JM, Wang ES, Lin JY, Vivek S (2009) Why do customers utilize the internet as a retailing platform? A view from consumer perceived value. Asia Pac J Mark Logistics 21 (1):144–160

4. Chu CW, Lu HP (2007) Factors influencing online music purchase intention in Taiwan: an empirical study based on the value-intention framework. Internet Res 17(2):139–155
5. Chu KM, Cheng S (2009) The construction model of customer trust, perceived value and customer loyalty. J Am Acad Bus Cambridge 14(2):98–103
6. Gu PF (2010) The relationship between customer perceived value and customer loyalty of C2C. Chinese doctor theses, Zhejiang University
7. Kolter P, Keller KL (2006) Marketing management, 12th ed. Prentice Hall, New York
8. Parasuraman A, Valarie Z, Leonard B (1985) A conceptual model of service quality and its implications for future research. J Mark 41–50
9. Peters C, Amato CH, Hollenbeck CR (2007) An exploratory investigation of consumers perceptions of wireless advertising. J Advertising 36(4):129–145
10. Pura M (2005) Linking perceived value and loyalty in location-based mobile services. Manag Serv Qual 15(6):509–538
11. Terblanche NS, Boshoff C (2010) Quality, value, satisfaction and loyalty amongst race groups, a study of customers in the South African fast food industry. South Afr J Bus Manag 41(1):1–9

Chapter 3
The Anthropometric Data of Chinese Male Taikonaut Candidates with 175 cm Stature

Xiaochao Guo, Duanqin Xiong and Yanyan Wang

Abstract All of Chinese male taikonaut candidates came from China Air Force pilots aged 25–35 years. The stature is ranged from 160 to 172 cm in medical selection while the upper limit may be shifted from 172 to 175 cm. The maximum stature is becoming one of the main constraints on taikonaut product design. The anthropometric data of Chinese male pilots with 175 cm stature was re-analyzed in comparison with that of pilots with 172 cm stature in the present paper, based on the databank of GJB4856-2003. The mean, stddev, and the 5th, 50th, 95th percentiles for the subsample were reported in 67 fundamental items according to GJB4856-2003 in statistics. The effects of bigger stature on production design were discussed.

Keywords Anthropometry · Taikonaut candidates · Male pilots · Stature · Product design

3.1 Introduction

The anthropometric data is fundamental to product design such as cockpit geometrical layout, eject seat and exit, personal protective equipment for pilots, and taikonaut. There may be different anthropometric characteristics with different user population. For example, the required stature (height) for NASA astronauts is 160–190 cm, but the stature should be 160–172 cm for Chinese male taikonaut candidates [1]. It is said that the upper limit of the taikonaut stature may be shifted from 172 to 175 cm in medical selection [2], and product design for them will adapt to change in the user population in the future.

In China, all of taikonaut candidates came from Air Force pilots aged 25 to 35 according to GJB4016-2000 [3]. The data of the latest anthropometric investigation

X. Guo (✉) · D. Xiong · Y. Wang
Institute of Aviation Medicine PLAAF, Beijing 100142, China
e-mail: guoxiaochao1965@163.com

© Springer-Verlag Berlin Heidelberg 2015

S. Long and B.S. Dhillon (eds.), *Proceedings of the 15th International Conference on Man–Machine–Environment System Engineering*, Lecture Notes in Electrical Engineering 356, DOI 10.1007/978-3-662-48224-7_3

in large scale for Chinese male pilots were presented in GJB4856-2003 [4], but the anthropometric characteristics of user population with 175 cm stature were not described. The anthropometric data of male taikonaut candidates of stature 175 cm was re-analyzed in comparison with those of 172 cm in the present paper based on the databank of GJB4856-2003, and the influences of change in the user population on production design were discussed.

3.2 Methods

3.2.1 Measurements

There were 67 fundamental items of anthropometry chosen from GJB4856. All data were collected by direct measurement with instrument precision of 1.0 mm.

3.2.2 Samples

There were 106 male pilots in Group 1 aged 25–35 in years with 172 cm stature, and 65 taikonaut candidates in Group 2 with 25–35 year-age but 175 cm stature from GJB4856-2003 databank. The averages of weight (body mass) were 67.72 ± 6.22 and 69.21 ± 7.20 kg for the groups.

Group 3 had all 107 male pilots with 175 cm stature in GJB4856.

3.3 Results

The data of 67 fundamental items in GJB4856 are given in Table 3.1. The Mean ($\bar{x}$), Std. deviation(s), and the 5th, 50th, 95th percentiles (P_5, P_{50}, P_{95}) are given.

3.4 Discussions

3.4.1 Two Samples with Different Stature

In GJB4856, the anthropometric items on head segment were coded in 1.1–1.53, the items with standing posture were coded in 2.1–2.162, the measurements were taken while subject sits were coded in 3.1–3.39, and the items on hand and foot segment were coded in 4.1–4.50.

Table 3.1 The data of 67 fundamental items for two groups of male taikonaut candidates (mm)

No	Code in GJB4856	Items	Group 1		Group 2					
			$\bar{x}$	s	$\bar{x}$	s	P_5	P_{50}	P_{95}	
1	1.1	Maximum head length	187.4	6.3	187.2	6.0	175.6	187.0	197.7	
2	1.2	Maximum head breadth	158.3	5.2	159.0	5.7	149.3	159.0	168.7	
3	1.3	Total head height	232.9	10.5	236.0	11.5	213.3	236.0	255.0	
4	1.35	Bitragion breadth	146.0	4.8	146.2	4.7	140.0	145.0	154.0	
5	1.40	Head circumference	572.9	12.9	574.3	13.5	553.3	575.0	597.1	
6	1.44	Sagittal arc	342.9	13.8	345.9	13.7	320.3	345.0	372.0	
7	2.1	Stature	1719.8	2.9	1749.1	2.9	1745.0	1749.0	1754.0	①
8	2.3	Eye height II	1604.0	9.6	1633.1	10.5	1615.3	1633.0	1651.0	①
9	2.4	Gnathion height	1486.9	11.4	1513.1	12.0	1490.6	1513.0	1536.7	①
10	2.10	Crotch height	793.4	22.5	807.3	18.7	773.1	805.0	839.0	①
11	2.11	Cervicale height	1468.9	12.0	1494.1	9.6	1476.2	1495.0	1510.4	①
12	2.15	Shoulder height	1398.7	14.4	1420.2	13.9	1397.3	1421.0	1440.0	①
13	2.22	Waist height	1042.9	15.5	1062.3	14.4	1043.3	1060.0	1088.7	①
14	2.25	Iliospinale height	938.0	19.9	957.9	20.8	919.1	958.0	987.0	①
15	2.26	Trochanter height	856.1	20.3	873.7	21.1	833.3	873.0	904.4	①
16	2.29	Radial height	1082.4	14.6	1095.3	16.9	1066.2	1092.0	1124.8	①
17	2.30	Radial stylion height	846.5	16.4	855.8	19.1	826.3	857.0	892.2	②
18	2.34	Middle finger tip height I	664.1	17.6	670.8	20.2	637.0	671.0	708.5	③
19	2.35	Middle finger tip height II (over head)	2156.6	28.8	2193.8	30.2	2153.6	2193.0	2245.9	①
20	2.39	Tibial height	440.0	12.5	448.2	12.1	423.5	447.0	468.1	①
21	2.43	Medial malleolus height	72.6	5.5	73.8	5.5	64.3	74.0	84.1	
22	2.52	Shoulder breadth	392.6	14.7	396.2	12.0	375.3	395.0	419.1	
23	2.61	Chest breadth	306.2	16.8	305.5	17.9	279.3	303.0	334.5	
24	2.65	Crista iliaca breadth	285.3	18.2	284.8	19.3	256.3	282.0	332.0	
25	2.67	Hip breadth	326.9	13.5	329.6	14.7	307.3	328.0	359.8	
26	2.77	Chest depth	225.4	14.2	226.2	15.5	203.3	226.0	255.2	
27	2.88	Length of upper extremity	734.6	18.7	749.4	20.0	716.4	750.0	782.1	①
28	2.90	Upperarm length	316.3	11.6	324.9	15.6	292.6	323.0	346.4	①
29	2.91	Forearm length	235.8	11.2	239.5	10.7	222.0	237.0	257.7	③
30	2.93	Thigh length	498.0	16.6	509.7	19.0	474.1	511.0	541.4	①
31	2.94	Leg length	367.4	12.9	374.4	13.2	352.6	374.0	398.4	②
32	2.97	Crotch length	696.2	40.0	701.9	41.2	640.9	700.0	777.7	
33	2.99	Dorsal length	441.0	17.0	446.5	14.2	420.6	449.0	469.1	③
34	2.102	Acromion-neck root length	138.8	10.4	139.7	10.7	124.3	140.0	159.1	
35	2.134	Neck root circumference	450.1	21.3	454.3	18.9	425.3	452.0	489.0	
36	2.135	Neck circumference	366.0	13.9	364.7	14.6	337.2	363.0	393.8	
37	2.139	Chest circumference I	923.1	43.7	923.6	49.2	845.1	917.0	1009.7	
38	2.141	Waist point circumference	792.3	65.2	782.0	60.6	696.5	777.0	908.4	
39	2.142	Waist circumference	807.2	62.0	798.0	61.8	715.8	796.0	921.4	

(continued)

Table 3.1 (continued)

No	Code in GJB4856	Items	Group 1		Group 2					
			$\bar{x}$	s	$\bar{x}$	s	P_5	P_{50}	P_{95}	
40	2.144	Hip circumference	932.7	38.9	938.2	42.0	880.6	935.0	1030.6	
41	2.148	Biceps circumference	286.8	19.3	289.1	22.3	256.5	285.0	330.7	
42	2.153	Forearm circumference	264.9	13.2	267.4	14.8	245.3	265.0	290.7	
43	2.155	Wrist circumference	166.0	6.0	167.0	7.3	157.0	166.0	178.7	
44	2.157	Thigh circumference	550.4	30.2	554.0	37.5	500.0	557.0	624.1	
45	2.161	Calf circumference	374.4	20.0	374.3	19.7	342.9	372.0	413.2	
46	2.162	Ankle circumference	220.9	10.2	221.0	9.9	206.0	220.0	241.5	
47	3.1	Sitting height	931.4	17.1	945.1	15.6	914.3	947.0	969.7	①
48	3.3	Eye height II, sitting	815.6	17.5	829.1	17.6	798.3	831.0	852.7	①
49	3.4	Shoulder height, sitting	610.4	18.1	616.2	19.6	579.4	617.0	645.1	③
50	3.8	Elbow height, sitting	262.4	22.4	262.6	22.6	221.5	263.0	301.7	
51	3.11	Thigh clearance height, sitting	154.7	12.2	153.7	12.8	135.3	152.0	179.2	
52	3.12	Knee height, sitting	506.8	12.2	515.4	12.1	497.0	514.0	534.7	①
53	3.13	Popliteal height, sitting	421.7	11.8	429.1	11.6	409.3	428.0	449.4	①
54	3.14	Acromial point-elbow distance	348.0	12.7	353.6	14.0	331.9	354.0	375.0	②
55	3.22	Buttock-knee length, sitting	569.4	13.6	578.8	13.2	557.6	578.0	601.7	①
56	3.23	Sitting depth	463.4	12.4	472.6	14.1	446.3	473.0	494.4	①
57	3.28	Elbow-elbow breadth, sitting	427.9	27.4	427.9	26.6	382.9	431.0	479.7	
58	3.29	Hip breadth, sitting	343.9	15.2	347.1	15.6	325.3	347.0	379.7	
59	4.1	Hand length	184.8	6.0	187.6	6.0	176.3	187.0	197.0	②
60	4.5	FINGER II length	70.5	3.3	72.2	3.2	67.3	72.0	77.7	②
61	4.10	Hand breadth at metacarpal	85.3	3.8	86.1	3.5	81.0	86.0	93.7	
62	4.13	Proximal knuckle II breadth	19.4	1.0	19.8	1.1	18.0	20.0	22.0	
63	4.14	Distal knuckle II breadth	16.6	1.1	17.1	1.1	15.3	17.0	19.0	③
64	4.21	Hand depth	30.8	1.9	31.3	1.7	28.3	31.0	34.0	
65	4.24	Hand girth I	205.7	8.5	206.8	8.5	193.3	207.0	222.1	
66	4.39	Foot length	253.6	6.9	257.9	7.2	244.3	258.0	267.7	①
67	4.43	Foot breadth	98.8	4.2	98.9	4.1	92.3	99.0	106.7	

① $P < 0.001$, ② $P < 0.01$, ③ $P < 0.05$ in this paper

172 cm stature covered the range of 1715–1724 mm in Group 1 and 175 cm stature ranged from 1745 to 1754 mm in Group 2 with reference to age stratification method in Annex E of GB/T 22187-2008/ISO 15535:2003 [5].

The percentiles for stature group in 172 cm were 56.20–64.30 % and the percentiles for 175 cm stature group were 77.60–83.00 % in 1739 cases of GJB4856 databank. This means that user population will extend from 64.12 to 82.86 % of total Chinese male pilot population in GJB4856-2003 when the upper limit of stature shifts from 172 to 175 cm for taikonaut selection.

3.4.2 The Anthropometric Characteristics of User Population with 175 cm Stature Relative to that of 172 cm Stature Group

3.4.2.1 Male Taikonaut Candidates with 175 cm and 172 cm Stature

For the measurements on head segment, no differences were found between Groups 2 and 1 in Table 3.1 for the six items in statistics.

For the measurements with standing posture, the differences were statistically significant between Groups 2 and 1 in 20 of 41 items (see Table 3.1). The items in height were larger in Group 2 with 175 cm stature than the same kind in Group 1 with 172 cm stature, except medial malleolus height coded as 2.43 in GJB4856. This meant that product design in height should enlarge the height of gate or exit may be adjusted from 172 to 175 cm approximately.

The greater the data of limb items is bigger in Table 3.1, the more the reachable area of pilots is expansive in Group 2. It is suggested that personal protective and life support equipment size may be different for male taikonaut candidates in Group 2 from that of Group 1 with the addition of dorsal length coded 2.99 according to GJB20A [6].

Space workstation design should also accommodate to standing operations for user population with 175 cm stature.

For measurements with sitting posture, there were 8 of 12 significantly different items between Groups 2 and 1 in Table 3.1. The design eye position (DEP) is a key point for workstation design such as aircraft with seated operations. If the upper limit of taikonaut stature shifts from 172 to 175 cm in medical selection, the DEP in sitting should be heightened approximately to 1.0 cm. Workspace and chair should be redesigned for target user population because of parameter changes based on anthropometric data such as (3.1) sitting height, (3.12) knee height, (3.23) sitting depth, and (3.13) popliteal height [7].

For the measurements on hand and foot segments, there were 4 of 9 measurements different between Group 2 and 1 in statistics. The differences may affect the design of personal protective equipment as well as controls operated by hands or feet.

3.4.2.2 Taikonaut Candidates to All Male Pilots with 175 Cm Stature

For the anthropometric data of all pilots of 175 cm stature, it seemed that only 2 of 67 items differed between Groups 3 and 2 in Table 3.2. The circumference items coded as 2.141 and 2.142 in GJB4856 were more in Group 3 than in Group 2 ($F = 4.32$ and 4.19, $\boldsymbol{P} < 0.05$). It suggested that Group 3 had approximately the same anthropometric characteristics as Group 2 on the one hand, and it was necessary to pay attention to the overweight trends with age [8] on the other hand, because there were more elder pilots (30.85 %) aged 36–53 in Group 3. The design

Table 3.2 The data of 67 fundamental items for all pilots with 175 cm stature in GJB4856 (mm)

No	Code in GJB4856	Items	Group 2		Group 3					
			$\bar{x}$	s	$\bar{x}$	s	P_5	P_{50}	P_{95}	
1	1.1	Maximum head length	187.2	6.0	187.4	5.9	177.0	187.0	197.6	
2	1.2	Maximum head breadth	159.0	5.7	158.7	5.2	149.4	159.0	167.0	
3	1.3	Total head height	236.0	11.5	235.1	10.9	214.8	236.0	254.0	
4	1.35	Bitragion breadth	146.2	4.7	146.5	4.7	139.4	146.0	155.2	
5	1.40	Head circumference	574.3	13.5	574.4	13.3	552.4	575.0	595.0	
6	1.44	Sagittal arc	345.9	13.7	342.7	15.0	320.0	340.0	371.2	
7	2.1	Stature	1749.1	2.9	1749.3	2.9	1745.0	1749.0	1754.0	
8	2.3	Eye height II	1633.1	10.5	1634.3	10.1	1616.4	1634.0	1651.0	
9	2.4	Gnathion height	1513.1	12.0	1514.1	11.5	1494.4	1513.0	1536.6	
10	2.10	Crotch height	807.3	18.7	809.6	18.4	778.4	808.0	839.0	
11	2.11	Cervicale height	1494.1	9.6	1497.1	10.7	1479.4	1497.0	1516.4	
12	2.15	Shoulder height	1420.2	13.9	1423.2	14.0	1399.6	1423.0	1445.0	
13	2.22	Waist height	1062.3	14.4	1062.4	15.1	1043.0	1061.0	1090.8	
14	2.25	Iliospinale height	957.9	20.8	957.1	19.1	924.8	957.0	987.0	
15	2.26	Trochanter height	873.7	21.1	874.7	20.2	837.6	874.0	905.6	
16	2.29	Radial height	1095.3	16.9	1098.5	16.8	1069.8	1100.0	1126.0	
17	2.30	Radial stylion height	855.8	19.1	859.1	19.2	827.0	859.0	892.0	
18	2.34	Middle finger tip height I	670.8	20.2	673.4	19.8	641.6	672.0	710.6	
19	2.35	Middle finger tip height II (over head)	2193.8	30.2	2195.8	28.5	2158.2	2196.0	2240.2	
20	2.39	Tibial height	448.2	12.1	449.4	11.9	427.4	449.0	469.6	
21	2.43	Medial malleolus height	73.8	5.5	74.8	5.3	65.4	75.0	83.2	
22	2.52	Shoulder breadth	396.2	12.0	396.3	12.8	375.4	395.0	419.6	
23	2.61	Chest breadth	305.5	17.9	307.8	18.8	280.4	309.0	339.6	
24	2.65	Crista iliaca breadth	284.8	19.3	289.7	21.4	256.4	287.0	332.6	
25	2.67	Hip breadth	329.6	14.7	331.8	14.4	307.4	331.0	357.6	
26	2.77	Chest depth	226.2	15.5	231.0	16.8	204.0	231.0	264.6	
27	2.88	Length of upper extremity	749.4	20.0	749.9	18.8	717.8	750.0	781.8	
28	2.90	Upperarm length	324.9	15.6	324.8	14.3	301.8	326.0	346.2	
29	2.91	Forearm length	239.5	10.7	239.4	10.6	222.0	239.0	257.0	
30	2.93	Thigh length	509.7	19.0	507.7	18.0	476.8	506.0	538.2	
31	2.94	Leg length	374.4	13.2	374.6	12.4	354.4	374.0	395.4	
32	2.97	Crotch length	701.9	41.2	708.8	40.5	647.6	708.0	776.2	
33	2.99	Dorsal length	446.5	14.2	450.4	16.6	420.8	450.0	480.0	
34	2.102	Acromion-neck root length	139.7	10.7	138.9	10.7	123.4	140.0	157.6	
35	2.134	Neck root circumference	454.3	18.9	456.6	20.7	425.0	453.0	493.0	
36	2.135	Neck circumference	364.7	14.6	369.5	16.3	340.8	370.0	400.8	
37	2.139	Chest circumference I	923.6	49.2	934.8	53.4	847.6	928.0	1034.0	
38	2.141	Waist point circumference	782.0	60.6	805.8	79.5	697.0	795.0	972.0	③
39	2.142	Waist circumference	798.0	61.8	821.5	78.9	716.4	815.0	990.6	③

(continued)

Table 3.2 (continued)

No	Code in GJB4856	Items	Group 2		Group 3				
			$\bar{x}$	s	$\bar{x}$	s	P_5	P_{50}	P_{95}
40	2.144	Hip circumference	938.2	42.0	945.8	43.5	881.4	940.0	1029.8
41	2.148	Biceps circumference	289.1	22.3	292.3	21.2	261.2	290.0	330.6
42	2.153	Forearm circumference	267.4	14.8	269.0	13.8	248.4	267.0	290.0
43	2.155	Wrist circumference	167.0	7.3	168.4	7.3	158.0	167.0	180.6
44	2.157	Thigh circumference	554.0	37.5	557.8	36.3	500.0	560.0	613.8
45	2.161	Calf circumference	374.3	19.7	376.1	20.0	343.2	377.0	411.2
46	2.162	Ankle circumference	221.0	9.9	222.4	10.2	206.0	222.0	241.8
47	3.1	Sitting height	945.1	15.6	944.5	15.3	915.0	945.0	969.6
48	3.3	Eye height II, sitting	829.1	17.6	829.5	16.6	801.4	830.0	852.6
49	3.4	Shoulder height, sitting	616.2	19.6	618.5	17.7	586.2	619.0	644.8
50	3.8	Elbow height, sitting	262.6	22.6	264.7	21.2	229.0	264.0	301.0
51	3.11	Thigh clearance height, sitting	153.7	12.8	154.2	12.2	137.0	152.0	175.6
52	3.12	Knee height, sitting	515.4	12.1	516.4	12.3	496.4	515.0	534.0
53	3.13	Popliteal height, sitting	429.1	11.6	429.3	11.6	409.4	429.0	449.2
54	3.14	Acromial point-elbow distance	353.6	14.0	353.8	13.4	332.2	354.0	375.0
55	3.22	Buttock-knee length, sitting	578.8	13.2	579.6	14.8	555.2	579.0	603.0
56	3.23	Sitting depth	472.6	14.1	473.6	13.9	447.4	473.0	495.0
57	3.28	Elbow-elbow breadth, sitting	427.9	26.6	434.9	28.9	383.2	437.0	490.0
58	3.29	Hip breadth, sitting	347.1	15.6	350.5	15.3	326.0	350.0	378.2
59	4.1	Hand length	187.6	6.0	187.4	5.7	177.0	187.0	197.0
60	4.5	Finger II length	72.2	3.2	72.2	3.3	67.4	72.0	78.0
61	4.10	Hand breadth at metacarpal	86.1	3.5	86.3	3.6	81.0	86.0	93.0
62	4.13	Proximal knuckle II breadth	19.8	1.1	19.7	1.2	18.0	20.0	22.0
63	4.14	Distal knuckle II breadth	17.1	1.1	17.0	1.1	15.0	17.0	19.0
64	4.21	Hand depth	31.3	1.7	31.4	1.7	29.0	31.0	34.0
65	4.24	Hand girth I	206.8	8.5	208.3	9.2	194.0	208.0	224.6
66	4.39	Foot length	257.9	7.2	257.7	7.2	245.4	258.0	268.6
67	4.43	Foot breadth	98.9	4.1	99.1	4.2	92.0	99.0	106.0

③ $P < 0.05$

of personal protective and life support equipment should adapt to elder pilots or taikonaut, especially in waist circumference if the designer is to take the elder user population into account.

3.5 Conclusions

If the upper limit of the taikonaut stature shifts from 172 to 175 cm, product design should adapt to changes in the anthropometric characteristics of target user population on the basis of the 67 items presented in Table 3.1.

References

1. Yan X (2014) China fulfil to build space station in 2022. http://society.people.com.cn/n/2014/0911/c136657-25637260.html. Accessed 13 Jan 2015
2. Yang L (2009) Chinese taikonaut stature limits broaden to 175 cm in selection. http://news.163.com/09/0708/11/5DMQBQIR00011MTO.html. Accessed 13 Jan 2015
3. GJB4016-2000 (2000) Medical standard for selecting astronauts. Military standard Publishing House of the General Armament Department, Beijing
4. GJB 4856-2003 (2003) Human dimensions of Chinese male pilot population. Military standard Publishing House of the General Armament Department, Beijing
5. ISO 15535:2012(E) (2012) General requirements for establishing anthropometric databases. International Organization for Standardization, Switzerland
6. GJB20A-2006 (2006) pilot personal protective and life support equipment size. Military standard Publishing House of the General Armament Department, Beijing
7. MIL-STD-1472G Human Engineering (2012) Department of Defense, pp 239–263
8. Guo X et al (2015) The Body Mass Index of taikonaut candidates. In: Long S, Dhillon BS (eds) Proceedings of the 14th international conference on man–machine–environment system engineering, Oct 2014. Springer, Berlin, Heidelberg, pp 129–134

Chapter 4
Selection of Basic Measurements on Head Segment of Male Pilots for 3-D Anthropometry

Xiaochao Guo, Lili Zhang, Duanqin Xiong, Linghua Ran and Yanyan Wang

Abstract The 3D digital design and flexible manufacturing technology on head protective products call for 3D anthropometric data from user population. ISO 20685:2010(E) recommended extracting the six measurements by head scanner while all standards relative to male pilot products were based on one-dimensional (1D) anthropometric databank with traditional measurement in China. The 35 basic measurements were selected by designer survey and data analysis of 904 cases with 53 measurement items in GJB4856 databank to confirm the technological design requirements for anthropometric data on head segment. And 40 landmarks on head segment were identified for 3D anthropometry. The basic measurements covered the main technological design usage for pilot products.

Keywords Basic measurements · Head · Male pilots · Anthropometry

4.1 Introduction

ISO 7250-1:2008(E) provides a description of anthropometric measurements which can be used as a basis for comparison of population groups in which there were six basic anthropometric measurements on head segment defined [1]. In order to use data from 3D scanners in compatible internationally databases, ISO 20685:2010(E) recommended extracting the six measurements by head scanner [2]. In fact, the technological design requirements for anthropometric data may be different for industries because of product and user population [3]. For example, GB/T2428 gave 41 head-face dimensions of adults, and there were totally 53 measurements on

X. Guo (✉) · L. Zhang · D. Xiong · Y. Wang
Institute of Aviation Medicine PLAAF, Beijing 100142, China
e-mail: guoxiaochao1965@163.com

L. Ran
China National Institute of Standardization, Beijing 100191, China

© Springer-Verlag Berlin Heidelberg 2015

S. Long and B.S. Dhillon (eds.), *Proceedings of the 15th International Conference on Man–Machine–Environment System Engineering*, Lecture Notes in Electrical Engineering 356, DOI 10.1007/978-3-662-48224-7_4

head segment listed in GJB4586 in China [4, 5]. So, the basic measurements on head segment for 3D anthropometry were chosen here on the basis of designer survey and analysis of user characteristic in GJB4856.

4.2 Methods

4.2.1 Designer Survey

The design requirements for anthropometric data were investigated in aviation industries which included six product designers for pilot such as aircraft cockpit of fighters, attackers, bombers, transports, helicopters, and trainers; and personal life support and protective equipment as well as pilot product standards.

4.2.2 Analysis of User Characteristic in GJB4856

There were 30 fundamental items with 1739 male pilots measured, and 23 recommended items with sample size of 904 pilots in GJB4856. The factor analysis and regression were therefore made for the 53 measurements on head segment in data reduction with SPSS software based on data of 904 male pilots.

The sample of pilots aged 32.46 ± 6.47 in years with stature (body height) of (170.87 ± 4.35) cm and body mass (weight) of (69.11 ± 8.62) kg.

4.3 Results

The main results of designer survey and analysis of pilot characteristic were in Table 4.1. There were 13 principal components extracted after Varimax rotation with Kaiser Normalization converged in 16 iterations and 70.57 % of total variance explained.

4.4 Discussions

4.4.1 Requirements from Designer Survey

There were 33 items selected as basic measurements if they were:

(1) listed as basic human body measurements in ISO 7250-1:2008(E), or
(2) presented both in GB/T5703 and GB/T2428, or

Table 4.1 Results of designer survey and factor analysis based on GJB4856 databank

Codes in GJB4586	Items	Designer desired	ISO7251-1	GB/T 5703	GB/T 2428	Components in rotated component matrix												
						c1	c2	c3	c4	c5	c6	c7	c8	c9	c10	c11	c12	c13
[a]1.1	Maximum head length	6	y	y	y				0.74									
[a]1.2	Maximum head breadth	6	y	y	y						0.75							
[a]1.3	Total head height	6		y	y													
[a]1.4	Auricular height	6		y	y									0.66				
[a]1.5	Vertex-pupil height	6																0.65
[a]1.6	Physiognomic facial length I	4			y	0.48												
[a]1.7	Physiognomic facial length II	5		y		0.68												
[a]1.8	Physiognomic superior facial length	4		y	y	0.49		0.66										
[a]1.9	Morphological facial length	4	y	y	y	0.75												
1.10	Morphological superior Facial length	2						0.74										
[a]1.11	Least frontal breadth	4			y						0.55							
[a]1.12	Bizygomatic breadth	6		y	y		0.51											
[a]1.13	Biocular breadth	6		y	y							0.74	0.54					
[a]1.14	Interocular breadth	6		y	y								0.85					
1.15	Breadth of palpebral	4										0.90						
[a]1.16	Interpupillary distance	6		y	y								0.77					
[a]1.17	Nose height	5		y	y			0.87										
[a]1.18	Nose length	5		y				0.86										
[a]1.19	Nose breadth	3		y	y												0.54	
[a]1.20	Median nose breadth	3														0.75		
[a]1.21	Nose depth	4			y													
[a]1.22	Median nose depth	3														0.55		
1.23	Median nose depth	2						0.57										
1.24	Midpoint-supramentale distance	3				0.79												

(continued)

Table 4.1 (continued)

Codes in GJB4586	Items	Designer desired	ISO7251-1	GB/T 5703	GB/T 2428	Components in rotated component matrix												
						c1	c2	c3	c4	c5	c6	c7	c8	c9	c10	c11	c12	c13
1.25	Midpoint-pogonion distance	1				0.84												
1.26	Midpoint-gnathion distance	2				0.86												
1.27	Mouth-supramentale distance	2				0.59												
1.28	Mouth-gnathion distance	4				0.71												
[a]1.29	Mouth breadth	6		y	y							0.53						
1.30	Height of mucous lips	3			y	0.58												
[a]1.31	Ear implantation	4			y										0.84			
[a]1.32	Ear implantation	5			y										0.86			
[a]1.33	Physiognomic ear breadth	5			y										0.52			
[a]1.34	Ear-ear breadth	6			y						0.74							
[a]1.35	Bitragion breadth	4		y	y						0.70							
[a]1.36	Bigonial breadth	3		y	y		0.54											
[a]1.37	Occiput-pronasale distance	5			y				0.78									
[a]1.38	Occiput-tragion distance	4			y				0.73									
[a]1.39	Occiput-eye protuberance distance	6							0.85									
[a]1.40	Head circumference	6	y	y	y				0.58									
1.41	Transversal circumference	4					0.62											
1.42	Gnathion opisthocranion circumference	4					0.51							0.49				
1.43	Vertex circumference	4					0.64											
[a]1.44	Sagittal arc	4	y	y	y					0.60								
[a]1.45	Transversal arc/bitragion arc	4	y	y	y									0.65				
1.46	Bitragion-frontal arc	2			y		0.48											
1.47	Bitragion-subnasal arc	3					0.74											
[a]1.48	Bitragion-gnathion arc	3		y	y		0.83											
[a]1.49	Bitragion-submandibular arc	4		y	y		0.82											

(continued)

Table 4.1 (continued)

Codes in GJB4586	Items	Designer desired	ISO7251-1	GB/T 5703	GB/T 2428	Components in rotated component matrix												
						c1	c2	c3	c4	c5	c6	c7	c8	c9	c10	c11	c12	c13
1.50	Bitragion-occipital arc	3		y					0.47									
1.51	Glabella-vertex-cervical I	4								0.89								
1.52	Glabella-vertex-cervical II	4								0.86								
1.53	Shoulder-vertex hemiarc	4								0.63								

[a]Selected as basic measurements for 3D pilot anthropometry

(3) desired by all the six aviation industries, or
(4) required by more than half the designers investigated and presented in either GB/T5703 or GB/T2428 withal.

4.4.2 Factor Analysis Based on GJB4856 Databank

It was found that there were 4, 4, 3, 5, 1, 4, 2, 3, 2, 3, 0, 1, and 1 items which were selected as basic measurements distributed in long dozen rotated component matrix in Table 4.1. The items coded 1.20 and 1.22 in GJB4586 were therefore picked while taking the component c11 into account.

The set of basic measurements on head segment included totally 35 items in final that are asterisked in Table 4.1.

4.4.3 The Basic Measurement as Predictor

The basic measurements could be used as predictors to calculate the values of other items listed in Table 4.1. The regression equations of other measurements were presented in Table 4.2 taking the 35 basic measurements as independents. All the linear models were significant by ANOVA ($\boldsymbol{P} < 0.001$).

4.4.4 Landmarks on Head Segment for 3-D Anthropometry

According to the definitions of the 35 basic measurements mentioned above in Annex C of GJB4856, there were totally 40 landmarks on head segment for 3D anthropometry which included glabella, pupilla (right and left), entocanthion (right and left), ectocanthion (right and left), eye protuberance (right), orbitale (right and left) with eyes; tragion (right and left), otobasion superius, otobasion inferius, retro-aurale, anteaurale, supra-auricular point, and infero-auricular point with ears; sellion, midpoint of nose, pronasale, subnasale, alare (right and left) with nose; cheil-angle point (right and left) and stomion with mouth; and vertex, trichion, euryon (right and left), frontotemporale (right and left), zygion (right and left), gonion (right and left), gnathion, opisthocranion, and inion on the other parts of head.

4.4.5 The Usability of the Basic Measurements

The basic measurements on head segment were presented in Table 4.3 which was cited by Chinese military standards in effect for technological design of pilot products [6–10]. They all were subsets of the set of basic measurements on head segment mentioned above.

Table 4.2 The effect of the basic measurements used as predictors in linear regression

Codes in GJB4586	Items	Regression equations	R^2	Residual
1.10	Morphological superior facial length	$y_{1.10} = 0.665x_{1.8} + 0.759x_{1.17} - 0.452x_{1.18}$	0.998	2.68
1.15	Breadth of palpebral	$y_{1.15} = 1.616x_{1.13} - 0.621x_{1.14}$	1	0
1.23	Median nose depth	$y_{1.23} = 0.340x_{1.8} + 0.142x_{1.14} + 0.516x_{1.17}$	0.994	2.67
1.24	Midpoint-supramentale distance	$y_{1.24} = 0.273x_{1.7} + 0.408x_{1.8} + 0.429x_{1.9} - 0.484x_{1.17} + 0.374x_{1.18}$	0.998	3.80
1.25	Midpoint-pogonion distance	$y_{1.25} = 0.310x_{1.8} + 0.763x_{1.9} - 0.242x_{1.17} + 0.168x_{1.37}$	0.998	4.01
1.26	Midpoint-gnathion distance	$y_{1.26} = 0.229x_{1.7} + 0.722x_{1.9} + 0.049x_{1.21}$	0.999	3.46
1.27	Mouth-supramentale distance	$y_{1.27} = 1.330x_{1.9} - 0.506x_{1.13} + 0.531x_{1.14} - 0.365x_{1.17}$	0.980	2.73
1.28	Mouth-gnathion distance	$y_{1.28} = 0.442x_{1.7} - 0.354x_{1.8} + 1.065x_{1.9} - 0.156x_{1.13}$	0.996	3.40
1.30	Height of mucous lips	$y_{1.30} = 1.245x_{1.9} - 0.420x_{1.12} + 0.406x_{1.14} - 0.243x_{1.36}$	0.978	3.06
1.41	Transversal circumference	$y_{1.41} = 0.126x_{1.7} + 0.098x_{1.12} + 0.154x_{1.40} + 0.349x_{1.45} + 0.273x_{1.48}$	0.999	16.34
1.42	Gnathion opisthocranion circumference	$y_{1.42} = 0.125x_{1.7} + 0.292x_{1.40} + 0.344x_{1.45} + 0.240x_{1.48}$	1	13.85
1.43	Vertex circumference	$y_{1.43} = 0.077x_{1.36} + 0.225x_{1.40} + 0.425x_{1.45} + 0.273x_{1.49}$	0.999	15.71
1.46	Bitragion-frontal arc	$y_{1.46} = 0.511x_{1.40} + 0.226x_{1.45} + 0.263x_{1.48}$	0.999	8.09
1.47	Bitragion-subnasal arc	$y_{1.47} = 0.108x_{1.12} + 0.309x_{1.40} + 0.583x_{1.48}$	0.999	7.10
1.50	Bitragion-occipital arc	$y_{1.50} = 0.044x_{1.22} + 0.184x_{1.34} + 0.593x_{1.40} + 0.179x_{1.44}$	0.999	11.13
1.51	Glabella-vertex-cervical I	$y_{1.51} = 0.178x_{1.3} + 0.412x_{1.40} + 0.410x_{1.44}$	0.999	14.35
1.52	Glabella-vertex-cervical II	$y_{1.52} = 0.160x_{1.3} + 0.518x_{1.40} + 0.322x_{1.44}$	0.999	14.53
1.53	Shoulder-vertex hemiarc	$y_{1.53} = 0.192x_{1.3} + 0.451x_{1.40} + 0.357x_{1.4}$	0.998	14.54

Table 4.3 The basic measurements cited by Chinese military standards

Codes in GJB4586	Items	GJB20A	GJB35B	GJB36A	GJB6851.1	GJB1471A
1.1	Maximum head length	y		y		
1.2	Maximum head length	y		y		
1.3	Total head height	y		y		
1.4	Auricular height	y				
1.5	Vertex-pupil height	y				
1.9	Morphological facial length	y				
1.11	Least frontal breadth	y				
1.12	Bizygomatic breadth	y				
1.13	Biocular breadth	y				
1.16	Interpupillary distance			y	y	
1.18	Nose length	y				
1.19	Nose breadth	y				
1.20	Median nose breadth	y				
1.21	Nose depth	y				
1.22	Median nose depth	y				
1.24	Midpoint-supramentale distance	y				
1.28	Mouth-gnathion distance	y				
1.29	Mouth breadth	y				
1.32	Physiognomic ear length	y				
1.33	Physiognomic ear breadth	y				
1.34	Ear-ear breadth	y				
1.36	Bigonial breadth	y				
1.37	Occiput-pronasale distance	y				
1.38	Occiput-tragion distance	y				
1.39	Occiput-eye protuberance distance		y	y	y	y
1.40	Head circumference	y				
1.41	Transversal circumference	y				
1.44	Sagittal arc	y				
1.45	Transversal arc	y				

4.5 Conclusions

The 35 basic measurements were selected by designer survey and data analysis based on GJB4856 databank. They could produce the values of the other items in GJB4856 and cover the main usage for technological design of pilot products. There were 40 landmarks on head segment to be identified for 3D anthropometry.

References

1. ISO7250-1:2008(E) (2008) Basic human body measurements for technological design—Part 1: Body measurement definitions and landmarks. International Organization for Standardi zation, Switzerland
2. ISO20685:2010(E) (2010) 3-D scanning methodologies for internationally compatible anthropometric databases. International Organization for Standardization, Switzerland
3. Guo X et al (2010) Design requirements of transportation industry for anthropometric data of pre-adult Chinese. In: Long S, Dhillon BS (eds) Proceedings of the 10th conference on man-machine-environment system engineering, Scientific Research Publishing, Inc., pp 65–68
4. GB/T 2428-1998 (1998) Head-face dimensions of adults. China Standard Press, Beijing
5. GJB 4856-2003 (2003) Human dimensions of Chinese male pilot population. Military standard Publishing House of the General Armament Department, Beijing
6. GJB20A-2006 (2006) Pilot personal protective and life support equipment sizes. Military standard Publishing House of the General Armament Department, Beijing
7. GJB35B-2008 (2008) Aircrew station geometry for fighter and attacker. Military standard Publishing House of the General Armament Department, Beijing
8. GJB36A-2008 (2008) Design and application requirements for human body templet of pilot. Military standard Publishing House of the General Armament Department, Beijing
9. GJB6851.1-2009 (2009) Cabin dimension series for bomber and transport-Part 1: Aircrew station basic geometry dimension for bomber and transport. Military standard Publishing House of the General Armament Department, Beijing
10. GJB1471A-2012 (2012) Aircrew station basic geometry dimension for military helicopter. Military standard Publishing House of the General Armament Department, Beijing

Chapter 5
Numerical Calculation and Analysis of Influencing Factors on Waist Force of Armored Vehicle Driver

Weiping Liu, Binhe Fu, Xixia Liu and Yi Jin

Abstract High morbidity of low back pain is a prominent problem for crews in operating armored vehicles. The low back pain of drivers has a direct relationship with the waist force under unreasonable postures. Based on the national military standard (GJB) and national standard (GB), this paper built a biomechanical model of armored vehicle driver's waist aiming at operation characteristics of armored vehicle driving. Influencing factors on waist force of armored vehicle driver were figured out. Numerical calculation was conducted and the change laws of influencing factors were analyzed, which gave out analysis conclusions. The results obtained from this paper can provide reference for improving waist force of armored vehicle drivers.

Keywords Waist force · Numerical calculation · Biomechanical model

5.1 Introduction

High morbidity of low back pain is a prominent problem for crews in operating armored vehicles. The investigation result from Ref. [1] indicates that the morbidity of low back pain of armored vehicle crews was 73.7 %. The morbidity of drivers, as high as 94.0 %, is the highest among all the armored vehicle crews and is 56–68 % higher than that of oversize vehicle drivers [2, 3]. The test data of typical operating posture of armored vehicle drivers from Ref. [4] indicate that the operating posture of armored vehicle driver was a form of compulsive position. The low back pain increases because of the long-time and single-posture repetitive operation under bad postures.

W. Liu (✉) · B. Fu · X. Liu · Y. Jin
Department of Mechanical Engineering,
Academy of Armored Forces Engineering, Beijing 100072, China
e-mail: Liuweiping@126.com

© Springer-Verlag Berlin Heidelberg 2015

S. Long and B.S. Dhillon (eds.), *Proceedings of the 15th International Conference on Man–Machine–Environment System Engineering*, Lecture Notes in Electrical Engineering 356, DOI 10.1007/978-3-662-48224-7_5

The low back pain of drivers has direct relationship with the waist force under unreasonable postures. It will provide helpful reference for relieving drivers' low back pain to analyze the influencing factors of waist force of armored vehicle drivers and help to change the laws. For this reason, regarding the typical armored vehicle driving cabin as the study object, this paper employed human biomechanical modeling and numerical calculation and analysis to analyze the influencing factors of the waist force of armored vehicle drivers, which concluded helpful results.

5.2 Mechanical Model of Driver's Waist Force

By analyzing several types of typical armored vehicles, it is concluded that hand control of armored vehicles are arranged in the same plane with driver's shoulder joint and motion track of hand operation. Therefore, this paper employs 2D multi-rigid body biomechanical model, as is shown in Fig. 5.1, for better analysis. Meanings of parameters are as follows:

G_1, …, G_5 are the mass of head, body, upper arm, forearm and hand separately. F_{my} is the back strength, which refers to the force from erector spinae. F_{ry} is the reacting force at the *L4*/*L5* against the top of sacrum and the bottom of lumbar intervertebral disc, which is exerted by sacrum and exert a force upon spinae. The component force of F_{ry} in the direction of spinae is the vertical pressure at the *L4*/*L5*, which is represented by F_{ryc}. The component force of F_{ry} in the vertical direction of spinae is the horizontal pressure at the *L4*/*L5*, which is represented by F_{rys}. F_c is the operating force exerting on arms when driving. α_Q is the angle between F_{ry} and vertical plane. α_A is the upper arm angle, namely, the angle between upper arm and vertical line. α_B is the forearm angle, namely, the angle between forearm and horizontal line. α_3 is the body angle.

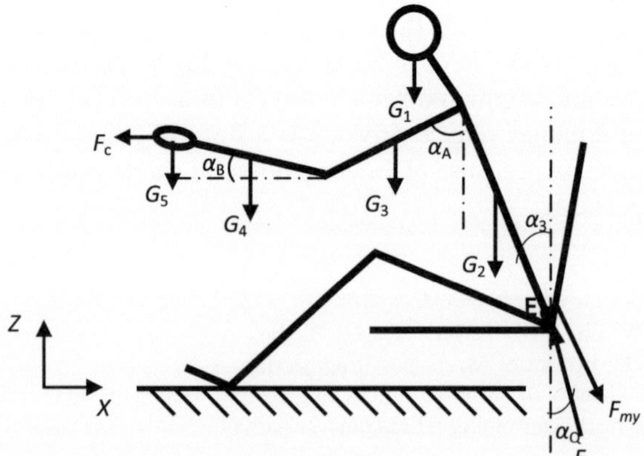

Fig. 5.1 Driver's 2-dimension biomechanical model

When body is in the state of balance, the sum of torque exerting on the *L4*/*L5* lumbar vertebra is 0. Based on the torque equilibrium equation $\sum M = 0$, it is defined as follows:

$$F_{my}X_b = \frac{\sum G_i X_i - F_c L}{5} \tag{5.1}$$

where X_i is the length of perpendiculars of barycenters at *L4*/*L5* separately, and X_b is the distance between erector spinae and spinae, which is 5 cm [5].

Relative mass of each link in human model and the parameters of barycenter relative position of each link in human model refer to date in GB/T 17245-1998. It is defined as follows:

$$F_{my} = \frac{H}{500}[G(0.87\sin\alpha_1 + 9.88\sin\alpha_3 + 0.58\sin\alpha_A + 0.2\cos\alpha_B) - F_c(0.29\cos\alpha_3 - 0.19\cos\alpha_A + 0.14\sin\alpha_B)] \tag{5.2}$$

Based on equilibrium of forces, $\sum F = 0$, i.e., $\sum F_x = 0$, $\sum F_z = 0$:

$$tg\alpha_Q = \frac{F_{my}\sin\alpha_3 - F_c}{0.5644G + F_{my}\cos\alpha_3} \tag{5.3}$$

reacting force at *L4*/*L5*:

$$F_{ry} = \frac{F_{my}\sin\alpha_3 - F_c}{\sin\alpha_Q} \tag{5.4}$$

vertical pressure:

$$F_{\mathrm{ryc}} = F_{ry}\cos(\alpha_3 - \alpha_Q) \tag{5.5}$$

horizontal pressure:

$$F_{\mathrm{srys}} = F_{ry}\sin(\alpha_3 - \alpha_Q) \tag{5.6}$$

5.3 Influencing Factors and Numerical Calculation

5.3.1 Main Influencing Factors Analyzing

According to Eq. 5.2, there are three factors influencing driver's waist force. First is the human parameter, which comes from the change of human geometric dimensions after mass distribution parameters of each link have been defined. Second is the parameter of cabin. From geometric model of driver's operating posture in Ref. [6], the change in parameters of cabin would influence parameters of driver's operating posture such as upper arm angle and body angle. Third is the operating force, which depends on performance of operating device.

5.3.2 *Numerical Calculation*

Based on geometric model of driver's operating posture built in Ref. [6] and driver's biomechanical model in this paper, a numerical calculation diagram of influencing factors in waist force of armored vehicle driver is presented shown as Fig. 5.2. Human geometric dimensions refer to data of Table 2 from Ref. [6] which regard the test data of armored vehicle crews' human dimensions given from

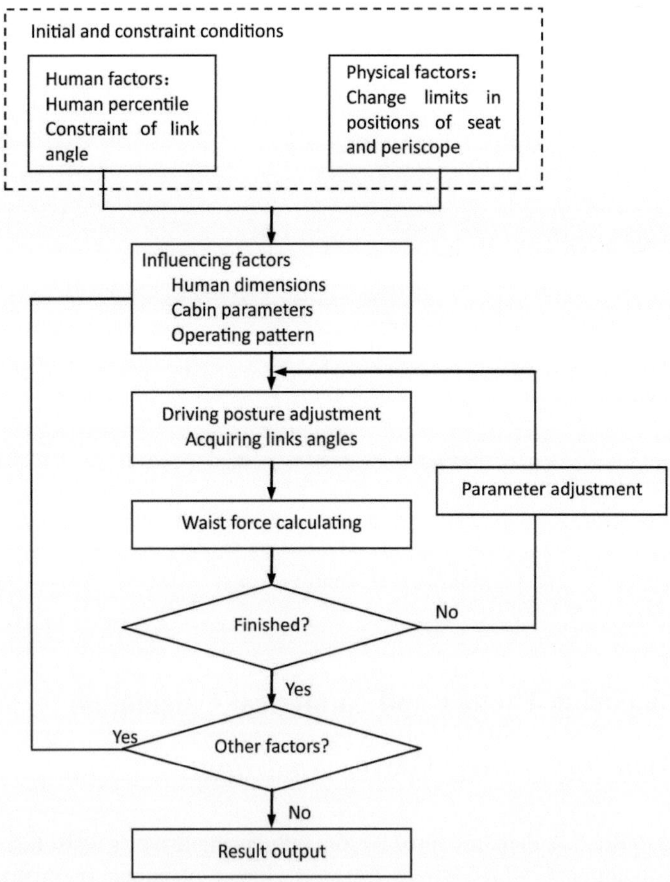

Fig. 5.2 Analyzing procedure in numerical calculation of influencing factors on driver's waist force

GJB1835-1993 as basic accordance. According to data in GB8420-87, human dimension of each link is revised by linear interpolation method.

To distinguish influence of each main factor, numerical calculation is conducted in three conditions as follows:

(1) Aiming at driving cabin of certain tank, simulated calculations of waist force in different percentile are conducted under conditions when windows are opened and closed. When windows are opened, each angle of driver refers to test data in Ref. [4];
(2) Taking 50th percentile human dimensions as study object, relationship between changes of cabin arrangement parameters and waist force of drivers by modifying positions of seats in driving cabin and periscopes within limits;
(3) Aiming at driving cabin of certain tank, operating force curve of sheering operating device is taken as input to conduct numerical calculations of driver's waist force in different percentile under conditions when windows are closed.

5.4 Result Analysis

5.4.1 Influence of Height on Driver's Waist Force

Figures 5.3 and 5.4 show analyses results of different percentile drivers under conditions when windows are closed and opened. Figure 5.3 shows waist force of different percentile drivers when windows are closed, while Fig. 5.4 shows waist force of different percentile drivers when windows are opened.

It is concluded from Figs. 5.3 and 5.4:

(1) No matter windows are closed or opened, waist force of high percentile driver is obviously bigger than that of low percentile driver. Take the window-closed driving as an example, waist vertical pressures of 50th and 95th percentile driver are separately 1.6 times and 2.4 times of 5th percentile driver, while

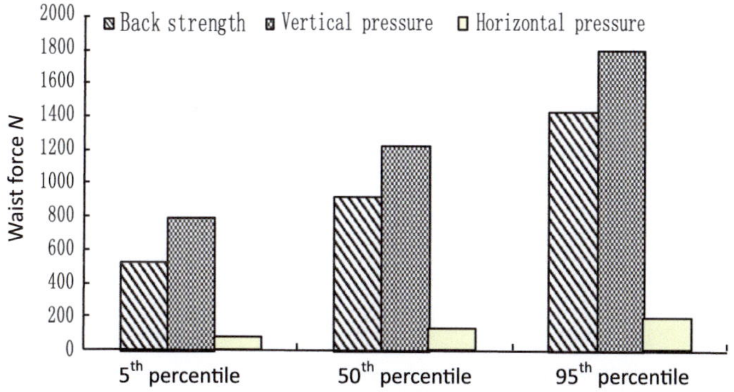

Fig. 5.3 Contrast of different percentile drivers' waist force during window-closed driving

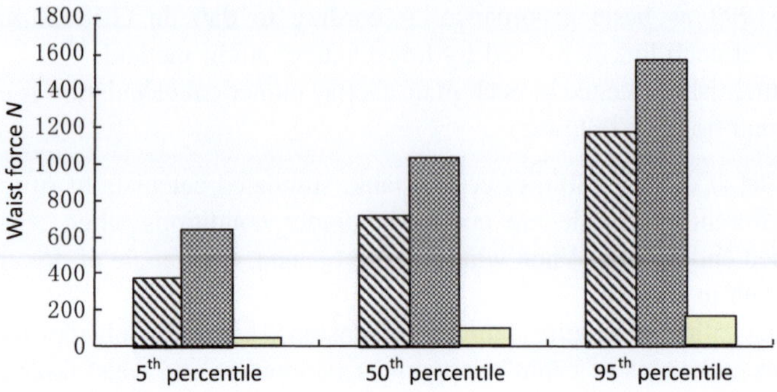

Fig. 5.4 Contrast of different percentile drivers' waist force during window-opened driving

horizontal pressures are separately 2.07 times and 3.37 times, which waist force of high percentile drivers are easy to feel fatigue. Therefore, armored vehicle drivers have to be selected to avoid high percentile people.

(2) Driver's waist force of window-opened driving is smaller than that of window-closed driving. During window-opened driving, there is less external observation constraint for drivers. Thus, there is more room for drivers to adjust postures and the waist force decreases because of the decreasing body angle.

5.4.2 *Influence of Cabin Arrangement Parameter Change on Waist Force*

Relationship between 50th percentile driver's waist force and driving seat is shown as Fig. 5.5, which shows influence of position change of driving seat in horizontal direction on waist force. Relationship between 50th percentile driver's waist force and position of periscope is shown as Fig. 5.6, which shows influence of position change of periscope in vertical direction on waist force.

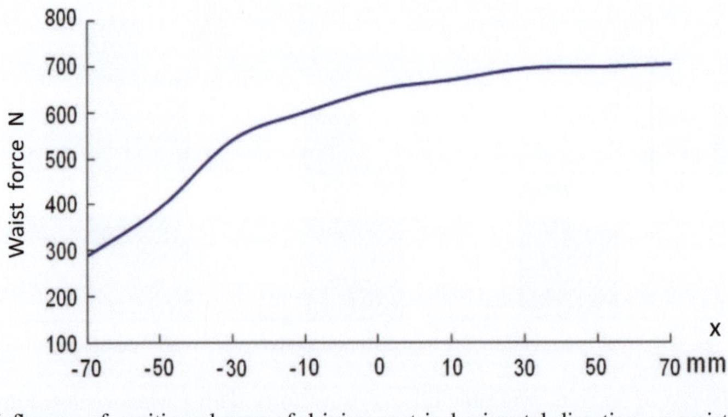

Fig. 5.5 Influence of position change of driving seat in horizontal direction on waist force

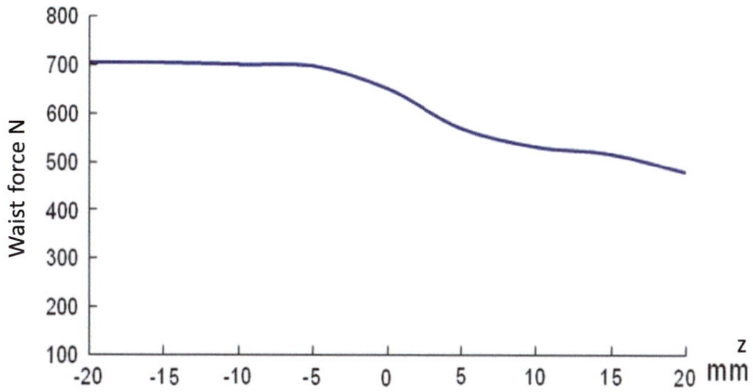

Fig. 5.6 Influence of position change of periscope in vertical direction on waist force

(1) It is concluded from Fig. 5.5 that driver's waist force increases as the relative position between seat and periscope increases. On the contrary, driver's waist force decreases as the relative position between seat and periscope decreases. The increase of relative position between seat and periscope results in the increase of body angle in sitting posture directly.
(2) It is concluded from Fig. 5.6 that driver's waist force decreases as the relative position of periscope increases. It is good in decreasing driver's waist force to increase the height of periscope within limits.

5.4.3 Influence of Operating Force on Waist Force

The relation curve between steering force of certain tank and rotational angle of joystick is shown as Fig. 5.7. Numerical calculation result of different percentile

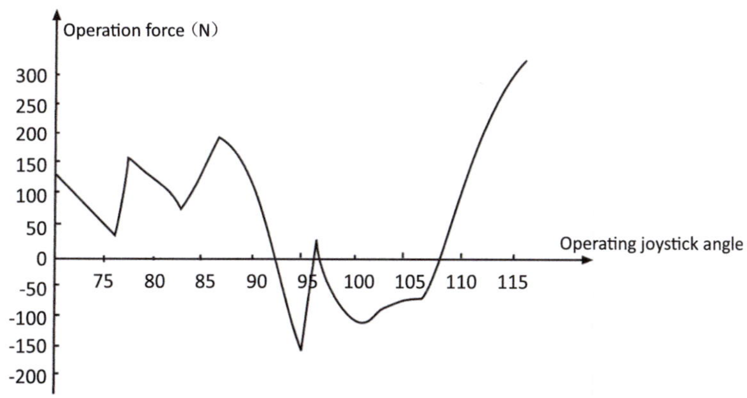

Fig. 5.7 Operating curve of operating joystick in certain armored vehicle

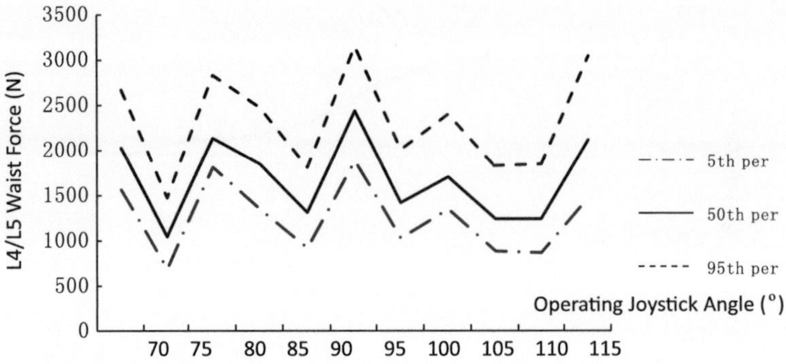

Fig. 5.8 Relationship between different percentile people *L4/L5* waist force and operating joystick angle

driver's waist force over the change of operating force is shown as Fig. 5.8 under conditions when windows are closed.

As shown in Fig. 5.8, driver's waist force varies with the rotation of operating joystick during actual operation. The increasing external force makes obviously increasing waist force.

It is indicated from the statistical result that more than 85 % operations of operating joystick are in the limit ranging from 70° to 105°. It is indicated from Fig. 5.8 that 5th, 50th and 95th percentile driver's waist force maximum are 1850, 2450 and 3167 N respectively. The corresponding horizontal pressure is 322, 524, and 710 N respectively. According to regulations from National Institute for Occupational Safety and Health (NIOSH), *L5/S1* lumbar vertebra pressure is supposed to be 3400 N and horizontal pressure to be 508 N. Pressure limit of *L4/L5* is similar to that of *L5/Sl*. Obviously, 50 % horizontal pressure of *L4/L5* driver's lumbar vertebra is out of the safe limit under this circumstance.

It is indicated from the result that operating force has significant influence on driver's waist force. Therefore, it is significant for relieving armored vehicle driver's low back pain to improve operating performance and decreasing force.

5.5 Conclusions

(1) External observation device of armored vehicle cabin constrains driver's operating posture. Therefore, high percentile drivers are unsuitable for operating in cabin and crews need to be selected;
(2) To decrease the influence of unreasonable operating posture on driver's waist force, it is suggested to open the windows during armored vehicle driving. Otherwise, it is supposed to adjust the seat position to decrease the relative horizontal position between periscope and driving seat. When the operating room is designed, the arrangement of periscope and driving seat are supposed

to satisfy the purpose of comfortable driver's body angle, and also increase the horizontal and vertical adjustment amount appropriately of driving seat;

(3) Operating device is supposed to employ power-assisting device and hydraulic device to decrease driver's waist force during armored vehicle driving.

References

1. Lu C, Ye Y, Yuan H et al (2002) Low back pain among the tank soldiers. Clin J Med Off 30 (1):34–35
2. Chen L, Hua G (2001) Evaluation and exercise therapy. Chin J Phys Med Rehabil 23(3):177–178
3. Magnusson ML, Pope MH, Wilder DG et al (1996) Are occupational drivers at an increased risk for developing musculoskeletal disorders? Spine 21(6):710–717
4. Liu W, Xie C, Wang F (2009) Test on armored vehicle driver's operation posture and its analysis. In: 8th international symposium on test and measurement. Chongqing, China
5. Wang S, Li L, Liu J (1994) Biomechanics analysis on waist of sitting work. Chin J Ind Hyg Occup Dis 12(1):29–31
6. Liu W, Xie C, Cao W et al (2008) Research on driving comfortableness for armored vehicle based on kinematics analysis. Veh Power Technol 3:20–23

Chapter 6
Peer Assessment Modeling and Design in Military Training

Cheng Jin, Zhibing Pang, Xiaogao Wang, Zhaofeng Luo and Quanliang Yin

Abstract Peer assessment software is useful as part of mutual assessment in military training to enhance the overall training level. We conducted experiments to obtain a scoring algorithm and the relevant weight on which the software should be developed to score each trainee. The basic algorithm and software procedure contribute to the final scoring of each trainee. By employing a large amount of experimental data, the algorithm and software may provide an accurate method for peer assessment in military training.

Keywords Peer assessment · Algorithm · Software

6.1 Introduction

In military training, peer assessment tends to be adopted in order to increase training enthusiasm and improve the training effect. In line with the unit's training conditions, peer assessment is used to target the post requirement of trainees and improve the trainees' organizational capabilities. Thus, the trainees may swap roles between trainee, organizer, and examiner. In traditional training, the organizer assesses the trainees and scores the subjects, which does not encourage the subjective initiative of the trainees. Lacking passion in training and study, trainees just simply prepare for their own examinations. The trainees' capability in different roles cannot be developed in a comprehensive way.

C. Jin (✉) · Z. Pang · Z. Luo · Q. Yin
Air Defence Forces Academy, Zhengzhou 450052, China
e-mail: fkb_mmes@163.com

X. Wang
Unit 73151 of PLA, Xiamen 361100, China

© Springer-Verlag Berlin Heidelberg 2015

S. Long and B.S. Dhillon (eds.), *Proceedings of the 15th International Conference on Man–Machine–Environment System Engineering*, Lecture Notes in Electrical Engineering 356, DOI 10.1007/978-3-662-48224-7_6

Mutual assessment makes trainees further involved in the military training. The mutual assessment modeling may improve the evaluation system as well as the trainees' capability to learn and guide the training.

6.2 Peer Assessment

Assessment with scoring plays the role of the baton in the course of instruction. It is the core of teaching quality management as well as the motivation for self-study [1]. From the instructor's perspective, assessment with scoring clarifies the learner's condition, which can be taken as the basis to improve the instruction mode. From the trainee's perspective, assessment with scoring allows the trainees to know themselves, which can be taken as the point of reference. Peer assessment is different from the traditional scoring mode, which is organized by trainers to score every aspect of trainees. Mutual assessment gives initiative to the trainees, who conduct regrouping and scoring. They have the ability to assess others and attain their own results. It may not only increase the trainees' enthusiasm but it can also examine their level of assessment, which can be another improvement for them to some degree [2]. Evaluation of scoring can give trainees a good command of their learning conditions, allow them to consider it as a reference resource to change their study methods.

6.3 Building the Model

After training and moving into the period of assessment (which also includes companions' evaluation of each other), we can divide every trainee's score into two parts: a teaching score and an evaluation score. These scores can then be combined according to their weightings, giving a final result.

6.3.1 The Trainee's Teaching Score

$$b_j = \frac{\sum_{i=1}^{n-1} N_i - N_{\max} - N_{\min}}{n-3}$$

In the above calculation, b_j is the jth trainee's assessment score; $\sum_{i=1}^{n-1} N_i$ represents the total score for the trainee; $N_{\max}$ represents for the trainee's highest mark; $N_{\min}$ represents the trainee's lowest mark; and n represents the number of the evaluated trainee.

6.3.2 The Trainee's Evaluation Score

$$c_j = K3 \times \frac{100\sum_{i=1}^{n-1}(1-\frac{|a_{ij}-b_i|}{b_i})}{n-1} + K4 \times \frac{100\text{m}}{C_{n-1}^2}$$

In the above calculation, n represents the number of the evaluated trainee; m represents the right number that the jth trainee gives to the other $n-1$ trainees in score order by comparison between any two means; a_{ij} is the score that the jth trainee gives to the ith trainee; b_i is the ith trainee's assessment score; c_j is the jth trainee's score for scoring accuracy; K3 and K4 are weighting coefficients; and $K3 + K4 = 1$.

6.3.3 The Trainee's Final Score

$$Z_j = K1 \times b_j + K2 \times c_j$$

In the above calculation, c_j is the jth trainee's score for scoring accuracy; b_j is the jth trainee's assessment score; K1 and K2 are the relevant weighting; and $K1 + K2 = 1$.

6.4 Determine the Weighting

6.4.1 The Experiment Subjects

Thirty-nine trainees who took part in equipment operation training for a week were the subjects of our experiment.

6.4.2 The Experiment Method

To formulate the corresponding weighted scoring table, each trainee was assigned to four weights according to their knowledge. After calculating the average and removing the outliers to take the average value, we compared the weighted scores of ten experts to determine the scores of the four weights [3]. The statistical weights after computing are shown in Table 6.1.

The value of K1 was consistent with the experts' scores when the highest and lowest scores were removed to get the average. K1 was determined to be 0.68. For the same reason, we determined that K2 is 0.32, K3 is 0.62, and K4 is 0.38.

Table 6.1 The statistical weight table

Weight		K1	K2	K3	K4
Trainee	The average of weight after removing 5 highest scores and 5 lowest scores	0.68	0.32	0.62	0.38
	The average of weight	0.67	0.33	0.62	0.38
The organizer		0.68	0.32	0.61	0.39

6.5 Design Software

Based on the algorithm formula [4], we used Matlab to read the data and design the types of the software. There are three main parts of the software: data input module, data processing module, and data display module.

6.5.1 Data Input Module

The data input module is as follows:

Open the file, the score table.xls. Input data in the sheet. The data input format is matrix A, where $A(i,j)$ represent for the score that the ith trainee give to the jth trainee.

6.5.2 Data Processing Module [5]

Open the pingfen.exe to continue the data processing module.

There are 13 procedures for the data processing module:

1. Input the score matrix A
2. Input $K1$, $K2$, $K3$, $K4$
3. Input the number of the trainees Num
4. Input the number of the trainees that score the ith trainee lieo_num
5. Get the number of trainees that have the ith trainee score hango_num(i)
6. Get all the possible situations for which the number of the trainees by comparison between any two means that the ith score $C^2_{\text{hango\_num}(i)}$
7. Get the correct possible situation for which the number of the trainees by comparison between any two means that the ith score m
   The procedure to know m is as follows: get the ith score for the people's scoring matrix $p\{i\}$; get the ith score for the people's scoring order matrix $pz\{i\}$; transform the $p\{i\}$ and $pz\{i\}$ individually to order matrixs $p1\{i\}$ 和 $pz1\{i\}$ which can compare between any two means, such as: $p\{1\} = [2 \quad 5 \quad 3 \quad 4 \quad 1\,]$. After transformation, it will be:

$$p1\{1\} = \begin{bmatrix} 2 & 5 \\ 2 & 3 \\ 2 & 4 \\ 2 & 1 \\ 5 & 3 \\ 5 & 4 \\ 5 & 1 \\ 3 & 4 \\ 3 & 1 \\ 4 & 1 \end{bmatrix}$$

Compare $p1\{i\}$ and $pz1\{i\}$ to get the number of correct rows $p1\{i\}$,which also is m.

8. Get the ith trainee's assessment score $b_i = \frac{\sum_{i=1}^{\text{Num}} A(:,i)}{\text{lieo\_num}(i)}$
9. The score that the jth trainee scores the ith trainee $a_{ij} = A(i,j)$
10. The correct scoring order score for the ith trainee $px(i) = \frac{m}{C^2_{\text{hango\_num}(i)}} \times 100$
11. The scoring accuracy of the ith trainee $\mathrm{d}f(i) = \frac{\sum_{j=1}^{\text{Num}} \left(1 - \frac{|a_{ij} - b_i|}{b_i}\right)}{\text{lieo\_num}(i)} \times 100$
12. Get the scoring score for the ith trainee $c_i = K3 \times \mathrm{d}f(i) + K4 \times px(i)$
13. Get the ith trainee's final score $z_i = K1 \times b_i + K2 \times c_i$

6.5.3 Data Display Module

Open the file for the score table.xls. The computing result is shown in the output data table [6].

6.6 Conclusion

The requirements for the military are the real solider, real equipment, real space, and real assessment. Every trainee should have a good command of knowledge and practice, repeatedly examining equipment and people for certain training subjects. Using peer assessment modeling, every trainee will interview and score other trainees objectively, thus making the abstract problem more clear, making qualitative evaluation more quantitative, making tedious calculations more procedural, improving the objectivity of evaluation, achieving rapid and accurate evaluation of the practice effect, deepening the practice of the military training, and realizing the transformation of the operator and the interview. When combined with peer assessment modeling, this modeling algorithm and outputting score via software greatly improve the ability of the trainees. In their future military careers, it will have vital significance by cultivating excellent professional supervisory talents.

References

1. Yang X (2012) An analysis of teaching quality monitoring and evaluation system status and improvement measure. J Railway Police Coll 2
2. Wang J (2012) University Teachers' teaching quality evaluation method research. Inf Technol 12
3. Zhao H, Xu S (1986) Analytic hierarchy process. Beijing Science Press, Beijing
4. Chen C (2011) MATLAB application examples succinctly image processing and GUI design. Electronic Industry Press, 02:260
5. Chen Y (2008) Proficient in MATLAB GUI design. Electronic Industry Press 2
6. Zhang Z (2002) MATLAB program design and application. Beijing Tsinghua University press

Chapter 7
An Analysis of Major Psychological Misunderstandings in Military Academy Course Design

Hua Li, Weixin Liu and Jie Xing

Abstract The significance and value of psychology in the military academy course design is not fully acknowledged and accepted, and the divergences in its theoretical cognition and practical exploration exist. Thus there are many psychological misunderstandings in course design. By wide research and theoretical demonstration, the article reveals the major psychological misunderstandings in the military academy course design, which include the misunderstandings of thinking, value orientation, emotion and attitude, and roles. Based on the analysis of these misunderstandings, the article finds out that the most fundamental way to eliminate the misunderstandings is to enhance the psychological quality of course designers. It is expected to provide effective psychological aid to the military academy course design.

Keywords Military academy · Course design · Psychological misunderstanding · Analysis

7.1 Research Purpose

The research and practice of military academy course design has not formed a complete system, and many differences still exist in the theoretical cognition and practical exploration. Most of all, the psychology in the military academy course design, as a new research perspective, has not been fully recognized and accepted for its significance and value. All these are the important reasons for the psychological misunderstandings in the military academy course design. To analyze and eliminate the psychological misunderstandings and to provide effective psychological aid to the military academy, course design is the basic expectation of this article.

H. Li (✉) · W. Liu · J. Xing
Air Defense Forces Academy, Zhenzhou 450052, China
e-mail: fkb_mmes@163.com

© Springer-Verlag Berlin Heidelberg 2015

S. Long and B.S. Dhillon (eds.), *Proceedings of the 15th International Conference on Man–Machine–Environment System Engineering*, Lecture Notes in Electrical Engineering 356, DOI 10.1007/978-3-662-48224-7_7

7.2 Major Psychological Misunderstandings in Military Academy Course Design

7.2.1 Misunderstandings of Thinking

The process of course design is a multi-objective decision-making, which is essentially a thinking process of problem-solving [1]. Thinking activities in problem-solving can be divided into four stages, which are characterizing the problem, designing the scheme, implementing the scheme, and evaluating the result. The extensity, profundity, creativity, and criticalness, flexibility and logicality of course designers' thought has a significant impact on the scientificity and practicality of course design. Misunderstandings of thinking are mainly manifested in two aspects.

The first is thinking-set. Thinking-set is a common failing generally existing in current military academy course designers. Set is a psychological state of preparation which was formed in previous activities, will influence follow-up activities. Thinking-set of course designer is used to manifest in solving the problem in a conventional way. Thinking-set will prevent people to discover and solve new problems. Take something for instance. The current structure of the military academy has undergone major changes; the original teaching plan has been developed for the teaching personnel cultivating program; the teaching syllabus has been replaced by the course standard; and the personnel training objectives model has been increased as a new structural levels [2]. Facing the new changes, the military academy course designers cannot completely grasp the spiritual essence of the teaching personnel cultivating program and the course standards, and cannot fully appreciate a series of influences of the new changes on teaching, so that they still design the personnel cultivating program and the course standard in the thinking mode of designing the teaching plan and the teaching syllabus, and they even "label" just the name on old syllabus and teaching plan.

The second is functional fixation. Functional fixation means that in solving problems, the individual often only focusses on certain general functions of things, but cannot have other latent functions in mind. For a long time, the military academy course designers only see the role of knowledge carrying and transferring of course programs, and overlook the potential role of shaping for values of emotional attitude and the ideals and beliefs of the cadets. The functional fixation is transferred to the course programs through course designers' decisions, so that attention will only be paid to the dominant part of the course other than the hidden part, and the intellectual part of the dominant part other than the nonintellectual part. These phenomena are not conducive to the cultivation of high-quality military personnel of a new type.

7.2.2 Misunderstanding of Value Orientation

The process of course design is also a process of value orientation. The value concept through the course design directly determines the results of the course design, and directly affects the confirmation of the goal of personnel cultivating, the design of the course standards, and the setting of the course. The value orientation of the military academy course design points to training qualified military personnel prepared for the military construction and military struggle. Qualified military personnel must comply with the "five phrases" requirements. They should have the faith, courage, and ability of "winning the battle" and "never degenerating," have comprehensive knowledge structure and adaptable ability, and have high-strength innovation spiritual, practical ability, profound scientific and humanistic attainment. They should also have a sound personality and correct world outlook, outlook on life and values, have patriotic spirit, moral qualities and aesthetic ability, and have the needs of life-long learning and healthy spiritual life in the world, which is the value orientation of the military academy course design. To accurately recognize the value orientation, apply it in the course design as guideline, and embody it in the design, is the primary factor of the military academy course design.

For a long time, the military academy course designers, driven by the motto of "knowledge is power," have been limited to think about how to teach cadets. Knowledge standard has become the gist of course design. And the cadets also take mastering abundant knowledge as their responsibility. This kind of value orientation makes the military academy course design pay overmuch focus on the construction of knowledge structure, and the gap between academy education and military training get wider and wider. Currently, the military academy course design guideline has changed to note the importance of capacity-building. The course concept of capacity-building proposes to effectively apply knowledge to solve practical problems. To acquire knowledge is to build and improve abilities, and to receive education is to be qualified for employment of the future. The course concept is a big step compared to the knowledge base, and it is more in line with the needs of social progress and personal growth. Ability standard has faced an embarrassing situation all along about how to effectively put knowledge into practical ability. Until now, teaching has not solved the problem. The ideology of "quality education" has become the consensus of current education. It advocates the integration of knowledge, ability, and quality, and its highest goal is to cultivate the national quality, reshape the national culture, uphold the national spirit, and perfect the personality. The personality-based value orientation thinks that the course objective should include knowledge and skills, processes and methods, attitude and values of these three areas. However, due to the deep-rooted sense of the knowledge standard, the course objective is bounded in the knowledge and skills aspect. The important teaching objective has been abandoned. The objectives should be to cultivate learners' ability to grasp the learning process and to master the science learning method, to create a good learning emotion, attitude and personality, and to establish a correct world outlook, outlook on life and values by leaning

7.2.3 *Misunderstanding of Emotion and Attitude*

Emotion and attitude is a factor that is often overlooked in the study of course design, while it has important effect. In the design process, course designers have not only a definite mind and value orientation, but also more or less a certain emotion and attitudes tendency.

Currently, the misunderstanding of emotion and attitude in the military academy course design is as following. As for teaching management staff, they just focus on the organization and planning of course design, and have no enthusiasm on specific design processing. As for teachers, they only concern about the task of their own, and always have a sense of completion of the task. Meanwhile, job burnout of course designers makes them not connect course design with their interests and values, so the course designers lack fiery passion and deepest wisdom. As for cadets, they have been free from the course design activities for a long time, so another problem is the serious lack of their initiative and enthusiasm of course design. In addition, the main body of personnel lack necessary contacts and interact with the cadets, therefore, it is difficult to take course design as a positive behavior.

7.2.4 *Misunderstanding of Roles*

The role is the product of human socialization. It is the unity of responsibility, right, and interests of people in the practice of social life. It is embodied by a series of role abilities of cognitive ability, role expectation, role attitude, and role playing [3]. How relevant personnel understand themselves and the status and role of others in course design affects the duty confirmation of designers and emotion and wills in course design. Two groups of course designer are the main personnel and participants, and there is a clear difference between them. The misunderstanding of roles, such as role positioning is not accurate, or ignoring the role, frequently appears in the process of course design.

Error positioning of the main part of course designers' roles can be summed up in two words: "maximize" and "minimize." For a long time, the main military academy course designers have a common notion with their own role and position, and they think that they do not play any direct role in the course design. What they mention often is that "we are the performers. We just perform what we are asked to. Personal roles are not embodied in the course design." So they fail to see the importance of their roles as designers, and to be aware of their responsibility and mission, and they ignore the cadets' role expectations about them. This is to "minimize" the course designers' own role consciousness. In the meantime, due to the deep-rooted ideology of academic authority in the process of course design, as well as the "official standard" of teaching management personnel, the course designers carry their academic authority and administrative authority up to the unquestionable degree in course design. This is to "maximize" the course designers' own role. Inaccurate positioning of the course designers' own role makes the course design lack democracy, interaction, and development.

As course design participants, the cadets' role and function has not been recognized since long. The military academy and teachers ignore the cadets' role, and the cadets themselves are unaware of their importance to course design. On the one hand, the management hegemony of executive authority, the course hegemony of academic authority, and teaching hegemony of teaching faculty has imprisoned the minds of the military academy course designers, so that cadets have not been included in the scope of participation of course design [4].

On the other hand, after enrollment, the cadets will follow the course program to start learning. Learn. What to learn, learn to what extent, and even how to learn, will not worry the cadets. What the cadets should do only is to attend class on time, listen carefully, remember key points, and get a high score, which has become a common way of all military academies to educate cadets. Cadets have become accustomed to it. So in the way of learning, they are used to receiving ready-made conclusions; in the management of understanding, they are accustomed not to violate the rules and regulations; in the ideal pursuit, they are accustomed to let it be; in the emotional tendency, they are accustomed to herd behavior. The subject consciousness of cadets is not clear. They do not have the desire to express their own wishes [5].

7.3 Conclusion

Knowing the misunderstanding and analyzing it aims to eliminate the misunderstandings. The effective way to eliminate the understandings is to enhance the psychological quality of the military academy course designers. Psychological quality is the core of the military academy course designers overall quality. The enhancement of psychological quality is beneficial to promoting the perfection of their own personality, the maintaining of a good working attitude, and the development of their potency, so that the quality of courses and the quality of education in military academy can be ultimately improved, and the goal of developing qualified military personnel of all-around development can be reached.

References

1. Alsubel etc. (1994) Education psychology—the cognitive view [M] (trans: Yu X). People's education press, Beijing
2. Huang G, Cai Q (2005) Course design—theory and practice [M]. Normal university press, Nanjing
3. Lu J, Wei Q, Li Q (2004) Psychology—the basic theory and education application (revision) [M]. Shanghai people's press, Shanghai
4. Zhu R, Dong H (1999) Modern military curriculum theory [M]. Military science press, Beijing
5. Wang Z, Jiang J (2005) Walk up to continuable development—the only way military education [M]. PLA's press, Beijing

Chapter 8
A Research on Operator Selection Criteria of a Certain Type of Equipment

Zhaofeng Luo, Zhibing Pang, Jun Chen, Hongyan Ou, Yu Jin and Cheng Jin

Abstract This paper is on purpose to increase the scientificity and reasonability of the selection and improve the training effect through the research on selection indices of certain equipment. It adopts the methods of consulting literature materials, expert marking, and practical testing to research the effects of physical and psychological indices on the selection of operators. The specific model of weapon has unique requirements on physiology and mental qualities of operating personnel. The physical and psychological parameter indices of operator are presented through data processing; the indices priority is determined and they are endowed with weights. Some of the physical indices are necessary for the selection, and the left physical and psychological indices have certain effect on the operator. The research result of this paper will provide feasible selection criteria for the operator selection which is of practical significance.

Keywords Operator · Selection · Selection criteria

8.1 Significance of Selection

The selection means choosing operators from serviceman group, who are qualified to the operating of specific model of equipment. Seeking the best combination between operator and equipment is the essence of selection [1]. This paper is defined as the purposes which can operators to be capable of their posts, it discusses the selection and evaluation through characteristics of equipment and person's operating. Thus, the working efficiency can be increased and best performance can be fully displayed. There are mainly three significances:

First, with the development of informatization, military equipment develop rapidly, military posts change from few to even more, division of labor become

Z. Luo (✉) · Z. Pang · J. Chen · H. Ou · Y. Jin · C. Jin
Air Defence Forces Academy, Zhengzhou 450052, China
e-mail: fkb_mmes@163.com

© Springer-Verlag Berlin Heidelberg 2015

S. Long and B.S. Dhillon (eds.), *Proceedings of the 15th International Conference on Man–Machine–Environment System Engineering*, Lecture Notes in Electrical Engineering 356, DOI 10.1007/978-3-662-48224-7_8

more detailed, and there exits great change of requirements to person. According to the situation of our army now, the selection also has some shortcomings. For example, there are more qualitative analysis and less quantitative analysis and the selection lack of scientific issue. The shortages of the soft power may become a bottleneck to the development of our army and become a "weak link" which should be developed greatly (Table 8.1).

Second, according to the combat power, selecting the personnel who can adapt to the demand of these posts better can shorten the cycle for equipment to be combat-ready and accelerate the combat ability. At the same time, the selected people have corresponding conditions and meet the basic quality of post demands. Finally, the human errors can be reduced efficiently, the integrity of equipment can

Table 8.1 Index system of operator selection

First-class indices	Second-class indices	Third-class indices
Physiological factors	Measurement of physical condition	Height
		Arm length
	Visual sense	Eyesight
		Visual directional search capability
	Auditory sense	Hearing
		Hearing positioning ability
	Response speed	Vision response speed
		Hearing response speed
	Physical strength	Arm power (pull)
		Arm power (push)
		Strength of right index finger
	Physiological adaptability	Endurance
		Anti-fatigue capability
		Anti-vertigo ability
	Overall coordination	Body balance capability
		Instantaneous stability
		Handedness
		Close the left eye tight
Psychological factors	Ability structure	Wisdom
		Observational ability
		Memory
		Attention
		Reaction time
	Character Structure	Braveness
		Self-discipline
		Stability
		Perseverance
		Anxiety

be ensured, the occurrence of accident can be prevented efficiently, and the non-combat loss can be reduced.

Third, from the angle of the economics, if the distribution of post is reasonable, people who meet the equipment post demands are selected to hold the post, the allocation human resources rationally and effectively can be realized; training time can be shortened and training costs will be decreased so that the whole economic benefits of military training will be increased.

8.2 Research of Building of Selection Model

Operator selection means the testing and judging of physiology and mental qualities [2] of candidates with the help of special testing device and application of scientific theory achievement according to different requirements of equipment, choose operators to be capable of their posts according to impersonal demands.

8.2.1 Build Index System of Selection

Building of a center type of equipment operator selection index system is a groundbreaking work in the field of specific post staff selection. On the premise of referring to plenty of domestic and international materials, this paper logically analyzes and arranges the selection on the basis of small-scale questionnaires, experimental study, expert consultation, and group discussions. From analysis and arrangement, some obvious repetitive and in-terembracing selection indices are eliminated preliminarily, each index is classified according to objective analysis method in order to screen indices [3, 4]. The preliminary index system follows:

8.2.2 Determine Indices Priority

This decision is taken by votes of experts. The priority 1 indices are the indices which are considered as necessary indices by more than 80 % of experts. There are 15 members in the experts group: 4 relevant experts from the academies, 3 main responsible leaders and counterparts staff officers, 2 direct-level officer, and 6 experienced operators.

From the votes of experts, some psychological indices can meet the requirement through corresponding training, while some physical indices are unable effectively to meet the requirement after training which are defined as priority 1. The indices which can be changed after training are defined as priority 2. The results are shown in Tables 8.2 and 8.3.

Table 8.2 Priority 1 indices

First-class indices	Second-class indices	Third-class indices
Physiological factors	Measurement of physical condition	Height
		Arm length
	Visual sense	Eyesight
		Visual directional search
	Auditory sense	Hearing
		Hearing positioning ability
	Response speed	Vision response speed
		Hearing response speed
	Overall coordination	Handedness
		Close the left eye tight

Table 8.3 Priority 2 indices

First-class indices	Second-class indices	Third-class indices
Physiological factors (U_1)	Physical strength (a_1)	Arm power (pull) (a_{11})
		Arm power (push) (a_{12})
		Strength of right index finger (a_{13})
	Physiological adaptability (a_2)	Endurance (a_{21})
		Anti-fatigue capability (a_{22})
		Anti-vertigo ability (a_{23})
	Overall coordination (a_3)	Body balance capability (a_{31})
		Instantaneous stability (a_{32})
Psychological factors (U_2)	Ability structure (a_4)	Wisdom (a_{41})
		Observational ability (a_{42})
		Memory (a_{43})
		Attention (a_{44})
		Reaction time (a_{45})
	Character Structure (a_5)	Braveness (a_{51})
		Self-discipline (a_{52})
		Stability (a_{53})
		Perseverance (a_{54})
		Anxiety (a_{55})

8.2.3 Determine the Indices Weight

On the basis of selected evaluation indices, determining the weight of indices is the key point. The determining of weight refers to the behavioral science which can not be obtained directly from mathematical analysis method. After expert scoring,

Table 8.4 Weight voting tables of experts

Experts num	Factors			Σ
	a_{11}	a_{12}	a_{13}	
1	0.333	0.333	0.334	1
2	0.25	0.25	0.5	1
⋮	⋮	⋮	⋮	⋮
15	0.2	0.45	0.35	1
Weight a_{1i} (i = 1, 2, …,)	0.29	0.338	0.372	1

opinions from experts will be centralized and corrected through practical test to determine the weight [5, 6].

First, make and hand out the questionnaires, determine the weight of each index through experts' weight marking.

Take a1 as an example, assess the relative important degree of 3 factors of a1; the predetermined weight grade interval is 0.05; the distributive weight sum of the 3 factors is 1 when every expert vote (Table 8.4).

Second, calculate weight vector. Take a_1 as an example, calculate and process the weight of 3 factors of a1.

From the above table and weight calculation method, the weight of the first factor a1 is:

$$A_1 = \frac{1}{15}\sum_{j=1}^{15} a_{1j} = (0.290, \quad 0.338, \quad 0.372). \tag{8.1}$$

Repeat this process, the weight of each index given by every expert can be obtained. Here, we suppose the weights given by each expert are same, so the average of the weight given by each expert is the final calculated weight of the index system. Analogously, the weights of a2, …, a5 can be obtained.

$$A_2 = (0.362, \quad 0.316, \quad 0.322)$$

$$\cdots$$

$$A_5 = (0.222, \quad 0.222, \quad 0.267, \quad 0.113, 0.176)$$

The same procedure may be easily adapted to obtain the weights of the 5 factors of *A* and 2 factors of *U*

$$A = (0.266, \quad 0.333, \ 0.401, \ 0.553, \ 0.447)$$
$$U = (0.567, \quad 0.433)$$

Third, determine the final weight. Summarizing the questionnaires, make use of the above method to obtain the weight of priority 2 indices of a certain type of equipment operator selection index system.

8.2.4 Measure and Assessment

According priority 1, weed out the candidates directly whose indices can not meet the demand of priority 1. In accordance with the priority 2, measure the corresponding indices of candidates, the score of each item of the three-level indices is 100, evaluation V is taken by the testing results or assessment, the desirable V always is given by assessment experts. Usually V ($V = \{V_1 \ldots V_n\}$) is equal to 100, 80, 60, 40, 20, which represent excellent, good, medium, poor, and very poor. The score formula of a candidate follows:

$$F_i = \sum V_i A^T. \tag{8.2}$$

Rank the candidates according to their scores to finish the operator selection. This method can also be used to assess the skilled operators to modify some indices and the weights (Table 8.5).

Table 8.5 Priority 2 indices

First-class indices	Second-class indices	Third-class indices
Physiological factors (0.567)	Physical strength (0.266)	Arm power (pull) (0.290)
		Arm power (push) (0.338)
		Strength of right index finger (0.372)
	Physiological adaptability (0.333)	Endurance (0.362)
		Anti-fatigue capability (0.316)
		Anti-vertigo ability (0.322)
	Overall coordination (0.401)	Body balance capability (0.467)
		Instantaneous stability (0.533)
Psychological factors (0.433)	Ability structure (0.553)	Wisdom (0.206)
		Observational ability (0.293)
		Memory (0.122)
		Attention (0.193)
		Reaction time (0.186)
	Character structure (0.447)	Braveness (0.222)
		Self-discipline (0.222)
		Stability (0.267)
		Perseverance (0.113)
		Anxiety (0.176)

8.3 Conclusion

This paper presents an index assessment system with different priorities through the building of operator selection system of certain equipment and opinions from experts and operators. Weights of priority 2 indices are endowed in this research which can provide some theoretical basis and concrete measures for the selection. After direct assessment toward the skilled operators and assessment toward the candidates, the testing data will not be opened, compare the testing data with actual training results after training and correct some indices and the weights in the system, so the reasonability of the selection system will be increased gradually. In the above index system, most indices can be measured correctly with the deep research on man machine environment system engineering. But there are no scientific and easy ways and means for some key indices which we urgently need, such as anti-vertigo ability, body balance capability, and instantaneous stability. We should do further research into these three indices, sum up the rules, and seek for the feasible and effective methods.

References

1. Pang Z (1992) Operator's skill. PLA Publishing Press, Beijing, pp 2–4
2. Pang Z (2000) Man-machine-environment system engineering of air defense forces. Zhengzhou Air Defense Forces Academy, Zhengzhou, pp 71–120
3. Psychology Institution of the Chinese Academy of Sciences (1993) Modified Cattell's chart of 16 personality factors.
4. The Army Military Training and Testing Program, Beijing. 2008:35
5. Li T, Pang Z, Yin Q, Fei H, Li H, Wu X (2011) Physiology quality and psychological traits of operators in multioperator single machine. In: The 11th man-machine-environment system engineering conference proceedings, pp 98–101
6. Qin L, Pang Z, Yin Q, Wu J, Ji C (2010) Physiology quality and psychological traits of operators in single operator single machine. In: The 10th man-machine-environment system engineering conference proceedings, p 110–116

Chapter 9
Operation Optimization of a Chariot Operator Using Remote Measurement of Heart Rate

Hongyan Ou, Honglei Li, Zhingbing Pang, Shuai Mu and Chenhui Li

Abstract In this study, remote measurement of heart rate was applied for qualitative and quantitative analysis, in order to optimize the process and improve the effectiveness of a chariot's operation. The methods used to investigate heart rate burden and time depletion of the operator included reality-based experiments, instrument measuring, comparison, and analysis. The heart-rate remote-measuring system was used in reality-based experiments; the experiments collected heart burden and time depletion before and after operation process optimization. After operation optimization, the heart rate of the operator decreased, as did the required operation time. These results may help to promote combat effectiveness and improve combating capability for members of the army.

Keywords Heart-rate remote measuring · Operator · Operation process · Optimization

9.1 Introduction

The principle of man–machine operation optimization was put forward according to the understanding that "the person's energy could not be stored; if the energy was not effectively used it would be wasted; and if the energy was excessively consumed the person would feel fatigued" [1]. The purpose of this principle is to shorten operation time, increase the useful workload, and decrease operation fatigue. Accordingly, operation staff are required to make full use of their operational abilities, and they also should harmonize the action of their hands and each part of their body. They are also required to improve complicated, slack, and superfluous actions, as well as to avoid unnecessary actions. In this way, an operation should become more scientific and effective.

H. Ou (✉) · H. Li · Z. Pang · S. Mu · C. Li
Air Defense Forces Academy, Zhengzhou 450052, China
e-mail: fkb_mmes@163.com

© Springer-Verlag Berlin Heidelberg 2015

S. Long and B.S. Dhillon (eds.), *Proceedings of the 15th International Conference on Man–Machine–Environment System Engineering*, Lecture Notes in Electrical Engineering 356, DOI 10.1007/978-3-662-48224-7_9

This study focused on a chariot in the command-and-control center of an advanced weapons system [2]. The chariot is an important firepower unit to promote the development of combat effectiveness in the armed forces. The equipment is a kind of individual weapon that requires qualified physical ability, skill, and intelligence [3]. Therefore, it is very important to reinforce the optimization analysis of the operator. It plays a crucial role in the effective work of the operator and combat effectiveness of the equipment.

9.2 Experiment

9.2.1 Subjects

The participating personnel were mainly of three types. The first type of personnel consisted of three operators of a certain chariot, with one operator at each position. The three operators were respectively called A, B, and C. Before the experiment, the three operators had mastered the operation of their positions, their physical condition was good, and their psychological status was normal. They had the ability to complete the experiment. The second type of personnel included ten experts who came from field units of the troops, military academies, research institutes, and military factories. These experts were responsible for designing the operation optimization of the operators. The third type was the personnel who organized the experiment: three instructors from military academies. They were responsible for adjusting and testing the instrument and equipment, as well as collecting and analyzing the experiment data.

9.2.2 Equipment

The equipment included a chariot that was used by the three operators. The heart-rate remote-measuring system "ACUMEN" was used for monitoring the operators' heart rate, physiology burden, and operating time. It also was used to collect experimental data and generate a heart-rate remote-measuring diagram, which provided a quantitative basis and data support for operation optimization analysis. One computer was used to analyze and process the experimental data.

9.2.3 Requirements

The three operators were required to have attained a well-qualified degree for the post operation prior to the experiment [4]. During the experiment, the three operators wore the heart-rate remote-measuring system as required, which ensured the credibility of the data. During the experiment, the three operators also conformed to operation requirements and test standards. If the operation appeared to violate the rules or a false phenomenon occurred, the data from the heart-rate remote-measuring system were not used.

9.2.4 *Contents*

The combat operation of the chariot was divided into different phases. Each phase was divided into a number of steps [5]. If a certain step was representative of the physical requirements and technical ability needed by the operators, the step was selected for the experiment. The experiment was divided into two stages: the first stage was the operation before optimization, and the second stage was the operation after optimization.

9.2.4.1 The Operation Before Optimization

At this stage, the three operators wore the heart-rate remote-measuring system. The operators carried on a normal operation process according to their peacetime training.

The operation's actions and process are shown in Fig. 9.1.

9.2.4.2 The Operation After Optimization

After the first stage of the experiment was completed, the expert team thoroughly discussed the action and operation flow of the operators. They eliminated any unnecessary actions, optimized the operation path, strengthened the coordination

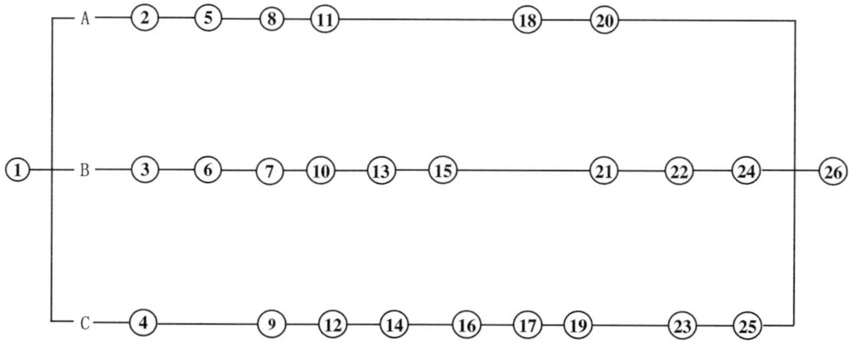

Fig. 9.1 Operation flow chart before optimization

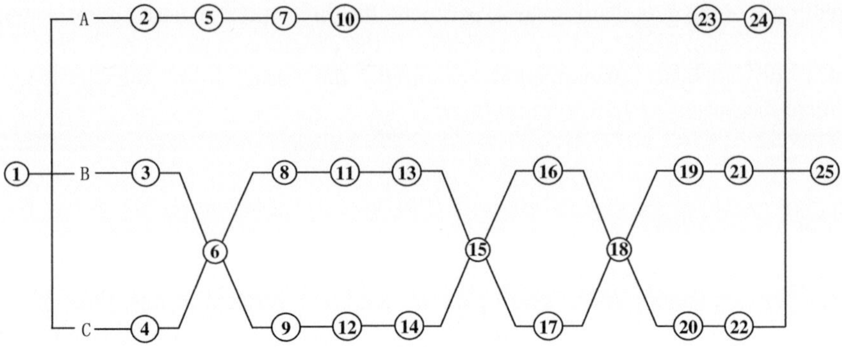

Fig. 9.2 Operation flow chart after optimization

among the operators, and constructed an operation flow chart after optimization. Then, the three operators again put on the heart-rate remote-measuring system and conducted an operation according to the flow chart after optimization. The three operators processed together and completed 25 operation actions in total and in sequence.

The operation's actions and process are shown in Fig. 9.2.

9.3 Experiment Data Analysis

9.3.1 Data Analysis Before Optimization

During the experiment, the heart-rate remote-measuring system recorded a diagram of the three operators' heart rates before and after optimization. By analyzing the data, we found that the three operators' heart rates showed different characteristics. The results for operator A and operator C were normal, but operator B showed an abnormality. We performed a detailed analysis of operator B as a typical case. The remotely measured heart rate results for operator B before optimization are shown in Fig. 9.3.

As shown in Fig. 9.3, before optimization, it took operator B 210 s to finish the whole operation. The heart rate of the operator B was mainly in the purple aerobic area. However, the heart rate reached the extreme limits area (red) three times, once exceeding 220 beats per minute.

When the heart rate reached the extreme limit area, it indicated that the operation at this time was heavy-loaded and the operator felt fatigue. Under this condition, the probability for error was large. If the operator remained in this condition for a long period of time, the operator could easily be injured [6]. Therefore, the operator's training should aim to avoid or reduce the occurrence of this phenomenon as soon as possible.

The main reasons for this phenomenon could include the following: First, the workload and/or difficulty of the operation was greater for operator B. Secondly, in

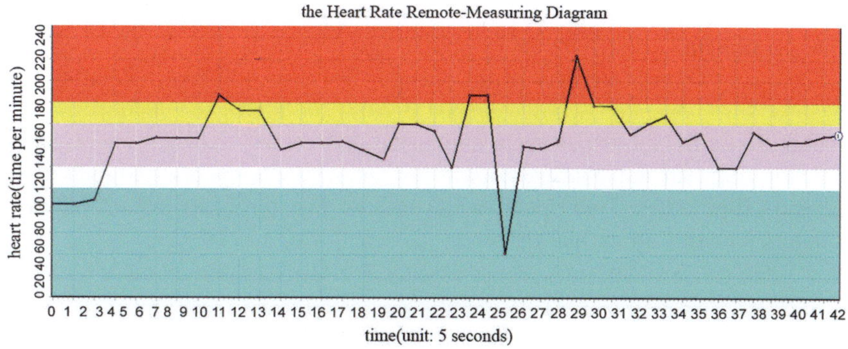

Fig. 9.3 Operator B's heart rate before optimization

the process of the experiment, operator B twice climbed to the top of the chariot from the ground, so the elevation gain was greater as it was 3.1 m high. To complete this task in the relatively short time, operator B had to improve the operating speed, which may make the heart rate burden worse.

9.3.2 Data Analysis After Optimization

Operator B's heart rate results after optimization are shown in Fig. 9.4.

As shown in Fig. 9.4, after optimization, it took operator B 175 s to finish the whole operation. The heart rate of operator B was mainly in purple aerobic area.

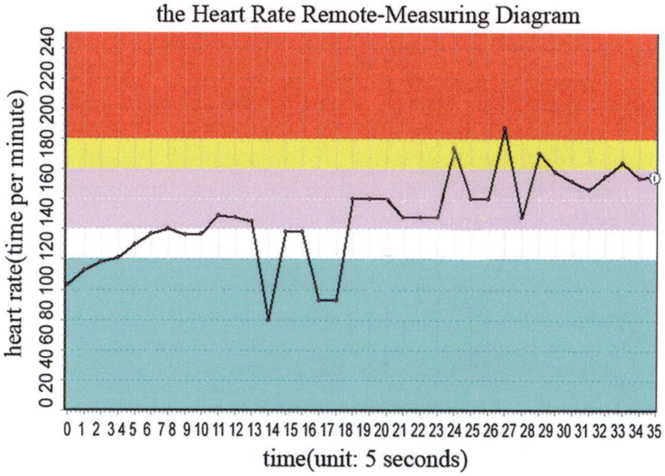

Fig. 9.4 Operator's B heart rate after optimization

Once, it reached the extreme limit area (red), but the maximum value was only 190 beats per minute.

There are two possible reasons for operator B's change in heart rate after optimization. First, operator B only climbed once to the top of the chariot from the ground. Secondly, the expert team eliminated unnecessary actions, optimized the operation path, strengthened the coordinator among the operators, and revised the operation flow chart. This optimization ensured that the heart rate of all three operators remained in normal range.

9.4 Conclusion

By comparing the operation flow chart with the heart-rate remote-measuring results before and after optimization, we can draw some conclusions. First, operator B independently completed nine actions before optimization. After optimization, operator B completed a total of ten actions: seven actions were independently completed and three actions were finished in conjunction with operator C. Compared with before optimization, operator B completed one more action after optimization. Second, operator B required 210 s before optimization and 175 s after optimization—a reduction of 35 s after optimization. Third, the heart rate of operator B was much higher before optimization, reaching the extreme limit area three times. The heart rate of operator B obviously decreased after optimization, only reaching the extreme limit area once—two times less after optimization.

Based on combat operation requirements and test standards, after operation optimization, the operating time was shortened, the operation burden was decreased, and the efficiency was enhanced. Therefore, the operation optimization process of this experiment was reasonable.

From the perspective of man–machine–environment systems engineering and research, we combined quantitative analysis with qualitative analysis to optimize the operation of a chariot. The man–machine operation optimization principle and a heart rate remote-measuring system were used in the study. These results can promote combat effectiveness and improve the combating capability of the army.

We hope this study encourages the development of man–machine operation optimization theory [5]. Because our data are limited and the sample size was small, we plan to complete a larger study in the future.

References

1. Guo X, Guo D, Lin Y (1991) Conspectus of the weapon operation subject. Tianjin Translate Publisher of Science and Technology, Tianjin, p 208
2. Ou H, Pang Z, Mu S, Li H, Yin Q, Zhang G (2013) Study on southeast and inshore area environmental worthiness of a certain equipment. In: Proceedings of the 13th international conference on man-machine-environment system engineering, Oct 2013

3. Ou H, Meng X, Pang Z, Liu D, Yin Q (2013) Study on man-machine integration operating power of a certain equipment. In: Proceedings of the 11th conference on man-machine-environment system engineering, Oct 2013
4. Pang Z (1992) Skill ability building of operator. Liberation Army Publisher, Beijing, p 8
5. Zhu Z (2004) Industrial psychology dictionary. Zhejiang Education Publisher, Zhejiang, p 3
6. Operation and Using Manual of A Certain Chariot (2008), p 05

Chapter 10
Analysis and Research of Grenadiers' Comprehensive Quality Based on Fuzzy Comprehensive Evaluation

Qian Shen, Haiyan Dai, Peng Han, Chunxin Wang, Chen Shen and Wenlong Zhang

Abstract To a great extent, how to make grenades function well and reduce accidents when throwing them depends on the comprehensive quality and ability of grenadiers. Subjective and objective factors should be considered when evaluating grenadiers' quality during training. In order to evaluate the quality of grenadiers, physical condition, operation ability, mental factor, and daily performance of grenadiers are analyzed and quantified as the base. This paper builds fuzzy comprehensive evaluation model of grenadiers' quality which is based on fuzzy comprehensive evaluation theory and adopts second-level fuzzy evaluation, also provides application cases, and obtains a result which is accordance with the actual condition.

Keywords Grenade · Grenadiers' quality · Fuzzy comprehensive evaluation

Grenade is a kind of small-sized throwing ammunition for attack and defense, used widely and largely as well. It can not only hit the effective target but also destroy tanks and armor vehicles. Grenades will be used largely to kill enemies under the high-tech condition in the future war [1]. This paper adopts fuzzy comprehensive evaluation to quantify the index of grenadiers' comprehensive quality, make standardized solution, and then offer the basic thinking of fuzzy comprehensive evaluation and modeling method, subsequently calculating specific example to get the same result with the facts.

Q. Shen (✉) · H. Dai · P. Han · C. Wang · W. Zhang
Air Defense Forces Academy, Zhengzhou 450052, China
e-mail: shenqian_0911@163.com; fkb_mmes@163.com

C. Shen
Xi'an Communication Academy, Xi'an 710106, China

© Springer-Verlag Berlin Heidelberg 2015

S. Long and B.S. Dhillon (eds.), *Proceedings of the 15th International Conference on Man–Machine–Environment System Engineering*, Lecture Notes in Electrical Engineering 356, DOI 10.1007/978-3-662-48224-7_10

10.1 Evaluation Index Construction and Quantization

10.1.1 Evaluation Index Construction

On the basis of expert knowledge and long-term practical experience [2], an excellent grenadier should have the following characteristics: healthy physical condition, skilled operation ability, steady mental quality, and excellent daily performance. Index and requirements are specifically shown in Table 10.1.

10.1.2 Evaluation Index Quantization

First of all, these indexes should be made standardized solution for comprehensive evaluation. On the basis of Delphi method, fuzzy statistics method is adopted to confirm. But expert experience method is usually adopted [3]. That is:

$$uV_j(u_i) = M_{ij}/p. \tag{10.1}$$

In this formula, M_{ij} is the frequency of $u_i \in V_j$; p is the numbers of expert estimator; $uV_j(u_i)$ is the membership function of $u_i \in V_j$.

10.2 Fuzzy Comprehensive Model of Grenadiers' Quality

10.2.1 Confirm Grenadiers' Evaluation Factor Sets

There are four factors which are physical condition, operation ability, mental quality, and daily performance to confirm grenadiers' quality according to Table 10.1. These four factors can make one factor set U.

$$U = \{U_1, U_2, U_3, U_4,\}$$

10.2.2 Build Evaluation Sets

On the basis of general standard of fuzzy comprehensive evaluation, excellent, above average, average and below average are represented by V_1, V_2, V_3, V_4 respectively, and then build evaluation sets $V = \{v_1, v_2, v_3, v_4\}$. For seeing much more clearly, evaluation level and evaluation will be represented by hundred-mark system. As shown in Table 10.2.

Table 10.1 Comprehensive quality evaluation diagram of grenadiers

Index (U)		Level factor (V)			
		Excellent (V_1)	Above average (V_2)	Average ($V3$)	Below average (V_4)
Physical condition (U_1)	Vision	2.0–1.5	1.5–1.0	1.0–0.6	≤0.6
	Age	29–21	21–18	29–35	≤18 or ≥35
	Fatigue period	12–14 (H)	10–12 (H)	10–7 (H)	≤7 (H)
	Total sick leave	≤10 (D/Y)	10–20 (D/Y)	20–4-(D/Y)	≥40 (D/Y)
	Physical fitness mark	100–8 5 (score)	85–75 (score)	75–60 (score)	≤60 (score)
Operation ability (U_2)	Education level	Above under graduate	Junior college	Senior high school or Junior college	Junior high school
	Subject theory	100–85 (score)	85–75 (score)	75–60 (score)	≤60 (score)
	Throwing distance	≥40 (M)	35–40 (M)	30–35 (M)	≤30 (M)
	Throwing accuracy	First-round hit	Second-round hit	Third-round hit	Above third-round hit
	Training duration	300	3–2 (Y)	2–1 (Y)	≤1 (Y)
	Obstacle-removing ability	Remove difficult obstacle alone	Remove simple obstacle alone	Remove simple obstacle with	Unable
Mental quality (U_3)	State of confidence	Best	Better	Good	Bad
	Emotion rhythm	Best	Better	Good	Bad
	Mental rhythm	Best	Better	Good	Bad
	Response time	≤0.13 (s)	0.13–0.14 (s)	0.14–0.16 (s)	≥0.16 (s)
Daily performance (U_4)	Award and punishment	Above merit citation class III or two citations	citation	None	Caution
	Leader's evaluation	Best	Better	Good	Bad
	Common satisfaction	≥90 %	90–80 %	80–60 %	≤60 %
	Cooperative ability	Have strong team cooperative consciousness	Have rather strong team cooperative consciousness	Have team cooperative consciousness	None
	Hobby	Love activities and get award	Often join in activities	Occasionally join in activities	None

Table 10.2 Mark level standard of evaluation index

Level	1	2	3	4
Evaluation	Excellent	Above average	Average	Below average
Mark	0.8–1.0	0.6–0.8	0.4–0.6	0–0.4

10.2.3 First-Level Evaluation

Each factor U_i (i = 1, 2,3, 4) will be evaluated comprehensively by initial model. In order to avoid the subjective problem from expert to confirm target weight, the combination method of AHP and Delphi will be adopted to confirm each index target. If getting all factors weight $W_i = \left(w_{ij}\right)_{N \cdot K}, (i = 1, \ldots N; j = 1, \ldots K_i)$ in U_i, each factor in U_i should build fuzzy mapping; generally adopt formula 1 to confirm membership of each index to each level, then obtain fuzzy relationship matrix $R_i = \left(r_{jm}\right)_{k_i \cdot p}, (j = 1, \ldots k;\ m = 1, \ldots p)$, and then obtain formula

$$B_i = (b_{i1}, b_{i2}, \ldots b_{im}) = W_i \cdot R_i, (i = 1, \ldots N; m = 1, \ldots P). \tag{10.2}$$

10.2.4 Second-Level Evaluation

If $U = \{U_1, U_2, \ldots, U_N\}$, choose B_i as single factor from U_i to evaluate, build fuzzy mapping, and then obtain fuzzy relationship matrix

$$R = \begin{pmatrix} B_1 \\ B_2 \\ \cdots \\ B_N \end{pmatrix} = \begin{pmatrix} b_{11} & b_{12} & \ldots & b_{1m} \\ b_{21} & b_{22} & \ldots & b_{2m} \\ \ldots & \ldots & \ldots & \ldots \\ b_{N1} & b_{N2} & \ldots & b_{Nm} \end{pmatrix}. \tag{10.3}$$

Suppose the weight distributions of factors in $U = \{U_1, U_2, \ldots, U_N\}$ are

$$W = \{W_1, W_2, \ldots, W_N\}. \tag{10.4}$$

And then obtain

$$B = W \cdot R = (W_1, W_2, \ldots, W_N) \cdot \begin{pmatrix} b_{11} & b_{12} & \ldots & b_{1m} \\ b_{21} & b_{22} & \ldots & b_{2m} \\ \ldots & \ldots & \ldots & \ldots \\ b_{N1} & b_{N2} & \ldots & b_{Nm} \end{pmatrix}. \tag{10.5}$$

After obtaining evaluation index from above formula, as the maximum membership degree principle, the evaluation result will be gotten from evaluation factors which correspond with maximum evaluation index in vector B; that is, fuzzy comprehensive evaluation conclusion. Meanwhile [4], comprehensive evaluation value can be obtained from the formula given below:

$$U = V_P \cdot B^T. \tag{10.6}$$

In this formula, $V_P = (1,\ 0.8,\ 0.6,\ 0.4)$.

10.3 Example Analysis

Second-level fuzzy comprehensive evaluation is adopted to evaluate grenadiers' comprehensive quality [5]. One grenadier's information is shown in Table 10.3.

Assume that ten experts will be chosen to mark relative importance of each index, on the basis of above result, AHP and Delphi methods are adopted to confirm each index weight, that is

$$W = (0.18,\quad 0.45,\quad 0.25,\quad 0.12), W_1 = (0.35,\quad 0.16,\quad 0.15,\quad 0.10,\quad 0.24)$$
$$W_2 = (0.10,\quad 0.12,\quad 0.35,\quad 0.22,\quad 0.08,\quad 0.13), W_3(0.24,\quad 0.26,\quad 0.15,\quad 0.35)$$
$$W_4 = (0.18,\quad 0.25,\quad 0.22,\quad 0.21,\quad 0.14)$$

3.1 first-Level

Fuzzy Comprehensive Evaluation to Grenadiers' Physical Condition(1) According to data from Table 10.3, choose 4 experts combined with data in Tables 10.1 and 10.2 to give quantification evaluation as formula (10.1) to one grenadier's physical condition U_1 and then confirm its fuzzy evaluation matrix.

$$R_1 = \begin{pmatrix} 0.8 & 0.75 & 0.75 & 0.8 \\ 0.8 & 0.75 & 0.8 & 0.75 \\ 0.8 & 0.75 & 0.8 & 0.75 \\ 0.6 & 0.6 & 0.6 & 0.6 \\ 0.8 & 0.75 & 0.8 & 0.75 \end{pmatrix}$$

(2) And then calculate fuzzy evaluation vector $B_1 = W_1 * R_1 = (0.78,\ 0.735,\ 0.7625,\ 0.7525)$ as formula (10.2).

10.3.1 First-Level Fuzzy Comprehensive Evaluation to Grenadiers' Operation Ability, Mental Quality, and Daily Performance

The same procedure is adapted to obtain grenadiers' operation ability U_2 whose fuzzy comprehensive matrix

Table 10.3 Information table of one grenadier

Name	Physical condition					Operation ability					
	Vision	Age	Fatiue period	Total sick leave	Physical fitness mark	Education level	Subjecttheory	Throwing distance	Throwing accuracy	Obstacle-removing ability	Training duration
	1.3	21	10	22	82	Junior college	90	35	Second-round hit	Remove simple obstacle alone	1 Year
Wang XX	Mental quality					Daily performance					
	State of confidence		Emotion rhythm	Mental rhythm	Response time	Award and punishment		Leader' s evaluation	Common satisfaction	Cooperative ability	Hobby
	Better		Good	Better	0.13	None		Better	75 %	Have team cooperative consciousness	None

$$R_2 = \begin{pmatrix} 0.8 & 0.75 & 0.75 & 0.8 \\ 1 & 0.9 & 1 & 0.9 \\ 0.8 & 0.75 & 0.8 & 0.75 \\ 0.8 & 0.75 & 0.8 & 0.75 \\ 0.8 & 0.75 & 0.8 & 0.75 \\ 0.6 & 0.6 & 0.6 & 0.6 \end{pmatrix},$$

$$B_2 = W_2 * R_2 = (0.798,\ 0.7485,\ 0.793,\ 0.7535).$$

Grenadiers' mental quality U_3 whose fuzzy comprehensive matrix

$$R_3 = \begin{pmatrix} 0.8 & 0.75 & 0.75 & 0.8 \\ 0.6 & 0.6 & 0.6 & 0.6 \\ 0.8 & 0.75 & 0.8 & 0.75 \\ 0.8 & 0.75 & 0.8 & 0.75 \end{pmatrix},$$

fuzzy evaluation vector $B_3 = W_3 * R_3 = (0.748,\ 0.711,\ 0.736,\ 0.723)$. Grenadiers' mental quality U_4 whose fuzzy comprehensive matrix

$$R_4 = \begin{pmatrix} 0.6 & 0.6 & 0.6 & 0.6 \\ 0.8 & 0.75 & 0.75 & 0.8 \\ 0.6 & 0.6 & 0.6 & 0.6 \\ 0.6 & 0.6 & 0.6 & 0.6 \\ 0.2 & 0.3 & 0.2 & 0.3 \end{pmatrix},$$

fuzzy evaluation vector $B_4 = W_4 * R_4 = (0.694,\ 0.5955,\ 0.6815,\ 0.608)$.

10.3.2 Second-Level Fuzzy Comprehensive Evaluation to Grenadiers' Quality

It is needed to consider all aspects evaluation to evaluate grenadiers' quality completely, that is, evaluate four subsets in evaluation factor sets $U = \{U_1,\ U_2,\ U_3,\ U_4,\}$ and then build fuzzy relationship matrix R

$$R = \begin{pmatrix} 0.78 & 0.735 & 0.7625 & 0.7525 \\ 0.798 & 0.7485 & 0.793 & 0.7535 \\ 0.748 & 0.711 & 0.736 & 0.723 \\ 0.694 & 0.5955 & 0.6815 & 0.608 \end{pmatrix}$$

to confirm each factor influence weight distribution vector $W = (0.18, 0.45, 0.25,$ and $0.12)$ and fuzzy evaluation vector $B = A{*}R = (0.76978, 0.718335, 0.75988,$ and $0.728235)$ in U.

On the basis of maximum membership principle, 0.76978 in B is equal to Max = (0.76978, 0.718335, 0.75988, and 0.728235), and its location corresponds with excellent level, thus, this grenadier's comprehensive evaluation is excellent. Meanwhile, according to fuzzy relationship matrix R and maximum membership principle, this grenadier's physical condition, operation ability, mental quality, and daily performance which are excellent, average, above average, average, respectively, can be confirmed; and it means that physical condition and mental quality of this grenadier are the best and he should enhance operation ability and daily performance.

10.4 Conclusion

The comprehensive quality of grenadiers is a fuzzy concept, which is not easy to quantify and has many influencing factors. This paper adopts fuzzy comprehensive evaluation to show the fundamental characteristic of decision thinking and reflect fully comprehensive quality of grenadiers, which has very good operational, scientific, and systematic abilities.

References

1. Cao H., Zhang Y. (2008) YuefeiGao. National Defense Industry Press, Weapons Instruction. Beijing
2. Editorial committee edits Shell and Ammunition (2010) Shell and ammunition. Aviation Industry Press, Beijing
3. Jijian X, Liu C (2006) Method and application of fuzzy mathematic. Huzhong Science and Technology University Press, Wuhan
4. Zheng X (2002) Radar effiency evaluation method based on fuzzy analytic hierarchy process. Modern Radar 55(2): 335
5. Ma L (2006) Foundation operational research course. Higher Education Press, Beijing

Chapter 11
An Analysis on the Screening Mechanism of the Manipulator Based on Game Theory

Chunxin Wang, Zaochen Liu, Qian Shen, Peng Han and Shuai Mu

Abstract There exist some problems in the recruitment process of the manipulator in the man-machine environment. We always hope to find high-quality personnel and make them bring greater benefits to the unit; however, high-quality personnel need more human resources cost. According to game theory, we have an analysis on recruitment screening mechanism from the perspective of benefit under the asymmetry information condition based on the signal mechanism between the cost of the benefit and the cost of HR. And we put forward some advice on how to determine a reasonable screening mechanism in accordance with the cost of human resources and the probationary period. We build a model to realize the conclusion and prove that it is feasible through the typical example. The conclusion can provide guidance for screening of the manipulator in all kinds of man-machine environment.

Keywords Game · Manipulator · Screening mechanism

11.1 Introduction

In the man-machine environment, choosing a manipulator is a key problem. For the screening of a manipulator in practical work there are the following two errors: The first is to pay attention to the candidates' qualifications, job title, and working ability, and the employer ignores the payable cost of candidates in the degree, academic title, and the corresponding ability. The second is to ignore the identification mechanism in the probationary period; the employer simply regards the probationary period as the training period during which a manipulator should be familiar with their work and do the related job well. In fact, the policy considering only the difference of degree and ability does not conform to the logic of

C. Wang (✉) · Z. Liu · Q. Shen · P. Han · S. Mu
Air Denfense Forces Academy, Zhengzhou, China
e-mail: chxinwang@163.com

© Springer-Verlag Berlin Heidelberg 2015

S. Long and B.S. Dhillon (eds.), *Proceedings of the 15th International Conference on Man–Machine–Environment System Engineering*, Lecture Notes in Electrical Engineering 356, DOI 10.1007/978-3-662-48224-7_11

economics; a well-educated personnel with strong abilities mean more benefits to the employer, but also high cost and high risk; so, we should establish reasonable screening mechanism to hire the job candidates with the better cost benefit, that is to say, comparison between wages and benefit is much lower. Scientific inspection mechanism in probationary period will eliminate information asymmetry to a maximum degree and ensure that the recruitment personnel has strong ability.

11.2 The Mechanism of the Cost Benefit

In the process of recruiting personnel, we always hope to find high-quality staff to bring greater benefit to the unit. Therefore, in the process of recruiting requirements, the recruiters must have certain qualifications; the staff can be recruited formally through a series of the screening mechanisms such as the written test and job interview. In fact, the screening mechanism is a kind of signal mechanism [1]. In the recruitment process, the employing units in the disadvantaged information position should take the initiative to design an effective mechanism to choose and recruit the personnel in the advantaged information. Thus the candidates automatically find the right position according to their ability, so it can not only ensure that the employing units can recruit the staff with strong ability, but also hire enough personnel.

We simply assume that the cost through which job candidates obtain the corresponding degree or pass the examination is distributed in [0, 1] interval. The cost through which the personnel with different quality obtain degree or pass the examination (time, energy and tuition) is usually different. The cost of low-quality personnel is much higher than that of high-quality personnel, so by comparing the costs and benefits, the personnel with different quality often choose different education level or can only examine different test result. Generally speaking, the personnel with the high cost have lower work benefit than the personnel with the low cost; so, in the recruitment process we tend to recruit those with lower human resources cost. We assume that the relationship between human resource cost and work benefit can be reflected through Fig. 11.1. Horizontal axis reflects the human

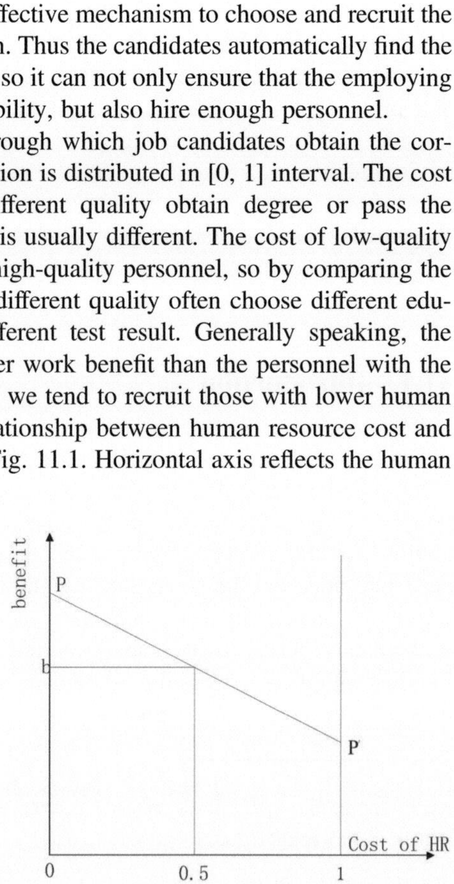

Fig. 11.1 Cost of HR and benefit

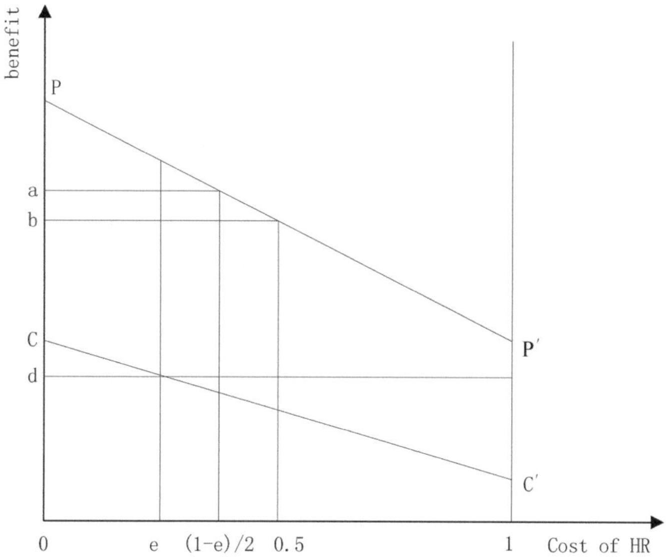

Fig. 11.2 The relation of the cost–benefit mechanism

resources cost through which the manipulator obtains the corresponding degree, from left side to right side the corresponding personnel's quality is from much higher to lower. Suppose PP′ line represents corresponding work benefit of the various cost personnel, namely the relationship between working benefit and the cost of human resources is a linear function relationship.

The manipulator of the cost and the corresponding benefit as shown in Fig. 11.1. If no signal mechanism, random selection of personnel, the average expected cost so the recruitment of manipulator is 0.5 averages expected and the its benefit is b, belong to the average. But usually we will use signal mechanism, such as on the candidates to mention certain qualification requirements and through the cost and the quality of personnel examination negative correlation, or further hypothesis is a linear function of, as shown in Fig. 11.2 CC′ line. Further assuming that if d is interested in getting hired, then the signaling cost is lower than that of the e personnel. But the cost is higher than the e personnel to signal the cost is higher than the signal of interest, so not to signal willingness. So, with the academic requirement as a screening mechanism, successfully will lower ability personnel exclude, average expected cost of hiring staff into (1−e)/2, the mean expected benefit manipulator a is higher than b.

From the above analysis we can find that it needs some steps to selected good ability of personnel requirements: (1) design different ability personnel signaling cost difference signal mechanism, if not the cost difference between the different ability of candidates to signal, could not identify personnel. This means that the academic requirements of the applicants cannot be too low, too easy to reach will not play a role; the academic requirements of the applicants cannot be too high, because when the requirement is too high due to very few people are willing to approach this signal, it cannot find a sufficient number of manipulator. Therefore, we can require the recruitment of personnel should have a bachelor's degree, but to engage in specialized technical posts in sports such as nursing, art, work, and academic requirements can be relaxed. (2) salary cannot give too high, because when the wages are too high, so that the D is higher than that of all the applicants ability to send out a signal cost (CC′lines all under D), low-ability candidates signal is also good, also can choose to send the same signal, this mechanism is also not discriminating manipulator ability.

11.3 The Screening Mechanism of Probation Period

Spence (1974) reveals the "signal effect" theory, due to the asymmetric information between the labor market, employers and candidates, the market requires a party to identify [2] a signal to help information deficiency. For employers, the candidate's education level plays a role in screening and instructions. Especially, in some units of this staff productivity is difficult to measure units, the use of this kind of signal is more important. But in practice, the Spence model in the presence of required separation equilibrium conditions of people of different abilities has great differences in ability. That is to say, only the ability of discriminating different diploma large effective (at present, because of the rampant fraud, make its effectiveness is limited), and for the ability to discriminate the education level of the same is helpless. At this time, an alternative screening method is a probation period screening mechanism. The basic approach is through the probation period screening during the probation period, paid to hire low wage rates, while in the period of probation after the end of the internal pay a higher wage rate, low ability of knowing who is difficult to hire during the probation period, Inner Mongolia mixed clearance, so at the end of the trial period was employed to obtain a higher wage rate, but during the probation period, take responsibility for the low wage rate losses, and will not participate in the recruitment [4], but the high ability of staff are confident to pass the probation period to prove their ability and would rather pay the probation period of low wage rate and obtain high wage rate in the period of employment. The separating equilibrium leads to the low ability of personnel are excluded from recruiting and job skills are higher personnel [3].

11.3.1 Analysis of Recruitment Model Under Asymmetric Information

In recruitment, some units and candidates are game players, the candidate's ability to theta type θ, it is personal information of the applicant Interest. Pure strategy unit is to select the length of the probationary period T and probation salary rate ω1 and after the trial period qualified personnel employment salary ω2; pure strategy candidates is the choice of "accept the trial" or "refuse to try". The following analysis of the game process, and according to the realistic situation given some reasonable assumptions:

① let $Q = f(L, \theta)$, Q here is output, θ is the operation of hand type, L is the quantity is employed for the type of personnel.

② let the wages rate is ω(θ) to setting unit for capacity personnel, the optimal employment benefit maximization problem as follows:$\max[f(L,\theta) - \omega(\theta)L]$. The first-order conditions for:

$$\frac{\partial f(L,\theta)}{\partial L} = \omega(\theta); \frac{\partial f(L,\theta)}{\partial \theta} = L\frac{\partial \omega(\theta)}{\partial \theta}$$

From this, we can get the type of hiring personnel is θ^*, and the number is L^*.

③ in the probation period T, if can identify the type of candidates $\theta \geq \theta^*$ and its probability is $P(T)$, it is a strictly increasing function.

④ now set the per unit utility of the candidates θ work in other enterprises work or out of work time is $U_0(\theta)$, T_0 is the expected employment cycle. Record type for the candidate's θ to participate in the trial of the expected utility is $U(\theta)$, here we use the monetary equivalent measure the level of expected utility, T_0 is equal in value to the time to participate in the trial personnel are employed and not employed the sum utility. That is:

$$U(\theta) = \begin{Bmatrix} \varpi_1 T + \varpi_2(T_0 - T)(1 - P(T) + U_0(\theta)(T_0 - T)P(T), \theta < \theta^* \\ \varpi_1 T + \varpi_2(T_0 - T)P(T) + U_0(\theta)(T_0 - T)(1 - P(T)), \theta \geq \theta^* \end{Bmatrix} \tag{11.1}$$

⑤ The assumption is willing to hire for the above personnel for $\theta \geq \theta^*$. The time discount factor candidates for 1 (the discount rate is 0). By the hypothesis can be seen, the employer's goal is to design a set of pure strategies enable $\theta < \theta^*$ select "reject trial", and the above $\theta \geq \theta^*$ selection "accept trial", which will not require low-ability personnel from participating in the recruitment of personnel of. Strategy design $(\varpi_1, \varpi_,, T)$ should satisfy formula:

$$\begin{Bmatrix} \varpi_1 T + U_0(\theta)(T_0 - T)P(T) + \varpi_2(T_0 - T)(1 - P(T) \leq U_0(\theta)(T_0), \theta < \theta^* \\ \varpi_1 T + \varpi_2(T_0 - T)P(T) + U_0(\theta)(T_0 - T)(1 - P(T)) \geq U_0(\theta)(T_0), \theta \geq \theta^* \end{Bmatrix} \tag{11.2}$$

If there is no positive lower bound of $U_0(\theta)$, then for any given strategy $(\varpi_1, \varpi_,, T)$, there are low enough θ to make the type (11.2) in the type (11.1) is not established, because the reservation utility is low enough for low-ability workers, given probation wage rate is attractive, So some very low ability will actively participate in the recruitment of staff. Similarly, when there is no positive upper bound of $U_0(\theta)$, for any given $(\varpi_1, \varpi_,, T)$, sure someone who has high ability do not content the second line of type (11.2), which are excluded from the outside in the recruitment. Therefore, if no bounds of $U_0(\theta)$, the probation period is not an effective screening mechanism. In fact, when existing at the same time over and lower bounds on model of $U_0(\theta)$, separating equilibrium exists and will distinguish personnel $\theta \geq \theta^*$ and $\theta < \theta^*$. Now, exist $M > m > 0$ and fit

$$\mathrm{m} \leq \mathrm{U}_0 \leq \mathrm{M} \tag{11.3}$$

Proved by below:

From type (11.2), can get:

$$\begin{aligned}&\frac{U_0(\theta_1)T_0 - \varpi_1 T}{(T - T_0)P(T)} - \frac{U_0(\theta_1)(1 - P(T))}{P(T)} \leq \varpi_2 \leq \frac{U_0(\theta_2)T_0 - \varpi_1 T}{(T - T_0)P(T)} \\ &- \frac{U_0(\theta_2)(1 - P(T))}{P(T)}\end{aligned} \tag{11.4}$$

For:

$$\frac{U_0(\theta_1)T_0 - \varpi_1 T}{(T - T_0)P(T)} - \frac{U_0(\theta_1)(1 - P(T))}{P(T)} \leq \frac{MT_0 - \varpi_1 T}{(T - T_0)P(T)} - \frac{U_0(\theta_1)(1 - P(T))}{P(T)} = A$$

$$\begin{aligned}&\frac{U_0(\theta_2)T_0 - \varpi_1 T}{(T - T_0)(1 - P(T))} - \frac{U_0(\theta_1)(1 - P(T))}{1 - P(T)} \geq \frac{mT_0 - \varpi_1 T}{(T - T_0)(1 - P(T))} \\ &- \frac{U_0(\theta^*)(1 - P(T))}{1 - P(T)} \\ &= B\end{aligned}$$

Set $\varpi_1 \in (0, m)$, if $\lim\limits_{T \to T_0} P(T) = P(T_0) > \frac{M - \varpi_1}{M + m - 2\varpi_1} > 0$, it has $B > A$, In fact, it equivalence to below:

$$\begin{aligned}&(MT_0 - \varpi_1 T)(1 - P(T)) - U_0(\theta^*)(1 - P(T)^2(T_0 - T) < (MT_0 - \varpi_1 T)P(T) \\ &- U_0(\theta^*)P^2(T)(T0 - T)\end{aligned} \tag{11.5}$$

When $T \to T_0$, type (11.5) left come to $T_0(m - \varpi 1)P(T_0)$, right comes to $T_0(M - \varpi 1)(1 - P(T_0))$

$$\begin{aligned}\text{When } \mathrm{T} = \mathrm{T}_0,\ &T_0(m - \varpi 1)P(T_0) > T_0(M - \varpi 1)(1 \\ &- P(T_0)) \text{ or } P(T_0) > \frac{M - \varpi_1}{M + m - 2\varpi_1}\end{aligned} \tag{11.6}$$

11.3.2 Confirm Probationary Period

We assume that during the probation period, personnel to establish a long-term cooperative relationship with the employer will try their best to work. With the total benefit function unit for $R = AK^{\alpha}L^{\beta}$, K is the cost of input is the human, A, α, β is constant. So the function of input is $K = c + V(t) + W(t)$. c is the cost of hire is the input of the training is the total wages. So the function of benefit is $N = R - K = AK^{\alpha}L^{\beta} - c - V(t) - W(t)$, it can meet the conditions below:

$$\begin{aligned} N'(t) &= 0 \\ N''(t) &< 0 \end{aligned} \tag{11.7}$$

11.3.3 Examples of Calculation

Benefit and elements of a production unit made at the 10 different recruitments in the input amount of historical data shown in Table 11.1, the recruitment of personnel recruitment plan in 2006 are as follows: 10 personnel, recruitment costs 8, staff recruited after the probation period per capita daily training costs 0.4, daily wages 0.2, determine the best period of probation.

We first determined according to the total benefit function of historical data, the total benefit of natural logarithm function on both sides, there is: $\ln R = \ln A + \alpha \ln K + \beta \ln L$

Through the calculation, will yield, capital and personnel data into the corresponding log data, and by using the least square method, can be calculated respectively, and the value of the parameter estimation. In equation form get the following estimation results:

Table 11.1 The historical data of recruitments between the benefit and input of a production unit

	R (Benefit)	K (input)	L (recruiter)
1	225	10	20
2	240	12	22
3	278	10	26
4	212	14	18
5	199	12	16
6	297	16	24
7	242	16	20
8	155	10	14
9	215	8	20
10	160	8	14

$$\ln R = 2.322 + 0.194 \ln K + \beta \ln L$$

On both sides to take against number of $\ln A = 2.322$. Get A = 10.2, so the total benefit function estimate of the situation for the:

$$R = e^{2.322} + K^{0.194} L^{0.878}$$

So the function of benefit is

$$N = 10.2K^{0.194} L^{0.878} - 8 - 0.4t - 0.2t \tag{11.8}$$

Type (11.8) content type (11.7) to meet the conditions so as to calculate the best for a trial period: 34 days.

11.4 Conclusions

Game analysis of manipulator recruitment screening mechanism, problems existing in the recruitment process, from the perspective of human resources cost is analyzed, the employer should signal mechanism of how to set up the recruitment of scientific, demonstrates the important role of probation period mechanism and how to calculate the optimum period of probation. Ensures that can attract higher quality candidates, and carried out the screening of candidates ability, provides some reference for the current forces in the fierce talent recruitment market on the correct use of screening mechanism to make scientific decisions. At the same time, in the evaluation of the candidates ability, can also use the fuzzy comprehensive evaluation method, considering the many other indicators, for example, knowledge, creativity, language ability, analyzing ability, organization executive ability, strain ability etc [5, 6]. So, we can help by the game analysis, to obtain useful talent person for unit.

References

1. Laffont J-J,TIROLE J (1987) Auctioning incentive contracts. J Political Econ 95:21–37
2. Spence A (1974) Competitive and optimal responses to signaling: an analysis of efficiency and distribution. J Econ Theory 8:296–332
3. Shiyu X (2002) Game of the economy. Fudan university press, Shanghai
4. Akerlof G (1970) The market for lemons: qualitative uncertainty and the market mechanism. Q J Econ Theory 39:251–269
5. Wang C, Dai F, Ge L, Li C (2009) A method of knowledge discovery based on the extension set. In: 2009 Third International Symposium On Intelligent Information Technology Application
6. Wang C, Dai F, Ge L (2009) Application of the grey fuzzy theory in evaluation of informationization risk. In: Third International Symposium On Intelligent Information Technology Application

Chapter 12
Research of Evaluation Methodology for Graduate Advisor Based on Learning—Guiding

Zhibing Pang, Rongzhi Yang, Cheng Jin, Hua Li, Chenhui Li and Honglei Li

Abstract Objectives, inspire advisor's enthusiasm on graduate student training, improve advisor's integrated quality in order to improve the quality of graduate level training. Methodology, combine factors from both student and advisor for advisor evaluation and construction of evaluation model. Results, constructed graduate student's capability and advisor's quality-based evaluation model, through analysis from both student and advisor. Conclusion, advisor is the key point to guarantee graduate student training, it's more comprehensive, reasonably and accurately to use student capability and advisor quality-based method on advisor evaluation, so that to enhance the level of advisor team.

Keywords Quality · Capability · Evaluation · Model

12.1 Introduction

The society increasingly concerns about the quality of graduate education, and the views about graduate advisor from both student and society changes with the rapid expansion of the scale of graduate enrollment [1]. Graduate advisor evaluation, as one of the most basic content of graduate education, has already been very popular in foreign universities. The development of graduate education in Chinese universities is relatively short. Although each university has initially established advisor selection rules, there's still no one operable system for graduate advisor evaluation, and the resulting deep-seated problems becomes increasing prominent. Previous researches mainly pay attention on advisor themselves, instead of coupling graduate students aspect for a more integrated advisor evaluation. This paper aims to propose a student-advisor coupling based evaluation system, and evaluate

---

Z. Pang (✉) · R. Yang · C. Jin · H. Li · C. Li · H. Li
Air Defence Forces Academy, Zhengzhou 450052, China
e-mail: fkb_mmes@163.com

© Springer-Verlag Berlin Heidelberg 2015

S. Long and B.S. Dhillon (eds.), *Proceedings of the 15th International Conference on Man–Machine–Environment System Engineering*, Lecture Notes in Electrical Engineering 356, DOI 10.1007/978-3-662-48224-7_12

graduate advisor in a more comprehensive, objective, and scientific way, by setting quality of education as the fundamental, encouragement for advisor as the purpose.

12.2 Factors of Evaluation System

Evaluation system analysis couples two factors from both graduate student and advisor, focusing on student–advisor coupled evaluation issue.

12.2.1 Graduate Student Factor

As a graduate in modern society, their own capability and quality are supposed to be followed the trend and social development. An outstanding graduate need to be evaluated from three aspects: knowledge, quality, and capability.

12.2.1.1 Knowledge

(1) Political Consciousness
Familiar with our party's latest theoretical achievements on route, guideline and policy, understand current international and domestic situation, maintain political character.
(2) Professional Knowledge
Proficient in and own passion on specialization, firmly establishes profession awareness, concern the development of their own profession, concern how their profession could meet social needs and adapt to trend of the times, actively seek change, be aware of professional activities of the academic community.
(3) Others
Diligent and good at reading, proficient on critical thinking, actively learn knowledge outside their profession, like humanity, geography, and social sciences, etc. Convert what they learned to innovative thinking, break conventional way of thinking, and enrich their knowledge system and connotation, fluent in foreign languages as well.

12.2.1.2 Quality

(1) Political Quality
Graduate students as the main force of technological power for our country, are supposed to have the correct political orientation, firm political stance,

distinctive political views, strict political discipline, mature political discernment, and strong political acumen [2], establish a correct worldview and outlook on life.

(2) Professional, Scientific and Cultural Qualities
Graduate students must have profound knowledge, master basic theoretical knowledge of discipline, be capable of independent scientific research and latest research achievements, be strive to occupy the forefront of disciplines, be able to make creative achievements in technical and professional disciplines.

(3) Innovation Quality
Innovation is the source for the development of modern society. Also it is the fundamental guarantee of building a harmonious society and achieving sustainable development. Therefore, graduate should not only be passive recipient of knowledge, but also should have a keen insight and judgment. They should keep scientific skepticism and achieve a comprehensive mental development, while intellectual growth, and ultimately have the ability to think and behave independently [3].

(4) Ideological and Moral Qualities
Future-oriented young people need a good sense of professionalism, initiative, social responsibility, historical mission, imagination, scientific spirit, humanities, etc. Graduate students in the modern society should establish good moral view, and correct hand the relationship between individual, collective, and country in such a complicated social.

(5) Humane Quality
Graduate students need to understand the nature of human society and its various literary, philosophical, and cultural tradition, while they training their scientific and cultural qualities. Also, they are supposed to understand points of views about economy, management, politics, as well as have communication writing skills.

(6) Physical and Mental Qualities
An individual who owns good physical and mental health is able to adapt to the social and natural environment, and then a career. Graduate students must have a sound body and a strong psychological endurance, to face both the joy of success and sadness of failure.

12.2.1.3 Capability

(1) Literature Retrieval
Literature retrieval needs the help of computers and networks, so graduates must be familiar with computer and network applications, especially several common search engines.

(2) Retrieval and Reading Ability in Foreign Language

Foreign language reading ability depends on a solid foundation of English, graduates should strength basic English, understand the situation of profession in foreign countries, broaden their knowledge.

(3) Independent Research and Design Capabilities
Graduates should take part in related research activities to enhance their academic and research capabilities, and try to publish papers on subject-related journals and conferences.
(4) Skilled Equipment Operation Capability
Graduates should be proficient and professional on equipment and related operation methods, be able to use laboratory testing software.
(5) Statistical Analysis and Statistical Software Ability
Graduates should have some knowledge of statistics, which related to subject design and data analysis, and be able to apply statistical theory into subject report and questionnaire.
(6) Summary and Thesis Writing Skills
Graduates should have solid writing skills, and the ability of converting what they achieve from research to written material.
(7) Ability to Capture the Tread of Social Development
This ability is related to personal development of graduates themselves. Graduates should develop such an ability to capture the trend of social development, and keep pace with the change [4].
(8) Life Planning Capability
Graduates should have the ability to plan for their life, know what they need to do, have a clear plan, so as not to get lost in the tide of social development.

12.2.2 Advisor Factor

The evaluation of advisor is mainly focus on ethics training, expertise, training guidance, and nature of teacher.

12.2.2.1 Personality Training

Advisors should affect students with their own thinking and attitude towards life, character, and personality. This section takes advisor's moral achievement, work attitude and life attitude as key factors.

Advisor's moral achievement is not only about their effects of moral quality, but also about personality shaping on students. For this purpose, ethics training should include following indicators: charisma, moral quality heritage, moral leadership, and well-guidance [5].

Work attitude is mainly about advisor's teaching attitude, professionalism, etc., which is more important for the growth of students than textbook knowledge.

The points of the work attitude should are being strict towards oneself, broad-minded towards others, teaching responsibility, tireless study spirit, willingness of teaching and care for students.

The effects from advisors to graduates are deep and benefit students for their whole lifetime. Attitude towards life should include the following indicators: life attitude and social attitude, which reflect advisor's broad mind, passion of life, passion of community, and willingness of contributing to students and society.

12.2.2.2 Expertise

Graduate advisor is a typical professional career, which indicts advisors have already achieved a comprehensive knowledge system and formed specialized educational content. As academic core activities, teaching and research are advisor's most essential work. This section takes teaching and research abilities as the two significant factors of "expertise".

Advisor's teaching ability is not only about cultural knowledge, but also about instructive ability and organization, management capabilities, which includes the level of teaching, teaching methods, teaching content, teaching quality and effectiveness. Embodied as follows: First, to ensure teaching content should be substantial and keep pace with ongoing international knowledge improvement, and to ensure that the amount of teaching hours and strict requirements. Second, be proficient in using modern teaching methods, improving teaching efficiency and quality.

Advisor's research ability is equally important, is one of the indispensable abilities for graduate advisor, including research projects and research achievement. Embodied advisors have the ability to obtain various research project funding and the ability to publish and write relevant papers and reports.

12.2.2.3 Training Guidance

The teacher called graduate advisor, calls for the establishment of "guidance" in the relationship between teachers and students, teaching methods vary by students' individual differences. This section takes training environment, training process and training quality as key factors.

Coaching skills is the most important aspect of the evaluation of the advisor, which requires advisors pay attention on students through their entire education process and various aspects of education, work, life, etc., train student to have the integrity and consistency. First, advisors should be able to provide students with learning, research environment and opportunities. Second, advisor's training plan should specifically include: emotional interaction with student, guiding students to implementation of projects, developing students' innovative learning attitude, innovation consciousness and communication skills.

12.2.2.4 The Nature of Teacher

For the three factors mentioned above, expertise is the starting point, training guidance is the basic point, personality training is the commanding point, but all the three will eventually be reflected in the advisor's quality [6]. The nature of teacher is the self-training of advisor's personality, which requires advisor to be able to couple moral education with intellectual instruction, able to give students professional ethics education, teaching seriously, respect for science, love truth, noble exotic, well-groomed, behavior properly, patient, self-discipline, etc. This also requires advisors have excellent teaching quality and sound personalities, set themselves as behavioral example for the students, so that affect students by their strong personality.

This section takes academic quality, knowledge structure, and interpersonal skills as indicators to measure "the nature of teacher". These three factors cover personality and communication with others, which are the sources to success and sustainable improvement. Academic quality is mainly about the academic moral quality, which is reflected in thinking. Knowledge structure consists of new knowledge and previous knowledge, and the deep knowledge base, the ability to acquire new knowledge, and innovative thinking are all very important. Interpersonal skills mainly reflected in the organization skill, coordination skill and communication skill, which is not only about the relationship between advisor and graduate, but also indicate their teamwork and organization skill as a team.

12.3 Model System Constructions

12.3.1 Model of Graduate Student's Capability

According to graduate's qualities and capabilities, with A, B, C … to represent each quality and capability, and given each item weights $m1$, $m2$, $m3$…, using X to represent graduate's integrate ability level, model is formed as following:

$$X = m1 * A + m2 * B + m3 * C + \cdots$$

12.3.2 Model of Advisor's Capability

According to advisor's qualities and capabilities, with a, b, c … to represent each quality and capability, and given each item weights $n1$, $n2$, $n3$…, using Y to represent graduate's integrate ability level, model is formed as following:

$$Y = n1 * a + n2 * b + n3 * c + \cdots$$

12.3.3 Integrate Evaluation Model

After constructing the models for both graduate and advisor, we assigned weights α and β (values of α and β vary by different disciplines) to these two items respectively, and eventually got the graduate-ability oriented evaluation model for advisor's performance, using Z to represent advisor's final integrate ability level, model is formed as following:

$$Z = \alpha * X + \beta * Y$$

12.4 Conclusions

Advisor evaluation is a significant job in university education; it's as well as a reflection of the graduate education quality. This paper takes both graduate and advisor's abilities into consideration, to evaluate advisor's performance. Using learning-guiding based evaluation model makes the evaluation method more convince and more comprehensive, which also combines graduate and advisor's achievements together, aligns graduate and advisor's abilities, couples learning and teaching, and ultimately enhances graduate education in Chinese universities, promote the overall goal of the quality of graduate education.

References

1. Li Jun, Zhang Na (2009) Thoughts on graduate quality evaluation system of teacher evaluation index. China High Educ Eval 1:69–72
2. Huifeng X (2010) On the responsibility of contemporary graduate. Chin Grad 1
3. Feng R, Wang G, Wu B (2002) On Ability of graduate. Chengdu Univ Tradit Med 2
4. Li D (1996) Cultivation of high-level personal from the Postgraduate Education Reform. Degrees and Grad Educ 2
5. Lu Y (2002) Research master instructor overall quality evaluation system. First Military Medical University Bachelor thesis, pp 9–10
6. Li L, Chen W (2008) Core factor tutor quality evaluation system analysis. High Educ Forum (3):187–190

Chapter 13
Technical Staff Job Engagement and Influencing Factors

Xuejian Sun, Yiqun Gan, Zhiyong Wang and Zhen Wu

Abstract The aim of this study was to examine Job Engagement concept as well as the scale for technology management in China. 230 subjects from a research institute in Beijing presented with questionnaires on job engagement, job burnout and job characteristics, and 20 of them were interviewed. Using SSPS software, the authors processed the data or information collected for psychological statistics analysis, exploring factor analysis, and regression analysis. The results indicated: the concept of job engagement can reflect the psychological status of the technical staff at work; UWES-C is reliable and valid; UWES-C and the job engagement prediction model can be helpful with safety management.

Keywords Job engagement · Job burnout · Job characteristics · Safety management

13.1 Introduction

At present, science and technology is rapidly developing and new knowledge explosion continued. Some enterprises carry out reform or development in China, and most technical employees have to take more time to work as well as learning new technology and skills. Accompanied by high working pressure, health problems or safety accidents have increased in recent years. According to the latest medical statistical data [1], there were more cases of psychological problems and mental disorders than ever in China. The safety department survey results [2] showed that 60–90 % of accidents were caused by human errors. This paper examines job-engagement and its influencing factors to explore the means of evaluating from a psychological perspective, so that it provides related technical support for safety management.

X. Sun (✉) · Y. Gan · Z. Wang · Z. Wu
Zhongguancun South Street, #31, P.O. Box 28, Haidian District
Beijing, People's Republic of China 100081
e-mail: sun15010164642@qq.com

© Springer-Verlag Berlin Heidelberg 2015

S. Long and B.S. Dhillon (eds.), *Proceedings of the 15th International Conference on Man–Machine–Environment System Engineering*, Lecture Notes in Electrical Engineering 356, DOI 10.1007/978-3-662-48224-7_13

13.2 Core Concepts

13.2.1 Job Engagement

Job engagement (JE) is defined as a positive, fulfilling, work-related state of mind. There are three dimensions to JE: (1) vigor (VI), invested high levels of energy in one's job and challenged difficulties with tenacity; (2) dedication (DE), a sense of significance, showed enthusiastic and proud at working; and (3) absorption (AB), characterized by being happily engrossed in one's job and not willing to get off the job. In 2002, job engagement was tested with the Utrecht Work Engagement Scale [3].

13.2.2 Job Burnout

Job burnout was a metaphor that was used to describe a state of mental weariness in workplace. It can be traced back to a survey of Maslach Burnout Inventory (MBI). Later, the original developed the new version that was called MBI-General Survey (MBI-GS). There were three dimensions in MBI-GS: exhaustion, cynicism, and efficacy [4].

In this paper, based or referenced on job burnout theories of Maslach and other scholars, the author edited Technical Staff Job Burnout Scale (TSJBS) using the data or information on psychological status at work, which were collected from interviewing and investigating some Chinese technical staff in Beijing [5]. In TSJBS, there are three dimensions: (1) mental fatigue (MF), a sense of psychological fatigue at work, feeling of difficulties in concentrating, memorizing, and thinking; (2) alienation sense (AS), reflected indifference to customers or a negative attitude towards job; and (3) inefficacy sense (IS), made a lower evaluation on one's occupational accomplishment, working performance was worse than that one would expect.

13.2.3 Job Characteristics

Job characteristics is defined as professional factors related to one's job. There were five dimensions in the Job Characteristics Scale (JCS): (1) treatment satisfied (TS), satisfied with payment, position, and incentive scheme; (2) job payload (JP), being demanded on the quality or quantity of jobs within a period; (3) working roles (WR), being expected to do two or more jobs in a period; (4) resource supply (RS), getting energy both physically and psychologically; and (5) resource matching (RM), working in good human–machine environment system, such as rational technical reserves for operating machine, good workplace to human fitness [5].

13.3 Interview and Results

20 subjects from a research institute in Beijing were interviewed by the author, who talked about their occupation experience. Some constituting elements of job engagement were found in the in-depth interview, which include: (1) liked their job and did jobs in high spirits; (2) enjoyed working and challenged difficulties; and (3) put their energy into job as much as possible. There were some issues found, such as pay benefits, job plans and roles assignment, significantly impacted job engagement.

13.4 Questionnaires and Results

13.4.1 Subjects

230 technical staff from a research institute in Beijing presented with self-reporting questionnaires on job engagement, job burnout, and job characteristics. There were 20 individuals interviewed among the subjects. In this survey, 199 test papers were received and the 190 papers were completed, of which the validly response rate was 82.6 %.

13.4.2 Instruments and Material

The following instruments and material were used:

- Utrecht Work Engagement Scale of Chinese Version (UWES-C)
  Utrecht work engagement scale (UWES) was made by Schaufeli originally, which was translated and revised by Chinese scholars [6]. There are 15 items which were grouped into three subscales and the items were scored on a seven-point scale (0–6). The subscales include VI, DE, and AB.
- TSJBS and JCS
  Both scales were revised by the author [5].
- Professional efficiency scale(PES)
  PES includes 9 items (e.g. “I can effectively complete the job and make an achievement”), which was edited by the author. All the items were scored on a four-point scale (0–3).

13.4.3 Subjects

Based on psychological principle of measurement and assessment, first, questionnaires were sent to subjects and taken back by the author. Second, data from this investigating was reorganized and some invalid papers or abnormal numbers were deleted. Third, the data or information was studied by using SPSS FOR WINDOWS software for psychological statistics or calculation.

13.4.4 Investigation Results

13.4.4.1 UWES-C

- Structure validity
  Using confirmatory factor analyses to the UWES-C, the three-factor model showed an acceptable fit to the data with $\chi^2 = 144.05$, df = 82; important model fitting parameters (GFI, CFI, etc.) reach or approach 0.9; RMSEA < 0.08. The results showed that this structure validity of UWES-C can meet psychometric requirements as the following Table 13.1.
- Internal consistencies
  Cronbach's (α) is displayed in Table 13.2. According to the relevant standard [7], it was good at the internally consistent.

13.4.4.2 The Scores (M and SD) and the Correlations

Subjects ($N = 190$) were assessed by UWES-C, TSJBS, JCS, and PES. The mean (M), standard deviation (SD) of the study variables, and correlations showed in Table 13.3.

13.4.4.3 Regression Analysis on the Influence Factors

JE was a dependent variable; the subscales of JCS and TSJBS were predictors (independent); TS, RM, JP, MF, and AS were predictors. The results showed in Table 13.4.

Table 13.1 Results of confirmatory factor analyses to UWES-C ($N = 190$)

Model	χ^2	df	GFI	AGFI	PGFI	NFI	CFI	RMSEA
3-factor	144.05	82	0.91	0.87	0.62	0.88	0.94	0.062

Table 13.2 UWES-C internal consistencies ($N = 190$)

Variable	VI	DE	AB	Engagement
α	0.7209	0.7677	0.7657	0.8984

Table 13.3 The scores (M, SD) and the correlations

Variable	Subjects		Correlations						
	M	SD	PES	JCS	TS	JP	RM	RS	WR
VI	19.70	6.26	0.704**	0.411**	0.374**	0.232**	0.385**	0.192**	−0.048*
DE	13.30	5.29	0.589**	0.482**	0.405**	0.326**	0.427**	0.206**	−0.037
AB	16.54	6.04	0.665**	0.402**	0.382**	0.304**	0.340**	0.162*	−0.088
UWES-C	49.58	15.93	0.721**	0.474**	0.427**	0.315**	0.423**	0.205**	−0.065
AS	12.76	6.69	−0.112	−0.094					
MF	14.58	5.14	0.283**	0.225**					
IS	7.21	3.76	−0.013	0.169 s*					
TSJBS	34.57	12.19	0.054	0.089					

Note $N = 190$; $*P < 0.05$, $**P < 0.01$ (Pearson correlation, two-tailed test)

Table 13.4 Coefficients on the subscales of JCS & TSJBS predicting JE

Model	Variable	Unstandarized coefficients	Std. error	Beta	T	Sig.	Tolerance
JCS	(Constant)	19.984	4.000	–	4.996	0.000	–
	TS	1.853	0.468	0.268	3.960	0.000	0.845
	RM	2.443	0.577	0.283	4.237	0.000	0.866
	JP	1.436	0.464	0.197	3.092	0.002	0.948
TSJBS	(Constant)	44.762	3.437	–	14.02	0.000	–
	MF	0.985	0.225	0.318	4.374	0.000	0.880
	AS	−0.748	0.173	−0.314	−4.323	−0.000	0.880

Notes IS, WR and RS were deleted; *Beta* Standardized regression coefficient

13.5 Discussions

13.5.1 The Meaning of Correlations

TS, JP, and RM could be important factors impacting JE because there were moderately positive correlations ($r > 0.3$, $p < 0.01$) between UWES-C subscales and JCS subscales (TS, JP, and RM) as the data displayed (Table 13.3), and their statistical validity is good [8]. The UWES-C positively correlated with mental fatigue ($r = 0.21$, $p < 0.01$), showing that the mental fatigue showed a person was indeed working hard. On the one hand, if there is no mental fatigue, it means that a person did not engage with the job. But, some employees, who were psychological and physiologically healthy, might have felt intense mental fatigue, meaning that they spent much energy or time in job (over-Job Engagement). If so, leaders of safety department should pay attention to the employees who might conduct unsafe activities or make troubles because they were too tired to work well.

The TSJBS and its subscale IS unrelated significantly with the score of UWES-C, which means both job burnout and job engagement are different in concept from each other, and inefficacy sense is not a significant influence factor to the job engagement.

13.5.2 Job Engagement Predicting Model in Reality Meaningfulness

Job engagement predicting model (JEPM) was built using standardized regression coefficients, which also called the path coefficient (Beta), as shown in the Fig. 13.1 (figures are numerical value of Beta). As Table 13.4 showed, in the model (JCS), some dimensions of job characteristics scale (TS, JP, and RM) attributed to job engagement in the same time, reflecting a level of organizational management; In the model (TSJBS), mental fatigue and alienation sense could impact (or mediate) Job Engagement.

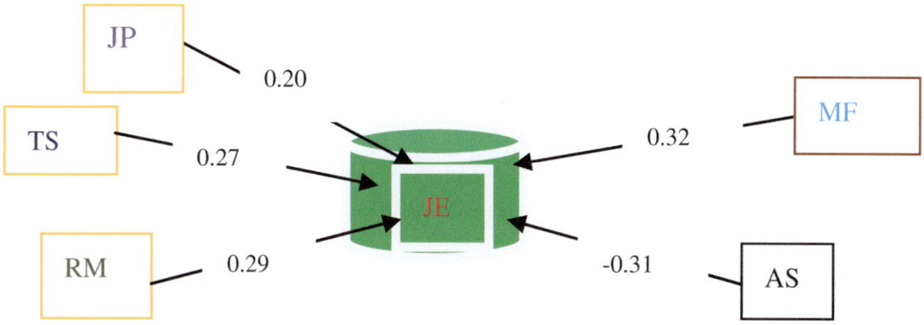

Fig. 13.1 JEPM

In the interview, there were two extreme cases of job engagement: insufficient and excessive. The former showed passive, distraction in day-dreams and so on; the later displayed one's unbalanced situation between work and life because of investing too much time into jobs, in which the individual did not enjoy eating and sleeping for many days so that the brain felt painful or disorder. Both of the cases could weaken safety management in the workplace, and even dangerous accident could occur in their working programs.

Job engagement is a psychological status to staff at work, which includes physical, cognitive, emotional perspectives etc. Job engagement plays an important role not only to staff's well-being, but also to organizational success and achievement [9].

Valuing human reliability and avoiding human factor accidents is an essential task for organizations to survive and develop. Thus, organizational managers should optimize Man-Machine-Environment Systems (MMES) at workplaces, and pay attention to staff's psychological and physiological needs. The job engagement predicting model can support organizational managing strategies, such as exploring good ways to resource matching, reasonable arranging job payload, reducing alienation sense with EAP [10], improving treatment satisfied by organizational culture, guarding against unsafe operation when technical staff might be in a case of mental fatigue.

To conclude, recognizing the significance of technical staff's Job engagement and influencing factors would be helpful with safety management or human resources program. R&D organizations or enterprises will be benefitted from safe and healthy environment so that they can more effectively do science and technology research and produce commodities for human society.

13.5.3 Job Engagement on Expending and PES Examined

Professional efficiency was an original subscale of MBI-GS Scale. However, some scholars consider it as a part of job engagement [3]. To examine the hypothesis, in this study, the professional efficiency scale (PES) was edited and operated by the author. PES was verified through using exploring factor analysis, and its internal consistencies were good at Cronbach's ($\alpha > 0.74$). The result displays that PES positively and highly correlated with UWES-C ($r = 0.72$, $p < 0.01$, Bilateral inspection), and reveals that professional efficiency not only has an impact on the job engagement but also could be an expending dimension of the job engagement scale.

13.6 Conclusions

With interviews and questionnaires, the data and information was collected and calculated by using SSPS software. This study's conclusions are as following:

- Job engagement became available to describe psychological status of the technical staff at work;
- UWES-C is reliable and effective, in terms of reasonably and objectively measuring technical staff's job engagement. A norm criteria of evaluating job engagement needs to be set up in future.
- UWES-C and job engagement prediction model can be helpful on research to human factor of safety programs as well as accident prevention.

Acknowledgments Heartfelt thanks to the Psychology Department of Peking University, interviewees and respondents of the questionnaire!

References

1. Yang S, Zhang H (2014) Study on the diagnosis and treatment of psychosomatic disease. Continuing Med Educ 28(7):64–66
2. Hu YC, Wu HH (2014) Research on accident model based on safety psychological factors. Value Eng 20:298–300
3. Schaufeli WB, Bakker AB (2004) Job demands, job resources, and their relationship with burnout and engagement: a multi-sample study. J Org Behav 25(3):293–315
4. Maslach C, Schaufeli WB (2001) Job Burnout. Annu Rev Psychol 52:397–422
5. Sun X (2012) Job burnout research of technical staff. In: Proceedings of 12th conference on man-machine-environment system engineering. 2012, Shanghai, China, pp 27–30
6. Zhang Y, Gan Y (2005) The Chinese version of utrecht work engagement scale: an examination of reliability and validity. Chin J Clin Psychol 13(3):268–270
7. Gan Y (2005) Psychological and behavioral science statistics. Peking University Press, Beijing
8. Robert MK, Dennis PS (2005) Psychological testing principes, applications and issues, 6th edn. Cengage Learn
9. Hu S, Wang Y (2014) Concepts, measurements, antecedents and consequences of work engagement. Adv Psychol Sci 22(12):1975–1984
10. Zhao R (2010) Employee assistance program—EAP consultant handbook. Science Press, Beijing

Chapter 14
Analysis on Professionalism and Ability Structure of Secrecy Sergeant

Hongbiao Wang, Huili Yu and Wei Zhang

Abstract One of the important contents of the sergeant system reform is to push forward the sergeant professionalism construction. According to the fact and teaching practice of military secrecy, the paper discusses the construction situation of secrecy sergeant professionalism, the equipped capability structure of secrecy sergeant professionalism, speeding up the occupational skill testing work and promotion training work of security sergeant, building the security system of secrecy sergeant professionalism, etc. in the ways of literature study, researching, expert interviewing in order to provide theory reference for pushing forward the process of security sergeant professionalism.

Keywords Secrecy sergeant · Professionalism · Capability structure

14.1 Introduction

As the skill application talent group, sergeants are the important protection of winning the war. The level of their qualities has direct influence on the generation of the fighting capacity of the army. It has become consensus of armies of all countries worldwide to take the professionalism construction of the sergeant group as the important content of the army professionalism. Our military has pointed out in *Plan on Implementing Military Talent Strategy Engineering* established in 2003 that: "sergeant professionalism system and optimized training system featured with characteristics of our military shall be formed by 2020." In 2011, it was again emphasized in *Outline of Military Talent Development Plan before 2020* that: "push forward the reform of sergeant professionalism." Secrecy sergeants shall refer to sergeants who hold the post of military confidential person and bear the duties and responsibilities of transferring documents and secrecy management and are a group

H. Wang (✉) · H. Yu · W. Zhang
PLA Institute of Special Operation, Guangzhou 510502, China
e-mail: 9058158@qq.com

© Springer-Verlag Berlin Heidelberg 2015

S. Long and B.S. Dhillon (eds.), *Proceedings of the 15th International Conference on Man–Machine–Environment System Engineering*, Lecture Notes in Electrical Engineering 356, DOI 10.1007/978-3-662-48224-7_14

of vital power in military secrecy. Pushing forward the professionalism construction of the secrecy sergeants has important meaning to advance the professionalism and normalization of the military secrecy career.

14.2 Analysis on Secrecy Sergeant Professionalism Status

14.2.1 A Group of Professional Group Is Necessary for the Development of Military Secrecy

The secrecy work develops by involving higher and higher science and technology content and deeper and deeper professional degree. The key point of well completing the military secrecy under informatization conditions is whether there is a group of high-quality secrecy talents who have strong political awareness, good work style, master technology and management skills [1]. However, the high-quality secrecy talent group needs occupational steadiness as protection to provide continual motive force of development.

14.2.2 Professionalism Shall Be Reinforced Due to the Specialty of Secrecy Sergeants

Comparing with other technological arms, the occupational failure of the secrecy sergeants may cause hidden, lagged, and comprehensive dangers and the losses caused hereby cannot be balanced by cash. Therefore, the occupational qualities of the group will directly influence the document transfer level and the safety management of confidential information of the military. From this point, the confidential sergeant group has higher professionalism requirements than that of other posts.

14.2.3 New Measures of State Secrecy Professionalism Provide References for Secrecy Sergeant Professionalism

Since 2012, state-relevant departments have gradually resolved the limitation factors such as vacancy of professional academic education, title series, and professional group in secrecy work that may influence the professionalism development of the secrecy work. First, Engineering Technology Senior Professional Post Assessment Committee of National Secrecy Administration of the People's Republic of China has been established and in further the secrecy system may automatically organize and assess senior technological title; second, the secrecy management major has been normally listed into major list of common undergraduate universities; third, state and some provinces have established association of secrecy workers [2]. The issuance of these measures has important meaning to push forward the professional development of the secrecy work as well as provide beneficial lessons and references for the military to develop professionalism of the secrecy sergeants.

14.3 Analysis on Ability Structure Necessary for Secrecy Sergeant Professionalism

Secrecy work is a kind of science of high comprehensiveness and has self objective law and suitable knowledge system and ability structure. Especially as the informatization construction of our military is developed deeper and deeper, higher requirements are proposed for the ability demand for secrecy sergeants: not only powerful political ideological qualities but also qualified professional skills; not only professional basic knowledge necessary for the post but also learning and innovative ability adaptive for new technology development. Therefore, focusing on the professionalism development of the secrecy sergeants, the ability structure shall be expressed in following aspects.

14.3.1 Political Ideology

Secrecy work is high-political work and professional awareness can be really strengthened only if the sergeants always have strong sense of responsibility and vision to the party and military career. It also decides that the secrecy sergeants shall have qualified political ideological qualities, which is the basis and precondition for being qualified for the post requirements. Political ideological qualities mainly refer to the requirements of ideological qualities and include political faith, political standpoint, political sensitivity, policy level and ideology and style, etc. To become a qualified secrecy sergeant, first, you shall have firm political faith and political standpoint, namely loyalty to the party and country and permanent persistence on taking the protection of the state safety and military benefits as individual task and understanding and handling various problems by consciously using the standpoint, viewpoint, and method of Marxism; second, you shall have highly political sensitivity and insight, meaning to be good at analyzing problems from political, correctly master the development trend of the secrecy work and complete the key work points of different periods and different backgrounds; third, you shall have highly policy level and can firmly carry out and perform the policies and laws of the state and military and correctly understand, master, and use relevant policies [3].

14.3.2 Professional Ethics

Professional ethics refer to certain behavior rules that people of certain occupation shall obey and are the materialization of social ethics in professional life. In order to enhance the professional ethics of all industries, various occupations shall make work principles, rules, and systems to behave people's behaviors and help them to form good professional ethics habit by starting from professional requirements and adapting the characteristics of the industrial workers. The secrecy sergeant shall obey and perform secrecy rules, security duties, and responsibilities and work rules of security office as an example, have high ethics qualities, treat fame and wealth and loss correctly, endure the loneliness, consciously resist various temptations and have the spirit of utter devotion and ethics of being an unknown hero; and shall establish service awareness and act as the bridge and link of up-and-down and inside-and-outside connection to push forward the smooth implementation of the secrecy work of the unit.

14.3.3 Basic Knowledge

Out of the complexity of the military secrecy, it demands that the personnel devoted to the career shall not only have good political qualities and spirit of selfless utter devotion but also own relevant basic knowledge. As a detail executor and pusher of the military secrecy work of one unit, whether the secrecy sergeant has consolidated basic knowledge is both the precondition of performing duties and responsibilities and the basis of further development. The basic knowledge mainly covers basic knowledge of culture, computer and internet, military secrecy, archival science, and basic laws and regulations of military secrecy, all of which will be used usually during the daily work of a secrecy sergeant. The master degree will directly influence the development of the work and even influence the safety of the military secrecy information.

14.3.4 Business Skills

The work of secrecy sergeant is fussy and meticulous and is a kind of detail work with special meaning and strong practice. Besides necessary basic knowledge, experience shall be accumulated continually during practice to enhance business skills and complete occupational work preferably. The secrecy sergeant shall familiarly master following business skills: daily management, document handling, stamp management, secrecy-based movement, storage and medium management, secrecy publicity and education and secrecy inspection of secrecy office. The first three businesses are generally completed by the secrecy sergeant independently and are as well the most important work content at present. It demands that the secrecy sergeant shall be extremely good at and proficient at this and can flexibly handle various circumstances they meet; the subsequent items are those that the secrecy sergeant can involve in. In early time, the secrecy sergeant may not be measured according to high requirements but shall familiarly master these contents as further develop and promote.

14.3.5 Study and Innovation

At present, as the construction of the army informatization goes deeper and further, the military secrecy stays at an important transition period and there are many new situations and new problems to be researched and resolved. The only way to get out and develop is to learn and innovate. Whether the secrecy sergeant has the learning and innovative ability and qualities is the ruler to balance the comprehensive qualities as well as the key factor on whether he can get succeeded. What learning and innovation reflect is a kind of spirit status. The motive power and practice cannot be generated without strong aggressiveness and responsibility sense. The learning and innovation of the secrecy sergeant shall be developed based on inheritance. Through mastering the historical law, the secrecy sergeant shall continually learn and innovate during practice, research and resolve new situations and problems met during the work and push forward the development of the secrecy theory and practice.

14.4 Conception of Accelerating Secrecy Sergeant Professionalism

Since the reform of sergeant system in 1999, our military has successively revised and improved relevant laws and systems such as *Sergeant Service Regulation* and *Sergeant Management Regulation* and initially established sergeant system with the characteristics of our military and expression of professionalism characteristics, which also built system basis for the professionalism of the secrecy sergeant. To accelerate the professionalism of the secrecy sergeant, emphasis shall be put on following aspects.

14.4.1 Carry Out the Professional Qualification Certification System of the Secrecy Sergeant

The sergeant's professional qualification certification is to make objective assessment on the theory knowledge and practice ability that the sergeant shall have and match the post. The secrecy sergeant shall be selected and promoted strictly according to the professional qualification certification and obtained job qualification of corresponding level. Insist on principles of "strict standard, strict requirements, first review and later use and work with certificate" and then conclude the sergeant into secrecy work group. The qualification certification of the secrecy sergeant can be implemented according to steps of selection, review, before-post training, qualification test, registration and record and supervision and management, which shall not only pay attention to the examination of knowledge level and intelligent structure but also emphasize the assessment on ethics, ability, professional ethics and work style and discipline during past work [4]. The access to secrecy sergeant can be safeguarded as long as the result of professional qualification certification is taken as the necessary condition for promotion and the effect of the professional qualification certification is strengthened to help to obtain professional development opportunity and establish firm basis for the professional development.

14.4.2 Reinforce the Promotion Training Degree of the Secrecy Sergeant

Promotion training: the sergeant shall join in training of corresponding level to be selected as professional technician sergeant or promoted up to professional technological sergeant and join in professional skill authentication exam. The sergeant who is qualified in the exam can be promoted. It refers to the "necessary exam for promotion and necessary training for promotion" usually mentioned. It is an effective measures suitable to the professional development and realizing the scientific development of the sergeant professionalism. However, according to the condition of secrecy sergeant's promotion training classes held in recent years, it still has difference from sergeant's promotion training of other majors. The reasons for that include not only objective factors such as the special requirements of the post and guarantee of frequently staying at post but also subjective factors such as the person has not sufficient attention to the promotion training. Therefore, active guidance shall be carried out and correct cognition shall be established on the height of advancing the professionalism of the secrecy sergeant to ensure the effective implementation of the promotion training and really enhance the professional level of the secrecy sergeant.

14.4.3 Construct Professionalism Protection System of the Secrecy Sergeant

The post characteristics of the secrecy sergeant decide the fundamentality and service of the work. Comparing with other posts, the secrecy sergeant cannot reach remarkable achievements. Although the secrecy sergeants are demanded to endure loneliness and poverty at the early period of the professional selection, emphasis shall be put on constructing protection system that can guarantee the secrecy sergeant's professionalism during system design. First, reputation incentive mechanism shall be established. Reputation awareness is certainly an important content of the professional awareness. The secrecy sergeant shall sufficiently know the importance of the secrecy work and understand that it is a kind of trust and important entrustment from the organization to be selected to devote to such profession. What is more, the idea that "it is achievement that problems cannot be found at this post" shall be established on awards implementation. The sense of reputation can be stimulated through timely awards. Strong sense of professional belonging [5] can be established only with strong sense of reputation. Second, post allowance shall be set. The post liabilities of the secrecy sergeant are vital. Any one of task cannot endure even the least bit error. Post allowance may slightly make material compensation to long-term high-strung spirit status and is also acceptance to the professional integrity. Third, professional technological sergeant system shall be explored. The cultivation period of the secrecy sergeant is long and effective measures are necessary to ensure the post steadiness. Professional technological sergeant system may help the secrecy sergeant devoted to secrecy work to have definite professional development expectation.

References

1. Lin JC, Qian HH (2007) Military security subject. Military Science Press, Beijing, pp 316
2. Huo R (2012) Pushing the process of security work professionalism. J Secur Work (10):4–5
3. Zhang Y (2008) Measures for managing national secrets personnel in Guangdong province. In: Learning materials about security. State Security Bureau in Guangdong Province, pp 208–209
4. Sun Y (2012) Thinking on reinforcing security records sergeant education. The PLA Records (2):18–22
5. Qiu F (2013) The idea about reformatting the system of sergeant professionalism for promotion. J Guilin Air Force Inst (1):16–17

Chapter 15
Research on Design of the Battlefield Soldier Physiological Status Monitoring and Analysis System

Chenhui Li, Zhibing Pang, Yanjun Zhang, Shili Chu, Xiaofei Zhai, Hualiang Xu and Zengjun Ji

Abstract After investigating the development of the physiological status monitor technology in China and abroad, the author studied the physiological status monitor technology for soldiers in the battlefield. According to the requirements for performance imposed by modern wars, the author analyzes the system composition, functional design, and data transmission of the Battlefield Solider Physiological Status Monitor System, so that it can analyze soldiers' operational intensity and fatigue procedure, and allocate combat missions properly on the basis of physiological information; estimate the condition of wounded soldiers in time and exercise evacuation; analyze comprehensively the physiological status of the personnel in combat units and thus provide a basis for command and control operations.

Keywords Information technology · Physiological status monitor · Radio frequency

15.1 Introduction

In the future information digital battlefield, soldiers fighting is the key factor to maintain the army integrated battle ability to win the war. More complex battlefield environments and advanced weapons and equipment require the highest requirements in physical fitness, skill, and intelligence of soldiers. Particularly, combat operation is fast and soldiers are always in a state of intense work, which will exert greater pressure and influence on their psychology and physiology. Personnel physiological states and exhaustion level may speed up the decline in their physical health status and directly affect their battle effectiveness, ultimately endangering their life safety and influence the completion of operational tasks. Most combat operations in future wars will be undertaken by small and frequent moving combat

C. Li (✉) · Z. Pang · Y. Zhang · S. Chu · X. Zhai · H. Xu · Z. Ji
Air Defense Forces Academy, Zhengzhou 450052, China
e-mail: lichenhui_zz@163.com

© Springer-Verlag Berlin Heidelberg 2015

S. Long and B.S. Dhillon (eds.), *Proceedings of the 15th International Conference on Man–Machine–Environment System Engineering*, Lecture Notes in Electrical Engineering 356, DOI 10.1007/978-3-662-48224-7_15

units. It will be important to master the unit soldiers' life and wound information; therefore, it is necessary to know the physiological state of soldiers and other information at any time. According to the statistical analysis of the US Army combat depletion, 20 % soldiers were killed in combat attrition, 2/3 injured soldiers had died in 10 min. Statistics of the Vietnam war show that 50 % of dead soldiers died from bleeding, 90 % of the injurious died before reaching medical institutions within 1 h of their injuries, 20 % of whom could be saved if timely emergency measures had been taken [1]. Studies have found that if the wounded soldiers timely take effective medical treatment within the first 1 h after their injuries, their chances of survival and recovery will increase. Hence, it is a prerequisite to monitor soldiers' vital signs for timely rescue in modern war; it needs to know soldiers' physiological and psychological state and casualties, as well as take effective measures to intervene and rescue. Therefore, to minimize casualties is of great practical value to commanders for operational deployment, battlefield ambulance, and reducing casualty rates.

15.2 The Present Situation in Research

In recent years, the "health care with their men in" slogan suggested that in all countries more and more attention be on study of single vital sign monitoring and rescue techniques, in order to develop future digital weapons of warrior system; for all soldiers vital signs be monitored through portable technologies or systems. In the US, NASA developed a wearable physiological measurement system to measure ECG, blood pressure, respiration, oxygen, and other physiological indexes of more than 30 kinds. The US MD biotech company developed a special eyepiece scanning probe, which according to the color of the capillaries in the eye can tell the content of oxygen in the body, determine the chemical contamination, and alert in a timely manner to take emergency measures to avoid casualties. The University of California developed a new type of underwear, equipped with biosensors, to record uninterrupted human physiological data that can be used to monitor the state of human life, and help medical staff to confirm the damage. Korea developed a physiological signal monitoring system based on helmets, which can detect soldiers' information, ECG, EOG, and EEG, monitor real-time the health of soldiers round the clock, providing commanders with the soldiers' physical condition information. Meanwhile, the system combines a global positioning system and rescue medical system that can significantly reduce rescue time for soldiers, especially for serious injuries and for those who are unable to send a distress signal. Professor Roy at Taiwan University launched the "Advanced Wireless Biomedical Health Monitoring System" using micro-nano-chip implanted in the skin surface to detect indicators such as breathing, heart rate, blood, and use the data through the remote, so as to monitor physiological indicators for early diagnosis and early warning. As early as the late 1980s of the twentieth century, the US military proposed a conception of individual vital signs monitored in the battlefield aimed at the entire battlefield information technology, so that every commander, soldier, and support personnel can get timely effective use of the required information, to always have a clear and accurate picture of the battlefield situation map, to ensure the battle plan, and the implementation of its mandate. The US Army developed the "Army land Warrior" program. Among them, the individual vital sign monitoring systems (Soldier Physiologic Status Monitoring, WPSM) is a US Army future force Warrior (FFW) part of the equipment. The US military planned to FFW in 2010, part of the equipment used will be completed in 2020. The US military developed chip Hospital systems, sensors, and medical injection devices integrated on a chip attached to the soldiers' skin; once sensor signal soldier is injured, it immediately initiates the emergency injection device. America also plans to implant new nanosensors to the soldiers' body to monitor the health of soldiers on the battlefields of the future, with rapid responses to the required state or providing drugs to the blood. In addition, the US Navy has developed the wisdom shirts project [2]; NASA developed the life Defender project [3] and battlefield wireless life monitored system (C-WHMS); France has FELIN plans [4]; United Kingdom has FIST plans, and Germany of future warrior [5], plans are joined on the individual life system monitoring technology, where through different sensors, physiological parameters of soldiers are obtained on life signs state changes for real-time monitoring, The purpose of achieving guiding operations is to rescue the soldiers.

A study on single vital sign monitoring started in China in late 2001. Li hongyi [6] researched individual physiological information monitoring system design; 2002, Guo jinsong [7] on individual life State remote monitored and the positioning system of research, using global positioning system GPS for positioning, achieved 1-km distance within life, and location information of transfer; 2005, Chen Jian [8] on individual life State monitoring system for has research, proposed has system program; 2009, Xin Li [9] Vital signs monitoring system based on ZigBee system design and research in 2010, Han Zhihai [1] on the sea, such as individual life and thoughts on development of condition monitoring system. In 2013, Xie Tai [10] signs monitoring system based on wireless RF technology research. Studies have shown that using vital signs monitor based on clinical technology is mature giving accurate human biological information, but on the field the combatants of movement are under long-term monitoring of vital signs and hence there are still some technical difficulties.

15.3 The System Design

Battlefield life monitor soldier vital signs are comprehensive monitoring, analysis and assessment, through the provision of advice and guidance, and risk factors for illness and injury is interfering in and controlling the whole process. Figure 15.1 shows the battlefield life monitoring of closed loop, first on soldiers for life signs of detection; then, through detection data of analysis and processing, achieved on its life State of assessment and risk forecast; according to decision information,

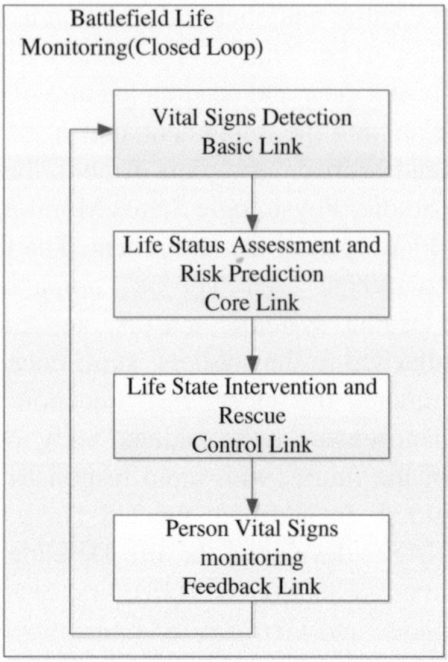

Fig. 15.1 Process of battlefield soldier vital signs monitoring

achieved on soldiers life State of intervention and ambulance measures; last, according to life signs of real-time monitoring, judge control links of actual effect, to achieved full of soldiers life signs monitoring process, and can and corresponding of combat command system interconnected, reached information data of shared.

Battlefield individual life signs monitoring system (BSPSMS) can real-time gets soldiers of life signs information and the injury situation, and sent to command center, achieved remote analysis, and injury assessment and Wei service resources scheduling, its construction target is improve soldiers of integrated combat capacity, increases its in combat action of combat effectiveness and survived chances, grasp forces overall fighting, effective implementation combat command, reached to minimum cost won war victory of target has important meaning.

15.3.1 System Features

BSPSMS is a new physiological parameter monitoring system with remote data transmission capabilities, is an advanced noninvasive sensor technology, information technology, satellite positioning technology and low-power wireless communication technology combined with a small device, can continuously monitor a soldier's vital signs and send people physical data and position reporting in real time. Modular design of the system, to facilitate agreement on unified standards, realizes all kinds of sensor Plug-and-play. In specific implementations, wearable structure should be taken, such as watches, mobile phone type, ring type, IC card, and to adapt to different environments, different levels of population.

Main functions of the system:

(1) Monitor vital signs. By soldiers, heart rate, blood pressure, pulse, respiration, body temperature and other physiological monitoring, analysis and monitoring of their damage.
(2) Fatigue testing. Using Visual methods recognized by the system's response to judge the State of fatigue characteristics, such as closed eyes, head, frequent yawning, etc. This non-contact method, it's easy, accurate, but susceptible to light and influence of individual factors.
(3) State intervention analysis and decision making. In the acquisition of soldier physiological signals and physiological response after signature, analytical processing, extraction and physiological parameters, control with basic physiology experts, determine the State of a soldier's life, sound the alarm or remote intervention. Meanwhile, base on the optimal deployment of combat missions of vital signs.

15.3.2 System Composition

By BSPSMS single depots and central station. Each of the soldiers wear a portable single depots, and Continuous Monitoring Cardiac, blood oxygen saturation, blood pressure, temperature, heart rate, breathing, posture, and environmental parameters, and the data store-and-forward to the central station for analysis, processing and display. Click alongside with the central site has a self-group of web features, click alongside to ID dynamically from the group internet access, central station at the Internet at the same time, it also has the battlefield communications network Internet functionality, and ultimately achieve multi-level raster the group to form a unified battlefield soldiers vital signs sensor system, as shown in Fig. 15.2.

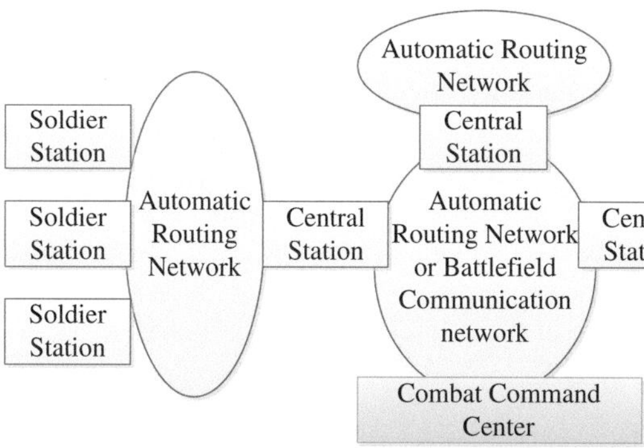

Fig. 15.2 Structure of system network

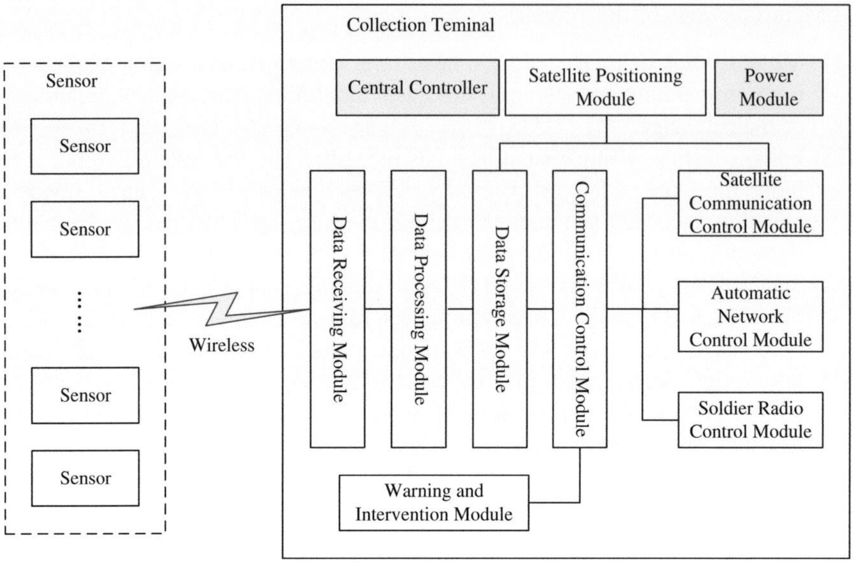

Fig. 15.3 Structure of soldier station

Click depot is a small, wireless sensor network, also known as the domain network (Body Area Network, BAN), which is a small removable, with communication of the sensor, and a collection of terminal (as shown in Fig. 15.3). Each sensor is to wear, to capture terminal BAN is the manager, is also BAN to an external network, such as battlefield communications network) the interface between data transmission and exchange. Click alongside the acquisition terminal and the sensor (including temperature probes, pulse detector, breathing detectors, blood pressure probe and the attitude Detector, etc.) BAN composition, receiving data and processing, while at the same time monitoring the environment parameters, such as chemical or biological environment detectors, life Signal Detectors, and sleep quality detector, satellite positioning systems, etc.). The sensor is in a portable monitoring unit, with real-time monitoring, vital signs, such as blood pressure, blood sugar, heart rate, breathing, blood oxygen saturation, and so on), making use of the wireless technology, such as Bluetooth, WIFI, ZigBee, etc.) with the acquisition terminal connected, and with a group of Web mode, in a given geographical area, access the central site.

Central Station, usually a unit-level command centers, to achieve a certain geographical area, a multi-center redundant configuration, you can also use battlefield communications network access charges at a higher level system, the commander informed with life of the soldiers, accurate judgment combat capability to provide viable basis.

15.3.3 Data Transmission

Monitoring system as a whole to achieve interoperability, and form a unified perception of vital signs, the key lies in data transmission technology. Therefore, when you design a data transmission module, it is important to take into account issues such as encoding, transmission systems, communication protocols, use as little as possible the amount of data delivered as much information as possible, the transfer process as shown in Fig. 15.4. Figure, apart from BAN mantel collecting physiological data, also can be used with individual Internet radio, to position it where necessary, automatically opens the radio sends the ambulance report information and State of life alarm intervention.

Due to the complexity of the soldier's own equipment and the unpredictability of the external environment, as well as the vital sign parameters measured different parts, using cable connecting the sensor will cause inconvenience to soldiers, so data transmission wirelessly must be taken. Vital sign monitoring project in a foreign country, General transfer information is divided into 3 stages, body area networks, local area networks and wide area networks. Body area network of individual stations stage between sensor and data acquisition terminals due to transmission distance, uses low power wireless transmission technology. Single LANs stage between the depot and the central station, according to battlefield environment with WIFI high speed transmission technology. Centers and wide area network between the combat command center stages, then used the battlefield communication networks, long-distance transmission of data. If necessary, vital signs information can also be classified using short messaging functions of Beidou satellite positioning system global transmission of critical information.

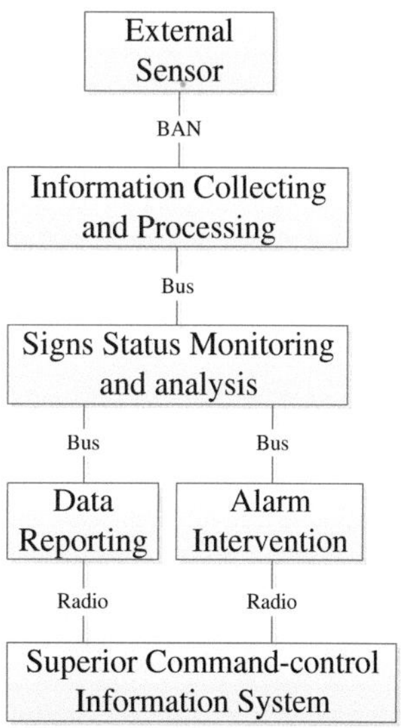

Fig. 15.4 Information translating process of the system

15.4 Conclusions

This paper researches the design method of battlefield soldiers' life signs monitoring system; a general method of the system construction is provided based on status of domestic and foreign system projects' construction; the composition structure, network mode and application of data transmission technology of the system are discussed. The following aspects should be researched continuously in the system design, including: (1) adapt complex and worse battlefield environment, especially coastal and plateau Alpine area and complex electromagnetic environment, etc., factors; (2) further improve sensors' sensitivity and reliability, make them timely and accurate to percept changes in sign parameters; (3) adaptability and effectiveness of data heterogeneous group network; (4) use short message function of Beidou satellite positioning system for data transmission; achieve unified battlefield soldiers' life signs perception, and can make first aid interventions to injured personnel and advantages deployment of combat missions timely according to signs state information.

References

1. Han Z, Li J (2010) Research on the naval solider physiological status monitor system. J Nav Gen Hosp 23(4), 12:232–234
2. Park S, Jayaraman S (2003) Enhancing the quality of life through wearable technology. IEEE Eng Med Biol Mag 22(3):41–48
3. Mundt CW, Montgomery KN, Udoh UE et al (2005) A multiparameter wearable physiologic monitoring system for space and terrestrial applications. IEEE Trans Inf Technol Biomed 9 (3):382–391
4. Chunfei Wang, Jianghong Shi (2007) Wireless sensor network in American War-fighter Physiologic Status Monitoring System. Chin Med Equip J 28(11):34–35
5. Liu L, Huang G et al (2013) Staff physiological monitoring and control system on special vehicle. Nat Defense Sci Technol 34(6), 12:33–36
6. Li H, Wang J et al (2001) Design of monitoring system of soldier physiological information. J Fourth Mil Med Univ 22(2):103–104
7. Jinsong Guo, Qinkai Deng, Jian Gong (2002) Development of apparatus for remote personal vital status monitoring and positioning. J First Mil Med Univ 22(4):320–322
8. Jian Chen, Xiuzhen Dong, Teng Jiao (2005) Research of soldier physiological status monitoring system. Chin Med Equip J 26(7):20–21
9. Xin Li, Xin Li, Jingwei Dong (2009) Vital signs monitoring system based on ZigBee. Inf Technol 8:174–177
10. Xie T, Liu X et al (2013) Research on physiological status monitor based on radio frequency technique. Chin Med Equip J 34(6), 6:4–6

Chapter 16
Analysis and Countermeasures of Aimer Operation Errors of a Certain Type of Anti-aircraft Artillery

Junlong Guo, Rongzhi Yang, Dapeng Wang and Zaochen Liu

Abstract Through analysis of aimer operation errors of a certain type of anti-aircraft artillery this essay finds the existing prominent questions and reasons, and takes effective measures to solve them. From training characteristics and operation errors reasons of aimers the essay deeply researches training matters of aimers using qualitative and quantitative analysis scientifically. In accordance with regularity in an orderly way and training scientifically the aimers can be gradually improved at operational level. According to the principle *from separate operation step to consistent operation, from simple operation to complicated operation and from simulated training to live operation* we should fully improve operational level of aimers and make them adapt training requirement of the live combat drill through scientific training, active changing training spot, and creating the live combat situation.

Keywords Operation error · Analysis · Anti-aircraft artillery · Aimer

16.1 Introduction

A certain type of anti-aircraft artillery is characteristic of fast firing rate, strong fire power and higher automation degree [1]. It mainly directs at attacking air raid weapons from low altitude and super low altitude, shouldering an important mission of countering air and missile threat. When we fire at air target making use of sighting device, the effect of fighting against target fully depends on operational skill and level of aimers [2]. When we have live fire drill some abnormal phenomenon that some aimers with excellent training performance and well-skilled operation procedure cannot completely develop themselves and have operational error usually occur. It plays an important significance for improving air combat

J. Guo · R. Yang · D. Wang (✉) · Z. Liu
Air Defense Forces Academy, East Jianshe Road no 24, Zhengzhou 450052, Henan, China
e-mail: 1137340834@qq.com

© Springer-Verlag Berlin Heidelberg 2015

S. Long and B.S. Dhillon (eds.), *Proceedings of the 15th International Conference on Man–Machine–Environment System Engineering*, Lecture Notes in Electrical Engineering 356, DOI 10.1007/978-3-662-48224-7_16

capability of a certain type anti-aircraft artillery to analyze and research operational characteristics, and discuss the aimers' operational error reasons.

16.2 Training Characteristics of Aimers

When using sighting device to fire, the main task of aimers is to research and track down aiming target in different direction and altitude, and step on the pedal to fire at air target [3]. The training characteristics of aimers are as follows.

16.2.1 Operational Essentials Are Standard

Operational procedure of aimers is complicated which includes receiving monitor's instruction, releasing lockage, searching and tracking down aiming target, changing the state of switch, implementing firing task, etc. During the training process the aimers must abide by operational regulation and conform to regularity produced by military skill [4], and step-by-step have operation according to standard training—well-skilled training—technique training procedure, and not make any mistake at all, otherwise this will cause damage to the equipment and staff.

16.2.2 Training Environments Are Complicated

Its objective existence and unchangeability affects the natural environment during military training. For example, some environmental changes such as geomorphic change, seasons shift, climate change, rotation of day and night will undoubtedly affect training effect. When combating in field conditions it is most difficult to make use of sighting device to research and track down aiming target. In different environments there exist huge differences for aimers' operational level; as long as they adapt all kinds of environments the operational level of the aimers can be normally developed.

16.2.3 Comprehensive Quality Is Highly Required

In order to counter air target of enemy effectively aimers' characteristics are fastness, steadiness, and precision. To attain the requirement the aimers must take on well-skilled operational skill, quick adaptability towards environmental change and well-prepared psychological quality, and combine ears hearing, eyes seeing, hands controlling, feet stepping and mind speculating with one another; the aimers have independent judging ability to make sure that firing occasion is controlled at suitable moment with precise aiming and an appropriate distance set, fully develop the whole efficiency and force of equipments.

16.3 Analysis of Operational Error Reason of Aimers

From teaching training for many years we have found that operational level of aimers and the exercise of their skills are closely related to skill familiarization, training environment, and training psychology. Some phenomena such as no searching targets, no steady tracking down target, and missing firing occasion is closely related to these factors during the live firing process.

16.3.1 Operational Skills Are not Well-Trained

Operational skills are the important bases which combines humans with equipment; as long as operational skills are well done we can combine humans with weapons and fully develop their combat efficiency. The aimer is one of the most important staff in the artillery squad, his operational skills are more required than other gunners, at the same time improving skills is a gradual process with steady advancement and we have no way to improve skills quickly. In the early training stage the aimer advances apparently; when skills come to a certain degree it is difficult to promote the skill level. The skills even come to a standstill and reach training bottleneck. The aimer needs quick response, excellent sight, flexible adaptability, and good psychological quality. The time interval of bottleneck varies in different people. The difference between excellent aimers and average aimers is whether bottleneck is quickly broken through; the aimers whose operational skills are not well grasped easily lose confidence and this leads to nervousness, deformation, and abnormal development of operational action.

16.3.2 Normal Training Environment Is Simple

Because of training equipment and training methods, normal training of a certain type of anti-aircraft artillery is limited to camp training spot; the training environment is relatively familiar and training means is very simple. However, live firing is often done in a new and strange area, uncertain factors increase, and it is difficult for aimers to adapt to the new environment, hence it is easy to lead to abnormal operation and make disastrous errors for the aimers.

During extremely cold weather gunners wear heavy uniform which cause operational inflexibility. During long-time operation it is easy to go numb for the human body, even extreme cold injury can occur. For fulfilling the same operation the body energy will consume more than under normal temperature and operational precision and rate will be greatly affected; in wet and hot weather with high temperature when the gunners are trained their body temperatures quickly escalate and more sweat will drain; all these factors can easily cause dizziness, sickness, spiritual wandering, even dehydrating, and much worse, lead to heatstroke. It is easy for human emotion to show irritation and anger. During sandstorm, heavy rain, and fog when the air is not clear and visibility is very low, it is difficult for gunners to observe everything, spirit is in high tension and it is easy for the gunners to fatigue; in extreme cold plateaus long-time training easily leads to plateau reaction and seriously affects training effect and does harm to human body to a certain degree because of high terrain, low air pressure, thin air, and low oxygen content.

In different environments the aiming precision and rate that same aimer conducts slightly differ (see Tables 16.1, 16.2 and 16.3). In accordance with the regulation of military training and test syllabus [5] we have a test towards the 4 aimers' grade under the three conditions. Before test the four aimers has arrived at the excellent level (finished within 6 s, errors within 5 mil), and they have fully warmed up. Through the test result we may conclude that under the different environments training performance greatly differs, some aimers are greatly affected from excellent level to average level even to disqualification.

Table 16.1 Test performance of aimer under the suitable condition (clear, breezeless, 25 °C)

Aimer	1st Aimer	2nd Aimer	3rd Aimer	4th Aimer
Aiming error	1 mil	No error	2 mil	1 mil
Target acquisition time	5″20	4″10	5″60	5″88

Table 16.2 Test performance of aimer under the windy and dusty condition (clear, fresh breeze, 25 °C)

Aimer	1st Aimer	2nd Aimer	3rd Aimer	4th Aimer
Aiming error (mil)	2	2	–	3
Target acquisition time	5″60	5″10	No acquiring target	6″40

Table 16.3 Test performance of aimer under the condition of high temperature (clear, breezeless, 36 °C)

Aimer	1st Aimer	2nd Aimer	3rd Aimer	4th Aimer
Aiming error (mil)	2	1	4	3
Target acquisition time	5″40	4″40	5″80	6″10

16.3.3 Psychological Change Under the Live Drill Is Abnormal

In live drill the training often combines tactic drill with live firing drill; the training conditions are poorer and are characteristic of more training program, high intensity and big difficulty, the psychological pressure of the trainees rapidly increases, it is more apparent for some important post such as aimers to receive training.

The first is to produce panic psychology. Under the live firing process the strong blast wave and special noise that artillery firing produce make aimers produce fear and uneasiness, their heartbeat is quickly increasing and concentration can't be focused, and this influences their spiritual state, even muscle quivering, thus seriously affect their operational performance. The second is to produce fatigue psychology. Continuously long-time training with high intensity gives aimers heavy spiritual burden, their body and emotion are extremely fatigue; strict safety regulation requires the aimers to concentrate on attention, it is easy to make their brains fatigue and thus leads to much lower firing precision. The third is to produce nervous psychology. During the usual training process some aimers' level is high, however, during the live firing process they are anxious to fire at target, thus this leads to psychological nervousness. An appropriate nervousness can develop potential capability of aimers, but extreme nervousness with heavy psychological burden may lead to functional imbalance of arms and legs, and cause serious barrier for development of operational skills.

16.4 Countermeasures

16.4.1 Training to Improve Aimer's Operational Level Scientifically

Through long-time training the aimers have been made to grasp tracking down, aiming and firing skills, however, because of different training intensity and methods the level of operational skills is different during the training process. So scientifically training is required to improve operational level of aimers gradually during the training process.

From the training effect it is very necessary to have a lot of repeated training and higher training intensity. But it isn't absolute to have a lot of repeated training and higher training intensity. A lot of repeated training and higher training intensity can cause aimers fatigue and lose interest, even make error operation occur frequently, thus reduce operational level. From training ways in accordance with the principle from separate operation step to consistent operation, from simple operation to complicated operation and from simulated training to live operation the training program orderly develops and gradually reinforce training difficulty, at last make operational skills of aimers become more perfect and the level is becoming much higher under the basis of regulation.

16.4.2 Change of Different Spot to Promote Aimer's Adaptability

Recalling the history, both home and abroad any war is done under the specific natural environment which plays immeasurable effect on combat process and result. Traditional aimers training is done in the training spot, much less considering training influence under other circumstance. So we must actively innovate training methods, comprehensively consider environmental factors, make aimers being trained under all kinds of environment and conditions, greatly improve aimers' adaptability.

Researching the influence of all kinds of natural environments towards aimers operation its aim is to solve the issues of training level. So we should fully make use of bad weather and worse situation, regard the training the aimers adapt under all kinds of conditions as the important content of normal training, completely improve their combat adaptability.

16.4.3 Creating Situations to Strengthen Psychological Quality of Aimers

The live training is the important path to improve combat ability during the peace time. If we need to combine military training with the war organically we must try to create the situation of live training, continuously strengthen psychological quality of aimers, and thus attain the real purpose like war.

The first is to create situation of tension. The aimers are relatively relaxed because the normal training environment is relatively loose and lack pressure, most of aimers are excellent during the normal training process, however, psychological nervousness for the aimers leads to deformation and error of operation during the live fire training, which will influence firing precision during the average situation, in critical time it will lead to firing failure, even more serious danger. So during the process of cultivation, nervous situation needs to be created and the education and training of psychological endurance ability should be trained all the time, through different ways such as non-scheduled test and check, reinforcement of training intensity psychological quality of aimers is reinforced. The second is to create the situation of combat spot. Through all kinds of modern means the battle situation will be deeply created by making use of sound, light, and smoke to simulate the roaring of air target, quake produced by explosion, big fire, gunpowder smoke, and so on. As a way of launching simulation bullets aimers are made in battle situation, gradually they reduce stress and fear toward battle and progress psychological endurance capability.

16.5 Conclusions

After analysis we conclude that the training of aimers is the characteristics of standard operational essentials, highly complicated training environment, and completely comprehensive quality. Operational level and skills and exercise of aimers are close related to skills level, the live combat training during the training circumstance, abnormal psychological change under the circumstance of live drill. Through experience, training, and the related practice we draw the conclusion that in order to acquire high training level and steady state of aimers we may improve operational level of aimers through training scientifically, promote aimers' adaptability through changing training environment actively, and also strengthen psychological quality of aimers through creating live combat situation. The research result of the essay will provide guidance and help for improving the training level and safety state of aimers and also provide experiences and lessons for other related operation and training.

References

1. Liu Z (2000) Twin 35 mm towed AA artillery weapons system. Artillery Mag 2:90–91
2. Song Y, Zhao H (1996) Brief firing analysis of 35 mm AA gunsight. Artillery Mag 4:84
3. Guo Junlong, Yang Qizi (2014) PG 99 type 35 mm AA artillery weapons and operations (operational manual). Liberation Army Press, Beijing, p 231
4. Pang Z (1992) Techniques generation of testers. Libertain Army Press, Beijing, pp 5–10
5. The Headquarters of General Staff of China's PLA Military Training and evaluating Syllabus of China's PLA Army. Liberation Army Press, Beijing (2008)

Chapter 17
A Quality Evaluation Index System of Group Leaders in an Aerospace Enterprise Based on a Fuzzy Evaluation Method

Yuchuan Bian, Man Zhang, Peng Qi and Jiang Yu

Abstract We investigated how to choose and appoint group leaders in an aerospace enterprise. The study analyzed the characteristics of group leaders' work, including their ability for management, operation, communication, and vocational morality, and analyzed their impact on the quality of the requirements for aerospace enterprise practitioners. The article built a structural model of group leaders in the aerospace enterprise, including 4 factors and 12 subsidiary factors. It established an index system of group leaders in the aerospace enterprise based on the basic theories of the fuzzy evaluation method and Delphi method. It examined the rationality of the index system through an example and offers a reference to choose and appoint group leaders in the aerospace enterprise.

Keywords Fuzzy evaluation method · Group leaders in aerospace enterprise · Quality evaluation

17.1 Introduction

With the development of the aerospace industry, there are a growing number of various types of spacecraft. In China, assembly, integration, and testing (AIT) of spacecraft in development are mainly carried out in groups. Groups are the most basic unit of the production and operation activities or management in aerospace enterprises. The quality of group leaders will directly affect the normal operation of the groups, along with the schedule, quality, safety, and other aspects of the work. Therefore, a scientific quality evaluation index system for group leaders should be established in order to guide the selection of group leaders in aerospace enterprises and meet the quality requirements for group leaders in new situations.

Y. Bian (✉) · M. Zhang · P. Qi · J. Yu
Beijing Institute of Spacecraft Environment Engineering, Beijing 100094, China
e-mail: byuchuan99@126.com

© Springer-Verlag Berlin Heidelberg 2015

S. Long and B.S. Dhillon (eds.), *Proceedings of the 15th International Conference on Man–Machine–Environment System Engineering*, Lecture Notes in Electrical Engineering 356, DOI 10.1007/978-3-662-48224-7_17

17.2 Characteristics of the Aerospace Enterprise and Quality Requirements for Group Leaders

The development of spacecraft is a work [1] of high technology, high risk, and large quality progress pressures, especially in the AIT development stage. In the mode of organizing production by units of groups, group leaders should properly arrange the work plans and labor division of personnel according to work content and communicate with personnel at all levels in the production process. In addition, because of the special status of spacecraft development in the national political and economic sectors, it also requires that group leaders should have high ideological qualities and professional morals. The quality of group leaders mainly manifests in the four aspects discussed in the following sections.

17.2.1 Management Ability for Group Leaders

Management is the main responsibility of group leaders—implementing all rules and regulations, rules of production, and operation of the enterprise, as well as accomplishing the working tasks and goals of the enterprise through management. It includes management of the group members, management of things, management of the process of production and operation (work), and management of information and time, among others.

17.2.2 Operation Ability for Group Leaders

As the leader of technical skills in the groups, group leaders should completely understand the related operation processes of their groups; master the necessary operation skills, business skills, and management techniques; have enough ability in the practical operation; and be able to discover and solve problems. This ability includes professional skills, discovering ability, emergency capacity, etc.

17.2.3 Communication Ability for Group Leaders

During the process of spacecraft AIT, group leaders need to communicate with group members, laboratory leaders, leaders of the project office, inspection personnel, scheduling personnel, quality personnel, and other group leaders, among others. Group leaders always become the focus of disagreements. Therefore, group leaders should have strong communication skills, good written expression and oral expression skills, and the necessary persuasion skills. Group leaders should fully, properly, and effectively communicate with people; win the understanding and support of subordinates, superiors, and peer group; properly manage conflicts; and conduct operational coordination across departments and levels.

17.2.4 Professional Morals for Group Leaders

Group leaders should be devoted to their duties, be conscientious, strive for perfection, perfect their work achievements, and continuously improve themselves. These abilities are mainly manifested in the responsibility and enterprise of work.

17.3 Structural Model of a Quality Evaluation Index of Group Leaders in Aerospace Enterprises

The appropriate choice and establishment of an accurate quality evaluation index are key to the effective evaluation of group leaders. There are many different kinds of system models. Commonly used models include the system structural model, simulation model, factor analysis, and principal component analysis [2].

The American psychologist David C McClelland stated that there is not much connection between intelligence test results and work success. The relationship between them depends on specific circumstances [3]. As a result, management ability, operation ability, communication skills, professional morals, and other requirements were incorporated in a quality evaluation index system for group leaders in aerospace enterprises. A system structural model has been established, as shown in Fig. 17.1.

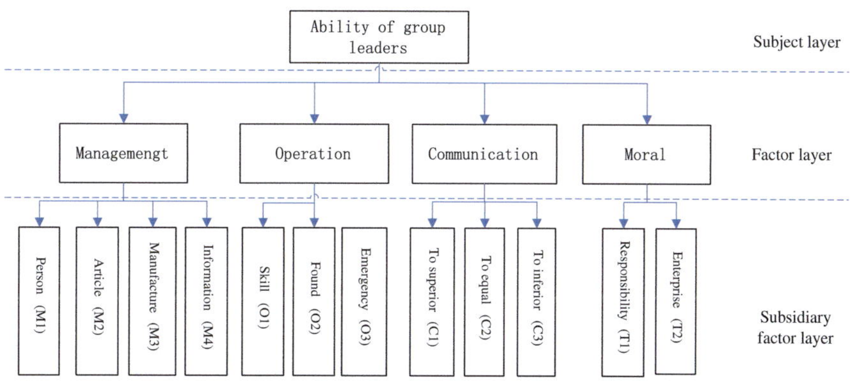

Fig. 17.1 The structural model of group leaders in the aerospace enterprise

As can be seen from Fig. 17.1, the structural model is divided into three layers: the subject layer, factor layer, and subsidiary factor layer. The subject layer is the final target of the structural model. The factor layer is the key factor to determine the subject layer. Subsidiary factors are the factors affecting every factor. They are also basal factors of the whole structural model. There are altogether four indexes in the factor layer of the quality evaluation system of group leaders in aerospace enterprises. There are altogether 12 indexes in subsidiary factor layer.

17.4 Determination of the Weight of Evaluation Indexes

In the evaluation model shown in Fig. 17.1, each factor in each layer has a different effect. There are differences between their importance, so indexes should be given different weights. In this study, the weight of each quality evaluation item and subweight of its evaluation index were calculated by the fuzzy evaluation method and Delphi method.

17.4.1 Establish a Judging Panel of Experts

According to the fuzzy evaluation method, a judging panel of experts needs to be established first. We selected a human resources director, research director, person of model project office, group members, and group leader from an aerospace enterprise to form the index evaluation group and determine the weight of each evaluation index.

17.4.2 Determine the Weight of Each Factor and Subsidiary Factor

Using relative weight, an evaluation index is taken as a unit and its percentage is calculated in the total evaluation index. Considering that the effect of the above indicators on the quality of group leaders cannot be normally quantitated, most of the factors should be secondarily quantitated; there also are effects of human subjective impact in weight distribution of indexes. As a result, the judgment analysis Table 17.1 should be listed after comparing evaluation items in accordance to the relative importance classes [4] of T.L. Saaty.

The relative importance between each evaluation item and the evaluation index of each item was evaluated using the Delphi method. This method gives 1 point to the least important factor, then compares the other factors with it to see their importance. The scores are completed as in Table 17.2 (selecting the evaluation

Table 17.1 Satty relative importance classes

Admeasurement	Meaning
1	Ci equals Cj
3	Ci important than Cj
5	Ci is more important than Cj
7	Ci is more important than Cj obviously
9	Ci is more important than Cj absolutely
2, 4, 6, 8	Ratio between the two interfacing classes

Table 17.2 Evaluation results of Expert W

Factors	Score	Subsidiary factors	Score
Management (M)	8	Person (M1)	7
		Article (M2)	4
		Manufacture (M3)	1
		Information (M4)	3
Operation (O)	6	Skill (O1)	5
		Found (O2)	6
		Emergency (O3)	4
Communication (C)	7	To superior (C1)	8
		To equal (C2)	2
		To inferior (C3)	7
Moral (T)	3	Responsibility (T1)	3
		Enterprise (T2)	3

results of expert *W* as an example). There were altogether ten experts participating in the consultation. The statistical result of scores are shown in Table 17.3.

A judgment analysis table of each layer was created according to the statistical results of the expert consultation. Table 17.4 shows the judgment analysis table of evaluation indexes for management capability. Other judgment analysis tables of the same layer can be made according to this calculation method. Table 17.5 shows the judgment analysis table for each factor item of the quality evaluation of group leaders in the aerospace enterprise.

The ratio of each evaluation item in Tables 17.4 and 17.5 can be calculated by dividing the average score of the evaluation item from expert consultation corresponding to the row of the ratio by the average score of the evaluation item corresponding to the column of the ratio. For example, the ratio of row O/column M of Table 17.5 is 1.02. This was calculated by dividing the average score 8.5 of the evaluation item corresponding to row O by the average score 8.5 of the evaluation item corresponding to column M. Therefore, the evaluation factor O (operation ability) is equivalent to 1.05 evaluation factor M (management ability).

The relative importance W_i can be calculated according to formula (17.1). Then, normalization should be carried out.

Table 17.3 Statistics of index system's factor

Factors	Total score	Average	Subsidiary factors	Total score	Average
M	85	8.5	M1	88	8.8
			M2	76	7.6
			M3	72	7.2
			M4	60	6
O	87	8.7	O1	79	7.9
			O2	66	6.6
			O3	52	5.2
C	72	7.2	C1	80	8
			C2	75	7.5
			C3	63	6.3
T	56	5.6	T1	55	5.5
			T2	48	4.8

Table 17.4 Judgment analysis of management ability

Factors/Ratio	M1	M2	M3	M4
M1	1.00	1.16	1.22	1.47
M2	0.86	1.00	1.06	1.27
M3	0.82	0.95	1.00	1.20
M4	0.68	0.79	0.83	1.00
$\sum_{j=1}^{n} a_{ij}$	3.36	3.89	4.11	4.93

Table 17.5 Judgment analysis of the factors

Factors/Ratio	**M**	**O**	**C**	**T**
M	1.00	0.98	1.18	1.52
O	1.02	1.00	1.21	1.55
C	0.85	0.83	1.00	1.29
T	0.66	0.64	0.78	1.00
$\sum_{j=1}^{n} a_{ij}$	3.53	3.45	4.17	5.36

$$W_i = \frac{1}{n}\sum_{j=1}^{n}\left(a_{ij}\Big/\sum_{j=1}^{n} a_{ij}\right) \quad (17.1)$$

Here, W_i is the weight of the evaluation factor (or subsidiary factor) i; n is the number of quality evaluation factors (or subsidiary factors) for the electronic group leader; i represents the row i; j represents the column j; and a_{ij} is the relative importance of evaluation factor (or subsidiary factor) of row i on the evaluation factor (or subsidiary factor) of column j.

Thus, it can be calculated that the weight set W of the evaluation items M, O, C, and T is (0.28, 0.29, 0.24, 0.19). The weight set W_m of the evaluation index of the management ability subsidiary factor is (0.30, 0.26, 0.24, 0.20). The weight set W_o

Table 17.6 Weight coefficients of (Subsidiary) factors

Factors	Weight coefficient	Subsidiary factors	Weight coefficient
Management (M)	0.28	Person (M1)	0.30
		Article (M2)	0.26
		Manufacture (M3)	0.24
		Information (M4)	0.20
Operation (O)	0.29	Skill (O1)	0.40
		Found (O2)	0.34
		Emergency (O3)	0.26
Communication (C)	0.24	To superior (C1)	0.37
		To equal (C2)	0.34
		To inferior (C3)	0.29
Moral (T)	0.19	Responsibility (T1)	0.53
		Enterprise (T2)	0.47

of the evaluation index of the operation ability subsidiary factor is (0.40, 0.34, 0.26). The weight set W_c of the evaluation index of the communication ability subsidiary factor is (0.37, 0.34, 0.29). The weight set W_t of the evaluation index of the professional morals subsidiary factor is (0.53, 0.47). The final calculated weights of factors (or subsidiary factors) are listed in Table 17.6.

17.5 Application of the Quality Evaluation Index Model for Group Leaders in Aerospace Enterprises

17.5.1 Choice of Evaluation Method

The specific evaluation method should be determined according to the evaluation purpose, object, and content. At present, the commonly used evaluation methods mainly include a written test, interview, psychological test, evaluation center, and other methods [5]. The scenario simulation interview method has been applied in the quality evaluation of group leaders in the aerospace enterprise. That is, a targeted scenario simulation question bank should be formulated according to each subsidiary factor of the quality evaluation model of group leaders before the interview. An example for the subsidiary factor of personnel management is "As a group leader, what kind of person do you think is qualified for part-time safety personnel in your group? What the duty should be?" An example of professional skills is "Please make a series of gestures used for commanding cranes." During the interview, candidates for group leaders who are being interviewed should extract questions item by item according to each subsidiary factor and answer. Expert interviewers should grade according to the answer and performance of the candidates for the simulation questions. After evaluating all the subsidiary factors, a final calculation and evaluation should be done according to the weight table of all factors (subsidiary factors) of quality evaluation, as shown in Table 17.6.

17.5.2 Creating an Evaluation Table

A quality evaluation table for group leaders should be made for the convenience of the expert interviewers to evaluate the group leader candidates. The table evaluates the candidates according to the 12 subsidiary factors of the quality evaluation structural model. Each subsidiary factor is divided into four grades: excellent, good, average, and poor. During the interview, expert interviewers should mark "√" on the corresponding grade according to the performance of group leader candidates. Only one grade is allowed for each item (Table 17.7).

When summarizing the results of quality evaluation, the probability of five grades for each evaluation index of the expert interviewers should be counted. That is, the grade probability of one index equals the number of experts evaluating that grade divided by the total number of experts.

When evaluating candidate F for group leader in an aerospace enterprise, there were 10 experts conducting an effective evaluation in the interview. The results should be counted and summarized according to the previously described method, as shown in Table 17.8.

Table 17.7 Evaluation table

Factors/Class	Excellent	Good	Average	Poor
Person (M1)				
…				

Table 17.8 Candidate F's evaluation

Factors	Subsidiary factors/Score	Excellent	Good	Average	Poor
Management (M)	Person (M1)	0.8	0.1	0.1	
	Article (M2)	0.8	0.2		
	Manufacture (M3)	0.9	0.1		
	Information (M4)	0.7	0.1	0.1	0.1
Operation (O)	Skill (O1)	0.9	0.1		
	Found (O2)	0.8	0.1		0.1
	Emergency (O3)	0.7	0.1	0.2	
Communication (C)	To superior (C1)	0.8	0.1	0.1	
	To equal (C2)	0.7	0.1	0.1	0.1
	To inferior (C3)	0.9	0.1		
Moral (T)	Responsibility (T1)	0.9	0.1		
	Enterprise (T2)	0.8		0.1	0.1

17.5.3 Comprehensive Evaluation of the Quality Evaluation Index

A comprehensive evaluation matrix can be calculated using a fuzzy change formula (17.2), for each fuzzy matrix of single factor from Table 17.8 and each weight set of the evaluation index in Table 17.6.

$$B_x = W_x \circ R \tag{17.2}$$

Here, x is the evaluation factor, R is the evaluation matrix, W_x is the weight set, and "o" is the Zadeh operator (∨, ∧).

Thus, the comprehensive evaluation matrix for the management ability of candidate F for group leader would be as follows:

$$\begin{aligned} B_m = W_m \circ M &= (\,0.30 \quad 0.26 \quad 0.24 \quad 0.20\,) \circ \begin{pmatrix} 0.80 & 0.10 & 0.10 & 0 \\ 0.80 & 0.20 & 0 & 0 \\ 0.90 & 0.10 & 0 & 0 \\ 0.70 & 0.10 & 0.10 & 0.10 \end{pmatrix} \\ &= (\,0.30 \quad 0.10 \quad 0.10 \quad 0.10\,) \end{aligned}$$

In the same way, we can get $B_o = (\,0.40 \quad 0.10 \quad 0.20 \quad 0.10\,)$, $B_c = (\,0.29 \quad 0.10 \quad 0.10 \quad 0.10\,)$, $B_t = (\,0.47 \quad 0.10 \quad 0.10 \quad 0.10\,)$.

Therefore, a comprehensive evaluation matrix consisting of each factor would be as follows:

$$B = (\,B_m \quad B_o \quad B_c \quad B_t\,)^T$$

The comprehensive evaluation matrix for candidate F for group leader can be calculated by fuzzy change with formula (17.2): $Z = (\,0.29 \quad 0.10 \quad 0.20 \quad 0.10\,)$. After normalization, we get $Z' = (\,0.42 \quad 0.14 \quad 0.30 \quad 0.14\,)$.

The overall quality of candidate F for group leader can be quantitatively evaluated. With a maximum score of 5, excellent, good, average, and poor correspond to 5, 4, 3, and 2, respectively.

$$F = Z' * (\,5 \quad 4 \quad 3 \quad 2\,)^T \tag{17.3}$$

It can be calculated that F = 3.76. Therefore, the overall quality of candidate F for group leader is between good and average. After being appointed as group leader for 1 year, F's performance was satisfactory and can basically ensure the normal operation of the group. However, he lacks any outstanding qualities.

17.6 Conclusions

According to the characteristics of the spacecraft development and aerospace enterprises, the quality of group leaders was investigated. A complete quality evaluation index system for group leaders in aerospace enterprises was established. Practical application has shown that a complete quality evaluation system is reasonable, practical, and can meet the requirements of aerospace enterprises for choosing group leaders.

References

1. Wang (2007) The risk management of aerospace project, 3rd edn. Project Management Technology
2. Bai S (2006) Systems engineering. Electronic Industry Press, Beijing
3. McClelland DC (1973) Testing for competence rather than for intelligence. Am Psychol 1:1–14
4. Wang C (2010) The study of quality evaluation based on professional manager. Master's degree thesis of Lanzhou University of Technology
5. Xu S, Wang J (2000) Ability evaluation. Enterprise Management Press, pp 100–102

Part II
Research on the Machine Character

Chapter 18
Rethinking the Mode of Modern Commercial Walking Street Planning and Design

Zhen Li and Fang Liu

Abstract In this paper, the modern, urban commercial walking street is the research object. The purpose is to promote the level of commercial walking street planning, and improve the quality of urban commercial shopping environment. Based on the updated mode of modern commercial walking street planning and design, a case analysis is performed. The planning ideas of the project of Ming Yang Commercial Walking Street in Xinyang is discussed and the important principles in planning and design of the modern commercial walking street from different aspects, such as the design concept, overall layout, landscape design and architectural modelling is summarized. Thus, this work makes a valuable reference for contemporary urban commercial walking street.

Keywords Commercial walking street · Planning · Space · Architectural design · Landscape

18.1 Introduction

The modern commercial walking street appeared in the 1950s. After World War II, economy in the western countries represented by the United States developed rapidly, and the urbanization process also moved forwards quickly. As a result, the standard of urban residents' living has increased greatly. So the requirement for the quality of the life is getting more and more diversified. At the same time, new urban

Z. Li (✉)
School of Civil Engineering and Architecture, Wuhan Polytechnic University, Wuhan, People's Republic of China
e-mail: 66994752@qq.com

F. Liu
The Infrastructure and Logistics Department, Huazhong University of Science and Technology, Wuhan, People's Republic of China
e-mail: 275724782@qq.com

© Springer-Verlag Berlin Heidelberg 2015

S. Long and B.S. Dhillon (eds.), *Proceedings of the 15th International Conference on Man–Machine–Environment System Engineering*, Lecture Notes in Electrical Engineering 356, DOI 10.1007/978-3-662-48224-7_18

problems appeared constantly, such as the urban expansion. Lots of urban residents have moved into the suburbs, the city centre has hollowed out [1]. Furthermore, the traditional narrow streets have been replaced by wide roads. Cars have dominated the urban life. It has caused the residents to travel with difficulty and feel lack of security.

Under the impact of this trend, a wave of suburbanization appeared in many developed countries. A lot of supermarkets and shopping centres have been built in new towns. At the same time, the traditional commercial blocks have been reconstructed and walking streets have become the main mode of shopping way in order to recreate the prosperity of the central region. Moreover, many commercial walking streets have appeared even in the new districts of cities. The emergence of the modern commercial walking streets effectively alleviates the traffic congestion in the central city. It has also reduced the air and noise pollution, which will greatly improve the customer's shopping experience [2]. And many facts have proved that it is a very successful pattern of urban design. The purpose of this paper is to promote the improvement of the urban commercial walking street planning's quality, and change the urban commercial environment. This study is explained with the example analysis. Here, the project of Ming Yang Commercial Walking Street is used as an example to discuss the new mode of contemporary commercial walking street planning and design.

The city of Xinyang is the famous tea producing area in our country. The project of Ming Yang Commercial Walking Street is located beside Shi River in the city. It has a strip of land with a wide square in the east, which is named, "Tea Rhyme". To the west of the land, there is an antique building named, "Ming Yang Tower". As a business project, it includes a tea-culture walking street, a museum on the subject of tea, office buildings, a supermarket and many more. The construction of this project will greatly improve the area's business and recreation, and enhance Xinyang's social lifestyle.

18.2 Planning and Design Concepts

18.2.1 A Reinterpretation of Traditional Tea-Culture

Xinyang's special location makes the blend of cultural exchange between the north and south, and forms the cultural characteristics in Central Plains, especially for the famous tea culture. Xinyang is known as the tea producing area since ancient times. The tea culture is the theme of this project; it combines Chinese traditional tea-culture with Xinyang's local characteristics, which is the unique way of this project's planning.

The core of Chinese traditional tea-culture is the four noble thoughts. They are "harmony", "peace", "happiness" and "truth". The four noble thoughts have particular value in modern society [3]. The current intense competition in the society has caused tension in people's physical and psychological lives leading to indifference in society. However, this problem could be resolved by the noble thoughts. For this, we introduce "the tea-ideas" in the planning, and try to show the charm of the traditional tea-culture from different aspects.

18.2.2 Creating a Funny Place for Shopping

Commercial atmosphere and shopping conditions are closely related. Monotony and lack of accessibility will lead to cold commercial environment. The planning tries to expand business area along the street. At the same time, we also design a walking street network, which can both increase the number of street stores and enhance the whole commercial street's equilibrium. This avoids the difference in commercial value between different sections. In addition, we design a special shopping mode, in which commodities of a same kind will be placed together so as to form a cluster effect and increase the walking street's appeal. Also, we introduce the functions of culture and entertainment. An urban stage is planned at the round square on the west side. Various plants for pavement landscape and a little theatre have been carefully arranged at the stage to form a warm atmosphere.

18.2.3 Building the Green Humanistic Landscape

To build a green environment is the important principle of the modern urban design [4]. The project is located at the foot of a hill and near to a river. One side, there is a natural undulating topography from west to east. On the other side, the clear charming Shi River is flowing by. This creates an excellent scope to form a green environment. The design implements the green and humanistic thought from the beginning, and rejects the idea of artificial modification. We try to preserve the nature as much as possible, pile up the rocks, decorate the stream and show the traditional thought of harmony between nature and human. In this way, we can achieve the aesthetic effect for shoppers.

18.3 The Overall Layout and Space Composition

Based on the theme of tea, we establish a combination of the points, the lines, and the planes. In this layout mode, the straight lines fit with the curves. The central commercial street is the developing axis of the whole land. In the south, business building is planned beside the city road, so as to show the regularity of the urban trunk road. In the north three buildings are arranged in a freely rotating fashion to resemble the shape of tea leaves. We have created different spaces by using various elements such as the supermarket at the end, the office buildings, the museum, the entrance square, and so on. This has reproduced the charm of the Chinese traditional architecture group's layout. As to the function, we have planned the business structure carefully. The free stores and the supermarket are combined. Moreover, the upper floors are designed into the apartment offices. The tea museum lies on the west square. It's designed like a fan, which implies a humble, natural philosophy of the tea. The supermarket and the office building are placed on the east side, in order to correspond with the central commercial walking street.

The diversity of the commercial walking street's spaces has been explored in the planning. It focuses on the following types of spaces:

Commercial space: Commercial walking street is not only a shopping place, but also a public environment. Thus, the design should be based on the human nature and respect for people [5]. On the other hand, the area of walking street is a window of the city too. It needs to inherit and continue the context of the local history, and use the modern vocabulary to construct the space environment with a local style. The design breaks the pure space in line state and form a rhythmic climax point by the comparison, with various bridges, and techniques such as the construction of blocks. In order to reflect the shopping's feature, we have increased the area along streets as much as possible. We have changed the design way of the building facade every a few units. The building's outer contour is avoided for uniformity. Some stores are wide and tall, while others are narrow and short. The whole walking street landscape needs a change in uniformity, in order to appear with continuous appeal for people.

Leisure space: Studies have showed that the modern shopping behaviour includes leisure activities. The quality of leisure space affects the shopping psychology directly. In the planning, we have made full use of the Shi River landscape nearby. We have modified the original land's condition rationally. The bank of the river is designed into different landscape platforms with different height. And besides, all kinds of recreation facilities have been arranged. And a series of squares provide space for various cultural activities' being carried out by their morphological changes. In this way, the warm and relaxing atmosphere is forming (Fig. 18.1).

Humanistic space: Successful design should embody the elegant cultural taste. We call it the materialized landscape of spiritual culture. It means that the soft elements of spirit infiltrate into a rigid design of solid material. So, we dig the essence of tea-culture and local characteristics. On one hand, I design the unique shopping and leisure space to show the thoughts of tea-culture's harmony and natural lasting. On the other hand, we learn the Chinese traditional house's layout and design the different node space into distinctive cultural space. Furthermore, we use poems on the rich tea-culture to name those spaces, such as "Spring tea flower", "Royal saplings" and so on. And besides, sculptures, cultural corridors and theme fountains are used together to create a strong cultural atmosphere.

Fig. 18.1 The design of leisure space

18.4 Landscape Design

The Shi River flows freely from the west and forms the central water street in the project, and runs through the whole area. Moreover, we have designed the bridges, pavilions and gravel path etc., to foil the faint tea culture. Also, for the design of waterfront we have used the techniques of Chinese traditional garden design, such as "surface water with back hill", "flowers foiled by trees". Green design gives priority to the colours of dark green and orange. The evergreens such as pine and spruce are chosen along with the deciduous plants. And the changes in the number of plants have also been paid attention to. The entrance square emphasizes the space's leading function and we choose the bright-coloured flowers to contrast the pavement's texture and colour in order to strengthen the function of the main entrance. The arrangement of plants appears to be complete and unified. It highlights the characteristics of conciseness and liveliness by the combination of the open lawn and bushes. Plants along the river focus on the edge's change combining with the bank line. The tea-culture is the clue of the overall landscape design. We have designed for the arrangement of three districts: the walking commercial area, entrance square and the waterfront recreational area. They contact each other by the main axis of the water street. As a result, the planned effect is successfully achieved. Modern architectural design is combined with the traditional building's details to create a light tea flavor (Fig. 18.2). We study the architectural form carefully and achieve the effect of a tea leaf lying on the wave . It forms a beautiful building contour line.

Fig. 18.2 Architectural appearance design

18.5 Conclusion

As a place that concentrates a city's history and culture, the commercial walking street is always being in a dynamic multidimensional space system. Therefore, the planning should handle many contradictions and coordinate the relationship between the business benefits and the urban functions. Only in this way, can the commercial walking street become the representative of a city's image as well as the centre of social activities [6]. Through this walking street planning, we deeply feel that the conventional mode of planning must be updated and be able to face the impact of the e-commerce and other advancements. This is the key to the planning.

References

1. Li H, Zhang X, Wang Y (2012) The ponder of urban commercial walking street plan: taking the JianLan commercial walking street of Lanzhou as an example. Huazhong Archit 12(4):109–112
2. Zheng S, Zidan W (2001) Explore people-oriented design of walking street planning. Fujian Archit Constr 3(141):7–9
3. Jiangping W, Yin L, Qin W (2005) Study on the modern commercial walking street construction. China Sci Technol Inf 8(15):291–292
4. Sekkei N, Kaisha K (2000) Sustainable architecture in Japan. Wiley, New York, pp 120–122
5. Ning L, Lin L (2008) Composition of traditional settlement and the analysis of its characteristics. Archit J 11:52–55
6. Rui Y, Xiaopeng B, Wengxiang W (2008) Building home of our spirits: good outdoor space design in community reconstruction under new situation today. Huazhong Archit 26 (12):109–112

Chapter 19
Knowledge Distribution and Text Mining of International Aviation Safety Research

Jie Li and Xiaohong Guo

Abstract Aviation safeties (AS) has become a hot topic in social media, for a number of AS issues have occurred in 2014. Generally research of AS were focused on narrow and specific topics. In this study, we try to analyze AS in a new sight, study the publications of the former scientists. The data was collected from the scientific database Web of Science and more than 1000 scientific papers were downloaded. Knowledge distribution and topics related to AS were discussed in this paper. The results show that the AS papers were mainly from the journal *Aviation Space and Environmental Medicine.* Geospatial analysis shows that NASA, USA and cities from the USA and the European Union are active regions in AS research. Topics analysis has shown that "aviat", "safeti", "accid", "human", "risk" and "factor" have high frequencies among the AS words. Words including "analysi", "human", "factor", "risk" and "aircraft" have high degree, been regarded as the information hub of AS research. Cluster of AS topics showed that AS mainly focus on five fields and can be summarized as outside and inside AS factors, risk assessment and human-factors etc.

Keywords Aviation safety · Aviation accident · Aviation incident · Citespace bibliometric · Co-words

19.1 Introduction

Aviation safety is one of the most important safety issues among safety related research, even it is not occurred frequently, and been regarded as one of the safest ways to travel. On the other side, the results of aviation accidents seem severer than

J. Li (✉) · X. Guo
Capital University of Economics and Business, School of Safety and Environmental Engineering, Beijing 100070, China
e-mail: lijie_jerry@126.com

© Springer-Verlag Berlin Heidelberg 2015

S. Long and B.S. Dhillon (eds.), *Proceedings of the 15th International Conference on Man–Machine–Environment System Engineering*, Lecture Notes in Electrical Engineering 356, DOI 10.1007/978-3-662-48224-7_19

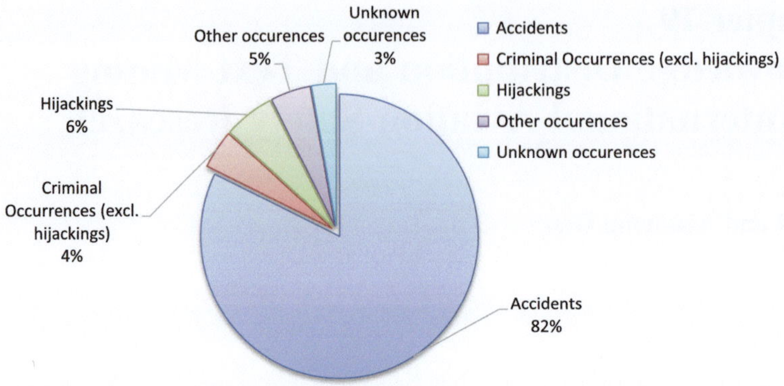

Fig. 19.1 Summary of Aviation safety issues from 1919 to 28th December 2014. The statistical data of aviation safety was collected from *ASN Aviation Safety Database* http://aviation-safety.net/database/

other traffic accidents, and have a bad effect on the public. Based on the statistics of the *ASN Aviation Safety Database*, there are totally 18,157 aviation hazards occurred from 1919 to 28th December 2014 (Incidents not included), for the composition of the aviation hazards see Fig. 19.1. And from 1948 to 2014, there are totally 89 missing aircraft, 27 on the land, 62 in the sea (Figure of Missing Aircraft from 1948 to December, 2014, can be download from http://aviation-safety.net/graphics/infographics/missingmapV2.jpg). Especially, in the year 2014, there are many severe aviation accidents which attracted public attention and made a significant impact on the society. For example, Malaysia Airlines Flight 370 has disappeared on Saturday, 8 March 2014, while flying from Kuala Lumpur International Airport, Malaysia, to Beijing Capital International Airport; the crash of Malaysia Airlines Flight 17 on July 17th, 2014, with 298 people dead, and also the Air Asia QZ8501 crash on December 28th, 2014.

Generally researches on aviation safety try to protect accidents or incidents in scientific ways, including developing new methods, techniques and also some phenomenological studies. it means that other scientists are always focused on a specific field of aviation safety issues, while our research analyze the aviation safety research in a global view, try to extract and organize the information of aviation safety from volumes of scientific publications.

19.2 Data and Methodology

The scientific data of aviation safety (AS) research was collected from Web of Science, including Science Citation Index Expanded (SCI-EXPANDED), Social Sciences Citation Index (SSCI) and Conference Proceedings Citation Index-Science (CPCI-S), which was developed by Garfield and is now a part of Thomson Reuters.

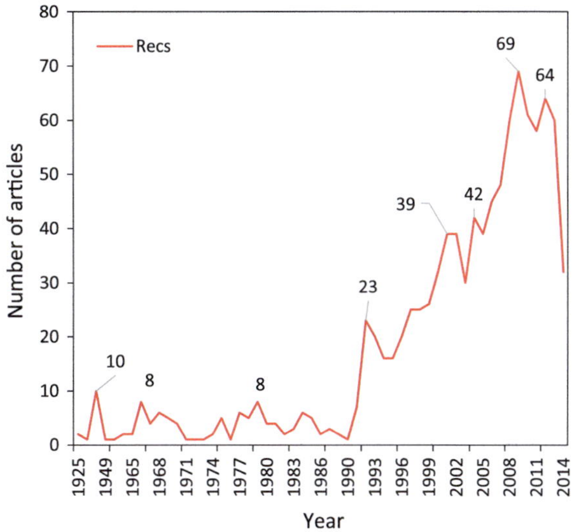

Fig. 19.2 Annual distribution of the data sample in aviation safety

The last update of the database is on 23rd July, while we downloaded the data on 25th July, 2014. The retrieval strategies were set as TOPIC: ("Aviation Safety" or "Aviation accident*" or "Aviation incident*"), Timespan was set from 1900 to 2014; totally 1005 records were retrieved from Web of Science. The annual distribution of the AS papers is shown in Fig. 19.2, and the publications were with a growth trend, especially from 1990 to 2013 (for the data in 2014 is not totally included), reflecting the increase of scientific attention in AS research. On the other side, the numbers of AS publications are not large. It may show the lack of knowledge cumulative in this area or scientists' publications in AS were undirected to safety issues, but maybe important to AS (for example Ergonomics, Medicine, or Engineering aspect).

Knowledge map methods were used in this study, which is a special kind of information visualization that exploits powerful human vision and spatial cognition to help humans mentally organize and electronically access and manage large complex information spaces [1]. The methods and contents in this research are discussed below,

1. Knowledge distribution research

In this study, knowledge distributions include two aspects, journals and geospatial distribution.

Journals map overlay method was used to reveal the knowledge distribution and transfer in journals level. This method was first developed in 2012 [2, 3] and combining the citing and cited map in 2014, which has been named journals dual-map overlays [4]. This method was used to give a global sight of a certain area. Geospatial distribution was discussed, including countries-/territories-level, institutions-level, and cites-level.

2. Words' frequencies and Co-words
Words' frequencies and co-words were used to analyze the topics of aviation safety research. Words' frequencies analysis uses natural langue process methods, by extracting the words and counting the number of occurrence of each word from the AS scientific documents. "Co-words" was first used in [5], and is also referred as semantic network as it can show the topics distribution, and the connection of the topics.

19.3 Results and Discussion

19.3.1 Distribution of the Aviation Safety Research

1. Knowledge distribution in journals level
There are 482 sources have published aviation safety (AS) papers, including important conference papers, *IEEE Aerospace Conference Proceedings*, *Conference on Aviation, Range, & Aerospace Meteorology*, and *DASC-AIAA/IEEE Digital Avionics Systems Conference* etc.. Journals from *Aviation Space and Environmental Medicine*, *International Journal of Aviation Psychology,* and *Aircraft Engineering and Aerospace Technology* As journals are mainly original sources of AS research more information on journals has been discussed. *Aviation Space and Environmental Medicine* (142), *International Journal of Aviation Psychology* (30), *Aircraft Engineering and Aerospace Technology* (22), *Journal of Air Transport Management* (19), and *Safety Science* (16) have published more than 15 AS papers.

The journals dual-map was created by CiteSpace[1] in Fig. 19. 3 (for detail steps on creating journals overlay refer Appendix A), on the left of the dual-map are the journals which have published AS papers; on the right side of dual-map are the journals that are cited by AS research. The left side can be considered as the knowledge carrier of AS research output, and the right can be seen as the knowledge bases of AS in journals level. Based on the dual-map, these papers are mainly from the cluster 1# mathematics, system, mathematical (left side) and obtained the knowledge mainly from 1# systems, computing, and computer (right side). And also knowledge flow from 3# (earth, geology, geophysics) to 3# (ecology, earth, and marine), 5# (health, nursing, medicine) to 2# (medicine, medical, clinical) and 7# (psychology, education and social) to 6# (psychology, education, health). On the left side, longitudinal axis of oval shows the number of publications in a certain journal, horizontal axis shows the number of authors in that papers. It shows that *Aviation Space and Environmental Medicine* has the largest publications and authors in AS than any other journals.

[1]One can freely use CiteSpace to create journals overlay map, the software can download from http://cluster.cis.drexel.edu/~cchen/citespace/.

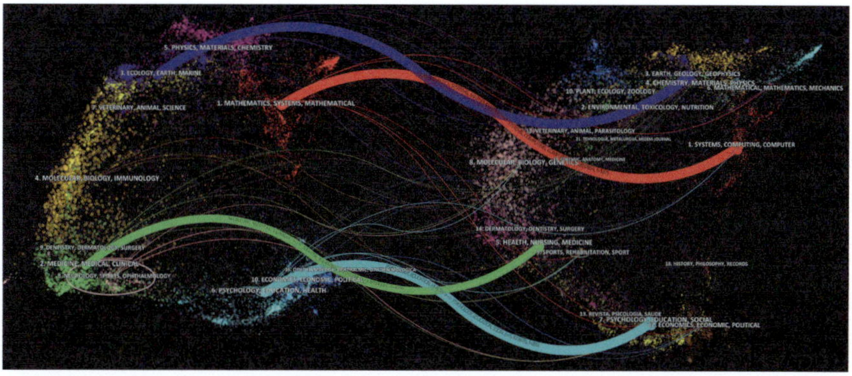

Fig. 19.3 Journals dual-map overlays of aviation safety, *Note* The *left side* of the map shows which journal has published the aviation safety articles, the *right side* shows journals that have been cited by aviation safety research. The online figure can be downloaded from https://drive.google.com/file/d/0B0HnDMi5NBF8X1VacmRqSU9qdGc/view?usp=sharing

2. Knowledge distribution in geospatial level

In this part, Countries/territories, institutions and cities were extracted from authors' address field, Totally 45 countries/territories, and nearly 670 institutions were extracted. USA has published 425 papers, ranking first in the countries/territories followed by UK (72), Peoples R China (63), Australia (39), and Germany (31). In the institutions-level, NASA (57), USA Federal Aviation Administration (56), Rutgers State University (19), Johns Hopkins University (18), Civil Aviation University of China (14), USN (14), University Illinois (12), USAF (12), and Cranfield University (11) have more publications than many other institutions. Furthermore, the name of cities were also extracted and mapped onto the world map, Fig. 19.4 shows the cities that are located in USA and European Union.

19.3.2 Topics Analysis

Top 100 normalized original keywords (or authors' keywords) were extracted from the documents using SCI2 [6], words' frequencies were listed in Table 19.1. And the network of the words was visualized by Gephi as shown in Fig. 19.5 [7]. The results showed that the words "aviat", "safeti", "accid", "human", "risk", and "factor" have the high frequencies. For these words we have used in retrieval process, and also they are much closer to our topic than other words. These words can be seen as basic words that authors used in his/her aviation safety (AS) research. Among the list "*aircraft*", "*air*", "*pilot*", "*flight*", "*airport*", and "*airlin*" have the similar meaning with the word aviation; authors also use these words instead of aviation in their AS research. The list of words, can also implicate

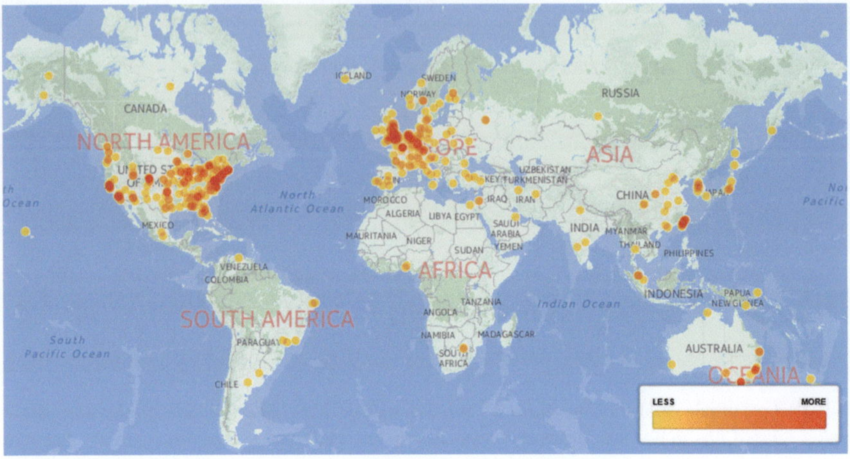

Fig. 19.4 Distribution of aviation safety research in cities-level

Table 19.1 Frequency of normalized original keywords

Words	*f*	Words	*f*	Words	*f*	Words	*f*	Words	*f*
Aviat	250	Decis	22	Remot	13	Head	10	Dynam	9
Safeti	217	Oper	21	Mine	13	Cultur	10	Medicin	9
Accid	108	Vision	20	Simul	13	Weather	10	Collis	8
Human	73	Mainten	20	Awar	12	Structur	10	Autom	8
Risk	63	Civil	18	Evalu	12	Lidar	10	Drug	8
Factor	60	Toxicolog	18	Train	12	Detect	10	Time	8
Analysi	58	*Airport*	17	Enhanc	12	Make	10	Sensor	8
Error	49	Traffic	17	Imag	12	General	10	Certif	8
Aircraft	49	Data	16	Prevent	12	Situat	10	Machin	8
Air	45	Process	16	Fuzzi	12	Visual	10	Qualiti	8
Pilot	40	Transport	15	Health	12	Fatal	10	Epidemiolog	8
Manag	35	Incid	15	Report	11	Crew	10	Administr	8
Model	35	Synthet	15	Volcan	11	Ceilomet	10	Behavior	8
Flight	33	Display	15	Ash	11	Postmortem	9	Support	8
Investig	32	Inform	14	Base	11	Program	9	Turbul	8
Control	31	Hfac	14	Design	11	Height	9	Bayesian	8
Assess	29	Classif	14	Sens	11	Crash	9	Text	8
System	28	Network	13	Scienc	11	Cloud	9	Runway	8
Hazard	24	Medic	13	Method	11	Theori	9	Injuri	8
Perform	23	Forens	13	*Airlin*	11	Reliabl	9	Standard	8

Note Normalized original keywords with lower case, tokenize, stem, and stop word text. Here *f* stands for frequencies of a certain word

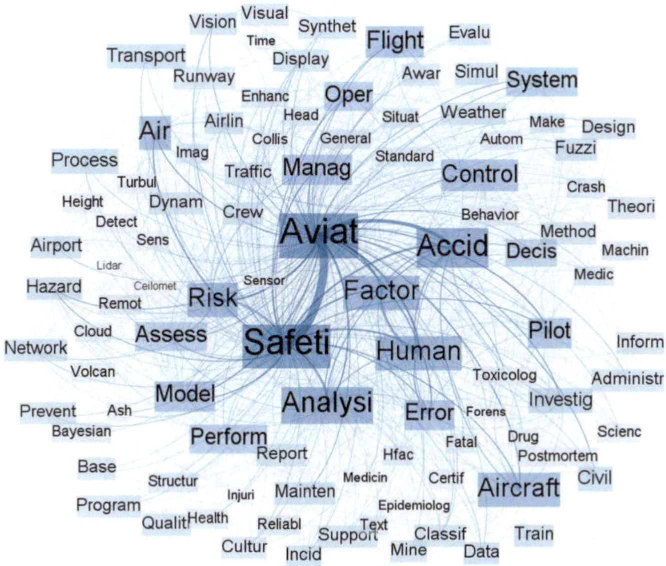

Fig. 19.5 Co-words network of aviation safety, Density [no loops] = 0.31898990, Average Degree = 31.58; *Note* The size of the node and font show the degree of the words, the larger the node and font, the higher the degree. The degree of a node is the number of links that it has with other nodes, node with higher degree act as information hubs in the network. Thickness of the edges shows number of co-occurrence between two words

that "human factors", "human error", and "vision" are hot topics in research. While "model", "simul", "fuzzi", "bayesian" show the key methods used in research.

After a deep analysis, the co-words network shows that "aviat", "safeti", "accid", "analysi", "human", "factor", "risk", and "aircraft" have high degree in the network, indicating that these words have more association with other words. These high degree words are the information hubs [8] in AS. Top ten values of the words pairs (co-occurrence) are "aviat-safeti" (168), "aviat-accid" (69), "human-factor" (49), "safeti-risk" (39), "aviat-human" (38), "aviat-factor" (37), "aviat-risk" (36), "error-human" (35), "safeti-accid" (35), and "safeti-human" (32). These important words have made the basic content of AS research and also the scientific focus of AS issues.

The co-words network was clustered by using VOSviewer,[2] including VOS mapping and clustering methods [9]. VOS mapping method was used to give the location for each word in two dimensional map, and VOS clustering method was used to give the color for each word to distinguish different clusters. For the details of each cluster see Table 19.2. Five clusters were obtained from the co-words network, and we named them: cluster #1 safety issues, outside factors effect

[2]VOSviewer can freely download from http://www.vosviewer.com/Home.

Table 19.2 Cluster of the co-words network

Cluster (number of words)	Words
#1(24) Safety issues (including outside factors effect aviation safety)	Safeti, Air, Model, Hazard, Transport, Airport, Process, Remot, Volcan, Simul, Sens, Ash, Imag, Dynam, Qualiti, Cloud, Structur, Health, Detect, Height, Turbul, Sensor, Ceilomet, Lidar
#2(22) Risk management and assessment, and also methods	Risk, Manag, Assess, Oper, System, Decis, Network, Fuzzi, Design, Airlin, Train, Evalu, Make, Bayesian, Support, Base, Method, Crash, Machin, Standard, Behavior, Autom
#3(19) Human factors and error, this is an inside factor that affects aviation safety	Human, Factor, Analysi, Error, Perform, Mainten, Incid, Classif, Hfac, Data, Mine, Cultur, Prevent, Inform, Report, Reliabl, Program, Theori, Text
#4(18) Accident investigation and rescue (including aviation medicine)	Aviat, Accid, Aircraft, Investig, Pilot, Toxicolog, Civil, Forens, Fatal, Scienc, Administr, Medic, Postmortem, Drug, Injuri, Medicin, Epidemiolog, Certif
#5(17) Air traffic control, factors, and vision related topics (including outside factors effect AS)	Flight, Control, Vision, Display, Synthet, Traffic, Head, Enhanc, Awar, Weather, Situat, Crew, Runway, Visual, General, Collis, Time

Note the coloued map of topics clusters can explored online https://drive.google.com/file/d/0B0HnDMi5NBF8ajVnMXRWTDVONzg/view?usp=sharing

aviation safety; #2 Risk management, assessment and methods; #3 human factor and error, related to ergonomics and reliability aspect, shown as the inside factor that affects aviation safety; #4 accident investigation and rescue; and #5 Flight control, environmental factors also included. These clusters mainly cover the aviation safety, while some newly emerged topics with low frequencies may not be extracted.

19.4 Conclusions and Discussion

Knowledge distribution and words related to aviation safety (AS) have been discussed in this research. The annual publications reflect the growth of the scientific research in AS, the reason is not just the more attention to this research filed, may be the growth of the science as a whole. Furthermore, some AS outputs were not published in journals and conferences, and also some journals and conference papers were not indexed by Web of Science. In this study, AS articles were mainly published in journal *Aviation Space and Environmental Medicine*, and the cluster of journals were mainly located in "mathematics, systems, mathematical", "ecology, earth, and marine", "medicine, medical, clinical" and "psychology, education, health", obtained the knowledge mainly from "systems, computing, and computer", "earth, geology, geophysics", "health, nursing, medicine", and "psychology, education, social". The USA is the most active country and so are the institutions, NASA and Federal Aviation Administration. Topics analysis from AS shows that Human related topics have emerged, including "human factors", "human error", and "human safety". In the network, the high degree words are "aviat", "safeti", "accid", "analysi", "human", "factor", "risk", "aircraft", and "air", andthese words are information hubs in the network reflecting their high connection with other words. The highly connected words are "aviat-accid", "aviat-factor", "aviat-human", "aviat-risk", "aviat-safeti", "error-human", "human-factor", "safeti-accid", "safeti-human", and "safeti-risk". These words were clustered into five factors, and can be summarized with outside factors, inside factors and risk evaluation.

Furthermore, this study is the first research using bibliography data and text mining methods to do AS research. Due to the limitation of data, results from these studies may not cover all the fields of AS research. Hence, more data should be available in future for the research.

Appendix A: Additional materials

How to create journal dual-map by using CiteSpaceIII

https://docs.google.com/presentation/d/12SFqI8elYdtNC_harUJXUfsZuH7yS7g92idtfKr5ZuE/edit?usp=sharing

References

1. Mane KK, Börner K (2004) Mapping topics and topic bursts in PNAS. Proc Natl Acad Sci 101:5287–5290
2. Leydesdorff L, Rafols I (2012) Interactive overlays: a new method for generating global journal maps from web-of-science data. J Inform 6:318–332
3. Leydesdorff L, Rafols I, Chen C (2013) Interactive overlays of journals and the measurement of interdisciplinarity on the basis of aggregated journal–journal citations. J Am Soc Inform Sci Technol 64:2573–2586
4. Chen C, Leydesdorff L (2014) Patterns of connections and movements in dual-map overlays: a new method of publication portfolio analysis. J Assoc Inform Sci Technol 65:334–351
5. Callon M, Courtial J-P, Turner WA, Bauin S (1983) From translations to problematic networks: an introduction to co-word analysis. Social Sci Inform 22:191–235
6. Team S (2009) Science of science (Sci2) tool, Indiana University and SciTech Strategies
7. Bastian M, Heymann S, Jacomy M (2009) Gephi: an open source software for exploring and manipulating networks. ICWSM 8:361–362
8. Song M, Chambers T (2014) Text mining with the Stanford CoreNLP. In: Measuring scholarly impact, pp. 215–234. Springer
9. Van Eck NJ, Waltman L (2010) Software survey: VOSviewer, a computer program for bibliometric mapping. Scientometrics 84:523–538

Chapter 20
Application Research on the Webpage Accessibility Based on the GOMS Model

Shuyu Shao and Hua Qin

Abstract In many studies, the majority scholars study webpage accessibility from the aspects such as psychology, consumer behavior, etc., but seldom studies focus on webpage accessibility from the aspect of quantitative analysis. In this paper, thinking four famous website as example, the formation of executive level quantitative model in the use of GOMS model, measure and analysis tasks quantitatively. At the same time, taking the time that participants spend as the dependent variable, intuitively compare and analysis the influence of webpage accessibility on website and page views, and thus could be able to provide suggestion for the websites so as to enhance web accessibility more target-oriented.

Keywords GOMS model · Webpage accessibility · Human–computer interaction · Behavior analysis

20.1 Introduction

As growing numbers of people access to the internet, a variety of websites are endless, and the speed is becoming faster, so that computer is playing a more and more important role in many aspects of human life. people also put forward higher requirements on the richness and accessibility of webpage. The design of Humanized interface, trends in web interface of user' interface and PC-based portable terminal are obviously the future developing trend of interface [1]. How to quickly attract users' attention, enhance the reputation of the website and improve website views, are the problems that each website operators must consider. From the perspective of users, evaluating the usability and efficiency by means of the webpage accessibility is the important way for user's experience evaluation of

S. Shao (✉) · H. Qin
School of Mechanical-Electronic and Automobile Engineering, Beijing University of Civil Engineering and Architecture, Beijing 100044, China
e-mail: ie1314@yeah.net

© Springer-Verlag Berlin Heidelberg 2015

S. Long and B.S. Dhillon (eds.), *Proceedings of the 15th International Conference on Man–Machine–Environment System Engineering*, Lecture Notes in Electrical Engineering 356, DOI 10.1007/978-3-662-48224-7_20

the website. How to analysis the web interface of human-computer relationship by using quantitative analysis and enhance the user's experience, is an area of focus for the researchers, and the GOMS model is exactly the analysis and modeling techniques for knowledge and cognitive process by using the user interacts with the system.

20.2 GOMS Model

The GOMS model is created by Card, Moran & Newall in 1983 in "The Psychology of Human Computer user model Interaction", and is widely applied in the field of human-computer interaction, which is by far the most successful human-computer interaction model. GOMS model is based on information processing theory in human problem solving. Through the task behavior that can be observed, the thinking process and the target structure, so as to reveal the cognitive skills that processing of complicated situations more effectively [2]. G(Goals) means the purpose of the user to perform the task. O(Operators) refers to the bottom of the cognitive processes and physical behavior, means the elementary perceptual, motor or cognitive actions that are used to accomplish the goals. M(Methods)what sequences of operators can be used to accomplish each goal or step. S (Selection rules)specify which method or judging criterion that should be used by the user to satisfy a given goal under a specific environment [3]. The potential application of GOMS is quite broad, the analysis of a task describes a structural knowledge that people needed in order to successfully complete the task. On this basis, when knowing the sequence of operation, it is possible to predict the execution time of task by quantitative prediction. Other aspects of analysis, such as error prediction and functional coverage and time for learning may also be accomplished by GOMS [4].

20.2.1 User's Behavior Analysis

The Keystroke-Level Model of GOMS method is a quantitative method on the behavior levels, the user's behavior is split into behavior modeling unit, once successful, and it is not necessary to measure each specific individual time, we can accurately predict the behavior sequence and the time required to complete the behavior sequence without a real user's intervention [5]. The interactive behavior of Keystroke-Level Model is decomposed into several basic action units,and each basic action unit's time is the average time through a lot of tests, as is shown in Table 20.1.

Table 20.1 The basic operation schedule

Name and mnemonic	Typical value (s)	Meaning
K(Keying)	0.2	Pressing the mouse or a button on the keyboard
P(Pointing)	1.1	Pointing with a mouse to a certain target that on the display
H(Homing)	0.4	Homing hands from keyboard to mouse or homing hands from mouse to keyboard
C(Change)	1.5	Switching between pages
M(Mentally preparing)	1.3	Psychologically preparation to enter the next step (representing pauses for routine activity)
R(Responding)	t	Waiting for computer system responding

20.2.2 The Quantitative Analysis of Task

The purpose of this experiment is to "Through the site navigation for sports navigation in La Liga to find the result of Real Madrid VS Levante ", according to the GOMS model. According to the GOMS model, the basic steps are obtained by analyzing the task operation from the aspects of level quantification, which is described in Fig. 20.1.

At the same time of completing the information behavior, we also need to complete a series of basic operations (KH and P), which are parts of GOMS model. How to recognize the time that a user stop by unconscious mental activity is a difficulty, which means the mentally preparing time (M) [6], the following is the list of the basic rules of orientation of users' mental activities:

1. M′ initial insertion: Insert M before K(keystrokes), insert M before P(Pointing) that is used to select a command, but do not insert M before P that is used to select the command parameters.
2. The expected M′ deletion: If the operator before M can fully forecast the operator behind M, then M will be deleted.
3. Delete M in the same cognitive unit: If a series of M-K(keystrokes) are belong to the same cognitive unit, remove all other M but for the first one.
4. Delete the overlapping M: If M and the response of computer time (R) overlap, then M will be deleted.

According to the related situation of the above rules 1 and 2, we can know that the initial value is HPKCPKCPKCP, the value of assessment revised through rule 1 is HMPKCMPKCMPKCMPKC, and we can figure out the essential operation steps according to rule1and 2 are: HMPKCPKCPKCPKC

Translated into the corresponding time is: 0.4 + 1.35 + 1.1 + 0.2 + 1.5 + 1.1 + 0.2 + 1.5 + 1.1 + 0.2 + 1.5 + 1.1 + 0.2 + 1.5 = 12.95 (s).

That is to say, in an ideal situation (means not considering the broadband speed and other external circumstances), the minimum time for visitors to browse the news is 12.95 s. This is the basic time, a time of basic actions that must be performed for users in order to complete a task. If visitors want to get news what

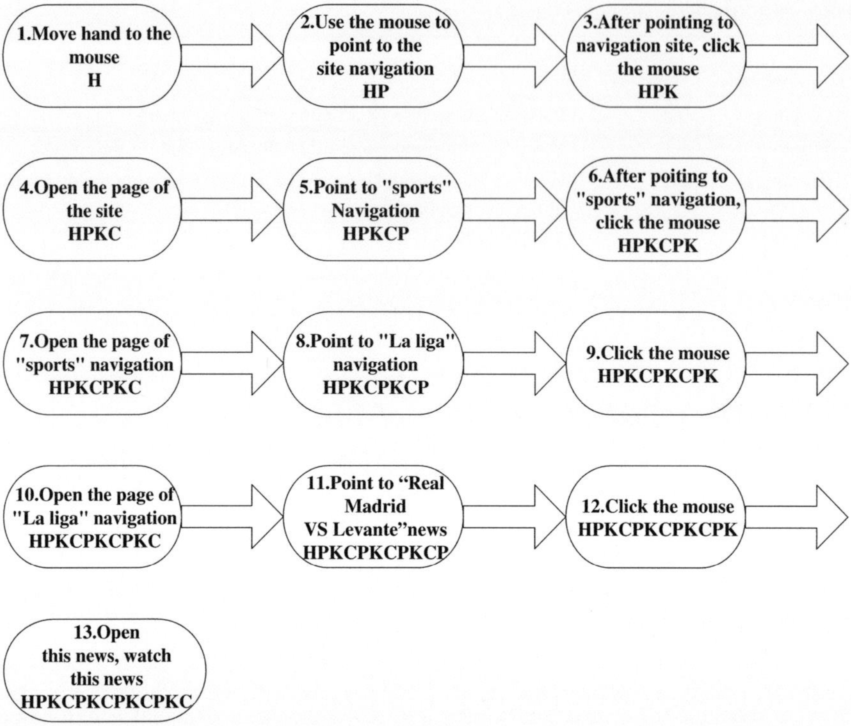

Fig. 20.1 GOMS executive level quantitative model

they want on the web page, website operators can meet this minimum time requirement, clicks to website and click rate will certainly be greatly improved.

20.2.3 Experiment

In the experiment, a total of 15 participants were recruited, they are all postgraduate students, and are very fond of sports, and often pay attention to the website of sports news, that means for participants, to find the sports news is a daily skilled operation on the site (Routine cognitive skilled behavior) [7]. In the smooth campus network environment, open navigation page, several well-known sites are in a prominent position of the navigation page, we asked participants to complete the task as described in 20.2.2, and use a stopwatch to measure the mission according to the quantitative step GOMS levels. The results are shown in Table 20.2.

As can be seen from the above results, the average time of participants who completed the task is about 16.3 s, more than the minimum time 12.95 s of GOMS model, taking the page jump delay into account, results are consistent with the prediction of the GOMS model to measure time. Because the functions of different sites

Table 20.2 The test time to complete tasks

Observed time(sec) Participant / Website	A	B	C	D
1	16.1	15.9	16.6	16.7
2	16.2	16.1	16.7	16.7
3	16.4	15.9	16.6	16.4
4	16.3	15.8	16.5	16.6
5	16.4	16.1	16.8	16.5
6	16.3	16.0	16.6	16.5
7	16.4	15.9	16.1	16.4
8	16.5	16.2	16.6	16.4
9	15.8	16.1	16.5	16.6
10	16.4	15.9	16.5	16.7
11	16.6	16.0	16.4	16.6
12	16.2	16.2	16.6	16.3
13	16.0	16.1	16.2	15.1
14	16.4	15.9	16.6	16.4
15	16.2	16.1	16.3	16.5
Average value	16.28	16.01	16.51	16.43

navigation are different, and navigation and reasonable position has a great influence on the website hits, so in the above steps, we should focus on the fifth step, namely, searching for the "sport" navigation. At the same time, the time of participants who completed the task in the B site is shortest,and C,D are longest. We consider the "sport" location of the four sites navigation Table 20.3, the result is consistent with Alex ranking data, that is to say the analysis of GOMS model is effective and feasible. In the subsequent analysis, website operators can pertinently improve Step 5 according to the specific circumstances. At the same time, in this step we can use MOD method to measure the time of the final step, finally to draw precise time.

Table 20.3 Position and click ranking of "Sports" navigation in webpage

Website	Sports position at the navigation position	"Sports" navigation page's click ranking at each website
A	Row 1 Column 4	5
B	Row 2 Column 1	2
C	Row 2 Column 8	8
D	Row 2 Column 9	8

20.3 Conclusion

Computer interface continuously into the life of modern people, people also put forward higher requirements to the quickness, accessibility and efficiency of the interface. People are paying more and more attention to the human-computer interactive interface design and usability evaluation. This paper based on the GOMS model, we quantitatively analysis the accessibility of webpage, taking the time to carry out performance measurement, measure the efficiency of human-computer interaction design in a quantitative way. Some researchers believe that most of the management information system (MIS) follow common interface design, but in the same time should consider the characteristics of commercial application and management orientation of the system and the system should be placed in the context of the organization and management of the design [8]. The target locations and target groups in different sites are different, this paper can give website operators inspiration, according to the strategy of the website navigation position, finally enhance the website's browses and improve the website's influence.

References

1. Li Y, Li L (2006) Graphical user interface development and classification. J Eng Geol Comput Appl 44(4):23–26
2. Li X (2011) On man-machine interface operation performance of vehicles based on BHR-GOMS Model. J Des Res 6(6):57–61
3. Kieras D (2003) GOMS models for task analysis [A]. Dan diaper. Handbook of task analysis for human-computer interaction . Lawrence Erlbaum Associates, Mahwah 83–116
4. Wang H, Dai G et al (2005) The eGOMS model of virtual reality human-computer interaction. J Huazhong Univ Science Technol (Natural Science Edition) 12(33):322–325
5. Luo S, Zhu S, Sun S (2002) Design of human computer interface. The mechanical industry press, San Deigo
6. Card SK, Moran TP, Newell A (1983) The psycholog of human-computer interaction. Erlbaum Associates, Hillsdale, New Jersey, USA
7. John B et al (1996) T he GOMS family of user interface analysis techniques: comparison and contrast. J ACM Trans Comput Human Interact 3(4):320–351
8. Zhang P (2005) Human-computer interaction research in management information system. J Manag Info Syst 22(3):9–14

Chapter 21
The Design and Application of the Air Purification System of the Meteorology Command Vehicle Based on the Photo-catalytic Technology

Kun Yang

Abstract This paper is focused on the mobile command vehicles of the meteorological department. It aims to study and design a set of air purification system which takes the photo-catalytic technology and the sensor collection technology as its core. The gas in the vehicle will be introduced into the gas inlet, going through three-layer mechanical filtration. The filtered gas will then go through a metal filter screen coated with catalytic materials. Under the ultraviolet light, the nanometer metal oxide (the catalytic materials) will chemically react with ultraviolet light to form a kind of powerful oxide which can sterilize and disinfect the filtered gas achieving the purification of the air in the cabin of the command vehicle and effectively guarantee the interior air quality and the health of the workers.

Keywords Photo-catalytic technology · Air purification · Sensor · Mobile command vehicle

21.1 Introduction

Nowadays, air pollution has become a hot spot issue around the globe and has done a certain degree of harm to people's health. As far as the indoor air pollution is concerned, there are all kinds of organics from house renovation and production, such as formaldehyde, benzene, xylene, etc. As for as the outdoor air pollution takes the pandemic PM2.5 (Fine particle matter, diameter of 2.5 μm or less) [1], for example, it can go directly to the alveoli of the lungs and is a major health hazard. Longtime exposure to particulate matter can cause diseases that hard to be cured. Therefore, the study on the management and control of the air pollutants has never ceased. This paper, targeted on the air pollution problem in the cabin of the mobile

K. Yang (✉)
Linyi Meteorological Bureau, Linyi 276004, China
e-mail: 99833964@qq.com

© Springer-Verlag Berlin Heidelberg 2015

S. Long and B.S. Dhillon (eds.), *Proceedings of the 15th International Conference on Man–Machine–Environment System Engineering*, Lecture Notes in Electrical Engineering 356, DOI 10.1007/978-3-662-48224-7_21

command vehicle, designs a set of air purification system based on the photo-catalytic technology. It can effectively filter the particulate matter in the air and play the role of sterilization and disinfection so as to guarantee the air quality and health of the working staff.

21.2 The Principle of Air Purification System

The air purification technology having been evolved over the two decades is a new and comprehensive technology [2]. The air purification system discussed in this paper is a set of system that integrates air filtration, sterilization and disinfection, and utilizes the air compressor to purify the gas in the cabin featuring high efficiency and thorough disinfection. In particular, it adopts photo-catalytic technology by utilizing the sterilization function of the ultraviolet rays and the catalytic materials, which makes the purification process fast and hi-efficient. The principle of the whole air purification system is shown in Fig. 21.1.

The gas sensor in the system collects the data of the air quality in the cabin. According to the results of the collected data and pollution extent, the control system will adjust the rotating speed of the blower and absorb the gas in the cabin through the air duct and the air intake of the vehicle into the purification system. After thorough purification and adsorption by going through three-tier special filter element and hi-efficient disinfection treatment through the catalytic action of nanocrystalline metal oxide, the fresh and clean air will go into the cabin through gas outlet.

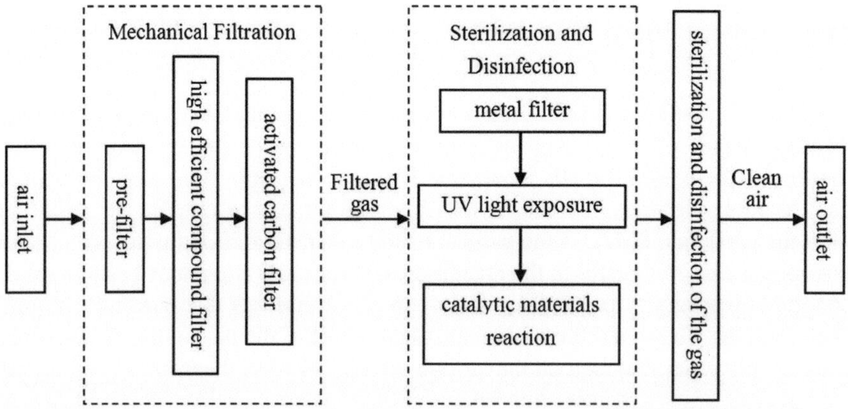

Fig. 21.1 Schematic diagram of the air purification system of the command vehicle

21.3 The Design of the Air Purification System

21.3.1 Overall Structure Design

The air purification system designed for the mobile command vehicle mainly fall into several parts, namely, gas cycle module, gas filtration module, gas collection module, gas sterilization module, and power module. The overall structure and procedure are shown in Fig. 21.2.

21.3.2 Air Cycle Module

The direct current blower is used to introduce the gas in the cabin to the air channel where the air purification system is installed. The control circuit of the blower has been designed as shown in Fig. 21.3. First of all, connect the power pin 1 with the power module to get the 12 vcd power; then, connect the blower pin 2 with ground connection, and pin 3 at the high battery state; finally, connect pin 4 with starting capacitance. After completing all these things, the operation of the blower can be controlled through optocoupler MOC3023. When the single-chip microcomputer sends low level signals to the optocoupler, the blower could be started. Meanwhile, according to the air condition in the cabin [3], the single-chip microcomputer can regulate the rotating speed of the blower through changing the duty ratio, thus adjust the system power.

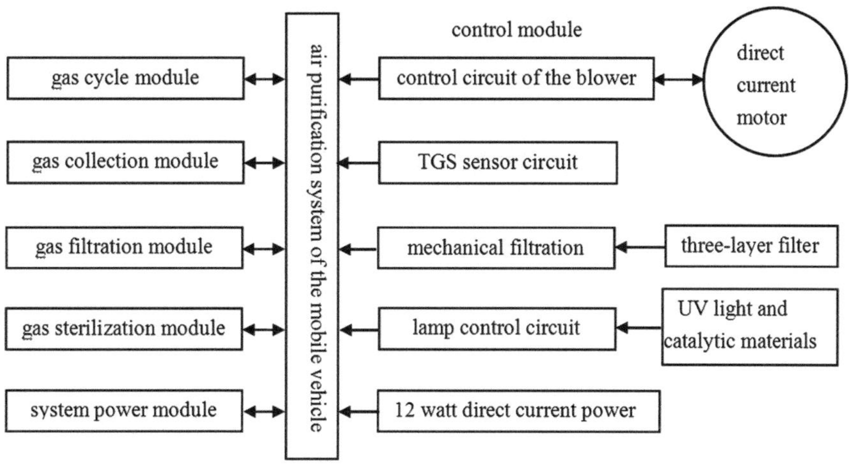

Fig. 21.2 Overall structure of the air purification system of the command vehicle

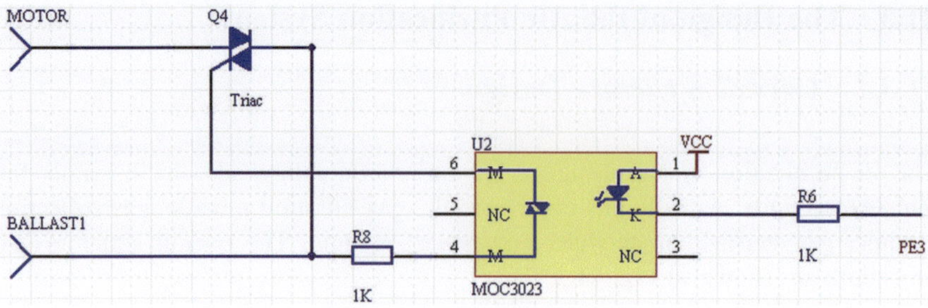

Fig. 21.3 Schematic diagram of the control circuit of the blower

21.3.3 Gas Collection Module

Controlling the operating power of the purification system is based on the gas collection. Through the gas sensor TGS800, we can get the quality and state of the gas in the cabin. If the air is seriously polluted, the blower will pick up rotating speed through the single-chip microcomputer, and vice verse. The TGS800 could access the gas composition through the internal materials and then regulate the voltage values of the input pin [4]. The principle of the control circuit of the sensor is shown in Fig. 21.4. The pins of sensor TGS800 will be linked to the power source, and the pins of SCM signal processing, respectively. The output connectors of the collected signals will be linked to the pins of PF1 and pb6 of the SCM. If the output voltage increases, it means that the air pollution is worsening, and the SCM will lift the rotating speed of the blower through computer program.

Fig. 21.4 Schematic circuit diagram of the signal collection of the gas sensor

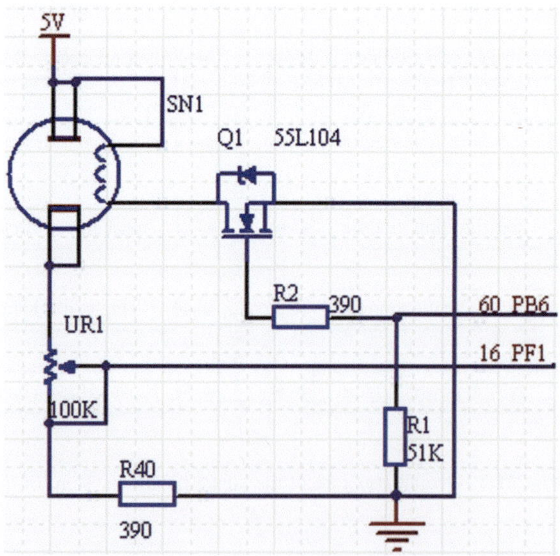

21.3.4 Gas Filtration Module

The filtration of the gas in the cabin of the meteorology command vehicle is a mechanical process in which the gas enters into the air channel of the purification system and firstly gets access to the air filter. It has used three-layer and high-efficient filter.

The first layer is called pre-filter which can filter out relatively large particles, such as, hair, pet fur, etc.

The second layer is called high-efficiency particulate air (HEPA) [5] which can remove the fine and small particulates with diameter of 0.3 μm and achieve purity of 99.97 %. The anti-microbial coating can effectively get rid of the dust and smog.

The third layer is called compound-activated carbon filter which is a unique 3D reticular absorber. It can effectively eliminate harmful gas and substance, such as the odor, formaldehyde, benzene, ammonia, bacteria, etc. (Fig. 21.5).

21.3.5 Gas Sterilization Module

This module makes use of the principle of ultraviolet sterilization. After being exposed to the UV light, the light will penetrate into the cytomembrane of the microbes, break the DNA structure of various germs, bacteria, parasite and other pathogenic bacteria, and damage their nucleic acids molecular bonds, which could cause immediate death and sterility, and thus, realize the sterilization and disinfection. First, coat the metal filter with catalytic materials (nanocrystalline metal oxides), and second, install a miniwatt UV lamps. Under a certain length of UV

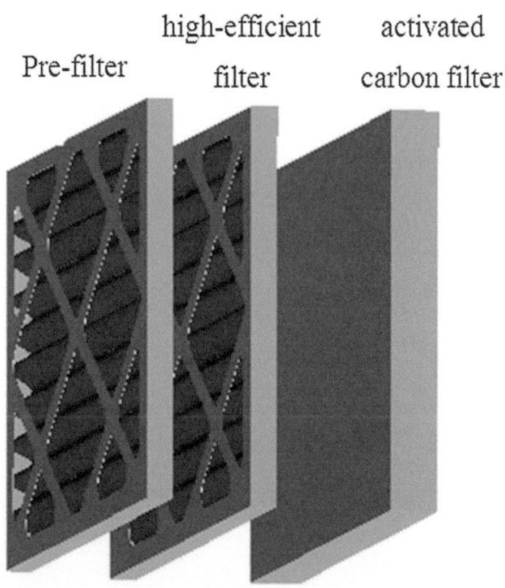

Fig. 21.5 Layout of the three-layer filter

light, the photo-catalysis will occur and the crystalline metal oxides will go through catalysis process and produce a large amount of strong oxides which will thoroughly sterilize and disinfect the filtered air.

UV lamp is a miniwatt lamp of 8 W. The drive and control circuit are similar to those of the blower, only lack of the power regulating function with the lamp. The structure layout is shown in Fig. 21.6.

21.4 The Application of the Meteorology Mobile Command Vehicle

Because of the global climate change and frequent extreme weather events, the meteorological department strengthens the construction of meteorology emergency mobile system to deal with significant social activities and public safety emergencies, in particular, the customization of the vehicle meteorology emergency system, in a bid to guarantee the meteorological monitoring and services and raise the control of meteorological disaster and hazard. The meteorological emergency command vehicle plays a major role in emergency mobile monitoring, emergency mobile communication, emergency supporting guarantee and emergency service and command.

The meteorology mobile vehicles have been widely used in meteorological departments. In order to guarantee the air quality in the cabins, the health of the workers in the vehicles and the precision of the important instrument installed in the vehicles, the command vehicles will be integrated with a set air purification system which utilizes the photo-catalytic technology. The layout is shown in Fig. 21.7. This kind of vehicle is refit on the basis of a coach. A section of air duct will be added to the original air duct of the air conditioning with air intake on one end and

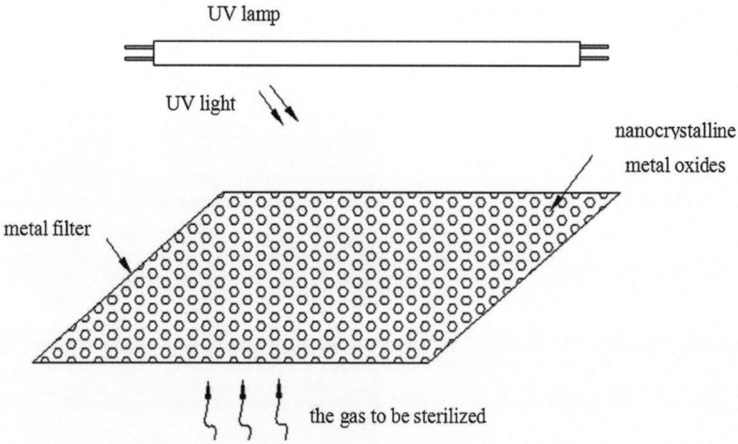

Fig. 21.6 Layout of the gas sterilization module

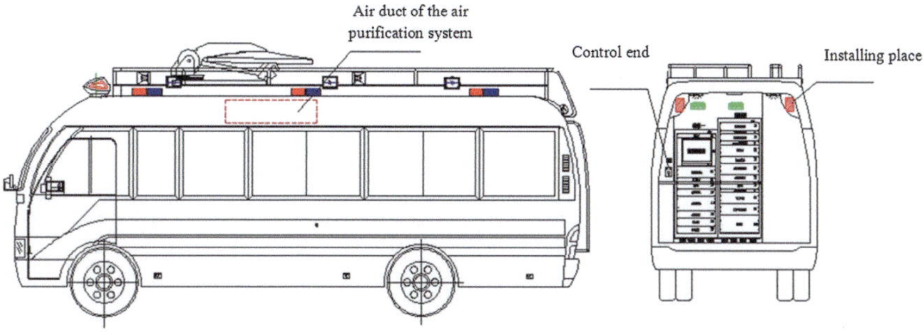

Fig. 21.7 The layout of the application of air purification system in the meteorology command vehicle

air outlet that is 500 mm away on the other. The air purification system and equipment will be installed in the added section of air duct. With this high-efficient air purification system, the gas in the cabin will be filtered, sterilized, disinfected rapidly and the health of the workers, and the safety and normal operation of the precise monitoring equipment will be effectively safeguarded.

21.5 Conclusion

This study on the air purification system of the meteorology vehicles is put forwarded on the premise of current increasingly serious air pollution. This purification system integrates scientific mechanical filter with photo-catalytic sterilization layer, which could not only achieve physical filtration, but also substantially and thoroughly eradicate the harmful elements in the air. Therefore, it is a relatively advanced method of dealing with the gas and adequately safeguards the air quality in the meteorology mobile command vehicles.

References

1. Zhao W (2013) Indoor Air pollution and purification technology. Chemical Industry Press, Beijing, pp 2–11
2. Bo Z (2010) Study and application of the new-type air purifier. The study on the static application technology 21(3):79–82
3. Wang X (2007) The control of the single-chip microcomputer of the electromotor. Beijng University of Aeronautics and Astronautics Press, Beijng, pp 156–167
4. Guo Z (2007) Study on the testing technology of the semi-conductor gas sensor and its array. University of Science and Technology of China, Hefei, pp 23, 24
5. Bi M (2011) The industrial signal producing and measuring device based on MSP430 series single chip micro controller. In: 6th international symposium on test and measurement 2011, vol 6, pp 668–671

Chapter 22
The Research of the Cluster of Modern Service Industry's Innovation Mechanism Under the Social Network Perspective

Yan Li, Jianping Jiang and Yingyi Wu

Abstract With the development of economic globalization, modern service industry has gradually become the main driver of economic growth, during which innovation has become the eternal topic, and clustering as well as networking have become the main trends in the development of modern service industry in the future. Based on the knowledge and funds, these two kinds of network resources acting as "bridge", the paper highlights the main parts of the knowledge sharing and fund raising to improve the cluster enterprise innovation. The features of the cluster network include three aspects: network centrality, network density, and structural holes. Knowledge sharing and fund raising have a significant impact on cluster innovation mechanism, and on the contrary, the characteristics of cluster network will affect knowledge sharing and financing as well, and what follows next is that it will affect the cluster innovation mechanism.

Keywords Cluster of modern service industry · Knowledge sharing · Innovation mechanism

22.1 Introduction

Social network is to put people or organizations as a "node", and through the interaction between these nodes, it can form a relatively stable relationship, one of which refers to the relationship existing in the circulation of knowledge, technology, funds, and other resources. The cluster innovation under the social network

Y. Li (✉) · Y. Wu
Energy Economy and Management Research Center, Xi'an University of Science and Technology, Xi'an 710054, China
e-mail: 632058171@qq.com

J. Jiang
School of Construction and Safety Engineering, Shanghai Institute of Technology, Shanhai 200235, China

© Springer-Verlag Berlin Heidelberg 2015

S. Long and B.S. Dhillon (eds.), *Proceedings of the 15th International Conference on Man–Machine–Environment System Engineering*, Lecture Notes in Electrical Engineering 356, DOI 10.1007/978-3-662-48224-7_22

refers to the main body of the cluster network that aims to improve their ability of innovating by contacting with other subjects interaction, including competition and cooperation, and which in turns affect the entire cluster consciousness and innovation. This paper mainly discusses the influence mechanism between modern service industry cluster network and innovation mechanism, with the help of the study toward the relationship among knowledge, funds, and the cluster innovation, so as to construct the conceptual model and explore a new method as well as some new ideas to deal with the modern service industry cluster innovation mechanism.

22.2 Research Hypotheses

22.2.1 *Variable Description*

Through the analysis of relationship between the modern service industry network and cluster innovation mechanism, the author has built a variable network including social network, knowledge flow, and funds flow, and the mechanism of cluster innovation.

(1) *Social network*: it refers to a social structure made up of multiple nodes, in the structure these nodes contact with each other and then form a variety of social relations [1]. Located in Xian's innovation of the interaction between the modern service industry cluster enterprises, and main characteristics includes the network centrality, network density and structural holes in three aspects.

(2) *Intermediate variables*: In order to have a better understanding towards the influence of the cluster network characteristics to the cluster innovation mechanism, the writer regards knowledge flow and funds flow as the intermediate variables in this paper [2].

(3) *Cluster innovation mechanism*: means that the cluster enterprise makes full use of various resources in the cluster in order to speed up the innovation and development of enterprises.

22.2.2 *The Relationship Between the Cluster Network, the Knowledge Flow, and Funds Flow*

(1) Network centrality:

Network centrality refers to the degree of centralization of the whole network, or refers to the core of the network [3]. It can be divided into three parts: degree centrality, intermediate centrality, and close centrality. Network centrality mainly affects the status and influence of nodes in the network. If in a cluster network, the enterprise has a high network centrality; it means that it can control more resources than others, thereby increasing the network knowledge flow and funds flow. Based on the above description, we can put forward the following hypotheses:

H1: The higher the degree of network centrality, the greater the knowledge flow can be produced.
H2: The higher the degree of network centrality, the greater the cash flow can be produced.

(2) Network density:
Network density refers to the close interconnection between the nodes. This paper argues that, if there is a widespread innovative cooperation between enterprises in the network, and then the cluster network has a higher density [4], this will have a great effect on the cluster innovation mechanism. Based on these findings, we can find that the degree of the network density can influence the communication between enterprises. So we can make some hypotheses:

H3: The larger the network density is, the more knowledge flow is;
H4: The larger the network density is, the more funds can be raised.

(3) Structural holes:
In some situations, there still remains a direct link between some nodes of social network, but they have no direct contact with the other nodes, so those who don't have direct contact or who have gap with other nodes can produce structural holes [5]. Studies suggest that structural holes prepare chances for the node to get interests. Namely, the node with more structure holes can stay in the center and get more resources [6]. So the importance of structural holes was not decided by the existence of the relation, but laid in that this kind of lack will influence the cross of the resources.

Based on the above description, this paper holds the point that in the network, the more a network has structural holes, the lower possibility that you get resources repeat, so the knowledge flow and the funds flow, so there are some hypotheses come up:

H5: The more structure holes, the more the knowledge flow.
H6: The more structure holes, the more the funds flow.

22.2.3 Knowledge Flow and Cluster Innovation of Modern Service Industry

Rich knowledge flows can bring a lot of innovative information, and it's important for enterprise to improve their innovation ability. Knowledge, therefore, can affect the circulation of nodes in enterprises in a very main way, and is of great significance to promote the development of the cluster innovation mechanism. As a result, the knowledge flow among enterprises in the cluster network can have a greater influence on the innovation ability of enterprises in the network and form a good innovation atmosphere. Based on the above reasons, the writer puts forward the following hypotheses:

H7: The greater the knowledge flow in the cluster network, the more benefit it will have to promote the cluster innovation mechanism.

22.2.4 *Funds Flow and Cluster Innovation of Modern Service Industry*

Cluster enterprise's innovation and development can't depart from the support of funds. In the process of cooperation with other enterprises, funds can get a rapid flow. This paper holds the view that the flow of funds will produce important influence to the innovation of the enterprise, and because of the network connection, funds flow can influence the innovation mechanism of cluster network as well. Through this, we can put forward the following hypotheses:

H8: In the cluster network the larger the funds flow, the greater the benefit to promoting the cluster innovation mechanism.

22.3 The Research Models and Analyses

22.3.1 *Methodology and Samples*

The author of this paper uses the form of the questionnaire. A total of 159 individuals participate in the survey, and among these people, 100 take the paper questionnaire and the other 50 receive the electronic document. After the investigation, we received 127 pieces of feedback and 113 of them are useful. When we finished stating these questionnaires, we use excels to make charts in order to make our investigation more clearly and vividly.

22.3.2 *Scale Examination*

This study used spss16.0 for scale examination of the samples. Because that the cluster network features, knowledge flow and funds flow, innovation mechanism, these four inventory has all show a value of 0.807 which is greater than 0.8, it shows the high internal consistency between these variables. At the same time, the author have a test using KMO test and Bartlett test of all variables, the result shows the data is suitable for factor analysis.

22.3.3 *Presentation and Results*

According to the variables such as cluster of the modern service industry's structure characteristics and the influence of the funds flow, knowledge flow, mechanism of cluster innovation model, and the author draw and set a SEM model based on AMOS7.0, import the data calculation and correct the results (Fig. 22.1):

22.3.3.1 The Analyzing Results of the Initial Model

First of all, the author establishes the initial path graph in AMOS and then calculates the estimation value; get the whole numbers of the SEM model. Among them, the CMIN value is 327.389, DF value is 178, CMIN/DF's value is 1.839, the number of *P* is 0.000, the degree of NFI is 0.809, and the value of RMSEA is 0.055. For the indicators meet the reference standard, it indicates that the hypothesis model and the data model have passed the test.

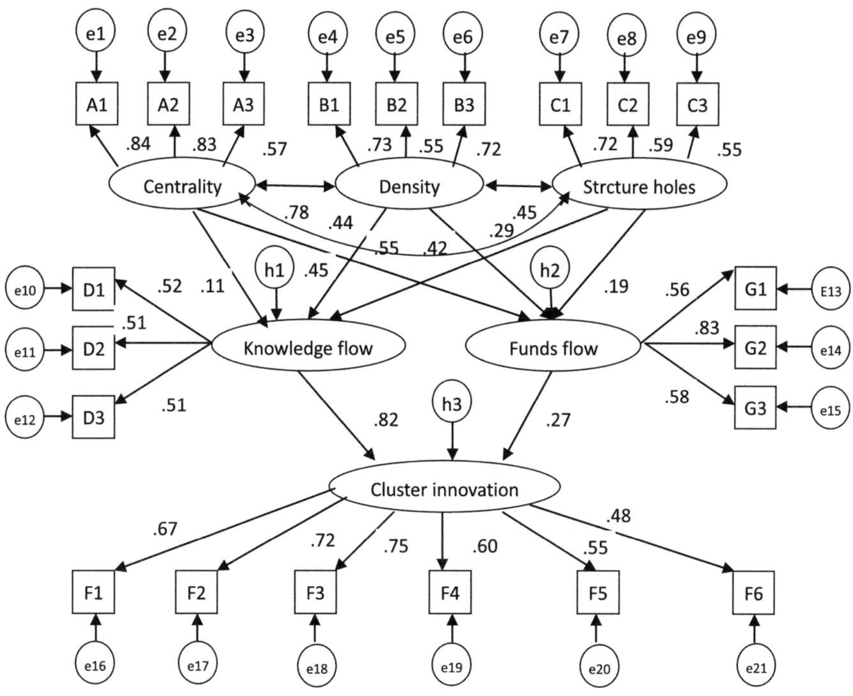

Fig. 22.1 The calculation results of the variables in SEM model

22.3.3.2 The Analysing Results of the SEM Model

This chart shows the relationship among the variables:

(1) Cluster network centrality's influence on the knowledge flow

The results showed that the network centrality of the modern service industry cluster network has no significant effect on the knowledge flow path, besides, as the specific data shows, the estimation value is 0.109, C.R is 0.528 and *P* in this chart is 0.598. From this data, it can be found that the critical value is less than 1.96, while the degree of *P* is more than 0.05, so we can make a conclusion that the quality of the network centrality has little effect on the knowledge flow. In other words, the result of H1 is false (Table 22.1).

(2) Cluster network centrality's influence on the funds flow

The results showed that the network centrality has positive influence on the funds flow, because as the specific data shows, the estimation value is 0.561, C.R is 2.118 and the number of *P* in this chart is 0.034. From this data, it can be found that the critical value is more than 1.96, while the degree of *P* is less than 0.05, so we can make a summary that the improvement of the network centrality can produce more funds. In other words, the result of H2 is true.

(3) Cluster network density's influence on the knowledge flow

As shown in the results, the network density has positive influence on the knowledge flow, because as the specific data shows, the estimation value is 0.461, C.R is 1.963 and the number of *P* in this chart is 0.050. From this data, it can be found that the critical value is more than 1.96, and the degree of *P* is the standard 0.05, so we can make a summary that the network density has a positive correlation with the knowledge flow. Generally speaking, the result of H3 is true.

(4) Cluster network density's influence on the funds flow

As it showed in the results, the network density has little connection with the funds flow, because as the specific data shows, the estimation value is minus 0.293, C.R is minus 1.112 and the number of *P* in this chart is 0.266. To sum up, the results show that the network density certainly has little connection with the funds flow. In short, the result of H4 is false.

Table 22.1 The relationship among the variables in SEM model

Route			Estimate	C.R.	P
Knowledge flow	<—	Network centrality	0.109	0.528	0.598
Knowledge flow	<—	Structural holes	0.422	2.708	0.007
Knowledge flow	<—	Network density	0.461	1.963	0.050
Funds flow	<—	Network density	−0.293	−1.112	0.266
Funds flow	<—	Network centrality	0.561	2.118	0.034
Funds flow	<—	Structural holes	0.186	1.190	0.234
Innovation mechanism	<—	Knowledge flow	0.815	3.628	***
Innovation mechanism	<—	Funds flow	0.265	2.353	0.019

*** $P<0.05$

(5) Structural holes' influence on the knowledge flow

The results showed that structural holes has positive influence on the knowledge flow, because as the specific data shows, the estimation value is 0.422, C.R is 2.708 and the number of *P* in this chart is 0.007. From this data, it can be found that the critical value is more than 1.96, while the degree of *P* is less than 0.05, so we can make a summary that structural holes in the network have a positive correlation with the knowledge flow. In a word, the result of H5 is true.

(6) Structural holes' influence on the funds flow

The results show that the effect of the structural holes on the funds flow in the cluster network is insignificant, because as the specific data shows, the estimation value is 0.186, C.R is 1.190 and the number of *P* in this chart is 0.234. From this data, it can be found that the critical value is less than 1.96, while the degree of *P* is less than 0.05, so we can make a summary that the effect of the structural holes to the funds flow in the cluster network is insignificant. In short, the result of H6 is false.

(7) The knowledge flow's influence on the innovation mechanism

According to the results showed in the chart, we can know that the funds flow in the cluster network has a significant effect on the innovation mechanism, because as the specific data shows, the estimation value is 0.815, C.R is 3.628 and the number of *P* in this chart is 0.001. From this data it can be found that the critical value is more than 1.96, but the degree of *P* is little, and shows a high connection, so we can make a summary that the funds flow has a significant effect on the innovation mechanism. Generally speaking, the result of H7 is true.

(8) The funds flow's influence on the innovation mechanism

As shown in the results, the network funds flow has positive connection with the innovation mechanism, because as the specific data shows, the estimation value is 0.265, C.R is 2.353 and the number of *P* in this chart is 0.019. From this data, it can be found that the critical value is more than 1.96, but the degree of *P* is less than standard number, it shows a high connection. To sum up, the results show that the funds flow's positive influence on the innovation mechanism. In short, the result of H8 is true. Comparing the data of hypotheses 7 and 8, we know that knowledge flow has a more significant influence on the innovation mechanism than funds flow, thus in order to improve the innovation mechanism we should pay attention to speeding up knowledge sharing.

22.4 Conclusion

According to the above study, the network centrality has a close connection with the funds flow, namely the higher the status of enterprises in the cluster network is, the more willing other enterprise is to build funds link with it. Network density has close ties with the knowledge flow but its relation with funds flow is unconsolidated. This is because that in the cluster network, the higher the density is, the richer the knowledge sharing is, and the closer the links between the two are;

Structural holes also has a close connection with knowledge flow, while it has a small influence on funds flow. All in all, knowledge flow and funds flow have significant positive influences on cluster innovation mechanism; moreover, the influence is more obvious for the knowledge flow.

Acknowledgments Supported by Project of Shaanxi Provincial Department of Education (11JK0061; 13JZ029; 14JK1445); Soft Science Project of Shaanxi Provincial Department of Science and Technology (2011KRM41); and Xi'an University of Science and Technology (JG1209; 2013SY01; 2014SX07; 2012SZ01).

References

1. Cross R, Liedtka J, Weiss LA (2005) Practical to social networks. Harvard Bus Rev 83(3):234–239
2. Huysman M, Wulf V (2006) IT to support knowledge sharing in communities, towards a social capital analysis. Inf Technol 21(1):69–74
3. Neelameghan A (2006) E-communities, social networks and technology management. Inf Stud 12(2):89–92
4. Chan K, Liebowitz J (2006) The synergy of social network analysis and knowledge mapping: a case study. Manag Dec Marking 7(1):74–79
5. Parise S (2007) Knowledge management and human resource development: an application in social network analysis methods. Adv Dev Human Resour 9(3):45–49
6. Ding Y (2011) Scientific collaboration and endorsement: network analysis of co-authorship and citation networks. J Inf 5:187–203

Chapter 23
Research on Applied-Information Technology with Trendy Utilization of Archival Information Resources

Shisheng Cheng, Yongqing Zhang and Haiyan Ma

Abstract In the big-data era, more than 80 % of information exist electronically. Without a doubt, paper and microfilm are the best, long-lasting storage forms for archiving. However, this is not practicable for the huge amount of digitalized information that will need to be archived in the near future. To some extent, it is even impossible because of the enormous amount of disk space that will be needed for big-data in different kinds of archives. Archival storage depends on the corresponding programs and technology for the eternal use of the digitalization archives' information resources. This chapter discusses some useful and practical functions and applied-information technologies for the utilization of archival information resources.

Keywords Applied-information technology · Archival information resources · Utilization · Functions and technologies

23.1 Automatic Classification

In two ways, more than 98 % of electronic information can be accurately classified in an automatic manner: First, the regulation for automatic classification must be based on electronic content management (ECM). Second, the organization

---

S. Cheng (✉)
Academy of Special Operations PLA, Guangzhou 510502, People's Republic of China
e-mail: chengshisheng731@sohu.com

Y. Zhang
Guangzhou Vocational School of Tourism and Business, Guangzhou 510280, People's Republic of China

H. Ma
Guangzhou Air-Force Automatic Commanding Workstation, Guangzhou 510071, People's Republic of China

© Springer-Verlag Berlin Heidelberg 2015
S. Long and B.S. Dhillon (eds.), *Proceedings of the 15th International Conference on Man–Machine–Environment System Engineering*, Lecture Notes in Electrical Engineering 356, DOI 10.1007/978-3-662-48224-7_23

classification must be clear and understandable. The remaining 2 % of information must be manually classified [1].

23.2 Automatic Tagging

Automatic tagging can improve the accuracy of information retrieval, which is very helpful for archive workers. However, almost 70 % of tagging work is completed manually, which reduces the efficiency of retrieval.

There are so many useful tools for automatic tagging on web sites such as YouTube and Facebook. These tools can help the end user to tag their own metadata in records, photos, and videos. In this way, these resources can be retrieved and enjoyed by others if approved in advance by the content owner.

23.3 Automatic Retrieving

Generally speaking, structured data is contained in the master database and its data bank, such as raw data. Non-structured data refers to records and files, such as office files, and resource records such as scanned images, paper, and microfilm.

An accurate ratio of retrieval is helpful for the location, search, and identification of archived files. Metadata is the key to the location and classification of information about the same topic, as well as to the rules to define whether the file should be archived or not. If this method is utilized correctly, the correct retrieval rate will exceed 90 %.

For researchers and end users, search is most commonly conducted using engines such as Google and Baidu, which is a blurry and less-than-accurate way of searching or retrieving [2].

23.4 Automatic Trapping

An automatic trapping program utilizes the service regulations to recognize or identify, copy, and retain e-mails and electronic files to the Element Management System (EMS). Meanwhile, useless records are deleted automatically, which helps to save disk space. In our information society, data occupy a huge amount of disk space. Storage space is a valuable resource that takes up many human resources and much physical space.

The technology of automatic trapping can acquire the correct files without the intervention of outside factor. It can catch the useful and correct information and distribute the corresponding records to the metadata bank, including the senders, receivers, the date, and the topic.

The leftover ratio can be nearly 0 % if the automatic trapping technique is used correctly.

23.5 Automatic Analysis

Analysis is part of more recent file management programs. The program's algorithm can complete functions such as browsing e-mails, office files, and other kinds of files. More importantly, the background relation is constructed among different kinds of files, which is helpful for automatic trapping and automatic classification.

The archive staff can use this method to decide if the files will be saved or stored as eternal records or to assign history metadata tags to the corresponding files.

23.6 Automatic Processing

Another newer function in file management applications is the automatic processing capability for information. This technology defines a conservation period for the files. The automatic processing software starts the processing program to decide if the files will be transferred to the archive or will be destroyed [3]. It also has other useful purposes, such as decreasing the tasks required by file management staff, transferring archival information to the archive, saving storage space, and deleting useless historical records.

23.7 Operation Flow Management

Operation flow management helps to automate manual operations. In this system, the management staff can see the whole process for the files. More than 90 % of the file management work can be implemented by procedures, including the sending and receiving of files, remote storage, and transferring to the archive. This method makes file management work more fluent and more efficient. In addition, the whole process can be supervised.

23.8 Merging Management for Structured and Nonstructured Information

A recent trend in the information management field is to manage original data files (so-called structured data) and office files (so-called non-structured data) using the same technology core. The related software, work flow, and practice work is in

Table 23.1 Utilization data in a district archive for 6 months

Contents		People (*n*)	Rolls (*n*)	Replications (*n*)	Printings (*n*)
Daily reception		25,621	22,511	146,746	5379
Weekend reception		2096	1437	8520	0
Among above	Trans-archive consulting	192	177	513	0
	In advance consulting	37	196	0	0
	Massive consulting	27	1076	3299	850
Total		27,973	25,397	159,078	6229

progress. This method may be more important in file and information management than archive management. Because this tool is so useful and so practicable, it can manage and supervise all the relevant files during the entire life cycle. In private domains, file management technology is used for archive appraisal and accession.

23.9 Data Analysis

The ultimate purpose for archived work is to be able to use it. Table 23.1 shows the utilization of archived files in a typical archive [4].

From the data in Table 23.1, it can be concluded that because the reception number is so large, it will take many archive staff a long time to finish it. However, if the task can be completed automatically—partly by using the applied information technology described in this chapter—the utilization efficiency will be higher and the reception number will be greater.

23.10 Conclusions

Classification, retrieval, analysis, trapping, and operation flow management are very important and useful tools for archive staff. In the near future, the price of these tools will decrease rapidly for the widely used departments and fields. In this way, archive staff must master different kinds of electronic storage technologies, especially applied-information technology. They also must cooperate with information technology and file management staff members to appraise and identify the files during their entire life cycle. Finally, the archive staff should use operation flow management software to define the correct way for accession and to ensure the correct digital storage method for different files and records, which will improve their own techniques and capabilities during the information life cycle [5].

References

1. Yao W, Moody K, Bacon J (2006) A model of OASIS rolebased access control and its support for active security. In: Proceedings of sixth ACM symposium on access control models and technologies. SACMAT 2001, pp 55–57
2. Seligman L, Rosentbal A (2005) XML impact on databases and data sharing. Computer 11–27
3. Yu S (2005) Building an open archive union catalog for digital archives. Emerald 17–27
4. Smith C (2006) Encoded archival context: authority control for archives. Info 663-techprocesses in libraries, pp 24–56
5. Luciana D (2001) Concepts, principles, and methods for the management of electronic records. Inf Soc 57–60

Chapter 24
Information Support for Real-Time Decision-Making Based on Big Data: Knowledge-Enabled Machine Activeness and System Efficiency

Xin Jin, Shiqiang Zong, Youjiang Li and Shanshan Wu

Abstract Real-time decision-making requires efficient information support that dynamically provides information useful to decision makers. Traditional fixed information service mode is challenged by Big Data environment, while search-based methods do not guarantee efficiency. To solve the problem, a method is proposed to improve machine activeness on information collection, allowing user to be focused on decision-making, so as to improve the efficiency of the whole decision-making system. During human decision-making process, machine keeps aware of the decision task context, dynamically recognizes user information requirement, automatically activates search process, based on domain knowledge previously built reflecting the latent relations between decision task types and the information required. It is proved by experiments that the method can effectively save user time cost on information requirement expressing and improve task relevance of collected information.

Keywords Information support · Decision-making · Big data · Knowledge enabled

24.1 Introduction

In many enterprise systems, along the operation process, decisions are made in real-time. For such decision-making, efficient information support is of great importance, which should be fast, dynamic, focused, and not distracting. In traditional enterprises, there are customized information services, which monitor decision-making process, and dynamically serves useful information to decision

X. Jin (✉) · S. Zong · Y. Li · S. Wu
Science and Technology on Information System Engineering Laboratory, Nanjing 210007, China
e-mail: venus_163@163.com

© Springer-Verlag Berlin Heidelberg 2015

S. Long and B.S. Dhillon (eds.), *Proceedings of the 15th International Conference on Man–Machine–Environment System Engineering*, Lecture Notes in Electrical Engineering 356, DOI 10.1007/978-3-662-48224-7_24

maker. The information flow is previously coded or configured. At what time, for which task, by which operation, which information should be got from which data source, and sent to which user, are all predefined [1, 2].

Recent years, Big Data has caused broad concern [3–5]. Increasing number of traditional enterprises start to utilize Big Data on Internet to support decision-making. As information distribution is unclear to users, customized information services is no longer fit. Instead, search engines have been vastly used. When user need some information for decision support, he firstly expresses his requirement into keywords. He needs to carefully choose keywords to avoid ambiguity and modify according to results. No matter how good is the performance of the search engine, in most cases, user needs time to manually filter search results. Normally, it is difficult for keyword-based method to achieve high precision, as machine does not understand user task, and user cannot express requirement precisely through keywords. Moreover, it follows a passive service mode, in that nothing returns if no keyword is submitted. In a word, Big Data increases complexity of information support, while keyword-based search method brings down the efficiency.

To solve the problem, a simple idea is to make the search process automatic, means machine actively does requirement expressing and information retrieval for human, so that human can be focused on decision-making. It is found that user information requirement during enterprise decision-making is much more regular than in personal manner. On making similar decisions, required information has similar content and scope. There are latent relations between decision task types and required information. If these latent relations are discovered and explicitly defined as domain knowledge, above automation is realizable. By this idea, a method is proposed, on how to prepare and utilize such domain knowledge to improve machine activeness and system efficiency. It is proved by experiment that the method can effectively reduce time cost on user information requirement expressing and improve information task relevance.

24.2 Related Researches

On helping user expressing information requirement, there are methods in IR (information retrieval) domain, e.g., keyword generation [6] and query recommendation [7]. These methods give suggestions on requirement expression, based on various kinds of knowledge, including user searching behavior analysis [8], term relationship [6], semantic relativity [9], and so on. They boosted the prosperity of search engine advertisement and got fast progress driven by its great commercial interest. However, they all require user to input some initial keywords as a hint and then give suggestions based on it. They do not care about user decision task, but about what user inputs in the search box. If the user chosen keywords do not precisely reflect his requirement, the recommended query and the returned search results are useless. Moreover, user still needs to distract part of his attention on initial keywords selection.

24.3 Method

By analysis of traditional customized information service mode, it is found when processing a same type of decision tasks, required information has similar scope and content, while differences mainly lie in detailed task features, such as related entities, space, and time contexts. There are latent relations between user decision task types and their INF-REQs (information requirements). Our method is to mine and define the relations into domain knowledge and then apply it to dynamically sense and recognize user INF-REQs.

24.3.1 Domain Knowledge Construction

Classify domain user decision tasks and divide until INF-REQs of two task types can be differentiated as variables. Take military operation command and control as an example, on making decisions about how to attack two targets of same type, their INF-REQs can be differentiated as a target name variable. But when making decisions about how to defend from target strike, INF-REQ totally changes, not as simple as a variable.

Six elements are suggested as a globally unique ID of task types, containing two groups: (1) <UserType, OperationType, ObjectType>, describing user decision-making operations on the user interface, i.e., who is currently doing what operation to which object on the user interface; (2) <ActorType, ActionType, TargetType>, describing the user currently concerned missions that happen in the real world, i.e., who is currently doing, or has done, or will execute what action to which target object. So the variable vector <U, O, O, A, A, T> can be used at task type ID.

For each task type, create an INF-REQ template. In this template, the scope and content of the required information are modeled, with variables defined in it. A model called TIREM (Task INF-REQ Express Model) is designed, as shown in Table 24.1, introduced in detail as follow. A template is constituted of an ID, and one or more TIREM models, described in "<REQ > … </REQ>" pairs. Variables are marked by "&()" operators. An example is shown at top right in Fig. 24.3.

LABEL is a literal description for human to understand the INF-REQ.USER, OPERATION, and MISSION define a decision task type from three dimensions: role of decision maker, his current operation on the user interface, and the user concerned mission that happens in the real world. SELECT and WHERE have the same meanings as in SQL. Difference lies in that each attribute following SELECT and WHERE is a "Subject.Property" pair. With subject as a limit, property semantics will not be misunderstood, thus improves the precision. For the required

Table 24.1 TIREM specification

```
<REQ>
  <LABEL>Literal description of the INF-REQ</LABEL>
  <USER>Information user type</USER>
  <OPERATION>User operation type</OPERATION>
  <MISSION>User concerned mission type</MISSION>
  <SELECT>Sub.prop1 &/|| Sub.prop2 &/|| …</SELECT>
  <WHERE>Sub.prop3 =/</> Value a &/|| Sub.prop4 =/</> Value b &/|| ……</WHERE>
  <FROM>Source 1 &/||/|- Source 2 &/||/|- ……</FROM>
  <MEDIATYPE>Type 1 || Type 2 || ……</MEDIATYPE>
  <WHEN>
    <START>Start time of the INF-REQ</START>
    <END>End time of the INF-REQ</END>
    <PERIOD>Period for refresh</PERIOD>
    <REPEAT>Period for repeated search</REPEAT>
    <EARLIEST>Earliest publish time</EARLIEST>
    <LATEST>Latest publish time</LATEST>
    <REALTIME>Real-time or non</REALTIME>
  </WHEN>
  <ORDERBY>1st rule |- 2nd rule |- ……</ORDERBY>
</REQ>
```

information, preferred sources, if there are, can be listed after FROM. Preferred types of the required information, e.g., text, image, video, audio, database, formatted message and so on, can be listed after MEDIATYPE. WHEN describes information timeliness. seven items are defined to describe timeliness requirement for both real-time and non-real-time information. ORDERBY defines how search results are ranked and ordered in the returned list.

There are two strategies to design templates: (1) by experience: invite domain experts to describe the required information according to their experiences, and translate into template files. (2) by learning: analyze user INF-REQs during task processing by aid of machine learning tools.

Finally, construct domain ontology. Ontology can solve the problem of cross-domain heterogeneity on describing INF-REQ semantics. So, it is suggested to construct user domain ontology and design templates using terms defined in the ontology. On source side, it is also suggested to construct ontology and describe its information metadata using terms defined in the ontology. OM (ontology mapping) tools [10] can be used to build alignment between different ontologies.

24.3.2 Domain Knowledge-Based Information Service

In running phase, on the user interface, functions are required to keep aware of the context information and dynamically assign values to the elements in <U, O, O, A, A, T>, so as to recognize current decision task type and features: (1) identify user by monitoring login operation, and assign value to variable UserType; (2) monitor user operations, and assign value to variable OperationType and ObjectType; (3) resolve formatted mission files, containing descriptions about the executor, actions and targets, and assign value to variable ActorType, ActionType, TargetType; (4) collect information about various kinds of entity IDs, area IDs, and system time, and form a feature variable set. Based on dynamically recognized <U, O, O, A, A, T>, load according template. For each variable in the template, assign with value of the according variable in the task feature variable set. After all variables have their values assigned, an INF-REQ is generated.

Before the generated INF-REQ is submitted to the search engines, its semantics should be described according to domain ontology, in order to get precise result feedback. For example, "warship:Lincoin" represents "Lincoin" aircraft carrier defined in ontology about warships and will not be mistranslated to other concepts like "people:Lincoin" or "car:Lincoin". In practices, we encounter synonym problems. For example, "wp:F-22" and "wp:Raptor" are 2 IDs about a same type of plane. To avoid low recall of search result, both IDs should be included in the INF-REQ. Similarly, "wp:FlyingRange" and "wp:Voyage" both mean the longest distance a plane could fly through and should both be included as well.

After above steps, an INF-REQ model is formed. Then it will be submitted to various search engines on Internet. According to value of REALTIME, the INF-REQ will be published to engines of real-time or non-real-time information sources. For durative INF-REQs, according to START and END, they will be made valid and invalid. If PERIOD has value, real-time information will be pushed to user periodically, while if REPEAT has value, the INF-REQ will be resubmitted to search engine periodically. For one-off INF-REQs, only the information published later than EARLIEST and earlier than LATEST will be returned.

Finally, the INF-REQ will be translated on data source side. For structured information in databases, it will be translated into SQL queries. For real-time information contained in messages pushed by P/S systems, it will be translated into subscription requests. For pure text information contained in webpages, it will be translated into keyword expressions. A set of translation rules have been defined to realize automatic translation from TIREM to SQL queries, subscription requests, and keyword expressions, which is not introduced in detail here.

24.4 Prototype System

The method has been implemented in a military command and control system, so as to build a prototype system for method verification and validation. During the whole process of operation command and control, commander needs information for decision support.

In the domain knowledge preparing phase, commander's decision tasks have been classified into 55 types, containing two-two combination of five user operation types (mission acceptance, situation analysis, operation planning, execution monitoring, and effect assessment) and 11 concerned mission types (fixed target striking, moving target interception, air attack defensing, anti-terrorism operation, etc.), forming a 5*11 matrix, each cell representing a task type. Templates have been designed for each task type according to commanders' experiences. User required information covers real-time, non-real-time, structured, and unstructured data. About 13 TB military data has been collected, distributed on 110 data nodes, in order to simulate Big Data environment, four domain ontologies have been constructed, including 84 concepts, 28 relations, and about 2 k instances.

Task awareness module has been developed and embedded in user interface, including time detection, login monitoring, formatted mission file analysis, work process monitoring, plan execution monitoring, and user interface operation listening. Functional modules have been developed to realize INF-REQ generation and submission, containing task analysis, template loading, template variable assignment, semantic description, and submission management. Result presentation module has been developed in user interface, including table, literal text, and map elements. On source side, INF-REQ translation modules have been developed for search engine, database, and P/S system, respectively.

24.5 Experiments

To verify the feasibility and performance of the method, experiments have been taken using the prototype system. Three commanders were invited to operate on user interface simultaneously. Each time a task in the matrix was triggered, a set of INF-REQs were generated, with search results returned automatically. In most time, the commanders needed not to input any keyword and were focused on decision-making. This is the main advantage of our method.

For comparison, each time a task was triggered, the three commanders were asked to input their INF-REQs manually, with time cost recorded, as shown in Fig. 24.1. Mean time cost for manual mode is about half a minute, while for automatic mode, about half a second. It is obvious that automatic mode can effectively save user time cost on requirement expressing.

All the manually input and automatically generated INF-REQs were submitted to search engines, with four result sets returned, which were compared on precision.

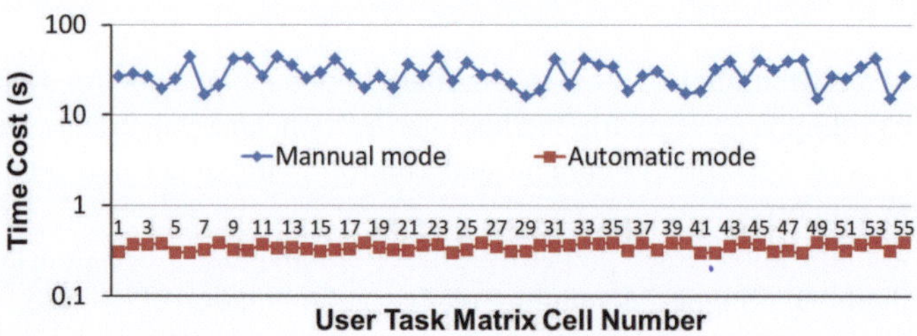

Fig. 24.1 Time cost comparison

Precision of a result set is the mean task relevance (evaluated by five human valuators) of all the search results. As shown in Fig. 24.2, mean precision for manual mode is about 38.7 %, while for automatic mode, about 60.5 %. Automatically retrieved information is more tightly focused on current decision task.

An instance is shown in Fig. 24.3. Accepted mission was to dispatch battle planes to intercept enemy planes, in order to defend some core facilities from attack. User current operation on interface was planning the flying path of the battle planes. At the top right is the dynamically loaded template for this detected task. Information is required on enemy planes' flying ability and real-time track within 10 min. Three variables were automatically assigned with values: EnemyPlaneType = "F-18", TargetArea = "Area Coordinate", CirremtTime = "14:00:00". One of the

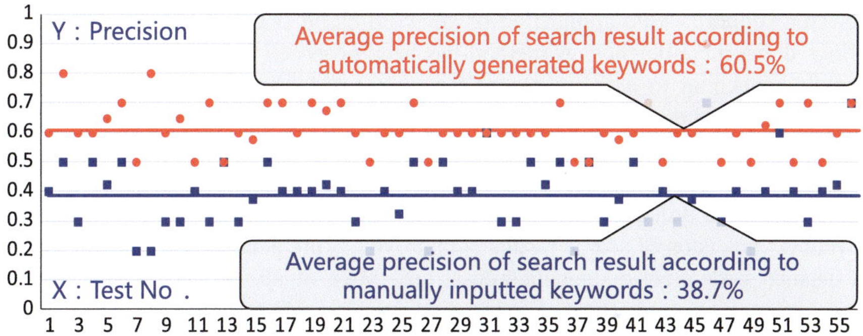

Fig. 24.2 Precision comparison

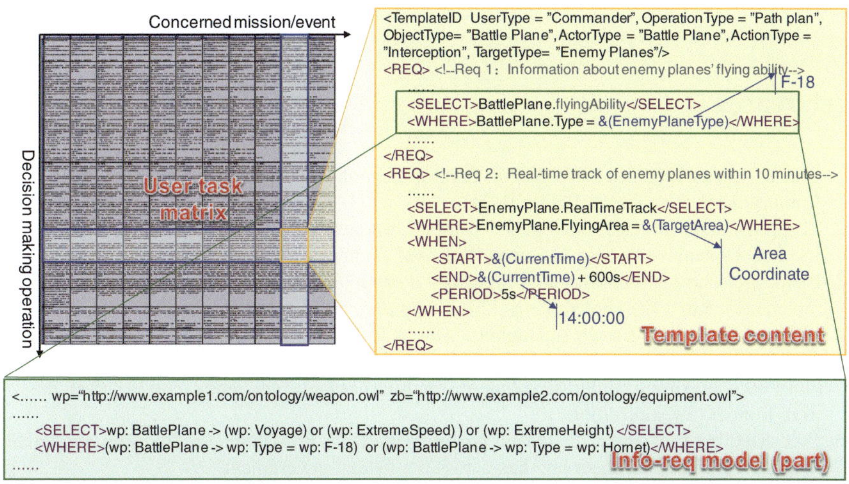

Fig. 24.3 INF-REQ model generation example

generated INF-REQs is shown at the bottom. Based on the domain ontology, the property "Flying Ability" was expanded with "Voyage," "Extreme Speed," and "Extreme Height"; and the value "F-18" was expanded with "F-18" and "Hornet." From the instance, one can see that normally the automatically generated INF-REQs are detailed, because lots of terms for restriction and expanding were added according to template and domain ontology. But for human, to input an INF-REQ with similar complexity is quite time-consuming. This also explains why automatically generated INF-REQs can get more precise search results.

24.6 Conclusions

Big Data challenges traditional customized information services for real-time decision support, while application of keyword-based search method significantly brings down the efficiency. A method has been proposed to solve the problem by enabling machine to actively sense, recognize, and express current information requirement, and do information retrieval automatically. Domain knowledge about latent relations from decision task types to commonly required information is applied to realize such automation. It is verified through experiments that: (1) efficiency of search process is improved as user time cost on requirement expressing is saved; (2) precision of search results is improved as the generated INF-REQ model is detailed with precise semantics described. Most of all, decision maker is free from information searching and focused on decision-making, so that the efficiency is improved.

Further researches are focused on knowledge construction. Template is a basic knowledge express pattern, and the next step is to mine rules form the gradually accumulated templates, so as to improve the method flexibility.

References

1. Chen ZX, Tan XQ (2013) Workflow Control technology for command information system. Command Inf Syst Technol 4(3):20–24
2. Zhang J, Chen ZB (2013) Intelligence distribution control requirements and realization of joint intelligence support system. Command Inf Syst Technol 4(2):33–36
3. Challenges and opportunities with big data—a community white paper developed by leading researchers across the United States (2013) http://cra.org/ccc/docs/init/bigdata/whitepaper.pdf
4. Manyika J, Chui M, Brown B, Bughin J, Dobbs R, Roxburgh C, Byers AH (2011) Big data: the next frontier for innovation, competition, and productivity. McKinsey Global Institute. http://www.mckinsey.com/insights/mgi/research/technology_and_innovation/big_data_the_next_frontier_for_innovation
5. Fact sheet: big data across the federal government (2012) http://digital-scholarship.org/digital-koans/2012/03/29/fact-sheet-big-data-across-the-federal-government/
6. Joshi A, Motwani R (2006) Keyword generation for search engine advertising. 6th IEEE international conference on data mining. IEEE Press, Hong Kong, pp 490–496

7. Ricardo B, Carlos H, Marcelo M (2004) Query recommendation using query logs in search Engines, current trends in database technology. Lecture notes in computer science. Springer, Berlin, pp 588–596
8. Google AdWords keyword tool (2013) http://google.com/select/KeywordSandbox
9. Abhishek V, Hosanagar K (2007) Keyword generation for search engine advertising using semantic similarity between terms. In: 9th international conference on Electronic commerce, Minnesota, p 94
10. Wei H, Yuzhong Q (2008) Falcon-AO: a practical ontology matching system. Web Semant: Sci Serv Agents World Wide Web 6(3):237–239

Chapter 25
Dark Pixels Law-Based Image Dehazing in UAV Reconnaissance

Qinlan Yue, Hongjie Zheng and Chengneng Jiang

Abstract The visible light band is an important band of UAV reconnaissance to acquire the information of targets. In this paper, degradation principles of haze images in UAV reconnaissance is first demonstrated by analyzing the effect of haze and fog upon UAV reconnaissance imaging. Then the entire atmospheric scattering model of haze images is set up straightly, and a realization of recovering a haze-free image in UAV reconnaissance using dark pixels law is discussed. Experiment results indicate that image dehazing based on dark pixels law can significantly improve the quality of reconnaissance tasks when there is a haze atmosphere around and a visible light imaging equipment onboard. The experimental results show that the method has obvious promotion effect on UAV reconnaissance equipment implementation of visible light imaging reconnaissance tasks in the haze situation.

Keywords UAV reconnaissance · Dark pixels law · Image dehazing

25.1 Introduction

UAV reconnaissance image refers to the visible image formed by processing electromagnetic wave energy, which is captured by the UAV reconnaissance equipment, from target reflection or radiation. The visible light band is an important band of UAV reconnaissance to acquire the information of targets. Under imaging condition of fine weather, reconnaissance image can be acquired with high quality in this band. In haze situation, due to the scattering of atmospheric particles, the edge of the image contour levels become fuzzy, relatively concentrated, poor contrast, thus the image will to some degree be degraded [1]. In this study, we analyze the influence of haze on the visible light image and the restoration method.

Q. Yue (✉) · H. Zheng · C. Jiang
Journal of PLA Academy of Special Operations, Guangzhou 510502, China
e-mail: qinlan_yue@sina.com

© Springer-Verlag Berlin Heidelberg 2015

S. Long and B.S. Dhillon (eds.), *Proceedings of the 15th International Conference on Man–Machine–Environment System Engineering*, Lecture Notes in Electrical Engineering 356, DOI 10.1007/978-3-662-48224-7_25

25.2 The Mechanism of Fuzzy Image Degradation in UAV Reconnaissance

First, the mechanism of haze generation is explained from the angle of physics, and then the analysis on the mechanism of fuzzy image degradation is carried on, proceeding from the fog imaging model.

25.2.1 *The Formation of Haze*

Aerosol mist is composed of tiny water droplets or ice crystals in the atmosphere, the average diameter of about 10–20 μm. It is relatively large compared with the visible light of wavelength 0.38–0.76 μm. The scattering of different wavelengths are basically the same, so the scene is milky white. Haze is formed by a large amount of suspended dust, sulfuric acid, nitric acid, organic hydrocarbons, and other particles which are uniformly distributed in the atmosphere, the average diameter of about 1–2 μm. Because of the Rayleigh scattering caused by the haze particles and the longer wavelength of the orange, yellow light, the scene looks slightly yellow. Despite the fog and haze formation principles are different, fog and haze both decrease atmospheric transmittance in the UAV reconnaissance image, thus affecting the imaging results.

25.2.2 *Diffuse Reflection Model*

The scattering of haze particles is reflected in two aspects: on the one hand, scattering on the reflected light of the surface features, which makes the arrival of airborne sensors attenuated; on the other hand, scattering on atmospheric ambient light and is transferred to the sensor [2]. According to the scattering theory, when the UAV is conducting visible light reconnaissance imaging in fog, light which reach the airborne sensor is mainly composed of two parts, one part is the incident light which has been weakened, the other part is the scattered ambient light [3]. The two parts constitute a complete model to describe the fuzzy image degradation mechanism, as shown in Fig. 25.1.

25.2.2.1 Incident Light Attenuation Model

Assuming that the incident light through the mist with the distance of dx, the incident light intensity of $E_0(\lambda)$, where λ is the wavelength of incident light. So at a distance of x, the attenuation of light can be expressed as:

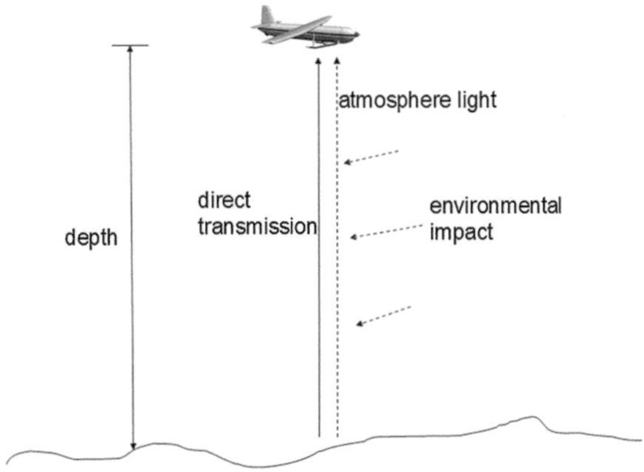

Fig. 25.1 Diffuse reflection model

$$\frac{\mathrm{d}E(x,\lambda)}{E(x,\lambda)} = -\beta(\lambda)\mathrm{d}x \quad (25.1)$$

The $\beta(\lambda)$ for atmospheric scattering coefficient which reflects the scattering ability of light unit amount of atmosphere on different wavelengths. Integral for x from 0 to d, we get the light intensity in d:

$$E_{\mathrm{dt}}(d,\lambda) = E_0(\lambda)e^{-\beta(\lambda)d}. \quad (25.2)$$

25.2.2.2 Environmental Incident Light Imaging Model

Larger particles in the atmosphere will produce light scattering on the ambient light, resulting in the partial involvement of lights in imaging. Assuming the atmosphere of light is the same, but just the direction, intensity and the intensity spectrum were unknown. The cone formed by the solid angle dw of imaging equipment and its optical axis, and the distance to UAV d is regarded as atmospheric light source, and the surface element volume in the distance of x for UAV:

$$\mathrm{d}V = \mathrm{d}wx^2\mathrm{d}x \quad (25.3)$$

If the strength of light distributed at any position is m, then the total luminous flux through the surface:

$$\mathrm{d}I(x,\lambda) = m \cdot \mathrm{d}w \cdot x^2 \cdot \mathrm{d}x \cdot \beta(\lambda) \quad (25.4)$$

There are:

$$\mathrm{d}E(x,\lambda) = \frac{\mathrm{d}I(x,\lambda)\mathrm{e}^{-\beta(\lambda)x}}{x^2} \tag{25.5}$$

So get the unit flux for:

$$\mathrm{d}L(x,\lambda) = m\beta(\lambda)\mathrm{e}^{-\beta(\lambda)x}\mathrm{d}x. \tag{25.6}$$

Integral:

$$L(d,\ \lambda) = m(1 - \mathrm{e}^{-\beta(\lambda)d}), \tag{25.7}$$

when the d approaching infinity, $L(\infty, \lambda)$ = m; when d equals zero, $L(0, \lambda) = 0$. So the light intensity reached UAV:

$$E_a(d,\lambda) = E_\infty(\lambda)\left(1 - \mathrm{e}^{-\beta(\lambda)d}\right). \tag{25.8}$$

25.2.2.3 Degradation Model of Fuzzy Image

According to the analysis of the theoretical model, the light intensity into airborne sensors consists of the incident light attenuation model and environmental incident light impact. That is:

$$E(d,\lambda) = E_0(\lambda)\mathrm{e}^{-\beta(\lambda)d} + E_\infty(\lambda)\left(1 - \mathrm{e}^{-\beta(\lambda)d}\right). \tag{25.9}$$

Because of the influence of visible light wavelength on the atmospheric scattering in the fog can be ignored, so the light intensity formula of sensor acquired can be simplified as the following form:

$$E = \mathrm{Re}^{-\beta d} + I_\infty\left(1 - \mathrm{e}^{-\beta d}\right). \tag{25.10}$$

Among them, R represents the original light intensity scene without the fog, in the other words the image after defogging. I_∞ represents the light intensity of infinite sky in the level direction, namely the global atmospheric light. While the $\mathrm{e}^{-\beta d}$ represents the transmittance of fog.

According to the mechanism of the fuzzy digital image degradation and the theoretical model, we can conclude the basic idea of the image dehazing method in UAV reconnaissance. By estimating the sky brightness and scene optical depth, to get the value of the ambient light, and then minus the value of ambient light from the degraded image, at last compensate the attenuation of the reflected light of scene, we can achieve the goal of image restoration.

25.3 Haze Image Recovering in UAV Reconnaissance Based on Dark Pixels Law

25.3.1 Calculating the Normalized Dark Pixels Law

According to Eq. (25.10) aforementioned, the haze image in UAV reconnaissance can be molded as:

$$\mathbf{I}(\mathbf{x}) = \mathbf{J}(\mathbf{x})t(\mathbf{x}) + \mathbf{A}(1 - t(\mathbf{x})), \tag{25.11}$$

where $\mathbf{I}$ is the observed intensity, $\mathbf{J}$ is the scene radiance, $\mathbf{A}$ is the global atmospheric light, t is the medium transmission describing the portion of the light that is not scattered and reaches the camera, and $\mathbf{x}$ is coordinate vector of some pixel in a image.

In a local patch, we consider that the transmission is constant, and the atmospheric light $\mathbf{A}$ is given. Then the dark pixels law of a particular color channel $c(c \in \{r, g, b\})$ can be calculated as:

$$\min_{\mathbf{y} \in \Omega(\mathbf{x})} (I^c(\mathbf{y})) = \widetilde{t}(\mathbf{x}) \min_{\mathbf{y} \in \Omega(\mathbf{x})} (J^c(\mathbf{y})) + (1 - \widetilde{t}(\mathbf{x}))A^c \tag{25.12}$$

here $\Omega(\mathbf{x})$ is the 15×15 pitch centered at $\mathbf{x}$, so the normalized dark pixels law is given by

$$\min_{\mathbf{y} \in \Omega(\mathbf{x})} \left(\frac{I^c(\mathbf{y})}{A^c} \right) = \widetilde{t}(\mathbf{x}) \min_{\mathbf{y} \in \Omega(\mathbf{x})} \left(\frac{J^c(\mathbf{y})}{A^c} \right) + (1 - \widetilde{t}(\mathbf{x})). \tag{25.13}$$

Therefore, we get

$$\min_{c \in \{r,g,b\}} \left(\min_{\mathbf{y} \in \Omega(\mathbf{x})} \left(\frac{I^c(\mathbf{y})}{A^c} \right) \right) = \widetilde{t}(\mathbf{x}) \min_{c \in \{r,g,b\}} \left(\min_{\mathbf{y} \in \Omega(\mathbf{x})} \left(\frac{J^c(\mathbf{y})}{A^c} \right) \right) + (1 - \widetilde{t}(\mathbf{x})). \tag{25.14}$$

25.3.2 Calculating the Initial Transmissivity Figure

From dark pixels law, we have

$$J^{\text{dark}}(\mathbf{x}) = \min_{c \in \{r,g,b\}} \left(\min_{\mathbf{y} \in \Omega(\mathbf{x})} (J^c(\mathbf{y})) \right) \to 0. \tag{25.15}$$

As the atmospheric light is positive, i.e., $A^c > 0$ which means that the first term in the right hand of (25.14) tends to be zero, we have

$$\min_{c \in \{r,g,b\}} \left(\min_{\mathbf{y} \in \Omega(\mathbf{x})} \left(\frac{I^c(\mathbf{y})}{A^c} \right) \right) = 1 - \tilde{t}(\mathbf{x}). \tag{25.16}$$

Thus, the initial transmittance can be acquired by the following formula:

$$\tilde{t}(\mathbf{x}) = 1 - \min_{c \in \{r,g,b\}} \left(\min_{\mathbf{y} \in \Omega(\mathbf{x})} \left(\frac{I^c(\mathbf{y})}{A^c} \right) \right). \tag{25.17}$$

25.3.3 Using the Cutout Method for Fine Transmittance Figure

In fact, there will be some halos and block artifacts in our recovered image, because the initial transmission is assumed as constant within a pitch size of 15 × 15. To take advantage of guided image filtering, which is believed to be a effective balance between matting quality and computational efficience, we can use it to capture the sharp edge discontinuities and outline the profile of the objects, to obtain a high-quality image with more less halos and block artifacts. As a particular application, the observed intensity **I** can be seen as a guide, and the initial transmission $\tilde{t}(\mathbf{x})$ is the filtering input, then the refined transmission $t(\mathbf{x})$ can be obtained using the guided filter.

25.3.4 Calculating the Atmospheric Light

A is assumed to be known in the formula derivation. Pick the top 0.1 % brightest pixels, among which the ones with highest intensity in the input image **I** can be selected as the calculate of atmospheric light **A**.

25.3.5 Calculating the Image After the Fog

After estimating $t(\mathbf{x})$ and **A**, put them into (25.11), then the scene radiance can be written as:

$$\mathbf{J}(\mathbf{x}) = \frac{\mathbf{I}(\mathbf{x}) - \mathbf{A}}{t(\mathbf{x})} + \mathbf{A}. \tag{25.18}$$

Since the transmission $t(\mathbf{x})$ can be very close to zero, it leads to the recovered scene radiance **J** is prone to noise. Therefore, a small positive constant t_0 is introduced to modify (25.18), i.e., to preserve a small amount of haze. So the output image can be expressed as:

$$\mathbf{J}(\mathbf{x}) = \frac{\mathbf{I}(\mathbf{x}) - \mathbf{A}}{\max(t(\mathbf{x}), t_0)} + \mathbf{A}. \tag{25.19}$$

25.4 Experiment Results and Analysis

In our experiment, the guided image filtering is introduced to refine the initial transmission [4]. Through repeated trials, the main parameters of guided filter are set as: pitch size is 15 × 15, guided radius is 60 pixels, haze coefficient is 0.98, upper bound of atmospheric light is 240, ε is 0.01.

In Fig. 25.2a is a UAV reconnaissance image with the top view of an outdoor scene covered by nonuniform haze. (b) is a result of the mist.

In Fig. 25.3a is the oblique view of an outdoor scene with uniform haze photographed by a UAV, which results in different thickness of haze while the depth of image varies. (b) is a result of the mist.

As we can see from the experiment results, by applying the dehazing method deviated from dark pixels law, we can recover a high-quality haze-free image from both vertical photographing with nonuniform haze and oblique photographing with uniform haze. For the method can simultaneously estimate the field depth, it is more preferred in an oblique photographing scenario. However, it fails when the image is degraded by a nonuniform haze or especially a very thick haze.

Fortunately, when dealing with a UAV reconnaissance haze image photographed vertically above some object, we can gain a high-quality image in most tests using dark pixels law, and the image definition is significantly improved.

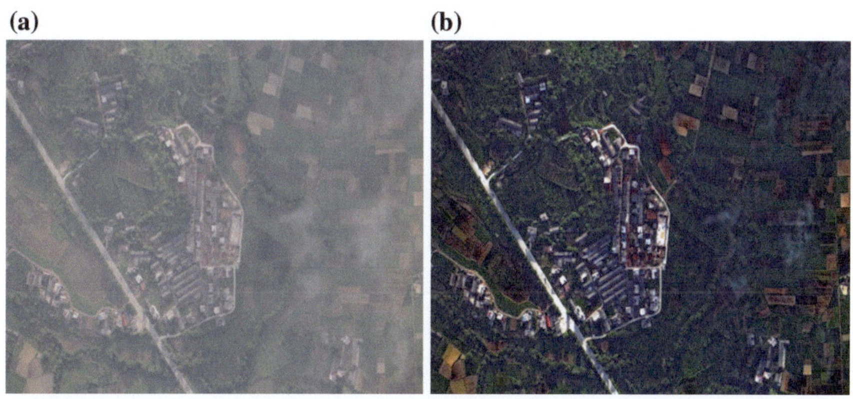

Fig. 25.2 Haze removal of vertical photographing. **a** Before going to the haze, **b** go after the haze

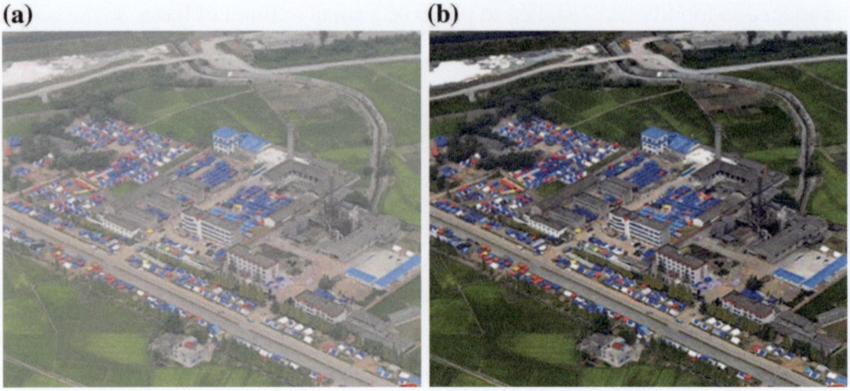

Fig. 25.3 Haze removal of oblique photographing with uniform haze. **a** Before going to the haze, **b** go after the haze

Acknowledgment Fund Project: Supported by The Natural Science Foundation of Guangdong Province (S2013010016141)

References

1. Huang X (2007) UAV reconnaissance imaging and intelligence processing. PLA Press, Beijing, pp 31–33
2. Yu J, Xu D, Liao Q (2011) Image defogging: a survey. J Image Graphics 16(9):7–22
3. Gan J, Xiao C (2013) Fast image dehazing based on accurate scattering map. J Image Graphics 18(5):583–590
4. He K, Sun J, Tang X (2013) Guided image filtering. IEEE Trans Pattern Anal Mach Intell 35 (6):1397–1409

Chapter 26
Analysis of Recognition Performance of SVMs Based on Three Types of Common Feature Datasets

Linbo Fang, Xiaoshan Song, Shuangyan Hu, Jing Yuan and Bei Wang

Abstract The recognition performance of a support vector machine (SVM) is seriously affected by the feature datasets it trains. SVMs with different type of feature datasets have different recognition performances. Three type of common feature datasets, including gray value dataset, Hu moment dataset, and principal component dataset, are studied by experiments in this paper; results show that SVMs based on the gray value dataset is applicable to the low-resolution image recognition; SVMs based on the Hu moment dataset is applicable to the high-resolution image recognition; SVMs based on the principal components dataset is too time-consuming to the real-time image recognition.

Keywords Support vector machine · Hu moment · Principal component analysis (PCA) · Gray value feature

26.1 Introduce

Before recognition with an SVM, the feature dataset must be constructed, by training which, the decision function model is generated. An SVM will be able to recognize the targets with the decision function model [1]. SVM with different type of feature dataset has different recognition performance. The dimension of some types of feature is so large that makes the SVM decision much time-consuming; the feature extraction of some other types is so time-consuming that affects the speed of the datasets construction and the real-time recognition; the feature extraction of other types may be much simple, and the dimension of the feature vectors is so low that makes a very fast recognition speed, while they may have low recognition accuracies. Three types of feature dataset are studied, which include gray value dataset, Hu moment dataset, and principal component dataset. Gray value of each

---

L. Fang (✉) · X. Song · S. Hu · J. Yuan · B. Wang
Xi'an Research Inst of High-Tech, Xi'an 710025, China
e-mail: fmailmail@163.com

© Springer-Verlag Berlin Heidelberg 2015
S. Long and B.S. Dhillon (eds.), *Proceedings of the 15th International Conference on Man–Machine–Environment System Engineering*, Lecture Notes in Electrical Engineering 356, DOI 10.1007/978-3-662-48224-7_26

pixel of an image is made as the feature value in the gray value dataset, which is the most basic feature of an image. PCA method, as a linear feature extraction method based on K-L transformation, can greatly decrease the dimension of the feature vector while keeping the main information I few new features. Hu moments feature is very classical in image feature extraction field, which is a type of global feature, and is insensitive to rotation, translation, and scale. All the three types are commonly used in the image recognition.

The dataset of an SVM must be constructed in particular format. "Libsvm" dataset format is an efficient format and common used in the SVM field [2]. Data shown in Table 26.1 are parts of the open test dataset "Vehicle" on the UCI website [2]; the data format is the "Libsvm" format. The first number in each line of Table 26.1 means the attribute (identity) of the line related. And the number 1, 2, 3, 4, respectively, indicate 4 types of vehicles including "bus," "van," "saab," "opel." Numbers in each line after the attribute value are features. In the format "*a*: *b*," "*a*" means index of the feature, and "*b*" means the feature value.

In this paper, three types of datasets including gray value, PCA, and Hu moments are constructed in "Libsvm" format, recognition performance of SVMs based on the three types of feature datasets are compared and analyzed.

26.2 Support Vector Machine

Given the training sample set: $\{(x_i, y_i), i = 1, 2, \ldots, 1\}$, in which, $xi \in R_n$, $y_i \in \{1, -1\}$means category, 1 means the total samples of the sample set, so the original optimization problem of SVM is:

$$\begin{aligned} &\min_{w,w_0} \frac{\|\boldsymbol{w}\|^2}{2} + C\sum_{i=1}^{l} \xi_i \\ &s.t.\ y_i\left(\boldsymbol{w}^T\boldsymbol{x}_i + b\right) \geq 1 - \xi_i\ , \xi_i \geq 0,\ i = 1, 2, \ldots, l \end{aligned} \quad . \tag{26.1}$$

In which, w is the normal vector of the SVM decision function model; ξi calls slack variable, is the sign of softening of the SVM classification interval, which means there can be misclassified samples by a hyper plane; $C > 0$ is a self-defined penalty factor, which controls the punishment degree of error classifying samples.

The dual optimization problem is got while the lagrangian multiplier method is introduced:

$$\begin{aligned} &\max\left[\sum_{i=1}^{l} \alpha_i - \frac{1}{2}\sum_{i=1}^{l}\sum_{j=1}^{l} \alpha_i\alpha_j y_i y_j K\left(\boldsymbol{x}_i, \boldsymbol{x}_j\right)\right] \\ &s.t.\ \sum_{i=1}^{l} \alpha_i y_i = 0,\ 0 \leq \alpha_i \leq C,\ i = 1, 2, \ldots,\ l \end{aligned} \quad . \tag{26.2}$$

Table 26.1 Parts of data in "vehicle" dataset

1	1:95	2:48	3:83	4:178	5:72	6:10	7:162	8:42	9:20	10:159	11:176	12:379	13:184	14:70	…
2	1:104	2:50	3:106	4:209	5:66	6:10	7:207	8:32	9:23	10:158	11:223	12:635	13:220	14:73	…
3	1:97	2:43	3:73	4:173	5:65	6:6	7:153	8:42	9:19	10:143	11:176	12:361	13:172	14:66	…
4	1:96	2:55	3:103	4:201	5:65	6:9	7:204	8:32	9:23	10:166	11:227	12:624	13:246	14:74	…

The decision function model can be got by solving the problem (26.2):

$$f(x) = \text{sgn}\left\{ \sum_{x_i \in SV} \alpha_i^* y_i K(\boldsymbol{x}_i, \boldsymbol{x}) + b^* \right\}. \tag{26.3}$$

We can see from (26.3) that the SVM is actually a linear classifier in the nonlinear space [3, 4]. The selection of SVM parameters can be through two steps [5]. First step is the selection of the kernel parameters, which can decide the mapping results to the sample space and the distribution of the feature vectors in the feature space; second step is the selection of the penalty factor C, which influences the misclassifying degree of the SVM decision function to the feature vectors, and the misclassifying degree becomes lower while the value of C is larger.

Before recognizing the target, the feature dataset of the target must be trained, though which generates the decision function. When recognizing the target, the feature vector of the target is substituted into the decision function and got the class value of the target by running the equation [6, 7]. Before training the feature dataset, the optimal model of the SVM must be determined, which means the kernel function and the parameters must be determined. The RBF kernel function is adopted in this paper. There are two parameters in a RBF-SVM which are the kernel parameter γ and the penalty factor C [8]. The parameters are selected by grid search method of 5-fold cross validation.

26.3 Image Feature Extraction

26.3.1 Hu Invariant Moments

Seven values is defined by Hu, which composed by the nonlinear composition of the central moment, which is invariant with position, scale, and direction of the target. The definition of the Hu moments is as follows:

$$\Phi_1 = \eta_{02} + \eta_{20}, \tag{26.4}$$

$$\Phi_2 = (\eta_{02} - \eta_{20})^2 + 4\eta_{11}^2, \tag{26.5}$$

$$\Phi_3 = (\eta_{30} - 3\eta_{12})^2 + (3\eta_{21} - \eta_{03})^2, \tag{26.6}$$

$$\Phi_4 = (\eta_{30} + \eta_{12})^2 + (\eta_{21} + \eta_{03})^2, \tag{26.7}$$

$$\begin{aligned}\Phi_5 =& (\eta_{30} - 3\eta_{12})(\eta_{30} + \eta_{12})\left[(\eta_{30} + \eta_{12})^2 - 3(\eta_{21} - \eta_{03})^2\right] \\ &+ (3\eta_{21} - \eta_{03})(\eta_{21} + \eta_{03})\left[3(\eta_{30} + \eta_{12})^2 - (\eta_{21} + \eta_{03})^2\right],\end{aligned} \tag{26.8}$$

$$\begin{aligned}\Phi_6 =&(\eta_{20}-\eta_{02})\left[(\eta_{30}+\eta_{12})^2-(\eta_{21}+\eta_{03})^2\right] \\ &+4\eta_{11}(\eta_{30}+\eta_{12})(\eta_{21}+\eta_{03})\end{aligned}, \quad (26.9)$$

$$\begin{aligned}\Phi_7 =& (3\eta_{21}-\eta_{03})(\eta_{30}+\eta_{12})\left[(\eta_{30}+\eta_{12})^2-3(\eta_{21}+\eta_{03})^2\right] \\ &+(\eta_{30}-3\eta_{12})(\eta_{21}+\eta_{03})\left[3(\eta_{30}+\eta_{12})^2-(\eta_{21}+\eta_{03})^2\right]\end{aligned}. \quad (26.10)$$

The value of the Hu moments is very small, which reach 10^{-20}, so the base 10 logarithm of the Hu moments is used in this paper.

26.3.2 Principal Component Analysis

Principal Component Analysis, PCA is a multiple sources statistical method which transforms several attribute indexes of the study object to a few irrelevant variables. The briefly description is as follows:

Suppose the input data matrix $\boldsymbol{X}_{n \times l}$ is given, in which l is the count of the vectors of the given matrix and n is the dimension of the vectors. The matrix is composed by the centralized sample data $\{\boldsymbol{x}_i\}_{i=1}^{l}$, in which, $\boldsymbol{x}_i \in R^n$ and $\sum_{i=1}^{n} \boldsymbol{x}_i = 0$. PCA can transform the input data vectors to the novel score vectors S_i using the Eq. (26.11).

$$\boldsymbol{S}_i = U^T \boldsymbol{x}_i \quad (26.11)$$

In which, **U** is a orthogonal matrix with $n \times n$, and the ith column u_i is the ith intrinsic vector of the sample covariance matrix C, the definition of C is

$$C = \frac{1}{l}\sum_{i=1}^{l} \boldsymbol{x}_i \boldsymbol{x}_i^T. \quad (26.12)$$

PCA solves the intrinsic problem shown by (26.13) at first.

$$\lambda_i \boldsymbol{u}_i = \boldsymbol{C}\boldsymbol{u}_i, \quad i = 1, \ldots, n. \quad (26.13)$$

In which, λ_i is a intrinsic value of C, $\boldsymbol{u}_i$ is the corresponding intrinsic vector (n dimension).

When the prior p intrinsic vectors **U** are used, $\boldsymbol{X} = \{\boldsymbol{x}_i\}_{i=1}^{l}$ is orthogonal transformed, as shown in (26.14):

$$\boldsymbol{S} = U^T \boldsymbol{X} \quad (26.14)$$

The novel vectors $\boldsymbol{S}$ ($p * l$) is called Principal vectors. When the prior few intrinsic vectors are merely used, the count of the principal vectors $\boldsymbol{S}$ will be reduced, then the dimension is reduced by PCA and the original data are compressed.

26.3.3 *Gray Value Feature*

Since an image can be described as a matrix, and each element in the matrix has a gray value, so the most direct and effective feature of an image is the gray values. All the gray values compose the feature vector of an image, which can describe all the details of the image [9].

26.4 Comparison of the Recognition Performance of SVMs

Three types of feature datasets including gray value, PCA, and Hu moments of two types of images including low-resolution image and high-resolution image are constructed, respectively. Images of four objects including cup, radio, desk lamp, and remote control are collected by the common camera. Thirty-six images are collected from each object, and the total number of images is 144. (1) Collection of low-resolution image. The distance between the object and the lens is 4 m; the pitch angle is 30°; the horizontal angle interval is 30°. The image of the cup is 60 × 103 pixels, and the other three images are 60 × 60 pixels each. Parts of images are shown as Fig. 26.1. (2) Collection of high-resolution image. The distance between the object and the lens is 0.5 meters; the pitch angle is 45°; the horizontal angle interval is 30°. The image of the cup is 200 × 343 pixels, and the other three images are 200 × 200 pixels each. Parts of images are shown as Fig. 26.2.

Feature datasets are constructed based on gray value, PCA, and Hu moments. The dimensions of gray value feature vectors of the 4 low-resolution images are 6180 (the cup) and 3600 (the other 3 objects), and which of the 4 high-resolution images are 68,600 (the cup) and 40,000 (the other 3 objects). The intrinsic values of the covariance matrix of each image are got in the PCA, which are sorted from large to small order, several largest values are selected to construct the feature vector. The number of the PCA feature dimension is 50 in this paper. Seven Hu moments are calculated to construct the Hu moments feature vector. The former two types of

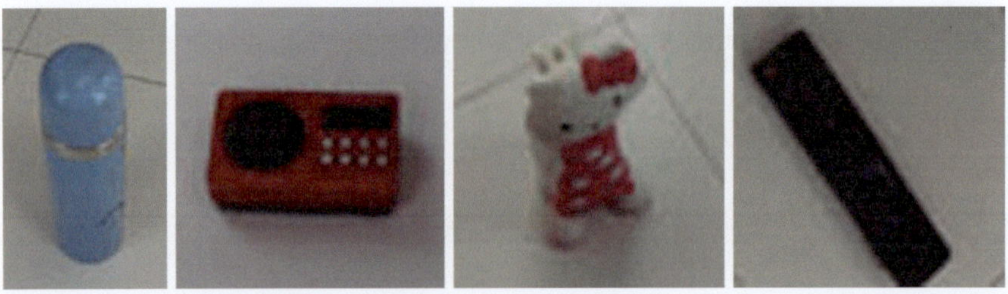

Fig. 26.1 Low-resolution images of 4 types of objects

Fig. 26.2 High-resolution images of 4 types of objects

feature vectors are too large to shown in this paper, so we show the Hu moments feature vectors in Tables 26.2 and 26.3. The "Libsvm" format are used [2], in which, the number 1, 2, 3, and 4 in the first column mean cup, radio, desk lamp, and remote control, respectively.

All the feature datasets are constructed through programming by MATLAB 7.1 in computer with 2.5 GHz CPU, 4000 M ROM. The time spending of construction of the 3 types of feature datasets are shown is Table 26.4, it is seen that the PCA feature dataset construction is most time-consuming, and the construction of Hu moments feature dataset is the fastest.

26.4.1 Comparison of the Recognition Accuracy to the Test Sets

Six test datasets are constructed to test the performance of SVMs: 3 low-resolution image feature sets and 3 high-resolution image feature sets, which contain 2 gray value datasets, 2 PCA feature datasets, and 2 Hu moments datasets. There are 68 feature vectors in each dataset which contains 4 classified images.

Before test, parameters selection and training to the 6 feature datasets must be done to get 6 decision functions. Then tests are made to the 6 test datasets, respectively. The results of the parameters selection and the recognition are shown in Table 26.5.

It can be seen from Table 26.5 that, to low-resolution images datasets, recognition accuracies of the SVMs based on the gray value dataset and the PCA feature dataset are 100.00 %, and recognition accuracies of the SVMs based on the Hu moments dataset are 79.41 %; to high-resolution images datasets, recognition accuracies of the SVMs based on the gray value dataset and the PCA feature dataset are 100.00 %, and recognition accuracies of the SVMs based on the Hu moments dataset are 98.53 %.

Table 26.2 Parts of the Hu feature dataset of the low-resolution images

1	1:2.727712	2:6.156892	3:11.028204	4:11.978735	5:23.525219	6:15.363979	7:23.854924
1	1:2.724618	2:6.153620	3:10.987145	4:11.818606	5:23.331982	6:15.143415	7:23.4210892
2	1:2.669382	2:7.468272	3:10.150558	4:9.844349	5:20.727475	6:14.214761	7:19.845511
2	1:2.688527	2:7.727749	3:9.967976	4:9.827159	5:20.985742	6:13.957358	7:19.725380
3	1:2.894398	2:8.860119	3:11.212009	4:11.176311	5:23.696711	6:15.931281	7:22.370955
3	1:2.900413	2:8.775412	3:11.126163	4:10.840807	5:22.170334	6:15.488955	7:21.873617
4	1:2.714436	2:6.606674	3:9.704559	4:10.413770	5:20.473812	6:13.806837	7:21.670202
4	1:2.690708	2:6.858658	3:10.427272	4:10.480518	5:20.949168	6:13.913519	7:21.525661

Table 26.3 Parts of the Hu feature dataset of the high-resolution images

1	1:2.726889	2:6.154043	3:11.024136	4:11.967784	5:23.509883	6:15.346867	7:23.822748
1	1:2.723469	2:6.150138	3:10.980186	4:11.839222	5:23.345621	6:15.178145	7:23.471157
2	1:2.668370	2:7.466069	3:10.146496	4:9.838472	5:20.690832	6:14.203097	7:19.835136
2	1:2.687584	2:7.715576	3:9.962536	4:9.821649	5:20.983454	6:13.944426	7:19.714370
3	1:2.899838	2:8.770728	3:11.127016	4:10.835524	5:22.161174	6:15.480622	7:21.866545
3	1:2.878240	2:8.610063	3:11.362029	4:10.810570	5:21.899506	6:15.203472	7:22.856028
4	1:2.699748	2:6.873056	3:11.136234	4:10.487159	5:21.482322	6:14.029132	7:21.420766
4	1:2.709123	2:6.717083	3:9.988203	4:9.758043	5:19.642383	6:13.123467	7:20.280184

Table 26.4 The time spending of 3 types of datasets constructions

	Gray value dataset	PCA dataset	Hu moments dataset
Low-solution images (s)	1.601	1079.851	0.187
High-solution images (s)	14.906	>9000.765	0.203

Table 26.5 Optimal parameters and recognition results

		γ	C	Recognition accuracies (%)
Datasets of low-resolution images	Gray value	0.587	0.236	100.00
	PCA	0.209	0.500	100.00
	Hu moments	125.333	7.780	79.4
Datasets of high-resolution images	Gray value	0.624	0.034	100.00
	PCA	2.700	0.342	100.00
	Hu moments	56.432	0.750	98.53

26.4.2 *Time Spending of Recognition to One Target*

There are 3 steps in a practical recognition: (1) collection of the target image by sensors; (2) pre-processing and feature extraction to the image; and (3) calculating the classification by substituting the feature vector to the decision function. So the time spending of recognizing a target includes the feature extraction time and the decision time. Time spending results of SVMs based on the 6 datasets to recognize one target are shown in Table 26.6.

It is seen from Table 26.6 that the SVM recognition based on the PCA feature dataset is most time-consuming, and the SVM recognition based on the Hu moments feature dataset is the fastest. To the low-resolution images, the SVM recognition based on the Hu moments feature dataset is most real time, but which has the lowest recognition accuracy; the recognition performance of SVMs based on the gray value feature dataset is very well, and the real-time performance can meet the need in practice, so it is applicable to the low-resolution image recognition. To the high-resolution images, the SVM recognition based on the PCA feature dataset is most time-consuming, and the SVM recognition based on the gray value

Table 26.6 Time spending of SVMs based on the 6 datasets to recognize one target

		Gray value	PCA	Hu moments
Time spending (s)	Datasets of low-resolution images	0.043	10.067	0.012
	Datasets of high-resolution images	1.512	>100	0.021

feature dataset is also time-consuming. So the two types of SVMs do not satisfy real-time requirements. And the SVM recognition based on the Hu moments feature dataset is most real time, which can satisfy real-time requirements, and the recognition performance can meet the need in practice, so it is applicable to the high-resolution image recognition.

26.5 Conclusion

Support vector machine is a classification machine with intelligent learning performance, which is a novel research focus following the neural networks. Recognition performances of SVMs based on 3 types of feature datasets are studied. It is concluded from the results of the examinations in this paper that: (1) recognition accuracy of the SVM based on the gray value dataset is the highest, but the real-time performance becomes low with the increase of the image pixels, so it is applicable to the low-resolution image recognition; (2) recognition accuracy of the SVM based on the PCA feature dataset is also high, but the feature extraction speed is too low to reduce the real-time performance; (3) recognition accuracy of the SVM based on the Hu moments feature dataset is fairly lower to the low-resolution image recognition, but which is high enough to the high-resolution image recognition, and the real-time performance is much high, so it is applicable to the high-resolution image recognition.

References

1. Dai K, Yu H, Ma X, Li Q (2014) Feature selection algorithms based on support vector machine. J Inf Eng Univ 15(1):85–90
2. Blake CL, Merz CJ (2003) UCI repository of machine learning databases [EB/OL]. http://www.ics.uci.edu/~mlearn/MLRepository.html
3. Hsu CW, Chang CC, Lin CJ (2003) A practical guide to support vector classification [OL]. http://www.csie.ntu.edu.tw/cjlin/papers/guide/guide.pdf
4. Vapnik V (1995) The nature of statistics learning theory. Springer, New York, p 1995
5. Jianchen Wang, Ganlin Shan, Qilong Zhang, Xiusheng Duan (2011) Research on feature selection method based on improved SVM-RFE. Microcomput Appl 32(2):70–74
6. Chapelle, Vapnik VN (2002) Choosing multiple parameters for support vector machines. Mach Learn 46(1–3):131–159
7. Joachims T (2000) Estimating the generalization performance of a SVM efficiently. University Dortmund, Dortmund
8. Wu K-P, Wang S-D (2009) Choosing the Kernel parameters for support vector ma-chines by the inter-cluster distance in the feature space. Pattern Recogn 42(5):710–717
9. Du J, Sun J (2013) Image classification based on feature weighting and support vector machine. Laser Infrared 43(3):315–318

Chapter 27
Study on the Aerial Target's Threat Degree Ordering Model Based on BP Neural Network

Songli Zhang, Yanfei Kong, Xiaohui Peng and Xiaogao Wang

Abstract This paper introduces the significance of air-raiding and anti-air-raiding for modern war and ordering of air attack targets to air defense, analyzes the influence index of aerial target's threat degree, and determines the normalized treatment of these influence indexes. The paper brings in BP neural network theory and illustrates its algorithm steps and establishes the model of BP neural network, training using the constructed network to sort the threat degree of aerial targets. Result shows that the established BP neural network model has a strong nonlinear mapping ability and self-learning ability, it provides a reference for the air defense unit to evaluate aerial target's threat degree, so could help commanders to make better decisions.

Keywords BP neural network · Threat degree · Air attack target · Air defense combat

27.1 Introduction

The modern war in recent decades shows that air-raiding and anti-air-raiding not only bear the brunt, but also throughout the global effect. Especially since the 90s in last century, air defense combat began to emerge as independent combat style, and air attack with anti-air attack has become the main combat style in modern war under the condition of informatization. How to carry out the operation decision quickly, have already become the key to play effectiveness of air defense weapon and to evaluate the threat degree of air raid target can help commanders to better

S. Zhang · Y. Kong (✉)
National University of Defence Technology, Changsha 410073, China
e-mail: kongyanfei09@sina.com

X. Peng · X. Wang
Army 73151, Xiamen 361000, China

© Springer-Verlag Berlin Heidelberg 2015

S. Long and B.S. Dhillon (eds.), *Proceedings of the 15th International Conference on Man–Machine–Environment System Engineering*, Lecture Notes in Electrical Engineering 356, DOI 10.1007/978-3-662-48224-7_27

target selection, rational distribution of fire, and maximize the effectiveness of weapon equipment.

There are many factors that affect the threat degree and not simple linear combination within factors. The traditional threat estimation often make a comprehensive with constant weight vector, which cannot solve the problem of difference between factors, the calculated results have a great different with the actual situation. The neural network model proposed in this paper, with well adaptive and self-learning ability, highly linear and nonlinear mapping ability, can be very good to mapping the complex nonlinear relation between the indexes, and learn to self correct.

27.2 Target Threat Degree Evaluation Index Systems

Threat refers to the enemy air raid weapons to defend objects and air defense weapons to threat degree. In air defense battle, choosing shooting target with biggest threat degree, can utmost defend the safety of us ground target.

Aerial targets' threat degree is decided by many factors, according to the actual combat, the influence index has roughly: target type, the height of the target, the target speed, shortcut, target distance, flying direction etc. If all the indicators are taken into account, the results may be more accurate, but is bound to generate data explosion, and make the problem complex. In a rapidly changing air defense fight, must race against time and decisive decision-making to win the battle initiative. So in order to balance, ignore some index, use the following five indicators to describe the target threat degree [1], as shown in Fig. 27.1:

(1) The target type, different target types, its use and target feature of radar detect and attack way are different, the threat degree is different. The target type will be divided into the following categories: small and medium-sized aircraft, large aircraft, and cruise missile.
(2) The target speed, for the same type of goal, with greater speed, penetration probability is greater, the greater threatening to us air defense forces.
(3) The target distance, the smaller distance between the air attack target and surrounded target, the time is more urgent for us to defense, and the bigger threat to the secured target.
(4) The route shortcut, shortcut of air target is refers to distance between the gun and target route level projection. The smaller of air target's shortcut, its attack intention is more obvious, and threat degree is greater.
(5) The target height, reducing fly height can make the probability of air raid weapons been found significantly reduces. Therefore, the height of the target is a very important factor to judge the target threat, it is usually divided into: high altitude, hollow, low altitude, and super low altitude.

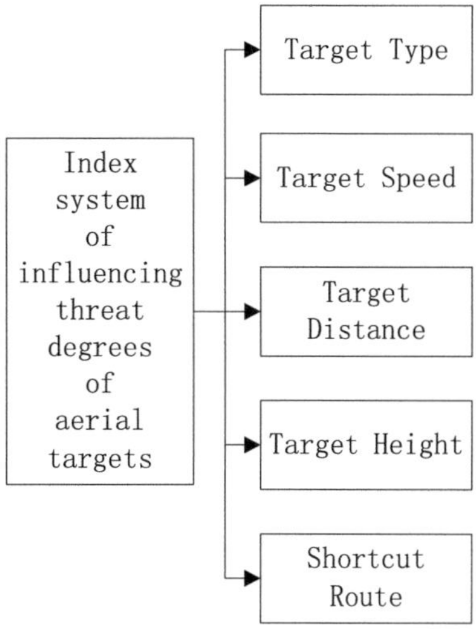

Fig. 27.1 Air attack targets threat degree index system

27.3 Model of Threat Degree for Judgment Based on the Neural Network

27.3.1 BP Neural Network Theory

Artificial neural network is nonlinear complex network system, widely connected by a large number of simple processing units, and has the ability of nonlinear function approximation, the very strong fault-tolerant, and learning ability.

Among them, the BP network is composed of input layer, middle layer (hidden layer), and output layer. The hidden layer may have several layers, each layer of neurons only accept the up neurons, so as to realize the nonlinear mapping from the input layer nodes to the output layer. The connection structure and weights can be obtained by learning. The information, neural network memorized, stored in connection weights between neurons, and from individual rights cannot see its stored information content, which is distributed storage mode [2].

Figure 27.2 is a typical sketch of BP neural network model. Its essence is built mapping from input layer parameters to the output layer parameters, which reflects the input parameters effect on the output of the form and amplitude, and reflects the essential relation between input parameters and the output results.

27.3.2 The Step of Algorithm

The BP model is a multi-layer perception mechanism. BP model with one intermediate layer, as shown in Fig. 27.3, which has n input nodes and k output nodes; their nonlinear activation function is sigmoid function:

Fig. 27.2 BP neural network structure

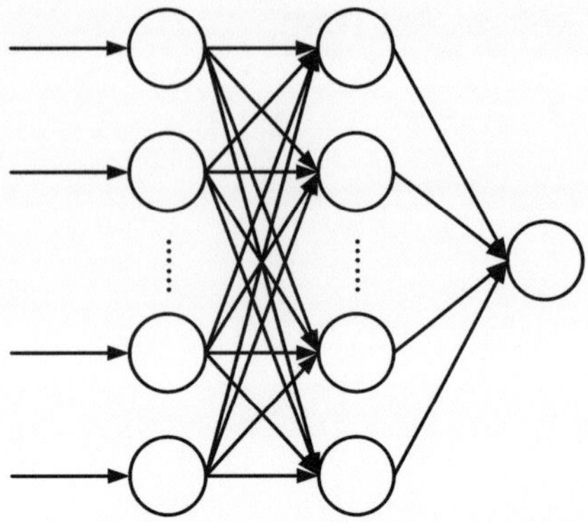

Fig. 27.3 BP neural network model

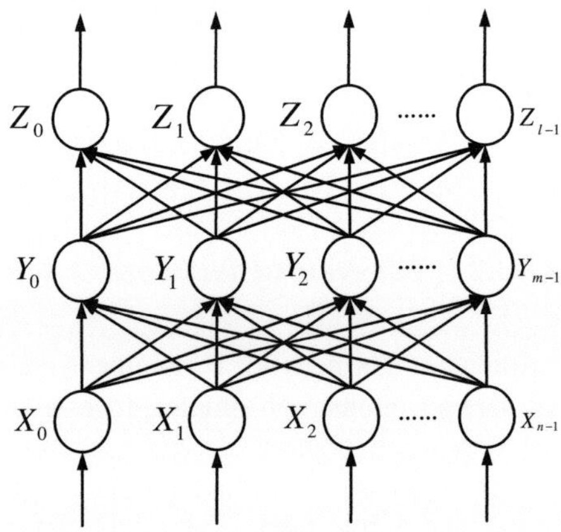

$$f(u) = \frac{1}{1 + e^{-u}} \tag{27.1}$$

Output of multi-layer calculating unit calculates according to Eqs. (27.2) and (27.3):

$$Y_j = f\left(\sum_{i=0}^{n-1} W_{ij}X_i - \theta_j\right) \quad j = 0, 1, L, m-1. \tag{27.2}$$

$$Z_k = f\left(\sum_{j=0}^{m-1} W_{jk}Y_j - \theta_k\right) \quad k = 0, 1, L, l-1. \tag{27.3}$$

In the formula: X_i is the ith node's output value of input layer; Y_j is the jth node's output value of the middle layer; W_{ij} is the weight coefficient from X_i to Y_j; W_{jk} is the weight coefficient between the jth node of hidden layer to the kth node of output layer; θ_j is the internal threshold of middle layer jth node; θ_k is the internal threshold of output layer's kth node; Z_k is the actual output value of the output layer's kth node.

Self-learning process of BP neural network is an iterative process. It is concrete algorithm as follows:

(1) Assigned a group of small random initial weights to the network, its value from 0 to 1, and make them inequal each other.
(2) Normalized the input data from 0 to 1, and determine the expected output signal (d_0, d_1, L, d_{l-1}).
(3) Computed the actual output value of neural network layer by layer.

$$f\left(\sum_{i=0}^{n-1} W_{ij}X_i\right) \rightarrow Y_j, f\left(\sum_{j=0}^{m-1} W_{jk}Y_j\right) \rightarrow Z_k(j,k=0,1,L,m-1). \quad (27.4)$$

(4) Started from the output layer, reverse adjustment weights, the adjustment formula is as follows:

$$\begin{array}{l} W_{jk} + \eta\,\delta_k Y_j \rightarrow W_{jk} \\ W_{ij} + \eta\,\delta_j Y_i \rightarrow W_{ij} \end{array}. \quad (27.5)$$

In the formula: $\delta_k = (d_k - Z_k)Z_k(1 - Z_k)$, $\delta_j = Y_j(1 - Y_j) \cdot \sum_{k=0}^{l-1} \delta_k \cdot W_{jk}$.
(5) Calculated the total error E, if $E \leq \varepsilon$, learn to stop, otherwise, go to (27.3) and recalculate:

$$E = 1/2\sum_{k=0}^{l-1}(d_k - Z_k)^2. \quad (27.6)$$

In the actual programming, if the step size is smaller, the learning speed is slower, but if the step size too much will cause the network appears to swing. In order to solve this problem, adding a momentum $\alpha(0<\alpha<1)$ to the formula (27.4), i.e.,

$$\begin{array}{l} W_{jk} + \eta\delta_k Y_j + \alpha \times \Delta W_{jk} \rightarrow W_{jk} \\ W_{ij} + \eta\delta_j Y_i + \alpha \times \Delta W_{ij} \rightarrow W_{ij} \end{array}. \quad (27.7)$$

In the formula: ΔW_{jk} is the difference between the two consecutive W_{jk}; ΔW_{ij} is the difference between the two consecutive W_{ij}.

27.4 The Example Analysis

27.4.1 *Quantitative Indicators*

Only the qualitative and quantitative indicators were quantified, can make the model easy to calculate and operate. Here take the idea of decomposition, the effect of every aspect of the index on target threat degree considered separately. And in order to eliminate the influence between the index class, usually made indicators normalized processing, here make the threat attribute values normalized to 0–1 [3].

(1) The target type k, according to the small aircraft (S), large aircraft (L), and cruise missiles (C) are quantified in 0.2, 0.4, and 0.6.
(2) The target speed v, is benefit type index, according to 0–1800 m/s for 0–0.9 of the corresponding quantization.

$$v = \frac{0.9v}{1800} \tag{27.8}$$

(3) The target distance s, is the cost indexes, according to 0–50 km for 1–0.1 of the corresponding quantization.

$$s = 1 - \frac{0.9s}{50}. \tag{27.9}$$

(4) The route shortcut j, belongs to the cost indexes, according to 0–5000 m for 1–0.1 of the corresponding quantization.

$$j = 1 - \frac{0.9j}{5000}. \tag{27.10}$$

(5) The height of the target h, accordance with the high altitude (H), middle altitude (M), low altitude (L), minimum altitude (MM), quantified in 0.2, 0.4, 0.6, and 0.8.

27.4.2 *Sample Collection*

Give 10 aerial targets, collecting every goal's corresponding values of the five indices, expectations of the target threat degree value given by the expert [4, 5] (Table 27.1).

Table 27.1 Sample data

Target number	Target type	Target speed (m/s)	Target distance (m)	Target height	Course short (m)	Expected degree
001	S	1000	500	H	100	0.60
002	L	500	600	M	800	0.38
003	C	800	200	L	2000	0.55
004	S	600	100	MM	300	0.50
005	S	1200	400	MM	800	0.45
006	C	800	800	M	1000	0.25
007	L	500	200	H	50	0.37
008	S	700	900	MM	800	0.40
009	L	800	1000	H	400	0.34
010	S	1000	600	L	100	0.80

27.4.3 Sample Training

The properties of 5 threat value as the input of neural network, so the first layer (input layer) with 5 nodes, the second layer (hidden layer) of 8 nodes, the third layer as output layer with 1 node, and characteristics output function with S type function, thus the output value is the threat degree of air attack target for the secured target.

Using a sample of 10, which is divided into two parts, one part is used for training, and the other part is used for validation. Target threat degree is given by experts, the values of some parameters: $\eta = 0.5$, $\alpha = 0.05$, $\varepsilon = 0.001$. According to the designed neural network model, use MATLAB programming for sample training [2, 6].

The former 7 groups are for training and validation with 8–10 group, the result of training samples as follows (Table 27.2):

According to the network weight of BP neural network training, use data of 8 h to 10 groups to validate the results. The verification results and error as follows (Table 27.3):

As shown in Fig. 27.4, the error between target threat degree value and the expected value is very small which is calculated through the neural network, and meet the requirements. It is thus clear that neural network has very strong nonlinear mapping ability and can be used for target threat assessment.

Table 27.2 BP neural network training results

Target number	Expected threat	Calculated threat
001	0.60	0.600856768
002	0.38	0.379054068
003	0.55	0.550802327
004	0.50	0.500204917
005	0.45	0.449727447
006	0.25	0.250326540
007	0.37	0.370748339

Table 27.3 Verification of BP neural network results

Target number	Expected threat	Calculated threat	Error
008	0.40	0.399997669	2.331E-06
009	0.34	0.340155334	−0.000155334
010	0.80	0.799651304	0.000348696

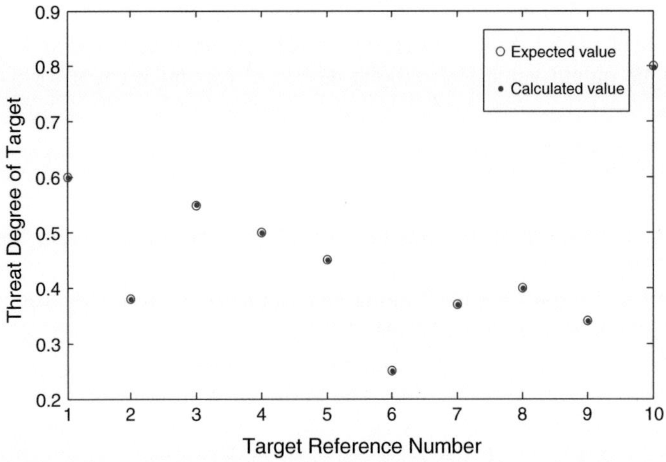

Fig. 27.4 Error between the calculated and expected threat degree

27.5 Conclusions

The informatization battle field varies from minute to minute, and opportunity is transient, only the correct judgment of enemy air raid targets threat degree for the target us air defense forces defend my, will grasp the key, solve the major contradictions, and made of victory of antiaircraft battle field. Threat sequencing model of air attack target provides reference for the air defense commander to correct selecting air raid target at a certain extent, shorten the command decision time, and make the best of the air defense weapon.

References

1. Peng J, Wang Z (2012) Study on the aerial targets threat ordering model. In: Ship electronic engineering, 9th
2. Zhang L (1992) The model of artificial neural network and its application. Fudan University Press, Shanghai
3. Wu W, Wang X (2009) Research for city of ground air defense planning. Military Science Press, Beijing, p 05
4. Li J, Yang G (2012) Ground target threat degree analysis based on neural network. Automation 03
5. Wang J, Lou S et al (2004) The neural network model of air attack targets threat evaluation. J Test Meas Technol 18
6. MATLAB application program neural network toolbox user's guide version 3.0. Math Work, 1998

Chapter 28
Research on Availability Simulation of Antiaircraft Gun Weapon Systems

Hao Liu, Hui Gu, Haimin Hu and Quanliang Yin

Abstract A simulation method that uses discrete event simulation for availability evaluation of antiaircraft gun weapon systems is presented based on the distribution characteristic of the RMS time of antiaircraft gun weapon systems. The simulation method can overcome the disadvantages of the analytic method, and can evaluate the availability of complex weapon systems, while RMS time follows any distribution. In the example, the availability of the antiaircraft gun weapon system is evaluated based on the simulation method presented in this paper. The example illuminated that the method is effective for availability evaluation of antiaircraft gun weapon systems.

Keywords Antiaircraft gun weapon system · Availability evaluation · Simulation

28.1 Introduction

A number of researches have evaluated the availability of weapon system based on Markov model. However, a review of these researches shows that the Markov model has a serious disadvantage, which is it needs the times of systems' lives; repair and management must follow exponential distribution and must be independent of each other. Markov model is an outstanding method when the evaluated system is a simple system, whereas antiaircraft gun (AA gun) weapon systems are composed of some subsystems which have other subsystems. Generally, the times of systems' lives, repair, and management do not all follow exponential distribution, and the failure of some subsystems is always dependent on each other. In this instance, the Markov model will not be fit to evaluate the availability of antiaircraft gun weapon systems.

H. Liu (✉) · H. Gu · H. Hu · Q. Yin
Air Defence Forces Academy, Zhengzhou 450052, China
e-mail: hmhu@163.com

© Springer-Verlag Berlin Heidelberg 2015

S. Long and B.S. Dhillon (eds.), *Proceedings of the 15th International Conference on Man–Machine–Environment System Engineering*, Lecture Notes in Electrical Engineering 356, DOI 10.1007/978-3-662-48224-7_28

Maintaining practice shows that preventive maintenance time usually follows normal distribution, corrective maintenance time usually follows logarithmic normal distribution, down time and logistic delay time usually follows exponential distribution, and administrative delay time usually follows Weibull distribution [1, 2]. Some papers point out that, when evaluating the availability of such systems, only simulation techniques can be useful to obtain credible predictions for availability parameters [3, 4]. The reason is that simulation methods can model the working behavior of system in great detail if it is thought necessary [5].

In view of this, a discrete event simulation method will be used to evaluate the availability of AA gun weapon systems.

28.2 Reliability Diagram of a Representative AA Gun Weapon System

AA gun weapon systems must perform searching and tracking aerial targets such as aircraft, helicopter and cruise missile, and launching shells to annihilate these aerial targets. A representative AA gun weapon system usually include one set of radar (i.e., a fire control system), n pieces of AA guns, etc. The reliability diagram of a representative AA gun weapon system is shown in Fig. 28.1.

28.3 Availability Evaluation Model

28.3.1 Maintenance Policy

The maintenance time and delay time are given based on the maintenance policy. There are three common uses of maintenance polices studied in this paper [6].

Fig. 28.1 The reliability diagram of a representative AA gun weapon system

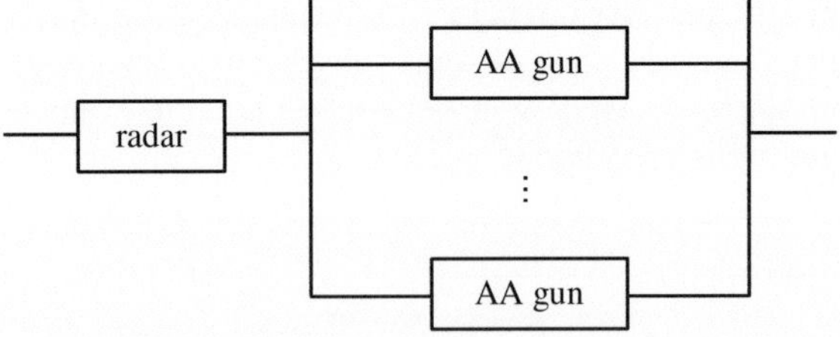

28.3.1.1 Policy I

Once the subsystem fails, it can be repaired by the appointed maintenance resource. There is a stated max-repair-time $T_{\max}$. If the real-repair-time exceeds $T_{\max}$, then the failed subsystem will be replaced by a new one, and a corresponding logistics delay-time will be needed.

28.3.1.2 Policy II

Once the subsystem fails, it can be repaired by the appointed maintenance resource. There is a stated max-down-time $T_{\max}$. If the real-down-time exceeds $T_{\max}$, the failed subsystem will be replaced by a new one, and a corresponding logistics delay-time will be needed. Once the AA gun weapon system can perform its scheduled function, the weapon system goes into operation.

28.3.1.3 Policy III

Maintenance action does not begin unless the weapon system is laid off. The weapon system shall halt when there are not enough subsystems to keep the weapon system working to perform its scheduled function, and all of the failed subsystems will be repaired. There is a stated max-repair-time $T_{\max}$. If the real-repair-time exceeds $T_{\max}$, then the failed subsystem will be replaced by a new one, and a corresponding logistics delay-time will be needed. The weapon system does not go into operation unless all failed subsystems are recovered.

28.3.2 Simulation Fundamentals

According to the real working process of weapon system, the first failed subsystem will be picked up. In other words, the weapon system fails when the weapon system performs its function for t_{Ri} which is the up-time of the ith subsystem where the up-time of this subsystem is the shortest. And the new up-time of other subsystems equals the result that the t_{Ri} was subtracted from the primary up-time. In the same way, the simulation should stop at the time of T which equals the sum of the up-time and down-time of weapon system. According to the probability theory, the up-time t_{Ri}, maintenance time t_{mi}, and delay time t_{di} of each subsystem can be known from the mean time between failures $(MTBF_i)$, mean time to repair $(MTTR_i)$, mean logistics delay time $(MLDT_i)$ of each subsystem. So, the availability of weapon system can be obtained as

$$A = \frac{T_{\mathrm{up}}}{T} \tag{28.1}$$

where A is the availability of the weapon system, T_{up} is the total up-time of weapon system, and T is the mission period of weapon system.

28.3.3 Basic Hypotheses

The assumptions in this paper are as follows:

(1) There is no correlativity between the failures of each subsystem. In other words, when one subsystem is failed, the failure rate of other subsystems will not change.
(2) The weapon system does not halt until there are not enough subsystems to keep the weapon system working for performing its scheduled function.
(3) Each subsystem has n spare parts for replacement. If there is no spare part to replace when a failed subsystem needs to be replaced, the subsystem will halt in the latter task duration.
(4) The system has only two states, working or halting, and each recovered subsystem is as good as new.
(5) The t_{di} of one subsystem is zero if it was recovered without spare part.

28.3.4 Three Times and Finding

The up-time t_{Ri}, maintenance time t_{mi}, and delay time t_{di} of each subsystem can be known from the $MTBF_i$, $MTTR_i$, $MLDT_i$ of each subsystem according to their distribution.

Through checking computations, the repair time of most electromechanical, electronic, and mechanical equipment follow logarithmic normal distribution, so, the logarithmic normal distribution is as the repair time distribution of the general products. Therefore, we assume the up-time t_{Ri} and delay time t_{di} of each subsystem to follow exponential distribution, and the maintenance time t_{mi} of each subsystem to follow logarithmic normal distribution. The three times of each subsystem can be obtained as

$$t_{Ri} = -MTBF_i \ \ln(r) \tag{28.2}$$

$$t_{di} = -MTTR_i \ \ln(r) \tag{28.3}$$

$$t_{mi} = e^{\mu+\sigma\sqrt{-2\ln r_1}\cos(2\pi r_2)} \quad (28.4)$$

$$\mu = \ln\ MTTR_i - 0.5s^2; \quad (28.5)$$

where r, r_1, r_2 are the variables which follow the uniform distribution on the unit interval, $\sigma = 0.5–1.3$.

28.3.5 Steps of Simulation

The steps involved in the working process simulation of weapon systems are as follows:

(1) Start simulation variables (that is to say: simulation clock, each subsystem status, mission period T, max-down-time or max-repair-time $T_{\max}$).
(2) To obtain up-time t_{Ri} maintenance time t_{mi} and delay time t_{di} of each subsystem by random sample according to their distribution.
(3) The up-time of weapon system is the minimum value of all subsystems.
(4) The failure appears with simulation clock and the up-time of weapon system increases. The subsystem which has minimum up-time failed when the simulation clock went to its up-time.
(5) If the failed subsystem is gun, switch to Step 9, else, switch to Step 6.
(6) If maintenance time t_{mi} does not exceed allowable $T_{\max}$, then $T_{\text{down}} = T_{\text{down}} + t_{mi}$, else there is a delay time, so, $T_{\text{down}} = T_{\text{down}} + t_{mi} + t_{di}$.
(7) Simulation clock goes to $T_s = T_{\text{up}} + T_{\text{down}}$.
(8) The weapon system goes to work again. The t_{Ri} of failed subsystem was subtracted from the up-time of other subsystems as the remaining up-time. Obtain up-time t_{Ri} maintenance time t_{mi} and delay time t_{di} of the recovered subsystem by random sample again.
(9) If the amount of failed gun does not exceed the threshold value, then switch to Step 11, else, switch to Step 10.
(10) To obtain the down-time of weapon system according to maintain policy, and to repair the failed guns. At the same time, $T_{\text{down}} = T_{\text{down}} + t_{mi}$ and $T_s = T_{\text{up}} + T_{\text{down}}$. When T_s does not less than T, switch to Step 12, else, switch to Step 8.
(11) To deal with the failed subsystem according to maintain policy.
(12) To obtain the availability of weapon system based on the total up-time and total down-time of weapon system.

Table 28.1 RMS attributes of the weapon systems

Attributes	Fire control system	AA Gun
MTBF (*h*)	80	90
MTTR (*h*)	0.4	0.3
MLDT (*h*)	0.5	0.5

28.4 Example and Analysis

28.4.1 Example

An AA gun weapon system includes a fire control system and six pieces of AA gun. The failure time and delay time of each subsystem follow exponential distribution, and their maintenance time follows lognormal distribution. The R&M&S parameters of each subsystem are shown in Table 28.1.

The allowable repair time T_{max} is assumed at 0.4 h, and there are enough spare parts for the maintenance of each subsystem.

28.4.2 Results and Analysis of Availability Simulation

The availability of the AA gun weapon system was obtained based on the simulation model presented in this paper, and is shown in Table 28.2.

From the results in Table 28.2, we know that there was no difference in the availability of weapon system between Policy I and Policy II; the availability value obtained in Policies I and II were greater than that obtained in Policy III when the mission period T exceeded 50 h, and the difference was greater with the mission period T prolonging. The results in Table 28.2 show that Policy III will induce a low availability of weapon system than Policies I and II, so, Policy III should not be the maintenance policy for this AA gun weapon system. Otherwise, we can know that the availability of this AA gun weapon system tends to decrease with increase in mission period T with all three maintenance policies, so the mission period T should be shortened to keep great availability.

Table 28.2 Availability of AA gun weapon system

T (*h*)	10	30	50	100	150	200	250	300	400	500
I	0.9970	0.9967	0.9964	0.9956	0.9952	0.9947	0.9945	0.9942	0.9939	0.9937
II	0.9970	0.9967	0.9964	0.9956	0.9952	0.9947	0.9945	0.9942	0.9939	0.9937
III	0.9970	0.9967	0.9963	0.9953	0.9944	0.9936	0.9930	0.9925	0.9918	0.9915

28.5 Conclusions

According to the working process and the characteristics of R&M&S time of AA gun weapon system, to overcome the disadvantage of Markov model, a simulation method that can fit any distribution of R&M&S time for availability evaluation of AA gun weapon systems has been presented in this paper. The simulation evaluation method can model the real working process of the subsystems of AA gun weapon systems, and can find the availability value by collecting the up-time and down-time of subsystems. This method can substitute the analytic method such as Markov model to evaluate the availability of complex weapon systems such as AA gun weapon systems.

References

1. Gang X, Tiantuo L (2003) Mento-Carlo method of system reliability analysis. Science Press, Beijing
2. Malwane MAA, Jinadasa G (2004) On steady state availability of a system with lognormal repair time. Appl Math Comput (150):409–416
3. Faulin J, Juan AA, Serrat C, Bargueño V (2005) SAEDES ++: determining complex system availability via simulation. In: Proceedings of the 2005 winter simulation conference, pp 1463–1472
4. Faulin J, Juan AA, Serrat C, Bargueño V (2008) Predicting availability functions in time-dependent complex systems with SAEDES simulation algorithms. Reliab Eng Syst Saf 93:1761–1771
5. Sha Jichang (1998) The most familiar complex system: the unrepairable system. Ref Nat Defence Sci Technol 19(3):1–2
6. Kingsley EF (1972) A General monte carlo simulation model for estimating large-scale system reliability and availability. Clemson University, Clemson

Chapter 29
The Strength Study of Thermal Power Unit Absorption Tower Desulphurization Pump Impeller Based on Fluid-Structure Interaction Calculation

Zhenjun Gao, Jianrui Liu, Chenxu Guo, Wei Fu and Xiaoke He

Abstract The study of impeller strength for thermal power unit absorption tower desulphurization pump with new material was performed based on fluid-structure interaction calculation. The numerical simulation results of flow field in thermal power unit absorption tower desulphurization pump under 0.8 Q operating point, 1.0 and 1.2 Q were imported to ANSYS Workbench, and the structure field of thermal power unit absorption tower desulphurization pump was studied to find the stress distribution law of impeller. In the end, the strength of impeller was checked. The results show that the stress distribution of main blades, back blades, and impeller–shroud is axial symmetrical, and the maximum equivalent stress is located in the junction between the impeller–shroud and main blade outlet, which means that this area has phenomena of stress focus. This paper was performed to provide reference and guidance for the design of thermal power unit absorption tower desulphurization pump.

Keywords Thermal power unit absorption tower desulphurization pump · Fluid–solid coupling theory · Impeller strength · Numerical study

29.1 Introduction

Thermal power unit absorption tower desulphurization pump is the key equipment of wet FGD. The main medium of thermal power unit absorption tower desulphurization pump is strong corrosive limestone-gypsum slurry, which make higher

Z. Gao (✉)
Binzhou Key Laboratory of Environmental Engineering and Technology, Binzhou 256600, China
e-mail: 570186276@qq.com

Z. Gao · J. Liu · C. Guo · W. Fu · X. He
Research Center of Fluid Machinery Engineering and Technology, Jiangsu University, Zhenjiang 212013, China

© Springer-Verlag Berlin Heidelberg 2015

S. Long and B.S. Dhillon (eds.), *Proceedings of the 15th International Conference on Man–Machine–Environment System Engineering*, Lecture Notes in Electrical Engineering 356, DOI 10.1007/978-3-662-48224-7_29

request for services life, operation efficiency, safety reliability, and maintenance [1]. The impeller is the key part of thermal power unit absorption tower desulphurization pump, which designs the hydraulic performance and reliability. The study of impeller strength can get the location and distribution law of stress–strain, which has important meanings for increasing the service life and safety reliability of thermal power unit absorption tower desulphurization pump [2, 3].

29.2 Fundamental Performance Parameters

Fundamental performance parameters of thermal power unit absorption tower desulphurization pump are shown as follows: flow rate $Q = 2440\ m^3/h$, head $H = 40$ m, rotational speed $n = 990$ r/min, efficiency $\eta = 81.5$ %, and match Power $P = 450$ kW.

29.3 Three-Dimensional Model and Grid Generation

The fluid-structure interaction calculation of thermal power unit absorption tower desulphurization pump involves flow field and structure field [4]. The flow field is the internal flow field of thermal power unit absorption tower desulphurization pump including volute, impeller, inlet water body, and outlet water body; the structure field is the impeller. The model of flow field is shown in Fig. 29.1. The

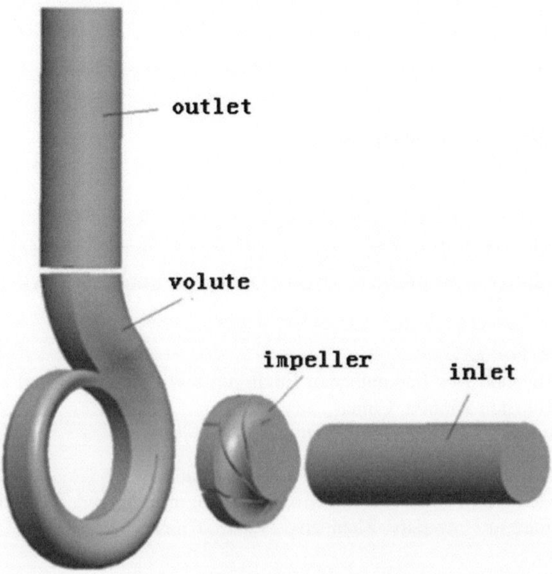

Fig. 29.1 The model of flow field

model of impeller is shown in Fig. 29.2. The impeller material is M26–23 V alloy steel.

The solution region of numerical calculation is the whole flow field. The unstructured tetrahedral grids were used in the numerical simulation, and the tiny structure parts were automatically encrypted [5, 6]. The grid of whole flow field is shown in Fig. 29.3. Automatic mesh generation of impeller in ANSYS software was performed in the model of impeller directly. The grids of impeller are shown in Fig. 29.4. The two solution regions should keep stable in their own equation solver to keep the corresponding relationship between the fluid interface and its contiguous solid interface [7]. The grid number of flow field is 1928609, and the impeller is 96452 with the grid-independent examination.

Fig. 29.2 The model of impeller

Fig. 29.3 The grid of whole flow field

Fig. 29.4 The grid of impeller

29.4 The Numerical Simulation of Flow Field

29.4.1 Numerical Methods of Flow Field

The velocity inlet and free flow outlet were adopted in calculation with SST k-ω turbulence model, no-slip boundary condition was for solid boundary, and the wall function method was adopted for dealing with the boundary condition at near wall region. The standard SIMPLE algorithm was used in pressure–velocity coupling, and the convergence precision was set as 10^{-4}.

29.4.2 The Reliability Research of Numerical Simulation

The hydraulic performance test of thermal power unit absorption tower desulphurization pump was done according to the national water pump test standard GB/T3216-2005 to verify the reliability of numerical simulation. The numerical simulation results and the test results of thermal power unit absorption tower desulphurization pump are shown in Figs. 29.5 and 29.6. As we can see from Figs. 29.5 and 29.6, the maximum error of pump head between calculation and experiment is 4.3 %, and the maximum error of pump efficiency between calculation and experiment is 3.5 %, which means that calculation results are in good agreement with experimental ones, so the calculation results is reliable, which can provide data basis for the calculation of structure field.

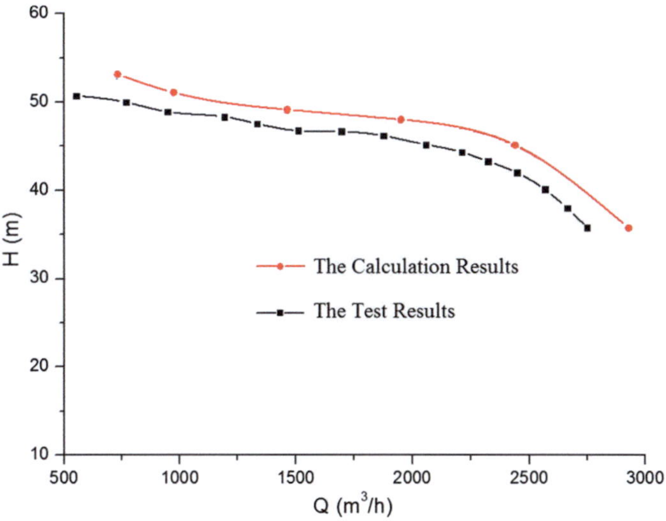

Fig. 29.5 The flow-head curve

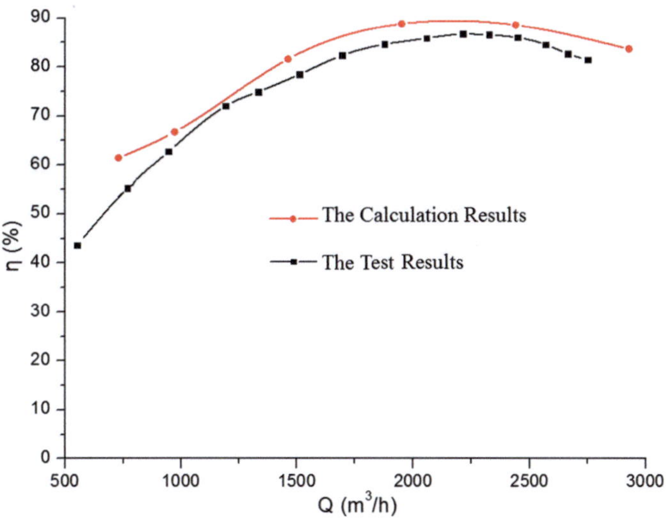

Fig. 29.6 The flow-efficiency curve

29.5 The Numerical Simulation of Structure Field

29.5.1 Numerical Methods of Structure Field

The numerical simulation of flow field was performed under 0.8 Q operating point, 1.0 t and 1.2 Q, then the numerical simulation results of flow field were imported to ANSYS Workbench; the corresponding pressures load under different operating points were loaded to impeller separately, the calculation began after choosing the right option of load and solving method.

29.5.2 The Results of Numerical Simulation

The distribution of equivalent stress and equivalent strain on the front and back of impeller was the research object; equivalent stress nephogram and equivalent strain nephogram of impeller under three operating points are shown in Figs. 29.7 and 29.8. As we can see from Fig. 29.7, the stress distribution of main blades, back blades, and impeller–shroud is axial symmetrical and the maximum equivalent stress is located in the junction between the impeller–shroud and main blade outlet which has the phenomena of stress focus. In Fig. 29.8, we can see that the maximum strain position of impeller is the sharp corner of extended part, the strain become smaller from sharp corner to impeller–shroud. The maximum strain under 1.0 Q is smaller than 0.8 and 1.2 Q. The strain of back blades and impeller–shroud is not obvious, and the strain of main blades has a symmetrical distribution. The static analysis results of impeller are shown in Table 29.1.

As we can see from Table 29.1, the maximum stress, the maximum strain and the maximum deformation are the smallest under design condition, which means that the pump should be operated under design condition as more as possible to increase reliability of thermal power unit absorption tower desulphurization pump.

29.6 The Strength Check

The designed impeller needs strength check to judge whether the impeller meets the strength requirement under the interaction of bending moment and torsion, the impeller bears the fluid reaction force, the dead weight of impeller, the rotating centrifugal force, and the friction between fluids. The strength of impeller was related to material, structure and the loads. The new material of thermal power unit absorption tower desulphurization pump is M26–23 V alloy steel whose allowable stress is 660 MPa and its ultimate strength is 862 MPa. The wear resistance, impact resistance, creep resistance, crocking resistance, and corrosion resistance of M26–23 V alloy steel are better than common steel. The material properties of M26–23 V alloy steel are shown in Table 29.2.

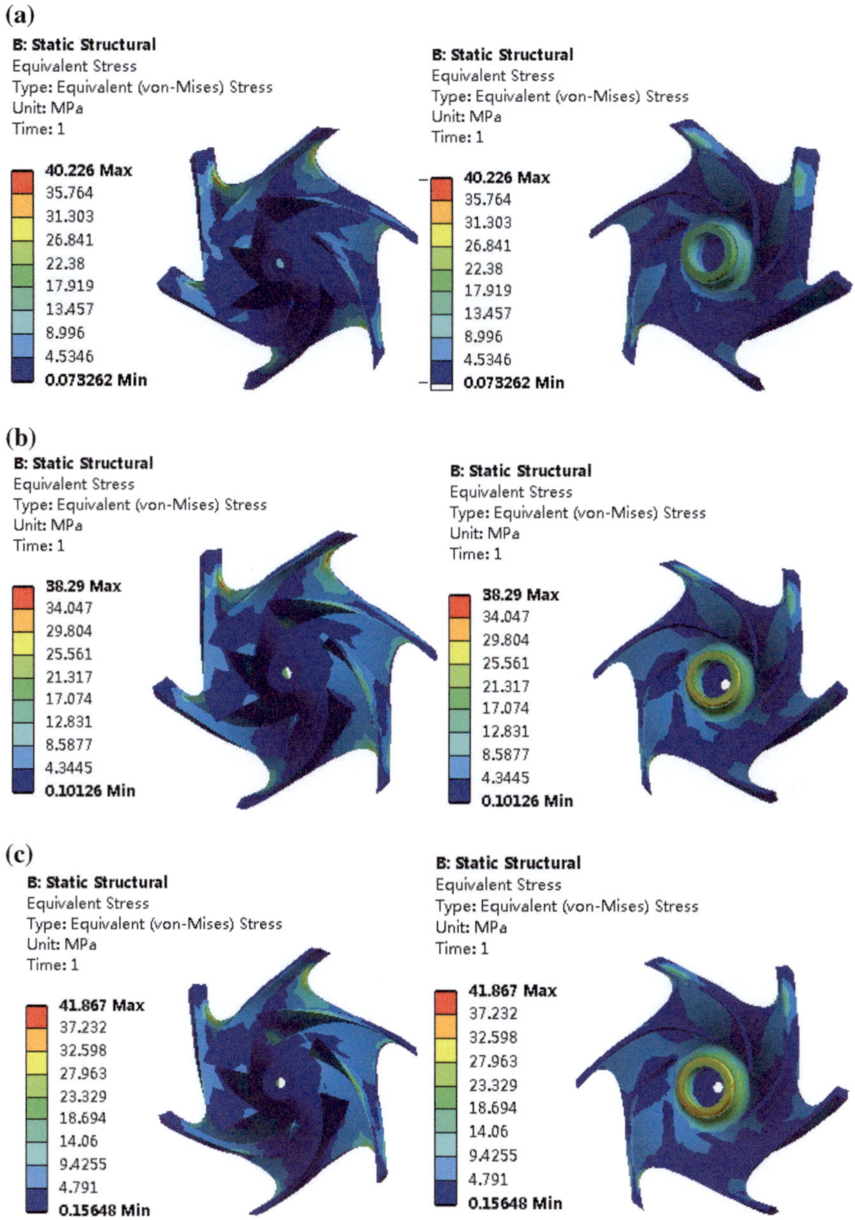

Fig. 29.7 Equivalent stress nephogram of impeller under three operating points. **a** Equivalent stress nephogram of impeller under 0.8 Q. **b** Equivalent stress nephogram of impeller under 1.0 Q. **c** Equivalent stress nephogram of impeller under 1.2 Q

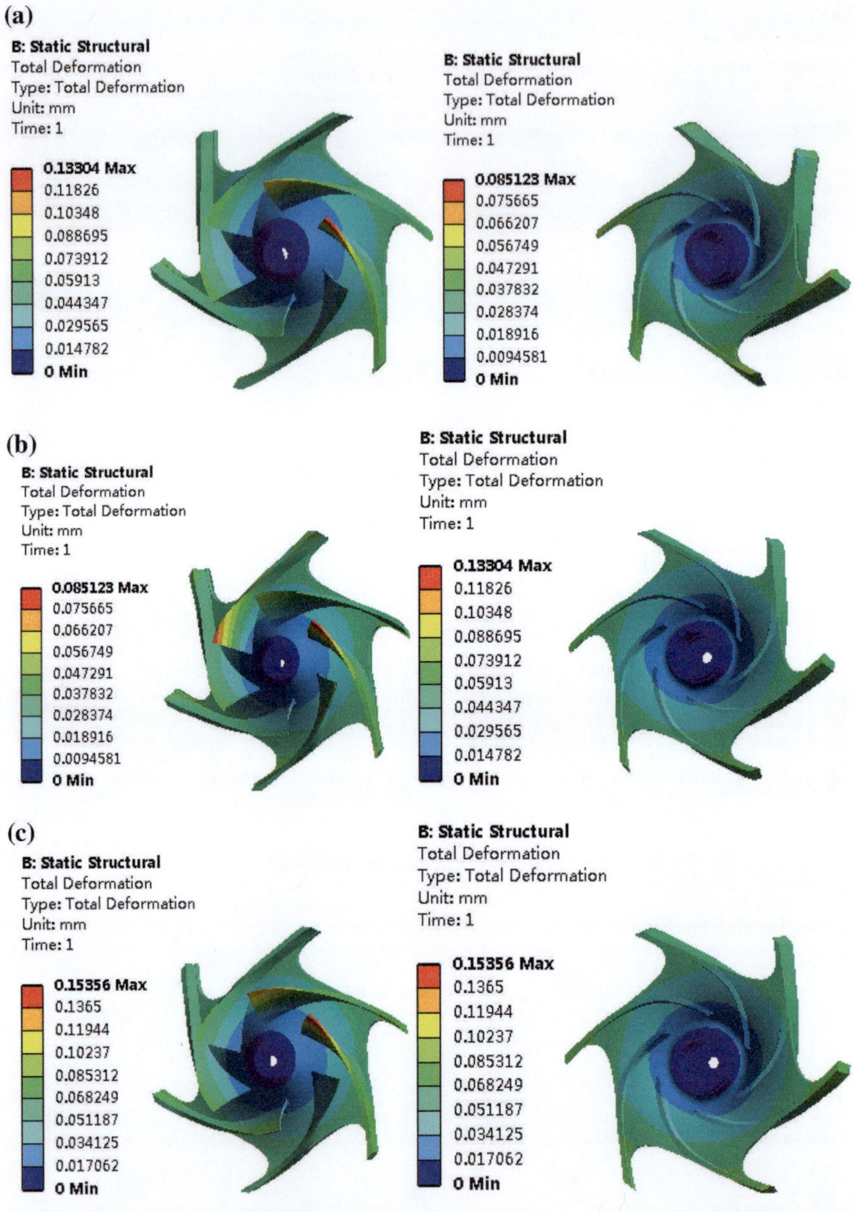

Fig. 29.8 Deformation nephogram of impeller under three operating points. **a** Deformation nephogram of impeller under 0.8 Q. **b** Deformation nephogram of impeller under 1.0 Q. **c** Deformation nephogram of impeller under 1.2 Q

Table 29.1 The static analysis results of impeller under three operating points

No	Flow	Max stress (Mpa)	Min stress (Mpa)	Max strain	Min strain	Max deformation (mm)
1	0.8 Q	40.226	0.073	2.01e-4	4.43e-7	0.133
2	1.0 Q	38.290	0.101	1.91e-4	6.51e-7	0.085
3	1.2 Q	41.867	0.156	2.15e-4	7.97e-7	0.153

Table 29.2 Material characteristic parameters of M26–23 V alloy steel

Density (kg m^3)	Elastic modulus (GPa)	Poisson ratio	Tensile strength (MPa)	Yield limit (MPa)	Allowable stress (MPa)
7863	216	0.32	680	1150	660

As we can see from Table 29.2, the maximum stress of impeller is less than allowable stress which means that the impeller meets strength requirement.

29.7 Conclusions

The fluid-structure interaction calculation of thermal power unit absorption tower desulphurization pump was done with ANSYS Software. The distribution law of stress and strain in impeller was analyzed, and the maximum stress point and the maximum deformation position were found. The strength of impeller was checked, which can provide reference for the design of thermal power unit absorption tower desulphurization pump.

Acknowledgments The research work was supported by Open Foundation of Binzhou Key Laboratory of Environmental Engineering and Technology No. bzhks201405 and Jiangsu Province Ordinary University Graduate Student Scientific Research Innovation Project No. CXZZ13-0674.

References

1. Hai-Yu Wang, De-Sheng Zhang, Wei-Dong Shi et al (2014) Analysis and optimization of structure in centrifugal pump based on fluid-structure interaction. J Drainage Irrig Mach Eng 32 (6):472–476
2. Xu X (2009) Experimental research on driving fatigue based on EEG analyzing. Tongji University, Shang Hai
3. Yang Wang, Wei-Jun Wang, Hong-Yu Wang et al (2013) Fatigue strength on stamped and welded impeller in multistage centrifugal pump. J Drainage Irrig Mach Eng 31(3):215–219
4. Han GAO (2008) Experimental study on the driver's physiological characteristics under simulated driving conditions. Harbin Institute of Technology, Harbin

5. Langthiem MA (2004) A numerical study of flow-induced noise in a two-dimensional centrifugal pump. Fluid Struct 19(3):349–368
6. Rong-Sheng Zhu (2014) Lin Peng (2014) Numerical simulation of solid-liquid two-phase flow in screw axial-flow pump. J Drainage Irrig Mach Eng 32(1):6–11
7. Bathe KJ, Zhang H, Wang MH (1995) Finite element analysis of in compressible and compressible flows with free surfaces and structural interaction. Comput Struct 56(1):193–213

Chapter 30
Manycore Polymorphic Computation Model for Complex Embedded System Tasks

Mei Luo, Linlin Li, Chengjian Li and Yongmei Luo

Abstract Tasks of complex embedded system have requirements on high performance computing and need to be changed with external environment. Considering the characteristics of complex embedded system tasks, we combine the advantages of polymorphic computing with manycore computing and introduce manycore polymorphic computation model. Meanwhile, we design three computing modes which is called as *computing state,* and three typical tasks are designed to verify the performance of our model. The result shows that no matter how data scale change, the model can work effectively in appropriate computing mode to meet the requirements of various tasks. The experiment also illustrates that each of the proposed computing modes can shorten execution time of different type of tasks.

Keywords Complex embedded system · Polymorphic computation model · Polymorphic computing mode · Task requirements

30.1 Introduction

In complex embedded system, there are many types of tasks are computationally intensive, which have strong real-time requirements. The response time of general computing platform is difficult to meet the strong time constraints of task requirements. With the change of external environment, computing manner of task is required to fit the change. Therefore, a model should be designed to dynamically select appropriate computing mode according to the needs of tasks.

According to the characteristic of complex embedded system tasks, we propose manycore polymorphic computing model (MPCM) and design various computing mode to fit different tasks.

M. Luo (✉) · L. Li · C. Li · Y. Luo
Xi'an Research Institute of Hi-Tech Hongqing Town, Xi'an 710025
People's Republic of China
e-mail: luom01-02@163.com

© Springer-Verlag Berlin Heidelberg 2015
S. Long and B.S. Dhillon (eds.), *Proceedings of the 15th International Conference on Man–Machine–Environment System Engineering*, Lecture Notes in Electrical Engineering 356, DOI 10.1007/978-3-662-48224-7_30

The rest of this paper, we introduce the development of polymorphic computing and manycore computing in Sect. 30.2. Section 30.3 describes the characteristics of complex embedded system tasks and designs manycore polymorphic computation model and its three polymorphic computing modes. Section 30.4 provides the test results. Finally, summary of the article is given in Sect. 30.5.

30.2 Related Work

The technology of polymorphic computing refers to support multi-level resource reconstruction in complex embedded systems. It also reflects an idea that software, hardware, and computational mode should adapt to the change of tasks' requirements. Currently, studies on polymorphic technology is focused on polymorphic architecture, resource allocation, task scheduling, etc [1–4]. The current studies show that the method of adapting to task's requirements is to build the polymorphic computation model which has the capability of reconfigurable, adaptive, and intelligent processing.

Manycore processor is suitable for intensive computing and attracts more and more attention nowadays. Currently, manycore processor is increasingly used in complex embedded system for a large-scale computing [5, 6]. Paper [7] describes multi-modes for manycore, but it does not combine polymorphic computing with manycore computing. Therefore, we propose a manycore polymorphic computation model for complex embedded system tasks and design three computing modes to fit the various computing requirements.

30.3 Manycore Polymorphic Computation Model

30.3.1 Logical Structure

Figure 30.1 shows the logical structure of manycore polymorphic computation model, it is composed by three parts: tasks layer, polymorphic service, and hardware resources.

Task layer is composed of a task set which has been configured explicitly computing mode and other parameters. We define $\tilde{T} = \{T_1, T_2, \ldots, T_n\}$ to represent a task set of system. $\forall T_i \in \tilde{T}$ is called a polymorphic task which can run on manycore. A polymorphic task is described in 5-tuple form $<\mathrm{TaskID}, \mathrm{priority}, \mathrm{r}_{\mathrm{num}_i}, \mathrm{Cmp\_mode}, \mathrm{Cmp\_data}>$, where parameter r_{numi} is the amount of resources T_i needs, Cmp_mode is computing mode of T_i, and Cmp_data is the data which need to be computed.

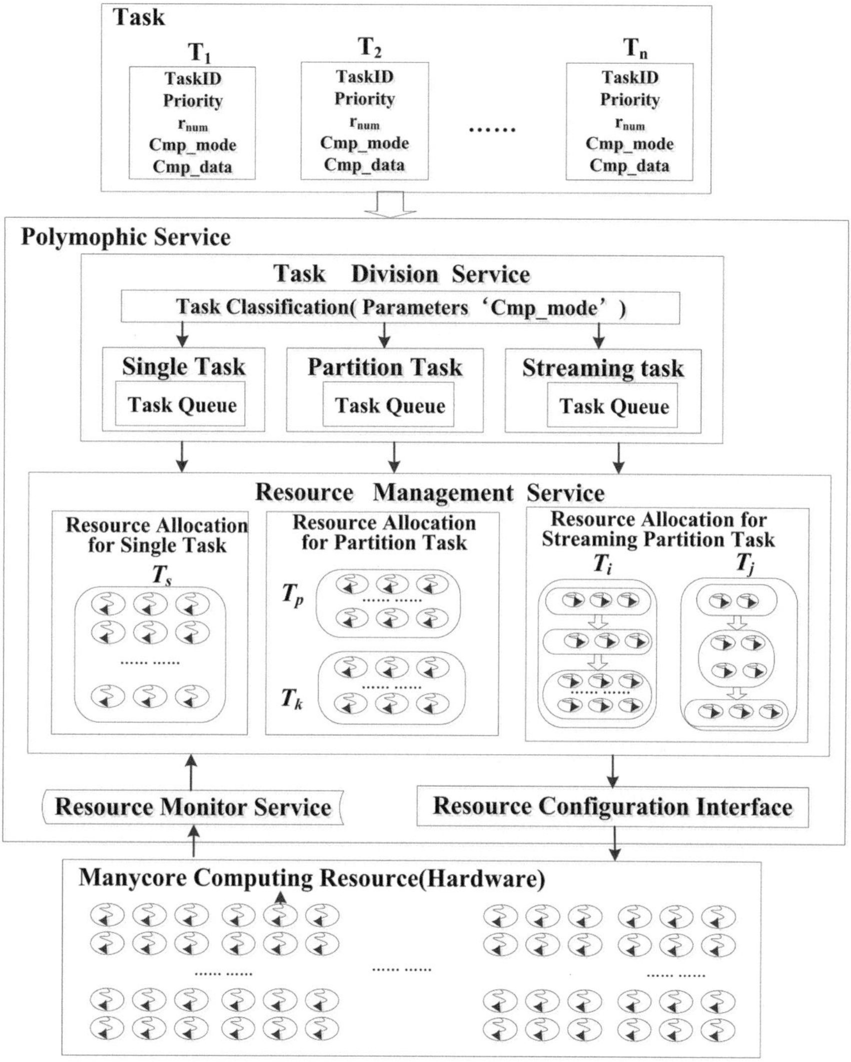

Fig. 30.1 Logical structure of manycore polymorphic computation model

Polymorphic service is the kernel of the model which is running on the host processor to provide 4 functionalities: task division, resource management, resource monitoring, and resource configuration. Task division service can sense real-time tasks and divide them to three types according to parameter Cmp_mode. Resource monitoring service can obtain resource status of hardware. Resource configuration service allocates tasks to manycore.

Hardware resources receive and compute data.

30.3.2 Polymorphic Computing Mode

Depending on the characteristics of complex embedded system tasks, we design three computing modes to fit the needs of different tasks. Different modes have its task queue, external commands, etc. When the task operation mode changes, the system will automatically schedule task queue to manage task and allocate resources.

In order to ensure the reliability and load balancing, we assume as follows:

1. Tasks of either mode are executed according to priority.
2. The three modes are mutually exclusive, that is, once one of computing mode is started, the other modes can not preempt it.

30.3.2.1 Single-Task Computing Mode

Single-task computing mode suits the task which can not be decomposed and has high computationally complexity and strong time constraints. If $\forall T_i \in \tilde{T}$ meets formula 30.1, T_i runs in single-task mode.

$$T_i \neq \{j \in [1, n] | T_{ij}\} \wedge T_i(r_{\text{num}}) \approx R \tag{30.1}$$

Formula 30.1 describes that task T_i can not be decomposed. Symbol $\wedge$ expresses that single-task meeting two conditions at the same time. R is free computing resources of manycore. Single-task mode awakes task from task queue which is ranked from highest to lowest priority. In this computing mode, all free resources are allocated to one task in the priority order.

30.3.2.2 Partition Computing Mode

This mode is for multiple tasks which are independent with each other and can be executed in parallel. $\forall T_i \in \tilde{T}$, $\forall T_j \in \tilde{T}$, If T_i, T_j satisfy formula 30.2, we call them independent that can be execute on different partition.

$$\begin{cases} \forall i \neq j, T_i(\text{Cmp\_data}) \neq T_j(\text{Cmp\_data}) \\ T_i \neq \{k \in [1, n] \mid T_{ik}\} \wedge T_j \neq \{s \in [1, n] \mid T_{js}\} \wedge T_i(r_{\text{num}}) + T_j(r_{\text{num}}) \leqslant R \end{cases} \tag{30.2}$$

In formula 30.2, tasks cannot be decomposed, and there is no data correlation between them. Independent multi-tasks are arranged by priority. In partition computing mode, all free resources are divided into paralleling running tasks in different partitions.

30.3.2.3 Streaming Partition Computing Mode

This mode suits the task which can be broken down into several calculation steps, if $\forall T_i \in \tilde{T}$ and $\forall T_j \in \tilde{T}$ fit formulae 30.3 and 30.4, we call them belong to streaming partition tasks.

$$\begin{cases} \forall i \neq j, T_i(\text{Cmp\_data}) \neq T_j(\text{Cmp\_data}) \\ T_i = \{T_{i_1}, T_{i_2}, \ldots, T_{i_m}\} \wedge T_j = \{T_{j_1}, T_{j_2}, \ldots, T_{j_m}\} \wedge T_i(r_{\text{num}}) + T_j(\text{r}_{\text{num}}) \leq R \end{cases} \tag{30.3}$$

$$\begin{aligned} &\forall i, j \in [1, n] \quad s.t. \\ &T_{i_m}(D_m) \rightarrow T_{i_{m-1}}(D_{m-1}) \rightarrow \cdots \rightarrow T_{i_1}(D_1) \\ &T_{j_m}(D_m) \rightarrow T_{j_{m-1}}(D_{m-1}) \rightarrow \cdots \rightarrow T_{j_1}(D_1) \end{aligned} \tag{30.4}$$

In formula 30.3, T_i and T_j have no data correlation and can be executed on different partition in parallel. Both T_i and T_j can be divided into many calculation step, m is the number of kernels. Formula 30.4 describes the relationship of these calculation steps, symbol $\rightarrow$ expresses data dependence, that is to say, data in the same streaming must be executed in order.

30.4 Performance Tests

The main idea of manycore polymorphic computation model is that according to task requirements, using different computing mode to implement task allocation and resource loading. Assuming adequate resources, our setting is following:

1. In single-task computing mode, we test target image recognition. As the target is more and more close, the accuracy of image is higher and higher. We record response time of task on multi-core CPU and manycore.
2. In multi-tasks partition computing mode, we test the adaption of manycore with the change of external environment. Three tasks are tested. When the external control change, data scale of tasks should change to meet the requirement of environment. We record the execution time of multi-core CPU and manycore.
3. In multi-tasks streaming partition computing mode, we test three tasks which can both be divided to three subtasks. The subtasks have data correlation, and every subtask can be executed on one kernel. We record the execution time of manycore.

The result is shown in Fig. 30.2.

Figure 30.2a illustrates that the task needs to occupy all free resources, with the data scale becoming bigger, the response time of manycore is less than multi-core CPU. Figure 30.2a shows that single-task computing mode can fit large-scale computing needs.

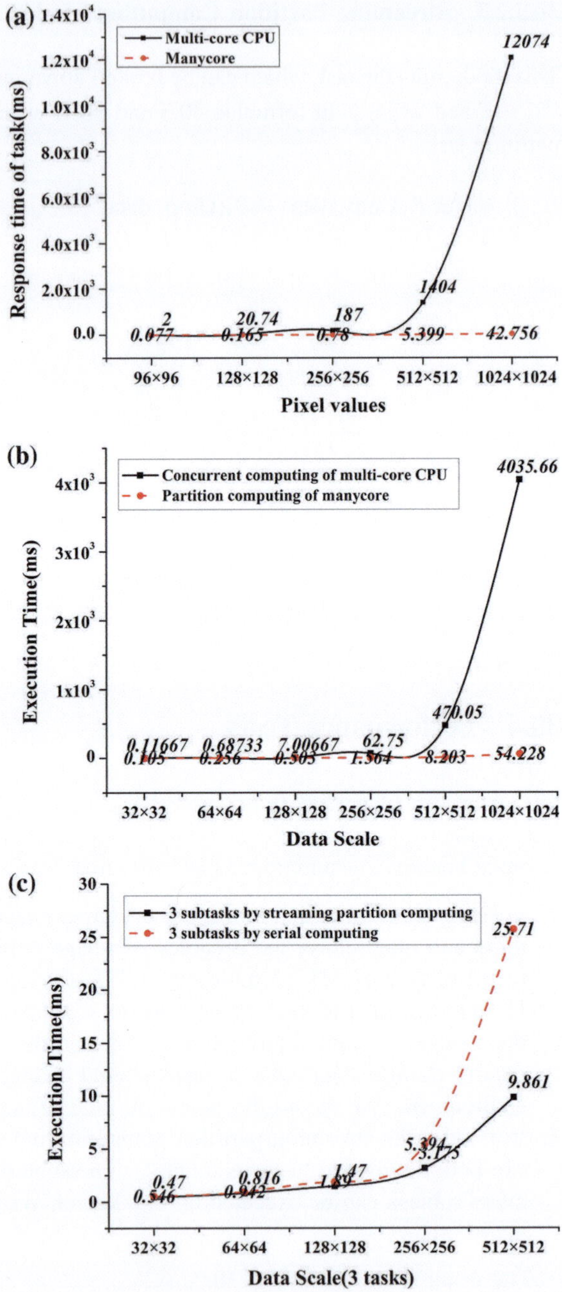

Fig. 30.2 Results of manycore polymorphic computing mode. **a** Single-task computing mode. **b** 3 tasks partition computing mode. **c** 3 tasks streaming partition computing mode

In Fig. 30.2b, partition computing mode can better meet the needs of multiple independent tasks than single-task mode. Whatever the computing scale changes, multi-task partition computing mode can reduce the computing time.

Tasks in Fig. 30.2c can be decomposed to subtasks. Although additional transmission and synchronization are needed, the execution time is less than serial computing mode. Therefore, we can see from Fig. 30.2c, streaming partition computing mode can also adapt to the tasks' requirements.

30.5 Conclusions

First, we conduct researches on polymorphic computing and analyze the advantages of manycore computing. According to the characteristics of complex embedded system tasks, we propose manycore polymorphic computation model, then design and test three polymorphic computing modes. The results show that each of computing modes can realize quick computation to ensure rapid response to different types of tasks.

References

1. Wu Y, Zhou X, Wu X et al (2013) An embedded real-time polymorphic computing platform architecture. In: 2013 international conference on mechatronic sciences, electric engineering and computer, pp 2326–2330
2. Luo M, Zhou X, Wang K et al (2013) Resource scheduling research of polymorphism task based on manycore architecture. In: 2013 5th intelligent human-machine systems and cybernetics (IHMSC), pp 62–65
3. Zhang Z, Ponpandi SD, Tyagi A (2014) An evaluation of user satisfaction driven scheduling in a polymorphic embedded system. In: 2014 IEEE 28th international parallel and distributed processing symposium workshops, pp 263–268
4. Hentrich D, Oruklu E, Saniie J (2011) Polymorphic computing: definition, trends, and a new agent-based architecture. Circ Syst 2(4):358–364
5. Chen R, Yong Y, Sun X et al (2013) An imaging processing technology for satellite-borne ScanSAR. Radar Sci Technol 11(2):136–140
6. Yu Y, Pang J, Shan Z et al (2012) Thread mapping model from CUDA to heterogeneous many-core architecture. Comput Eng 38(9):282–284
7. Wang K, Wu X, Luo M (2013) Construction for multi-mode computing system based on many-core processor. Mod Electron Tech 36(22):1–4

Chapter 31
A New Automatic Quality Control System for Fiber Ring Production

Xinfeng Li and Haoting Liu

Abstract Fiber optical gyro (FOG) as one of the most important component in Fiber inertial measure unit (FIMU), the production quality of which will affect the accuracy of FOG and FIMU; through decade years improvement in craftwork design, the main target has shifted to quality control promotion during production. This paper has proposed a new methodology for automatic production quality control; the method uses the computer vision technology to apply a system which can distinguish real-time fiber ring production images, applying pattern recognition of data mining technology for understanding of the type of fault production image. Pre-processing of computer vision has treatment to distil the image from real-time noised image for feature achievement, upon the qualified conducted result; a fast pattern recognition method support vector regression (SVR) has fast convergence which has utilized delightful result.

Keywords FOG · FIMU · Fiber ring · Median filter · Pattern recognition · Support vector regression (SVR)

31.1 Introduction

Fiber inertial measurement unit (FIMU) is the navigation and guidance component in missile and other aircraft units. In carrier flight test, the sensitive parameters of angular velocity and apparent acceleration for relative inertial coordinate (launching coordinate) will be sent through standard electrical interface to control system for gesture control and guidance calculation. Compare to mechanical gyro, fiber optical gyro (FOG) has many advantages, such as high reliability, long-usage life, wide dynamic range, quick start, small size, light weight, and direct output, especially for

X. Li (✉) · H. Liu
Hangtian Dianzi, No. 1 Fengying East Road, Beijing 100094, China
e-mail: yueyang2004@yeah.net

© Springer-Verlag Berlin Heidelberg 2015

S. Long and B.S. Dhillon (eds.), *Proceedings of the 15th International Conference on Man–Machine–Environment System Engineering*, Lecture Notes in Electrical Engineering 356, DOI 10.1007/978-3-662-48224-7_31

strap inertial measurement unit. Through decade years development, FIMU has been widely used in military and civilian territories.

Over decade years, the employment of FOG and FIMU products has desire of quality improvement of key components; fiber ring is one of the most important components in FOG; the craftwork in design and production has direct affection to the accuracy of FOG and FIMU. Researchers have perceived that the old manual way in production recognition is not suitable for fast technical improvement. A new methodology of automatic quality control system has been presented in this paper.

Fiber ring production has one main step which is fiber ring rolling; the seamless rolling is fundamental step in production; two main problems in fiber ring production are cracking and overlaying, in order to recognize the problem patterns; fast and reliable pre-processing image algorithm is traditional corrosion and median filter to distil the significant information from noise.

As the important step in image pre-processing, lots of works have been applied on image segmentation. These works can be divided into three major method categories: threshold [1, 2], color clustering [3, 4], and statistical model [5]. Another important pre-processing method is the segmentation based on the area of texture, however, as the optimal descriptor is not always clear. Various authors [6–10] attempt to find critical criteria to describe a specific texture. Moreover, in order to classify the problem images for the rolling problems varieties, support vector regression (SVR) as one of the most popular approaches has been presented in this paper.

The structure of this paper is comprised as follow: first part is general information of fiber ring rolling production theory to describe the whole production steps, emphasizing the key problem may appear during production. Second part is description of related technology, which reviews the recent expertise. Third part is illustration of manual production quality control and the one using computer vision technology, by which, the fiber ring quality comparison has presented at last to express the experiment result images for the advancement of the algorithm.

31.2 Research Background

Industrial production problem identification and quality monitoring are important topics of research in the area of industry safety assessment and quality monitoring. Especially for minor production in high accurate requirement, for example, the fiber-related production, to identify the existence of cracks and overlaying in structural or mechanical production and evaluate the affect of the damage caused by the cracks.

Recent research in auto industrial quality recognition has comprised electronic, metallurgy, automobile industry, in which the appearance defect type inspection has expressed the difficulties and varieties as big challenge with additional value. Compare with automobile and electrical industry, the fiber inspection particularity requires specific algorithm development; automatic production quality monitoring and problem recovery can be achieved by combining with electronic control system.

31.3 Methodology

The quality of fiber ring production is monitored through production image processing in real time. In which, the first step is pre-processing using median filter operator, and Harr feature will be achieved before the key pattern recognition is applied, in which fast pattern recognition contained rapid and precise cross validation method for parameter setting.

A image filtering

The real-time production images will been effected by unavoidable noise which includes impulse noise pulse noise and Gaussian noise etc. The full of target feature is enclosed in boundary; however, the real-time image of fiber rolling production is reflected by the fuzzification. The enhancement of image boundary for recognition of target features will be considered as fundamental step in image pre-processing. In this paper, the methodology applies meridian filtering as the filtering and image smooth method to let the low-frequency signal pass to remove the effect of noise, the result is well restored with boundary information and reasonable adaptive ability. Regard to the processing speed of all control system, a 3 * 3 master plate is used as smooth method in meridian filtering.

B the boundary tracking inspection

The boundary features are described by direction and amplitude, the gray change is smooth through the boundary change, the pixel which is vertical to boundary direction will have cleared visible change. Therefore, the gradient direction and local maximum are calculated as follow for boundary detection:

$$\frac{\partial f}{\partial r} = \frac{\partial f}{\partial x} \cdot \frac{\partial x}{\partial r} + \frac{\partial f}{\partial y} \cdot \frac{\partial y}{\partial r} = f_x \cos\theta + f_y \sin\theta$$

in which, $\frac{\partial f}{\partial r}$ has the situation

$$\frac{\partial(\partial f/\partial r)}{\partial \theta} = 0 \quad \text{and}\, f_x \sin\theta_g + f_y \cos\theta_g = 0$$

to have maximum value, therefore

$$\theta_g = \tan^{-1}(f_x/f_y)$$

The max gradient which is

$$g = (\frac{\partial f}{\partial r})_{\max} = \sqrt{f_x^2 + f_y^2}$$

known as gradient pattern, the properties has decided the boundary detection functionality. And the gray transformation direction which can be seen as boundary direction, the result can be obtained by

$$\theta_g = \tan^{-1}(f_x/f_y).$$

In this paper, the Canny method has been applied in boundary detection, it is comprised by first derivative of Gaussian function which is also seen as the optimization of product of Signal to noise ratio (SNR) and location.

C problem pattern recognition

Apply Harr feature to detected boundary image, which include horizontal and vertical features, diagonal feature, and pointing feature. The weight of coefficient is selected to enhance the extracted different Harr feature during production. Meanwhile, support vector regression method is applied to classify the problem pattern which is fiber cover and rolling gap.

Through using support vector regression method to classify the undetected boundary pixels which is wide applicable and the accuracy is adjustable through controlling method parameters.

31.4 Experiment and Results

See Figs. 31.1, 31.2, 31.3, 31.4, 31.5, 31.6, and 31.7.

In order to compare the accuracy and time efficiency for our algorithm with normal fiber production, the comparison is shown as table below:

Algorithm	Time (per rolling) (s)	Accuracy (%)
Normal manual production	10	75
Production involved with proposed method	2.3	98

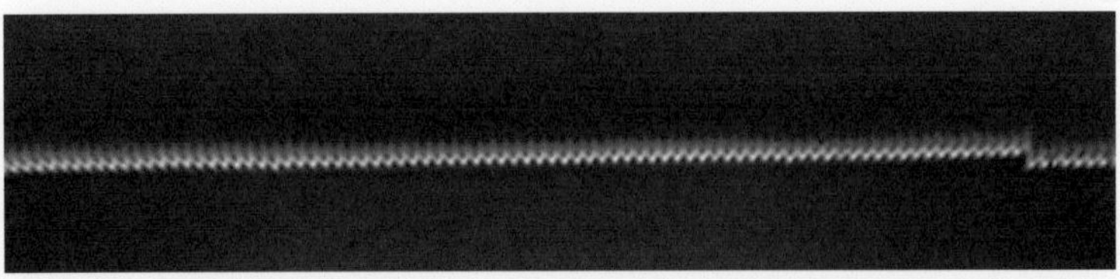

Fig. 31.1 Normal production results

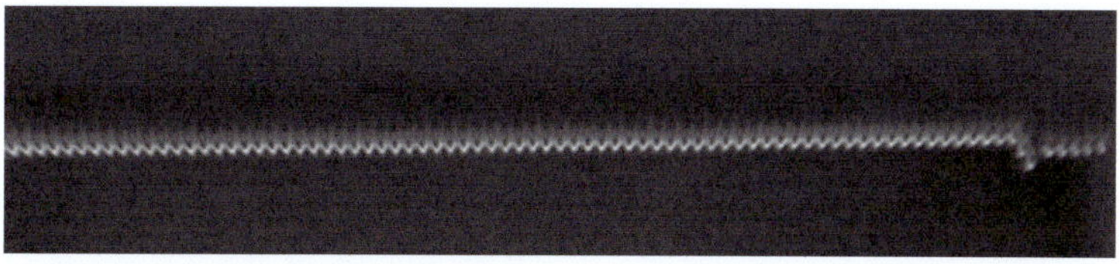

Fig. 31.2 Production overlaying

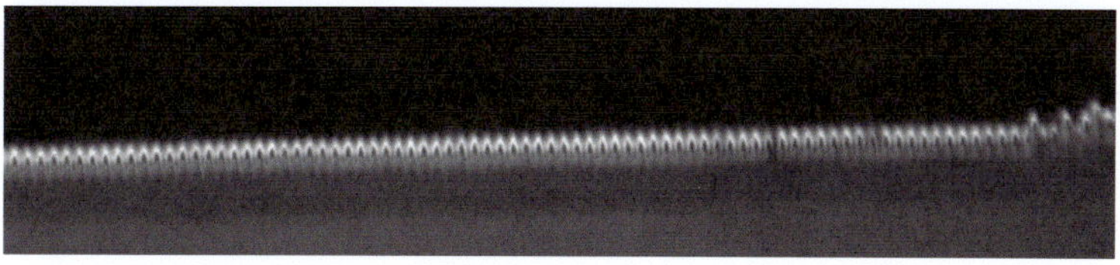

Fig. 31.3 Production cracking

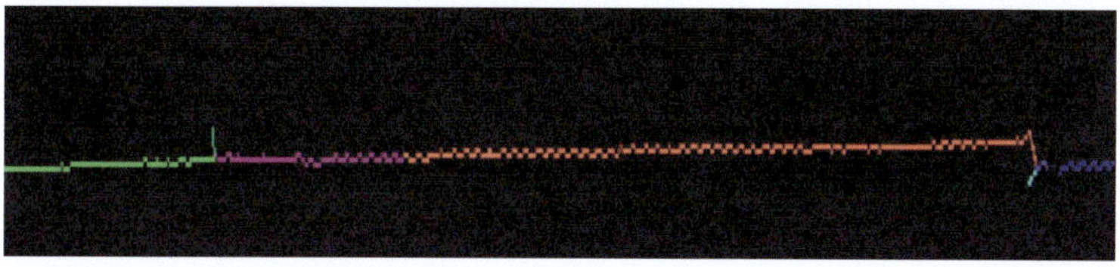

Fig. 31.4 After boundary detection

Fig. 31.5 Problem image processing

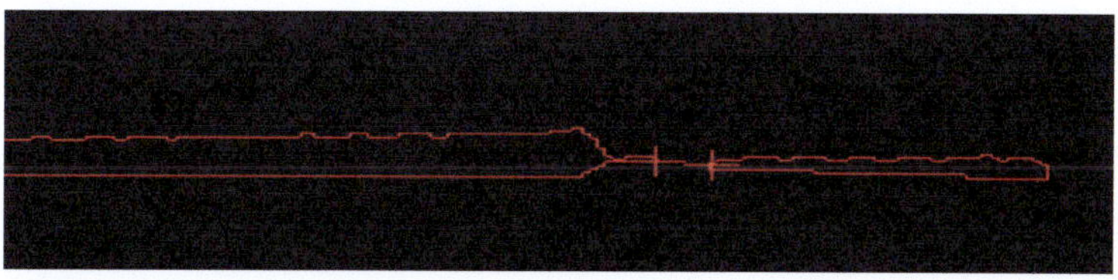

Fig. 31.6 Problem location processing

Fig. 31.7 The problem represented by Harr feature

31.5 Conclusions

This paper has presented a automatic quality control system with computer vision and data mining method to segment and recognize production quality from real-time images with complex background. This paper compares the proposed approach with normal manual production methods. Experimental results demonstrate that the boundary tracking method obtains better segmentation results and can effectively achieve and distinguish problem feature vectors from images with complex background.

References

1. Otsu N (1978) A thresholding selection method from gray-level histogram. In: IEEE Trans Syst Man Cybern 9(1):62–66
2. Niblack W (1986) Introduction to digital image processing. PrenticeHall, Englewood Cliffs, pp 115–116
3. Song Y, Liu A, Pang L, Lin S, Zhang Y, Tang S (2008) A novel image text extraction method based on k-means clustering. In: Proceedings of seventh IEEE/ACIS international conference on computer and information science, pp 185–190
4. Gllavata J, Ewerth R, Stefi T, Freisleben B (2004) Unsupervised text segmentation using color and wavelet features. In: Proceedings of the 3rd international conference on image and video retrieval. Dublin, Ireland, pp 216–224
5. Gao J, Yang J (2001) An adaptive algorithm for text detection from natural scenes. Comput Vis Pattern Recogn 2:84–89
6. Duvernoy J (1984) Optical–digital processing of directional terrain textures invariant under translation, rotation, and change of scale. Appl Opt 23(6):286–837
7. Sezer OG, Ertüzün A, Ercil A (2004) Independent component analysis for texture defect detection. Pattern Recogn Image Anal 14(2):303–307
8. Sobral J (2005) Optimised filters for texture defect detection. In: International conference on image processing (ICIP), pp 3165–3168
9. Sebe N, Cohen I, Grag A, Huang T (2005) Machine learning in computer vision. Springer, New York
10. Gonzalesand RC, Woods RE (2002) Digital image processing, 2nd edn. PrenticeHall, Englewood Cliffs

Chapter 32
A New Hybrid Genetic Algorithm for Parameter Estimation of Nonlinear Regression Modeling

Jinshan Chen

Abstract A new algorithm for parameter estimation of nonlinear regression modeling based on hybrid genetic algorithm (HGA) is proposed with a constraint condition by defining n dimensional column vector $X = (x_1, x_2, \ldots, x_n)^T$; $a_i \leq x_i \leq b_i$ ($x_{ji} = \text{IND}[j]\ .\ \text{chrom}[i] \times (b_i - a_i) + a_i$; $j = 1, 2, \ldots, N$; $i = 1, 2, \ldots, n$) and founding individual fitness function $F(X, t) = [S(\theta, t) + \varepsilon]^{-1}$, and using the linear ranking selection strategy. According to the fitness value of population member size from good to bad arrangement of $X_1, X_2, \ldots, X_N$ and then through a series of operations of transform, crossover, and mutation, the HGA is terminated by the terminal criterion of algorithm. The new algorithm can rapidly converge to the global optima by performing global search and local search alternatively, and achieve better performance than those of traditional algorithms. The experiments verify generalization and effectiveness of the algorithm.

Keywords Genetic algorithm · Nonlinear parameter estimation · Least-squares estimation

32.1 Introduction

Nonlinear regression analysis has been widely applied in the fields of voice signal processing, biological medicine, image processing, and modeling cognition. However, due to its complexity of nonlinear model and difficulty in obtaining parameter estimation, thus it is limited its application and development greatly. Traditional nonlinear regression model parameter estimation algorithm includes simulated annealing algorithm [1], Nelder-Mead algorithm [2], Gauss-Newton

J. Chen (✉)
Department 3, Institute of Special Operation, Guangzhou 510500, China
e-mail: js63@163.com

J. Chen
PLA Institute of Special Operation, Guangzhou 510502, China

© Springer-Verlag Berlin Heidelberg 2015

S. Long and B.S. Dhillon (eds.), *Proceedings of the 15th International Conference on Man–Machine–Environment System Engineering*, Lecture Notes in Electrical Engineering 356, DOI 10.1007/978-3-662-48224-7_32

algorithm [3], variable metric algorithm and homotopy algorithm [4], etc. These algorithms only have effect on some type of particular issues and require models with continuous, derivable, and single-peak features. Fang [5] proposes the sequential method which does not need auxiliary information of differential coefficient and works as a “mountain-climbing” searching method in principle. Moreover, the efficiency of algorithm is limited to the selection of initial fields to a large extent and fining precision in each time of iteration. Genetic algorithm is a whole-situation optimization algorithm, but it is the issue of slow convergence [6]. Thus, combining genetic algorithm with traditional optimization-selecting algorithm (mountain-climbing algorithm, simulated annealing algorithm, and Newton algorithm) can improve the efficiency of algorithm on the basis of algorithm universality. Renders et al. [7] proposed an algorithm combined quasi-Newton’s algorithm and traditional GA, used for nonlinear function optimization, which improves the convergence speed of algorithm. However, due to the first-order derivative of quasi-Newton algorithm required function, the universality of the method was limited to some extent. This paper proposes a new hybrid genetic algorithm for parameter estimation of nonlinear regression modeling. The new algorithm only utilizes function value for searching, thus it has wider application scope.

32.2 Nonlinear Regression Modeling and Least-Squares Estimation

The general model of nonlinear regression may be expressed as $Y = f(X, \theta) + e$, in which “e” subjects to normal distribution $N(0, \sigma^2)$, $X \in R^m$, $Y \in R^1$, $\theta \in R^P$, and P represents the number of parameters.

Set up the known watched value $\{(x_i, y_i); i = 1, 2, \ldots, n\}$, the issue of nonlinear regression modeling parameter is to obtain the least-squares estimation of θ, *that is,* calculate $\hat{\theta} \in R^P$, making any $\theta \in R^P$ with $S(\hat{\theta}) \leq S(\theta)$.

$$S(\theta) = \sum_{i=1}^{n} [y_i - f(x_i, \theta)]^2. \tag{32.1}$$

This paper takes asymptotic regression model (32.2) as an example, to discuss the parameter estimation under the least-squares estimation of nonlinear regression model. In fact, this method is similarly applicable to other nonlinear regression models.

Asymptotic regression model

$$f(x) = \alpha - \beta\gamma^x \qquad 0 < \gamma < 1 \tag{32.2}$$

32.3 Parameter Estimation Algorithm of Nonlinear Regression Modeling

Parameter estimation algorithm of nonlinear regression modeling based on hybrid genetic algorithm is described as follows.

32.3.1 *Code*

The essence of coding is to set up a mapping between the solution space and searching space of algorithm. While using coding of real number, each individual shall be expressed with n-dimensional real vector, that is $X = (x_1, x_2, \ldots, x_n)^T$; $a_i \le x_i \le b_i$ $(i = 1, 2, \ldots, n)$. To define any individual x_{ji} = IND[j] . chrom[i] × $(b_i - a_i) + a_i$;$j = 1, 2, \ldots, N$ in population, then the vectors meeting this relationship $X_j = (x_{j1}, x_{j2}, \ldots, x_{jn})^T$ can meet the boundary restriction conditions of optimization issue.

32.3.2 *Individual Evaluation*

According to the optimization requirement of model (32.2) nonlinear parameter estimation, set up individual fitness function $F(X, t) = [S(\theta, t) + \varepsilon]^{-1}$, ε standing for the positive number small enough. As for the individuals in population, calculate and record the fitness value in each individual and select the optimized individual X_H and worst individual X_L.

32.3.3 *Selection*

To avoid over-early convergence and stagnation phenomenon in algorithm, adopt nonlinear rank selecting strategy. First, hypothesize group members listed from good to bad as per the fitness values $X_1, X_2, \ldots, X_N$, then allocate selection probability P_k according to a nonlinear function.

$$P(X_k) = P_k = [c - d \times k/(N+1)]/N,\ k = 1,\ 2\ldots N \tag{32.3}$$

Of which, c, d are constants; the values of c and d shall be met

① $\sum_{k=1}^{N} P_k = 1$; ② As for any $k = 1, 2, \ldots, N$ then $P_k \ge 0$, and $P_1 \ge P_2 \ge \ldots \ge P_N$. Evidently, set up $1 \le c \le 2$ and $d = 2(c - 1)$, then the above requirements can be met.

32.3.4 Transform

Transform the worst individuals as follows.

(1) Calculate the centroid of mating pool species $X_{C.}$
(2) Reflect X_L: $X_R = X_C + r(X_C - X_L)$, r represents function.
(3) If $F(X_R, t) > F(X_L, t)$, then execute expansion operation. $X_E = X_C + q(X_R - X_C)$, q stands the expansion coefficient.
(4) As for any point X_W in a polygon apart from X_L, then $F(X_R, t) < F(X_W, t)$, execute the shrinkage operation: $X_s = X_C + s(X_H - X_C)$, s stands for the shrinkage coefficient.
(5) If $F(X_R, t) < F(X_L, t)$, then all points approximate to X_H point: $X_W = X_H + 0.5 (X_W - X_H)$
(6) Make the best points in X_R, X_E and X_S to replace $X_{L.}$

32.3.5 Crossover

To maintain the diversity and in population and avoid premature convergence, adopt neighbor-pairing principle [8]. Not only can this pairing method avoid superior mode from expanding, but also meet the fine grits parallel mode of genetic algorithm, easily acquiring larger parallel mode.

Provide $P(X) > P_c$ (P_c stands for crossover probability); two parent individuals Parent [1] and Parent [2] as neighbor-pairing principle, with their child individuals Child [1] and Child [2], adopt integral algorithmic crossover. First, generate the random number $a_1, a_2, \ldots, a_n$ from $n(0, 1)$ section, then

Child [1] . chrom[i] = a_i * Parent [1] . chrom[i] + (1 − a_i) * Parent [2] . chrom[i]
Child [2] . chrom[i] = a_i * Parent [2] . chrom[i] + (1 − a_i) * Parent [1] . chrom[i]
in which $i = 1, 2, \ldots, n$.

32.3.6 Mutation

According to formula (32.3), calculate the selection probability $P(X_k)$ of each individual; as for individual Parent[j] of variation probability $P(X_k) < P_m(P_m)$, randomly select a gene i for mutation.

$$\text{Child}[j]\,.\,\text{chrom}[i] = \text{Parent}[j]\,.\,\text{chrom}[i] + \delta$$

In which, $\delta = \lambda \times N(0, 1)$, where $N(0, 1)$ stands for standard normal random variable, $\lambda = C_M/t$, C_M stands for a constant, and t stands for the number of iteration.

32.3.7 Terminal Criterion of Algorithm

Set up a maximum evolution number G; when the evolution number reaches G or the results without evident improvement within 50 generations, the algorithm terminates, that is, $|S(\theta,t) - S(\theta,t+50)| \le \varepsilon$ (ε stands for a small positive), or $t = G$, algorithm ends.

32.4 Examples and Results

Calculation examples aims for calculating the parameter $\theta = (\alpha, \beta, \gamma)$ of asymptotic regression model (32.2), among which $0 < \gamma < 1$. The parameter of hybrid genetic algorithm is taken its value as: population size $N = 50$; crossover probability $P_c = 0.5$; mutation probability $P_m = 0.03$; maximum revolution number $G = 1000$, $\varepsilon = 0.000001$.

The experiment dataset is illustrated as in Table 32.1. Table 32.2 shows the results of nonlinear regression parameter estimation using the method of HGA, AIDEA [8] and HSA [9]. Here lists the average result obtained from 100 times of random calculation.

Experiment data selection

Dataset1 = {(2,18.6), (3,22.6), (4,25.1), (5,27.2), (6,29.1), (7,30.1)},

Dataset2 = {(0,20.518), (1,21.138), (3,21.734), (5,22.218), (7,22.286)}

Table 32.1 $\hat{\sigma}^2$ is the estimation of variance, defined as $\hat{\sigma}^2 = \sum_{i=1}^{n} \frac{(y_i - \hat{y}_i)^2}{n-3}$.

Table 32.1 shows that, as for two given data, HGA estimation results approximate to AIDEA [8] and HSA [9] estimation. This demonstrates the effectiveness of algorithm.

Table 32.1 Results of estimation

		HGA	AIDEA	HSA
Dataset1	α	33.7909611	33.8022682	33.8023
	β	26.6905654	26.6979828	26.698
	γ	0.7527665	0.7529937	0.752995
	$\hat{\sigma}^2$	0.0349082	0.03491553	0.03491537
Dataset2	α	22.4855205	22.487058	22.487
	β	1.9574835	1.958616	1.9586
	γ	0.704972	0.705544	0.705539
	$\hat{\sigma}^2$	0.00587683	0.0058764825	0.005876535

Table 32.2 Time of estimation(s)

	HGA	AIDEA	HSA
Dataset1	4.81	6.51	6.33
Dataset2	4.82	5.72	6.32

Considering each iteration of three algorithms has large calculation variance. This paper selected algorithm convergence time as index to compare with the efficiency of three algorithms. When estimation of variance $S(\hat{\theta})$ is smaller than the given thresholds, The algorithm ends. Set up a same thresholds for the three algorithms. Their convergence time is listed in Table 32.2.

Table 32.2 shows that HGA algorithm has the highest efficiency of estimation. The estimation of AIDEA is closely related to selection of initial fields and each iteration the precision of refinement neighborhood. HSA has slow iteration due to the influence of simulated annealing initial temperature and annealing rate. Thus the convergency time is long.

32.5 Conclusions

Hybrid genetic algorithm boasts the advantages of simple description, generalization, and less restriction on initial conditions and optimization in the whole situation. While applying in the parameter estimation of nonlinear regression model, it overcomes the disadvantages of traditional algorithm on strong restriction conditions and bad generalization. It shows good application prospect.

References

1. Chen H, Cui D W, Yan TS et al Race exponent based simulated annealing rank selection operator. ACTA Electronical Sinica 37(3):586–589 (in Chinese)
2. Nash J C, Walker-Smith M (1987) Nonlinear parameter estimation. Marcel Dekker Inc., New York, Basel
3. Ma N, Wei G (1999) Research on nonlinear modeling of time series. J Electron 16(3):200–207 (in Chinese)
4. Wei G, Man KF, Kwong S (1997) Homotopy theory based nonlinear modeling for time series. Chin J Electron 6(3):41–45 (in Chinese)
5. Fang KT, Zhang JT (1993) A new algorithm for calculation estimates of parameters of nonlinear regression modeling. Acta Mathmeticae Applicatae Sinica 16(3):366–377 (in Chinese)
6. He QH, Wei G, Lu YQ (1998) Evolution of genetic algorithm. J Electron 26(10):118–112 (in Chinese)
7. Render JM, Flasse SP (1996) Hybrid methods using genetic algorithms for global optimization. IEEE Trans Syst Man Cybern (Part B) 26(2):243–258
8. Liao F, Mao MH, Zhao W (2013) Nonlinear parameter estimation based on average information differential evolution algorithm. Math Pract Theor 43(19):123–128 (in Chinese)
9. Guo DL, Xia HM, Zhou YQ (2010) Hybrid simulation annealing and evolution strategy algorithm in non-linear parameter estimation application. Math Pract Theor 40(22):91–98 (in Chinese)

Chapter 33
The Quantum Cryptography Communication and Military Application

Xiaoli Zheng and Digang Jiang

Abstract The quantum cryptography communication became a new communication technology, with the developed of quantum computing. Quantum cryptography used some specific properties of quantum physics, and it solved some difficult problems of classical cryptography. It is a reliable technology for ensure the security of network communication that in future ages of optical communication. Based on the quantum key distributions protocol, and in recent years the United States, Germany, Sweden, Japan, and other western developed countries in the progress in research of quantum cryptography communication from the research results, the quantum cryptography communication has entered the practical stage in developed countries; our country also should be more in-depth and extensive research in this aspect; the assay analyzed the application perspective of the quantum cryptography communication in military communication.

Keywords Quantum cryptography communication · Cryptographic algorithms · Protocol

Quantum cryptography technology is a newly emerging interdisciplinary science combined quantum physics and cryptography. It solves the issues that modern passwords are intercepted and decoded during the process of delivery. Nowadays, Chinese army still lags behind in the construction of informationaztion. Particularly, there are some potential dangers in the information security field. Thus, it is of great significance to research quantum cryptography communication technology to improve Chinese army communication guidance security.

X. Zheng (✉)
PLA Institute of Physical Education, Guangzhou 510500, Guangdong, China
e-mail: Pig_100@163.com

D. Jiang
Zhongshan School of Medicine SYSU Guangzhou, Guangzhou
510080, Guangdong, China

© Springer-Verlag Berlin Heidelberg 2015

S. Long and B.S. Dhillon (eds.), *Proceedings of the 15th International Conference on Man–Machine–Environment System Engineering*, Lecture Notes in Electrical Engineering 356, DOI 10.1007/978-3-662-48224-7_33

33.1 Distribution Protocol of Quantum Secret Keys

The most important part of Quantum cryptography is quantum secret key distribution, whose purpose is to make the two parts of communications utilize the unreliable channels to finish the negotiating generation of secret keys [1].

33.1.1 BB84 Quantum Secret Key Distribution Protocol

BB84 protocol principle is to utilize the indeterminacy principle of single photon and quantum information channel. Alice randomly select 4 photon polarization (0, $\pi/4$, $\pi/2$, and $3\pi/4$) to send Bob, forming the photon status series with some polarization status and record the corresponding basis vector type of photon status. After receiving the signals sent by Alice, Bob starts to receive the photo status series sent. Bob selects one from two measurement basis vectors for each photon to measure, then records the results of measurement and keeps it secretly. After receiving and measuring the polarization status photon series sent by Alice, Bob shows the basis vector or measurement types used in the measuring process to Alice. Alice compares them and tells Bob the comparison results about what are correct and what are wrong. According to the comparison results, Alice and Bob translate all the correct photon polarization status as binary system bit string according to the predetermined agreement, so as to obtain the required secret keys.

33.1.2 B92 Quantum Secret Key Distribution Protocol

The principle of B92 protocol is to utilize the undistinguishable principle of non-orthogonal quantum states, which is decided by indeterminancy principle. First, select the any two sets of conjugated measurement base (here we take polarization direction 0° and two sets of linear polarization state of 90, 45, and 135°). Define 0 and 135° representing quantum bit "0." 45 and 90° representing quantum bit "1." Legal user Alice randomly launches polarization state (here take 0 and 45°). Bob randomly uses polarization state (here take 90 and 135°) for synchronizing measurement. By utilizing the above features, Alice and Bob realize B92 quantum secret key distribution as the following steps.

(1) Alice randomly takes 0 and 45° photon polarization state to launch the photon pulse selected to Bob.
(2) Bob randomly selects the 90 or 135° basis-deflecting check. If Bob's checking direction is vertical to Alice's selected direction. Detector cannot receive photons completely. While checking in the angle of 90°, it is with 50 % of probability that photons can be received. Once Bob measures photons, he can speculate the polarization status of photon sent by Alice.

(3) Then, Bob informs the conditions Alice receives photons through the public channel and but publicize the measurement basis. Both parties give up the unmeasured data. If no bugging or disturbance, both Alice and Bob can jointly possess a set of same random number series.
(4) Bob transforms the received photons as photon bit string then.
(5) Bob randomly publicizes some bits and has Alice to determine with errors or not.
(6) After Alice confirms that no persons are bugging, the left bit string can be established as code book.

This method is simpler than BB84 protocol, but the transmission rate is decreased by a half. Only 25 % of photons can be received.

33.1.3 E91 Quantum Secret Key Distribution Protocol

E91Protocol principle is to take EPR effect, with the following communication process.

(1) The photon pairs produced from EPR resources, respectively, are sent to legal users Alice and Bob in the direction of $\pm Z$. Alice randomly selects checked biased basis (linear polarization basis or round polarization basis) measuring a photon 1. The measurement results shall be determined by EPR correlation.
(2) Meanwhile, Bob also randomly adopts checking polarization basis to measure another photon 1 of the received EPR correlation pair and record the measurement result.
(3) Then Bob randomly uses the measurement basis through public channel (not publicized measurement results). Alice tells Bob that those checking polarized basis are selected rightly. Then the reserved results kept by the two parties and are transferred as quantum bit string. Then secret book is established.

Which is different from BB84 that what the data kept by the two parties are using Bdl inequality formula to check. If violating inequality formula, it demonstrates that quantum channel is safe, not bugged. If meeting the inequality formula, it demonstrates channels are in troubled, existing eavesdroppper. In a word, the security originates from Bell principle. The protocol is safe according to the quantum mechanics principle.

33.2 Research Test of Quantum Secret Key Distribution Protocol at Home and Abroad

Quantum cryptography test is one of the fastest developments of quantum information processing field. Currently QKD tests mainly include two aspects, QKD in optic fibers and in free space. QKD in optic fibers has reached the conditions of realistic application. QKD in free space must be realized QKD above 2 kms in ground surface test. The current testing results also approximate the practical using conditions.

The first QKD test was completed by Bennett et al., with the transmission distance of 30 cm [2]

Currently quantum cryptography can pass 30 km optic fiber channel at the rate of 1 k bit/s, using 1.3rim semiconductor laser and having secret key exchange safety through BB84 protocol. Further improvement of checking technology is expected to realize secret key exchange over 100 km at the rate of 20 k bit/s. British F3T lab is working hard at research and exploring the possible encoding technology with all kinds wavelength.

B92 secret key distribution protocol has conducted the safe secret key distributing test on 15 km optic fiber channel.

Barnett et al. also proposed data-declining protocol, jointly or independently used with BB84 protocol. Ekert proposed another EPR protocol based on totally different physical phenomena according to Bell maxim. These are suggestions of proposing testing system, but not realized.

BT lab has explored the quantum secret key distribution technology in optic fiber network and proposed the secret key distribution scheme of broadcasting tree-shape two-level broadcasting.

Geneva University of Swiss transmitted 1.3 um-wavelength quantum optic signals on 1.1-m-long optic fiber on the basis of polarization encoding scheme of BB84 protocol in 1993, with the error code rate only 0.54 %. In 1995, Geneva University completed the test in 23-km-long civil optic cable laid on the bed of Geneva Lake, with the error code only 3.4 % [3].

In 1999, Sweden and Japan cooperated to completed quantum cryptography communication test on 10-km-long optic fiber.

America Los Alamos Lab successfully realized 48-km quantum secret keys system running for 2 years. They used QKD system in the free space in 2000, with the transmission distance of 1.6 km.

In October 2002, University of Munich, Germany and British military research agency cooperated to transfer photon secret key between Zugspitze and Kalvin Del Peak with the distance 23.4 km between Germany and Austria successfully. The successful test made it possible to pass secret key through approximate satellite and establish global password sending network.

In August 2003, the scientific researchers of American National Standard and Technological Institute researched a detector to explore the single-pulse light. Meanwhile, it can lower the wrongly deleted or "leaking testing" rate.

In March 17, 2004, Japan NEC Company created the new record 150 km of quantum cryptography delivery distance, which makes the practicality of quantum cryptography possible.

In May 2003, USTC made breakthrough in experimenting research of quantum entangled stage purification. This research not only resolves the difficulties of remote communication, but also it can greatly promote the research of error tolerance quantum calculation.

Due to all kinds of reasons, Chinese research on quantum cryptography communication technology started late. In 1995, CAS Physics Institution first made the demonstration test in China based on BB84 protocol scheme. In July 2003, the experts of USTC Quantum Information Key Lab laid a 3.2-km-long cryptography communication system based on quantum secrets in the campus. In November 2003, ECNU researched the first quantum cryptography communication system successfully.

33.3 Military Application of Quantum Cryptography Communication

Quantum cryptography boasts the advantages of unconditional security and monitoring the buggers, which endows its great applicable value in military communication field [4, 5].

33.3.1 Concealment of Using in Military Communication

Wireless communication is to utilize electromagnetic wave to transmit signals, requiring variant antenna launching and strong electromagnetic wave radiation. Laser communication also requires strong optic waves of radiation. This strong electromagnetic and optic wave would expose the communication and destroy the hidden conditions of military communication. Quantum communication does not have electromagnetic wave radiation or optic wave radiation, which provides "Electromagnetic silence" environment of hidden communication. The enemies do not know whether communication exists or the position of communicators, which makes the communication really hidden.

33.3.2 Preventing Confidential Information from Bugged

In military communication, if adopting quantum communication, buggers intend to measure the polarization status of quantum but do not know which coordinate system shall be selected. Then his measurement action would make the measuring data of receivers producing 25 % errors. By monitoring error rate, buggers can be checked. While protecting the useful information, the bugger can be fooled and detected.

33.3.3 Quantum Cryptography Communication Can Utilize Computers Network to Deliver Information

Security confidentiality regulations regulates that military information shall not be transmitted through Internet in consideration of the potential threat in Internet. Thus, equipment connected with Internet cannot be used in military field. However, if adopting use quantum cryptography transmitting information, it is possible to transmit information on the Internet. Quantum teleportation utilizes the EPR particles' that remote connection to realize the transmission of unknown quantum sub-status, separating particle 1 information as classic information and quantum information to the receivers. Quantum information is the rest unrestricted information during the measurement by senders, which is transmitted through entangled status of EPR particles (called as particle 2 and 3).

During the process of measurement, particle 2 and 3 relieve the entanglement and entangles with particle 1. Particle 3 receives partial information of particle 1. Classic information is obtained by measuring particle 1 and 2. Using classic information, the receiver can recover the status of particle 1 through a orthogonal transformation corresponding the classic information on the hand-held particle 3, so as to transmit particle 1 to the receivers' hand. Considering that the third party cannot bug the information, users can receive the encrypted information from the encrypted quantum flow on the Internet, so that it is not necessary to lay out an exclusive wire route. This technology can be successfully used for the Internet transmission of military data. While guaranteeing the transmitting speed and quality, it can guarantee the security of information and reduces the military fee expenditure of vast military communication network.

33.3.4 Quantum Cryptography Communication Can Realize the Super Optical Speed Communication of Military Information

While transmitting information of quantum in quantum communication, it is unnecessary to know whether the receiver is located. Quantum information transmitting process will not be separated by any barriers, thus the teleportation status of quantum is called as the super space transmission of quantum states. The delivery speed of quantum information depends on the shrinking speed of quantum status. This shrinking speed has exceeded the velocity of light. Therefore, the transmission speed of quantum information exceeds the velocity of light. All information passing quantum channel transmission can realize the superluminal quantum teleportation, realizing the superluminal quantum communication. This means can advance the military communication capacity and realize leaping development of Chinese Army communication.

33.3.5 Used for Information Guarding

Thanks to its detectable feature for being bugged, quantum cryptography communication will make the network information security epoch-making revolutionary. Quantum encrypted equipment can integrate with the current optic fiber communication equipment, so as to make the fiber communication terminal equipment. Quantum communication does not send electronic magnetic radiation or strong light radiation. The enemy has difficulty in intercepting and destroying the other party's passwords, which can guarantee the security of information transmission. This technology can be used to improve the transmission security of military optic cable network information, so as to improve the information protecting ability.

References

1. Zeng G (2010) Quantum private communication. Higher Education Press, Berlin
2. Wang W (1998) The principle of quantum cryptography and its prospect. Commun Priv 1998 (1)
3. Zhang J (2005) New progress on experimental quantum cryptography. Physics 2005(10)
4. Gui Y (2002) Quantum key distribution. Prog Phys 2002(4)
5. Li J (2003) Quantum communication. Opt Commun Technol 2003(7)

Part III
Research on the Environment Character

Chapter 34
Mass Concentration of Atmospheric Particulate Matter and Pollution Characteristics of Heavy Metals in Southern Tailings Reservoir

Yingyun Liu, Pei Xiong and Danfeng Han

Abstract Total suspended particle (TSP) and fine particulate matter ($PM_{2.5}$) samples were collected continuously using an atmospheric particulate matter sampler at sampling points in a tailings reservoir in the south from October 28 to November 3, 2013. The mass concentrations of particulate matter in the air, the contents, and morphological analysis of heavy metal elements (Pb, Cd, U, Cu) were also studied. The results indicate that the air particulate matter concentrations exceed the mass concentration limits of national Grade II ambient air quality standard (GB3095-2012); The heavy metals content in air particulate matter from high to low order is: Pb > Cu > U > Cd and heavy metals are easier to be concentrated on fine particulate matter ($PM_{2.5}$); heavy metal elements (Pb, Cd, Cu) exist mainly in the exchangeable morphology, especially the element Cd.

Keywords Tailings reservoir · Particulate matter · Mass concentration · Morphological analysis

34.1 Introduction

This tailing reservoir is one where the tail sand contains heavy metals such as uranium, lead, cadmium, etc. At present, in the treatment project of the decommissioned tailings, the comprehensive treatment of the dam slope has been completed and the beach is the next management procedure. In order to provide the foundation for the tailings dam beach management project, the mass concentration of air particulate matters and characterization of associated heavy metals pollution were analyzed at sampling points in a tailings reservoir beach in autumn of 2013.

Y. Liu · P. Xiong (✉) · D. Han
School of Environmental Protection and Safety Engineering, University of South China, Hengyang 421001, Hunan, China
e-mail: gxylyy0924@126.com; 1072799825@qq.com

© Springer-Verlag Berlin Heidelberg 2015

S. Long and B.S. Dhillon (eds.), *Proceedings of the 15th International Conference on Man–Machine–Environment System Engineering*, Lecture Notes in Electrical Engineering 356, DOI 10.1007/978-3-662-48224-7_34

34.2 Experiment and Methods

34.2.1 Instruments and Reagents

Intelligent high-capacity total suspended particulate sampler TH-1000C type (Tianhong Intelligent Instrument Factory of Wuhan), smart large flow TSP sampler 2031 Type (Institute of Applied Technology in Laoshan, Qingdao), trace uranium analyzer MUA type (Beijing Research Institute of Uranium Geology), inductively coupled plasma atomic emission spectroscopy ICAP6300MFC, electric oven blast 101-2A type, electronic balance, muffle furnace, quartz cell, graphite crucibles, glass fiber filters, high speed centrifuges, PH meter, etc.

Hydrofluoric acid, nitric acid (Guaranteed reagent), distilled water, magnesium chloride, etc.

34.2.2 Sample Collection

A sampling point is located outside a pump room of the tailings dam area. Two samplers collected simultaneously total suspended particles (TSP) and fine particulate matter ($PM_{2.5}$) samples. Daily, 24 h samples were collected continuously from October 28 to November 3, 2013 and the sampling time is 10:00 (Beijing time). Six data respectively for TSP quality and $PM_{2.5}$ quality (total twelve data) were obtained and numbered in chronological order [1]: TSP numbered 1–6 and $PM_{2.5}$ numbered 7–12. At the same time, the meteorological data was also recorded.

34.2.3 Sample Monitoring and Analysis Methods

Mass concentration monitoring of total suspended particulates (TSP) and fine particulate matter ($PM_{2.5}$) was carried out according to "Ambient air-Determination of total suspended particulates-Gravimetric method"(GB/T15432-1995) and "Technical Specifications for gravimetric measurement methods for $PM_{2.5}$ in ambient air" (HJ 656-2013). Analysis of the uranium amount was conducted by "Analytical method of trace uranium in air-laser fluorescence method" (GB12377-90). The analysis and calculation of heavy metal element content in air were completed with reference of "Ambient air-Determination of lead-Graphite furnace atomic absorption spectrometry" (HJ539-2009).

34.3 Results and Analysis

34.3.1 Weather Conditions of Sampling Time

This paper records the weather conditions in the sampling period, including temperature, relative humidity, wind, etc., as shown in Table 34.1.

34.3.2 The Mass Concentration of Air Particulate Matter (TSP and $PM_{2.5}$)

Mass concentrations of TSP and $PM_{2.5}$ air particles at the sampling points in autumn are shown in Table 34.2. TSP concentration range is 421.6–154.9 ug/m^3, with average concentration of 318.2 ug/m^3. $PM_{2.5}$ concentration range is 318.9–78.5 ug/m^3, with an average concentration of 216.3 ug/m^3.

Table 34.2 shows that the mass concentration of atmospheric particulate matter in the tailings reservoir in autumn changes largely with the weather conditions. Under sunny dry weather lasting for several continuing days, the average 24 h maximum concentration of $PM_{2.5}$ and TSP are 421.6 ug/m^3 and 318.9 ug/m^3. They are both higher than the concentration limits of national Grade II ambient air quality standard (GB3095-2012) which are 300 ug/m^3 and 75 ug/m^3, respectively, for TSP and $PM_{2.5}$; Among them were the concentration of $PM_{2.5}$ reaches 4.3 times of the national standard in developed countries.

During the sampling period, average concentrations of TSP and $PM_{2.5}$ exceed the national Grade II and the $PM_{2.5}$ concentration exceeds the standard everyday. Because of rain, the increase in air relative humidity and other factors, the mass concentrations of atmospheric particulate matters TSP and $PM_{2.5}$ both decline.

The mass concentration ratios of $PM_{2.5}$/TSP at monitoring points are calculated, which are 50.7–75.7 %, with an average of 65.4 %. According to Table 34.1, the mass concentration ratio of $PM_{2.5}$/TSP is also large when TSP value is large; the mass concentration ratio of $PM_{2.5}$/TSP is also small when TSP value is small. It can

Table 34.1 Weather conditions in the sampling period

Date	Weather conditions	Relative humidity (%)	Temperature (°C)	Wind power
10.28–29	Sunny	41–75	17–26	Breeze
10.29–30	Sunny	29–70	18–24	Breeze
10.30–31	Sunny to overcast	46–80	16–19	Two grade
31–11.1	Overcast to light rain	53–90	17–21	Two grade
11.1–2	Overcast	75–87	17–20	Breeze
11.2–3	Overcast	79–91	15–19	Breeze

Table 34.2 Mass concentration of atmospheric particulate matters (TSP and $PM_{2.5}$)

Sample number	Mass concentration of TSP (μg/m^3)	Sample number	Mass concentration of $PM_{2.5}$ (μg/m^3)	Mass concentration ratio of $PM_{2.5}$/TSP (%)
1	416.9	7	283.6	68.0
2	395.2	8	274.4	69.4
3	421.6	9	318.9	75.7
4	317.9	10	226.7	71.3
5	202.3	11	115.9	57.3
6	154.9	12	78.5	50.7
Mean	318.2	–	216.3	65.4

Table 34.3 Concentration of metal elements in atmospheric particulate matters (TSP and $PM_{2.5}$) (ng/m^3)

Element name	Uranium (U)	Lead (Pb)	Cadmium (Cd)	Copper (Cu)
Content of the TSP sample	109.5	441.5	44.8	161.9
Content of the $PM_{2.5}$ sample	76.8	339.7	32.0	104.6
The percentage of $PM_{2.5}$/TSP sample content (%)	70.1	76.9	71.4	64.6

be inferred that small rain and other weather conditions are more conducive to the purification of fine particulate matter.

34.3.3 Air Uranium and Heavy Metals Concentrations

This paper respectively collected TSP and $PM_{2.5}$ samples in air and analyzed the content of heavy metal elements (Pb, Cd, U, Cu) with average mass concentration, as shown in Table 34.3.

According to Table 34.3, the distribution of metal elements contents in atmospheric particulate matter at the tailings reservoir is: Pb > Cu > U > Cd. The concentration of Pb element is highest among these elements, respectively, 441.5 and 339.7 ug/m^3 for TSP and $PM_{2.5}$.

Comparing Tables 34.2 and 34.3, it is shown that the average mass concentration ratio of particulate matters $PM_{2.5}$/TSP is 65.4 %; however, the average ratios of the mass content of heavy metals Pb, Cd, and U in $PM_{2.5}$/TSP increase and reach, respectively, 76.9, 71.4 and 70.1 %. The average ratio of the Cu mass content in $PM_{2.5}$/TSP is very close to the average ratio of $PM_{2.5}$/TSP. These results show that it is easier for heavy metals to be concentrated on $PM_{2.5}$.

34.3.4 Morphological Analysis of Heavy Metal Elements in Atmospheric Particulate Matters

In order to understand the mode of occurrence of heavy metal elements in atmospheric particulate matters of the tailings reservoir, this paper analyzes existing forms of trace heavy metal elements Pb, Cd, Cu in particulate matters TSP and $PM_{2.5}$ samples, using Tessier morphological classification and extraction [2]. The results are shown in Table 34.4.

In this table, F1 stands for exchangeable form, F2 for carbonate phase, F3 for iron (manganese) oxide bound, F4 for organic state and F5 for residual. At the same time, the percentages of different chemical forms are also calculated and showed as histograms in Figs. 34.1 and 34.2.

In the five kinds of chemical speciation using Tessier morphological analysis, heavy metals in the form of exchangeable speciation (F1) easily enter the environment through ion exchange and are absorbed by plants and humans [3]. In F1 speciation, percentages of the three heavy metals Pb, Cd, Cu in TSP, and $PM_{2.5}$ particulate matter are all high in this paper; element Cd is the highest, respectively, 80.8 and 78.3 % in TSP and $PM_{2.5}$; Pb and Cu also reach up to 48.1–55.2 %. It can be illustrated that exchangeable speciation is the main form of heavy metals Pb, Cd, Cu in air particulate matter in the tailings reservoir.

Heavy metal elements of Carbonate fraction (F2) can be released by mild acids [4]. In this thesis, the distribution of F2 is: Pb > Cu > Cd.

Heavy metal elements of Fe (Mn) oxide bound state (F3) can be deoxygenated when the pH value is reduced or redox potential is lowered. Therefore, conversion occurs easily when the external conditions change, increasing the activity and bioavailability which has potential hazards [5]. In this thesis, the elements Pb accounts for a large percentage, 21.5 % and elements Cd, Cu account for small percentage, only 3.6 to 5.6 %.

Heavy metal elements of Organic state (F4) can be released only under strong oxidizing conditions. The distribution of F4 in the tailings particles in air is Cu > Pb > Cd; the element Cu represents a percentage of 11.2–14.9 %, Cd only about 1 %.

Table 34.4 Matter concentrations of different chemical speciation of heavy metals in air particulates (ng/m^3)

Element	TSP					
	ρ(Total)	ρ(F1)	ρ(F2)	ρ(F3)	ρ(F4)	ρ(F5)
Pb	441.4	212.2	106.7	95.1	16.4	11.0
Cd	44.7	36.1	5.1	1.6	0.6	1.3
Cu	163.5	80.9	21.9	9.0	18.3	33.4
Element	$PM_{2.5}$					
	ρ(Total)	ρ(F1)	ρ(F2)	ρ(F3)	ρ(F4)	ρ(F5)
Pb	339.7	187.6	85.1	43.3	10.2	13.5
Cd	32.2	25.2	4.8	1.1	0.3	0.8
Cu	104.7	52.0	13.0	5.9	15.6	18.2

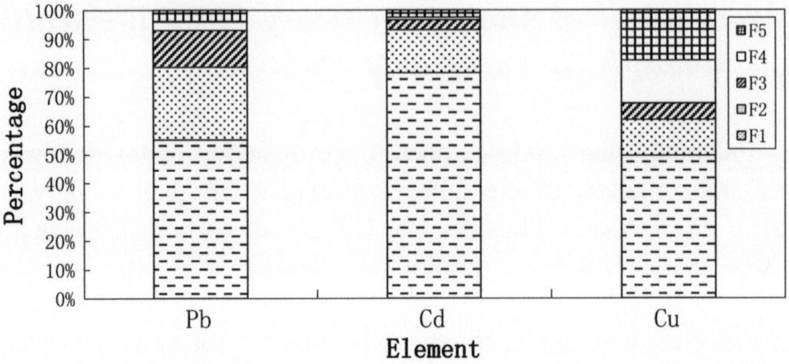

Fig. 34.1 Percentages of different chemical speciation of heavy metals in air particulates TSP in autumn

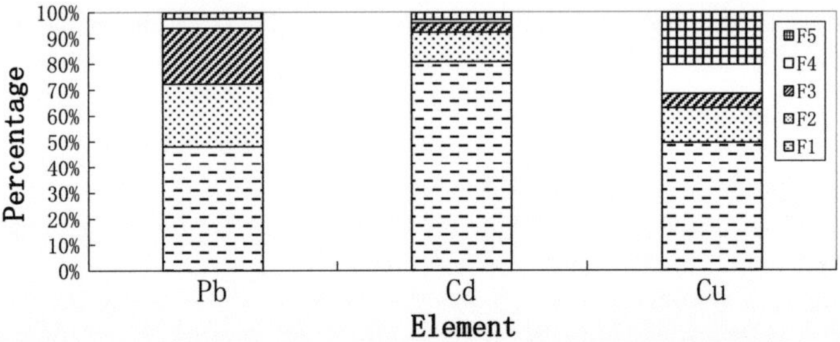

Fig. 34.2 Percentages of different chemical speciation of heavy metals in air particulates $PM_{2.5}$ in autumn

Heavy metal elements of residual fraction (F5) are stable and relatively safe to the environment. The element Cu represents a large proportion of 17.4–20.4 %, elements Pb and Cd only 2.5–4 %.

34.4 Conclusions

(1) TSP and $PM_{2.5}$ particulate matter concentrations in the tailings reservoir in autumn exceed the concentration limits of national Grade II ambient air quality standard (GB3095-2012) and the exceed situation of $PM_{2.5}$ is more serious.
(2) The order of the heavy metal element contents in particulate matters is Pb > Cu > U > Cd; uranium and heavy metal elements are more likely to be concentrated on $PM_{2.5}$.
(3) Heavy elements Pb, Cd, Cu in atmospheric particulate matters exist mainly in the form of exchangeable speciation which can easily cause harm to plants and humans, especially for the elements Cd.
(4) The tailings reservoir need the beach surface management in order to reduce particulate air pollution.

References

1. Feng X, Dang Z, Huang W (2008) Pollution level and chemical speciation of heavy metals in $PM_{2.5}$ during autumn in Guangzhou City. Environ Sci 29(3):569–575
2. Shuman LM (1985) Fractionation method for soil microelements Soil Sci 140:11–22
3. Zhai Y, Chen L (2010) The speciation and pollution evaluation of heavy metals in TSP of Changsha in Autumn. J Hunan Univ 5(37), 78–82
4. Lei M, Liao B, Qin P, Zhou X (2007) Morphologic analysis of Pb, Cd, Cu and Zn in the contaminated soil of a mine and study of their biological activity. Ecol Environ 16(3):807–811
5. Vangronsveld J, Cunningham SD (1998) Metal contaminated soils: insitu inactivation and phytorestoration. Springer, New York, pp 1–15

Chapter 35
Illumination Effect Estimation Based on Image Quality Assessment

Haoting Liu, Jin Yang, Wei Wang, Zhexi He and Wenpeng Yu

Abstract An image analysis-based illumination effect estimation method is proposed. This technique is developed for the illumination device design and application of space station. To estimate the illumination effect, first, the visible light camera is utilized to capture the image of illumination region. Second, the illumination direction and its distribution are estimated by the image analysis methods. The Mojette transform is used to estimate the illumination direction; while the image quality metrics are employed to evaluate the illumination distribution. Third, to improve the computational effect further, both the subjective and objective image quality evaluation methods are employed. Many experimental results have shown the validity and the correctness of our proposed technique.

Keywords Illumination direction · Illumination distribution · Image quality · Space technology · Visual ergonomics

35.1 Introduction

With the fast development of space technology, China will build her own space station [1] in the next a few years. Constructing a space station is a complex system engineering task; many subsystems will be built, including the space craft subsystem, the space environment and life safeguard subsystem, the power supply subsystem, the space robot subsystem, and the ExtraVehicular Activity (EVA) subsystem, etc. Different to the application on ground, the space station works in an extreme hard environment [2], in which it will face the vacuum, the cosmos radiation, and the alternative light change, etc. The astronaut will live in the

H. Liu (✉) · W. Wang · Z. He · W. Yu
Beijing Institute of Aerospace Control Device, Beijing, China
e-mail: imkyran@hotmail.com

J. Yang
Astronaut Research and Training Center of China, Beijing, China

© Springer-Verlag Berlin Heidelberg 2015

S. Long and B.S. Dhillon (eds.), *Proceedings of the 15th International Conference on Man–Machine–Environment System Engineering*, Lecture Notes in Electrical Engineering 356, DOI 10.1007/978-3-662-48224-7_35

space station and accomplish a series of scientific experiments. Thus, the design of space station should consider not only the function design but the ergonomic design [3].

The illumination device [4] will provide a proper environment light for video surveillance application of space station. It will affect the working state of most of the subsystems in that station. The EVA, the tele-operation robot application, or the state monitor of the formation flying satellite, will all need a sophisticated illumination environment. Because the space station will enter and leave the earth shadow periodically, the environment light source (including the light from the sun, the moon, and the earth, etc.) is relatively simple; in addition, the shapes or the surfaces of the working regions of space station are known. Thus, it is possible to estimate the environment illumination change accurately in orbit. The luminometer or even a camera can be used to evaluate the environment illumination.

In this paper, an image analysis-based illumination effect estimation method is proposed. The LED lamp will work with a visible light camera together, thus it can provide a reliable illumination environment for the distinct imaging of the camera. To achieve that target, first the camera will be used to capture the image of the target region. Second, several image analysis algorithms will be employed to estimate the illumination condition. The Mojette transform [5] will be utilized to estimate the light source direction; while the Image Quality Evaluation Metrics (IQEMs) [6] will be employed to analyze the illumination distribution. Third, in order to improve the computational precision, both the subjective and the objective image quality evaluation methods [7] are developed.

In the following sections, first the illumination system design of space station application will be presented. Second, the technique details about illumination evaluation will be introduced. Finally some experiment results will be given.

35.2 Space Illumination Applications

Figure 35.1 shows us the environment light analysis and a kind of illumination device design of the space station application. Image (a) shows us the environment light source of the space station. Image (b) is a case of illumination device design. From Fig. 35.1 we can see that: first the illumination device is needed not only when the space station enters the earth shadow region, but in the case of flying out of that region. Because the space station will move very fast, the main light source, i.e., the sun light, will also change its illumination direction and intensity quickly. As a result, we have to use the illumination device to make a light compensation in order to get a better video imaging effect. Second, both the visible camera and the LED lamp are utilized in the illumination device. When using the illumination device, a robotic arm can be used to send this device to a wanted position and implement the photograph or video monitor task.

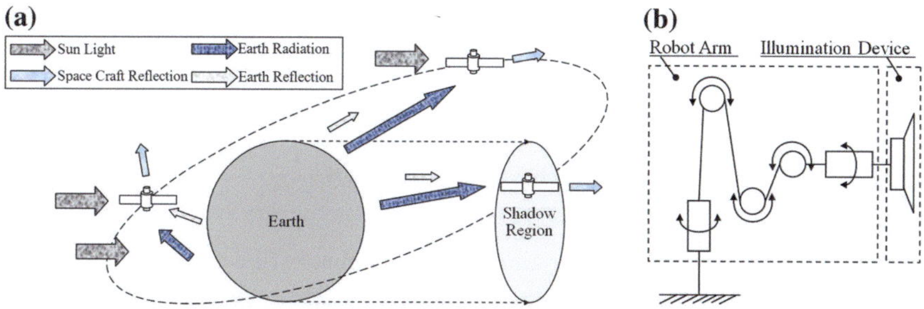

Fig. 35.1 The application prototype of the proposed space illumination device

35.3 Proposed Method

35.3.1 Methodology

Figure 35.2 shows us the computational flowchart of the proposed illumination estimation method. Both the illumination direction and its distribution are calculated. When estimating the illumination direction, the texture analysis-based Mojette transform is utilized. The Mojette transform calculates a series of projection relationships in a finite set of the observation angle, thus it is sensitive to the direction change of image texture. As for the space station application, the surface texture of the space shuttle is a kind of metal; the texture can reflect the illumination direction distinctly. The IQEMs are employed to assess the illumination distribution of that image region. Some of the no-reference IQEMs are utilized to evaluate the image brightness distribution. The no reference IQEMs are independent with the contents of image, thus it can be used to evaluate the brightness objectively. To

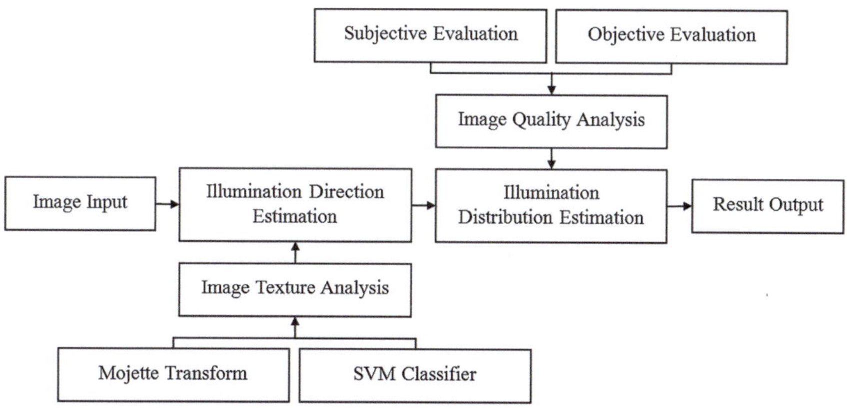

Fig. 35.2 The proposed illumination effect estimation method

improve the computational effect of IQEMs, both the subjective and the objective image quality evaluation techniques are utilized.

35.3.2 The Illumination Direction Estimation

The Mojette transform is used to estimate the illumination direction. Let us consider an image block *I(i, j)* and an observation angle θ_k: for each θ_k, we can create a corresponding projection direction and compute a cluster of the projection results by accumulating the gray values in that direction simply. We can realize that process by calculating the Eqs. (35.1) and (35.2). In (35.2) symbol n_k is the element number of a projection. The Mojette transform is a texture analysis-based method; thus it is needed to choose a proper image region when using this tool. After we have used the Mojette transform to collect the texture feature of the illumination region, we can employ the Support Vector Machine (SVM) [8] to identify the illumination direction because of its high classification precision.

$$M_{\theta_k}(x,y) = \sum_{x=1}^{M}\sum_{y=1}^{N} I(x,y)H(o_k - yp_k + xq_k) \quad (35.1)$$

$$n_k = (M-1)|q_k| + (N-1)|p_k| + 1 \quad (35.2)$$

where *I(x, y)* is the intensity of an image. Symbols *M* and *N* are the height and the width of an image. If $x = 0$, $H(x) = 1$; otherwise, $H(x) = 0$. Symbol θ_k is the observation angle, symbol $\tan\theta_k = q_k/p_k$, $p_k, q_k \in Z$, $GCD(p_k, q_k) = 1$. Symbol o_k is the intercept of the projection line *k*.

35.3.3 The Illumination Distribution Estimation

The IQEMs are utilized to evaluate the illumination distribution. On one hand, the flood fill algorithm [9] is utilized to segment the initial illumination regions. The flood fill algorithm is an image region processing technique, which can combine the pixels with similar intensity. On the other hand, both the image brightness metric and the contrast metric can be utilized to compute the illumination intensity in the computed illumination regions. The computational equations of these IQEMs are shown by (35.3) and (35.4). The image brightness metric M_{IBL} can reflect the illumination intensity of the image block directly, while the image contrast metric M_{ICL} can only describe the illumination effect. It means that M_{ICL} can evaluate the illumination distribution indirectly.

$$M_{\mathrm{IBL}} = \left(\sum_{k=1}^{N} h_k \cdot (I_k)^p\right) \Big/ N \tag{35.3}$$

$$M_{\mathrm{ICL}} = \sum_{k=1}^{N} \left(\left(I_{\max}^k - I_{\min}^k \right) / \left(I_{\max}^k + I_{\min}^k \right) \right) \Big/ N \tag{35.4}$$

where symbol h_k is the pixel number of the gray value I_k in image histogram; symbol N is the pixel number of the sample block; and symbol p is a control parameter, here $p = 3$. Symbols $I_{\max}^k$ and $I_{\min}^k$ are the maximum and minimum gray values of the kth block.

The ergonomic experiment method is utilized to build the relationship between the subjective evaluation levels and the objective computation results of the illumination image IQEMs. First, numbers of accumulated illumination images are collected to build a database. Second, we design an ergonomic experiment to ask subjects to evaluate the illumination effect of that database above according to their subjective cognition results. For example we can inquire the subjects to score the illumination effect into five levels. Then we can get a subjective illumination classification of different illumination environment. Third, we use the equations in (35.3) and (35.4) to calculate the corresponding images in that database again; then we can get the objective evaluation result of that image dataset. Finally, we can get a relationship between the subjective evaluation level and the objective computation results by the Canonical Correlation Analysis (CCA) [10] method. Thus the design method of the objective IQEMs can be improved by that relationship.

35.4 Experiments and Discussions

A series of simulation experiments are carried out in our PC (Intel(R) Core (TM) i5-3337U CPU, 1.8 GHz, 4.0 GB RAM). In the following parts, first some evaluation and simulation results will be presented. Then a discussion will be given.

35.4.1 Evaluation of Illumination Estimation Method

To show that our proposed method can estimate the illumination effect, a series of experiments are designed. Figure 35.3a, b shows us a test image sample and its evaluation result. In order to evaluate the illumination direction estimation technique, i.e., the computational effect of Mojette transform, the PhoTex texture database (http://www.macs.hw.ac.uk/texturelab/resources/databases/photex/) is used here. Figure 35.3a shows us an image sample in that database. The texture data is "1.acc.0.60.270". Figure 35.3b shows us the computational result of the Mojette transform in 0°–180°. From this result, we can see that the illumination direction can be estimated correctly by that method. Figure 35.3c presents an image sample for the subjective and

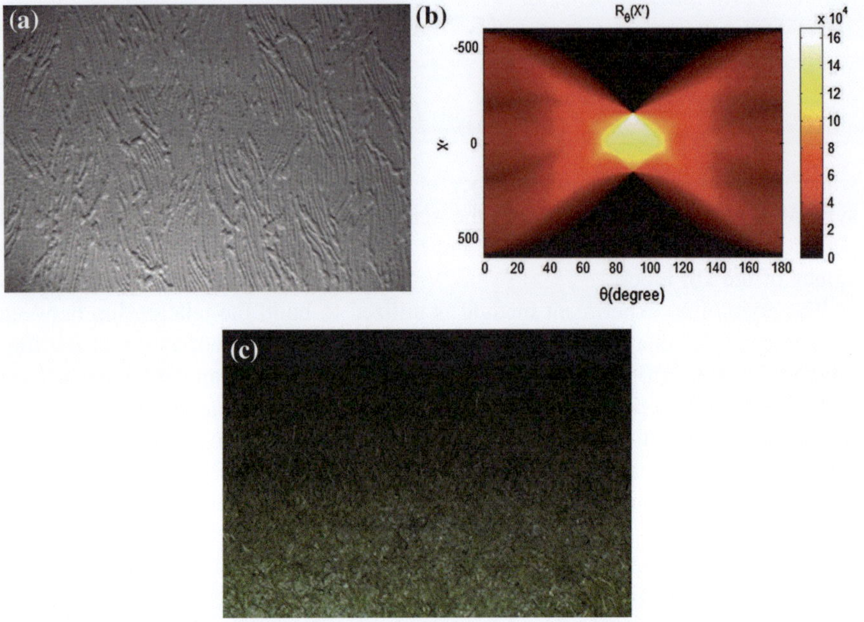

Fig. 35.3 The test image samples and the computational result of Mojette transform

the objective image quality evaluation methods. As for image (c), the subjective evaluation level is 4 (the total subjective classification levels are 5); while the objective image quality metrics, i.e., the M_{IBL} and M_{ICL}, are 113.667 and 0.325, respectively. Thus we can build the connection between the subjective and the objective image quality evaluation methods by CCA technique. And the computation effect of the objective evaluation metric can be improved by CCA.

35.4.2 *Simulation and Application*

To evaluate the computation effect of our proposed method, a simulation program is designed to test the computation effect of illumination estimation. Figure 35.5 shows us the simulation interface and its corresponding illumination simulation results. Image (a) is the simulation interface, in which two light sources are set to a surface texture. The intensity of light source and its position can be set arbitrarily in that program. Images (b) and (c) are its simulation result. The ray tracing method [11] is used to simulate the illumination effect. In (b) the light source intensity and its position are 3 and (1, 1, −2), respectively; in (c) the light source intensity and position are 8 and (3, −3, −2). The target region in Fig. 35.4a is an uneven plane with some textures. This program can be used to simulate the space equipment maintenance task and its illumination effect for a space station application.

(a) (b) (c)

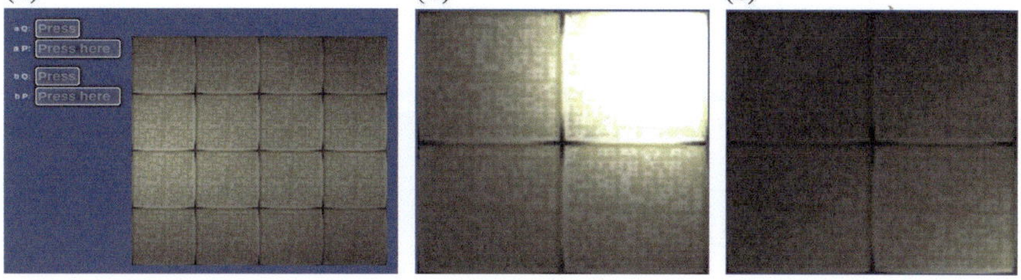

Fig. 35.4 The interface of an illumination effect simulation program and its simulation results

Table 35.1 The comparison results between the setting parameters of simulation program and the image evaluation results of our proposed method

Images	Light source direction	Illumination intensity	Illumination distribution
(b)	30/36.7	3/3	Sole/sole
(c)	45/55.1	5/5	Multiple/multiple

Table 35.1 shows us the comparison results between the setting parameters of that simulation program above and the evaluation results of our proposed method for image (b) and (c) in Fig. 35.4. In Table 35.1, the data in the right side of the oblique line is the setting value; while the data in the left side of the oblique line is the estimated value. Currently, the direction index of the estimated light source is only a 2D vector. The illumination distribution can be estimated by the Gaussian Mixture Model (GMM) [12]. In this paper, only two issues are considered, i.e., the sole GMM distribution and the multiple GMM distributions. From those results, we can see our method can identify the environment illumination effect to some extents.

35.4.3 Discussions

The evaluation of illumination environment is a significant task in many application areas. The remote medical application, the engineering operation, the marine exploration, etc. all have their own illumination environment requirement. In our research work, we classify the illumination problem as the illumination evaluation issue and the illumination control one. Currently, even the illumination evaluation issue cannot be solved well because of the complexity of the environment light. The light almost comes from everywhere, thus it is very difficult to judge the light source and its intensity. To make this problem simple, we can simplify the number of light source and use its shadow to analyze the direction of light source in future.

Figure 35.5 shows us the related technique and its application prospect of the intelligent illumination system. Different to the traditional intelligent illumination

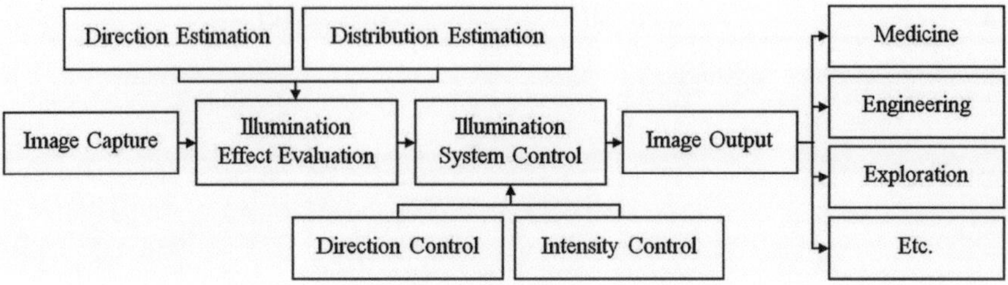

Fig. 35.5 The related technique and application prospect of the intelligent illumination system

system, which only uses the photosensitive resistance or the infrared sensor to detect the environment change, we propose to use the visible light camera to estimate the environment light. Because the visible light camera has a wide light wave response field, more particular analyses can be made by this kind of light sensor. Both the 2D and the 3D environments light distribution should be reconstructed in future. The no-reference image quality metric can be used as a powerful tool to estimate the illumination environment.

35.5 Conclusion

In this paper, an image analysis-based illumination effect estimation method is proposed. In contrast to other system or application, both the Mojette transform and the image quality metrics are utilized to estimate the light source direction and the illumination distribution. An integrated image quality evaluation method which considers both the subjective and the objective image quality evaluation technique is proposed. The advantage of this method is its fast computational speed; while its disadvantage is its limited processing ability. Currently, it can only be used when the illumination surface is simply or known. In future, the real-time 3D environment light estimation technique will be developed for our system.

References

1. Chen Y, Yang Y, Qian H, Xu Y (2012) The design of robot for space station operation. In: IEEE international conference on robotics and biomimetics, pp 247–252
2. Kennewell JA, Vo B-N (2013) An overview of space situational awareness. In: International conference on information fusion, pp 1029–1036
3. Liu H, Li J, Huang W, Han D, Xu F, Yang W, Zhou S (2010) Ergonomic evaluation and design of the VDT system for the hermetic cockpit. In: IEEE international conference on computer-aided industrial design and conceptual design, pp 629–632
4. Wu L, Zhang T, He S, Zhang B, Li N, Liu S (2010) Junction temperature measurement on the light-emitting diodes lamp for space applications. In: International conference on optoelectronic mircoelectronic material devices, pp 131–132

5. Recur B, Desbarats P, Domenger JP (2010) Mojette reconstruction from noisy projections. In: International conference on image processing theory tools applications, pp 201–206
6. Liu H, Cheng J, Lu H (2010) Abnormal change detection of image quality metric series using diffusion process and stopping time theory. In: Pacific rim conference on multimedia, pp 272–283
7. Winterlich A, Zlokolica V, Denny P, Kilmartin L, Glavin M, Jones E (2013) A saliency weighted no-reference perceptual blur metric for the automotive environment. In: International workshop quality multimedia experience, pp 206–211
8. Patle A, Chouhan DS (2013) SVM kernel functions for classification. In: International conferences on advances technology engineering, pp 1–9
9. Bhargave N, Trivedi P, Toshniwal A, Swarnkar H (2013) Interactive region merging and objective retrieval method using mean shift segmentation and flood fill algorithm. In: International conferences on advances computing communications, pp 157–160
10. Liu H, Li J, Wang Z, Chen J, Lu H, Zhao Y (2010) Image quality feedback based adaptive video definition improvement for the space manipulation task. Int J Image Graphics 11 (2):153–175
11. Underwood JP, Gillsjo D, Bailey T, Vlaskine V (2013) Explicit 3D change detection using ray-tracing in spherical coordinates. In: IEEE international conference on robotics and automation, pp 4735–4741
12. Liu H, Wang W, Li X, LI F (2012) Defect segmentation of fiber splicing on an industrial robot system using GMM and graph cut. In: IEEE international conference on robotics and biomimetics, pp 1968–1972

Chapter 36
An Analysis on Jamming Threat of Air Raid Targets

Honglei Li, Hao Lu, Meng Kang and Jiang Luo

Abstract Modern air raid operations are featured by large scale, high intensity and diversified aircraft platforms with complex jammers of strong jamming capability, various jamming methods and broad jamming frequency spectrum. It is actually fairly difficult to make judgment on the threat of air raid targets, as there are many factors that need to be taken into account. This paper analyzes the main factors that may affect the jamming threat of air raid targets, and sets up the analytical model of jamming threat of air raid target by joining altogether nine factors of platform and radar in accordance with multi-attribute decision theory. The model is then applied in the calculation of three jammers and the result agrees fundamentally with the actual situation, therefore providing an effective approach to the jamming threat judgment for land-based air defense operations.

Keywords Air raid target · Jamming · Threat

36.1 Introduction

The rapid development of enemy electronic equipment, electronic jamming method and system make possible to reduce aircraft loss and increase hitting accuracy. In the meantime, hostile jamming in the conditions of such high and new technology system is posing increasing threat on our electronic devices like radar, without which finding, tracking, and guiding become impossible and the air defense combat effectiveness shall be greatly reduced. Recently, local wars have proved that air raid operation is scaling up, jammers of various new systems and new technologies are emerging one after another, and aircraft platforms for jammers are developing faster and faster. It is thus an issue that needs urgently to be addressed to evaluate the threat of hostile jamming source quickly and effectively, and bring electronic

H. Li (✉) · H. Lu · M. Kang · J. Luo
Air Defence Forces Academy, Zhengzhou 450052, China
e-mail: fkb_mmes@163.com

© Springer-Verlag Berlin Heidelberg 2015

S. Long and B.S. Dhillon (eds.), *Proceedings of the 15th International Conference on Man–Machine–Environment System Engineering*, Lecture Notes in Electrical Engineering 356, DOI 10.1007/978-3-662-48224-7_36

devices like radar into full play. In this paper, the method to evaluate threat of air raid targets is studied with mathematical models whose practicality is illustrated with an example.

36.2 The Comprehensive Modeling of Air Raid Target Jamming Threat

36.2.1 Main Factors that Affect Jamming Threat

Modern air raid operations are featured by large scale, high intensity, and diversified aircraft platforms with complex jammers of strong jamming capability, using various jamming methods, and broad jamming frequency spectrum. It is therefore necessary for the defensive side to take resistant measures against jamming, or to strike the air-based threat source directly. Threats of various jamming sources on the defensive side are different.

Main factors that may affect the jamming threat of a radiation source include the velocity and the height of the jammer platform, range between jammer and radar, jamming power [1], type of jamming, jamming frequency coverage, radar type and system, anti-jamming capacities of targeted radar, and so on.

36.2.2 Determining the Analytical Model of Main Attributes

36.2.2.1 The Factor of Velocity of Jammer Platform

As a general rule, a faster jammer platform means that more offensive and greater jamming threat. Therefore the velocity attribute E_1 of the jammer platform can be expressed as:

$$E_1 = \frac{V_i - V_{\min}}{V_{\max} - V_{\min}} \tag{36.1}$$

In the above equation, v_i represents the velocity of the jammer platform number i, while v_{min} and v_{max} refer respectively to the minimum and maximum velocities of all jammer platforms.

36.2.2.2 The Factor of Height of Jammer Platform

Under normal circumstances, a lower jammer platform is less likely to be detected and therefore is of greater threat. Thus the height attribute E_2 of the jammer platform can be expressed as:

$$E_2 = \frac{h_i - h_{\min}}{h_{\max} - h_{\min}} \tag{36.2}$$

In the above equation, h_i represents the height of the jammer platform number i, while $h_{\min}$ and $h_{\max}$ refer respectively to the minimum and maximum heights of all jammer platforms.

36.2.2.3 The Factor of Range Between Jammer Platform and Radar

In general, a jammer platform closer to ground targets is of greater threat. Therefore the range attribute E_3 of the jammer platform can be expressed as:

$$E_3 = \frac{R_{\max} - R_i}{R_{\max} - R_{\min}} \tag{36.3}$$

In the above equation, R_i represents the range between the jammer platform number i and the ground target, while $R_{\min}$ and $R_{\max}$ refer respectively to the minimum and maximum ranges of all jammer platforms.

36.2.2.4 The Factor of Jamming Power

The jamming power is determined by tactical demands. Usually, it is required that in the minimum jamming range, the jamming power be greater than the echo signal at the input terminal of targeted radar receiver, with a SIR of 10–13 dB. At present, CW jamming power is several hundred watts, or even as high 1–2 kw; PW jamming is several kilo watts, or even as high as tens of kilo watts with a duty ratio of 2–10 % [2].

If we use E_4 to represent the power factor to evaluate the influence of receiver SIR upon jamming capability [3], it is expressed as:

$$E_4 = \begin{cases} 0, & P_{\mathrm{rj}}/P_{\mathrm{rs}} \le 0.5\,K_j \\ 1 - \frac{2}{3}\left(\frac{P_{\mathrm{rj}}/P_{\mathrm{rs}}}{K_j} - 0.5\right), & 0.5K_j < P_{\mathrm{rj}}/P_{\mathrm{rs}} < 2K_j \\ 1, & P_{\mathrm{rj}}/P_{\mathrm{rs}} \ge 2K_j \end{cases} \tag{36.4}$$

In the above equation, P_{rj} stands for the jamming signal power received by radar, P_{rs} for the echo power, K_j for the SIR for radar to work normally. A greater value of E_4 means greater jamming capability. The SIR $P_{\mathrm{rj}}/P_{\mathrm{rs}}$ is calculated as:

$$\frac{P_{\mathrm{rj}}}{P_{\mathrm{rs}}} = \frac{P_j}{P_s} \cdot \frac{4\pi R^2 G_j}{\sigma G_A F_A} \tag{36.5}$$

In the above equation, P_j and P_s refer, respectively, to the transmitting powers of jammer and radar, R is the range between radar and jammer, G_j is the gain of jammer on radar antenna, σ is the scattering area of the self-defense aircraft, G_A is the radar main lobe gain, F_A is the equivalent loss of jamming power.

36.2.2.5 The Factor of Type of Jamming

The type of jamming poses immediate influence upon the effect to some extent, on condition that other factors are the same. The optimum type of jamming is usually determined according to the system, operating mode and performance parameters of the targeted radar. The type of jamming is categorized into combined jamming, suppression jamming and deception jamming [4].

Suppression jamming means to transmit strong jamming signals through electro-magnetic devices to disturb or destroy the normal working of radar of the counterpart, which is of noticeable effect on all radars and is therefore frequently applied in ways such as spot jamming, barrage jamming, and frequency-swept jamming.

Deception jamming is employed to transmit fake signals to targeted radar receiver in order to increase the targeted radar's work burden and distract it from real targets; it is divided into range deception jamming, angle deception jamming, and speed deception jamming.

Combined jamming, as is usually seen very effective, is understood as a combination of suppression jamming and deception jamming.

On the basis of the jamming mechanism cited from document [2], and combining the effects of various jamming methods, we set the range of threat as [0, 7] and obtain Table 36.1 as follows:

In practical application, the resultant jamming capability of each jammer is approximately equal to the sum of all jamming capacities. Therefore the factor of type of jamming E_5 is expressed as:

$$E_5 = \sum_{i=1}^{7} U_i C_i \tag{36.6}$$

In the above equation, U_i is assigned the value of 0 or 1, namely, $U_i = 1$ if a designated type of jamming is adopted, and $U_i = 0$ if the corresponding type of jamming is not adopted; C_i stands for the value of the threat on radar when the type of jamming number i is adopted.

Table 36.1 Threat on radar with various types of jamming

Type of jamming	Spot jamming	Barrage jamming	Frequency-swept jamming	Range deception jamming	Angle deception jamming	Speed deception jamming	Combined jamming
Value	2	2	2	2	2	2	6

36.2.2.6 The Factor of Jamming Frequency Coverage

The jamming frequency spectrum must cover the frequency spectrum of the targeted radar, whose typical value is 0.5–18 GHz and is now expanding to MMW (35–40 GHz).

Take E_6 as the frequency attribute factor to evaluate the matching degree of transmitting frequency and receiving frequency. If the range of radar frequency is $[f_{r1}, f_{r2}]$, the range of jammer frequency is $[f_{j1}, f_{j2}]$, and $d\{[f_{r1}, f_{r2}]\cap[f_{j1}, f_{j2}]\}$ represents the intersection of two frequency ranges, then:

$$E_6 = \begin{cases} 1 - \frac{\min(f_{j2}, f_{r2}) - \max(f_{j1}, f_{r1})}{f_{r2}, f_{r1}} & d\{[f_{r1}, f_{r2}]\cap[f_{j1}, f_{j2}]\} = 0 \\ 1, & d\{[f_{r1}, f_{r2}]\cap[f_{j1}, f_{j2}]\} \neq 0 \end{cases} \quad (36.7)$$

A greater value of E_6 means better matching of radar frequency and jamming frequency and therefore better jamming effect.

36.2.2.7 The Factor of Type of Targeted Radar

Modern ground-to-air radar includes early warning radar, height finding radar, fire control radar, guiding radar, observation post radar and so on [4]. The value of radar type can be assigned by expert scoring method. Early warning radar, guiding radar, and fire control radar are assigned comparatively greater values because early warning radar is the primary guarantee for air defense units to find air situation, while guiding radar and fire control radar show that there is firepower against aerial target. The threat on height finding radar is comparatively less; as observation post radar is employed against enemy aircraft low altitude penetration, it is least prioritized and is assigned he smallest value. By expert scoring method, the values are respectively 0.7, 0.6, 0.9, 0.9, 0.5, and the factor of type of targeted radar E_7 is thus obtained.

36.2.2.8 Anti-jamming Technical Capacities of Targeted Radar

On the basis of the values of influences of technical measures upon anti-jamming capacities given in document [5], and combining relevant characteristics of radar anti-jamming technologies, we list the influences of jamming upon ten types of anti-jamming measures as is shown in Table 36.2:

In practical application, the resultant jamming capability is approximately equal to the sum of all technical measures that are jammed. Therefore the factor of anti-jamming technical capacities, E_9, is expressed as:

Table 36.2 Influences of jamming upon anti-jamming technical measures

Technical measures	Frequency agility	MTD	Pulse compression	Polarized variability	Broadband-limiter-narrowband	Variable frequency repetition	Variable frequency sweep	Frequency diversity	Side lobe suppression	Complex signal processing
Value	2	4	4	4	6	6	6	6	6	6

$$E_9 = \sum_{i=1}^{10} U_i C_i \tag{36.8}$$

In the above equation, U_i is assigned the value of 0 or 1, namely, $U_i = 1$ if a corresponding system is adopted by the targeted radar, and $U_i = 0$ if not; C_i stands for the value of the influence upon radar anti-jamming capacities when the anti-jamming measure number i is adopted.

36.2.3 Determining the Evaluation Scheme Set

The evaluation scheme set V is the collection of jamming sources of three jammers, i.e., $V = \{V_1, V_2, V_3\}$. We need to conduct comprehensive analysis on the aforementioned nine attributes with the precondition that the jammer is carried in the same environment, so as to obtain the evaluation value of jamming threat of various schemes.

36.2.4 Determining the Weight of Each Evaluation Attribute

The weight of each evaluation attribute shows its significance for the overall jamming capacity; as the preference information of the attribute, the weight shall pose considerable influence on the final decision-making. There is many a method to determine the weight such as Delphi method, AHP method, link relative method, interval estimation method, and so on. We adopt the 1–9 scale in the AHP method according to the opinions of experts and set up the matrix for the importance of those nine attributes to be compared between any two as follows:

$$B = \begin{bmatrix} 1 & 1 & 1 & 1/5 & 1/7 & 1/5 & 7 & 9 & 5 \\ 1 & 1 & 1 & 1/3 & 1/7 & 1/5 & 7 & 9 & 5 \\ 1 & 1 & 1 & 1/3 & 1/7 & 1/5 & 7 & 9 & 5 \\ 5 & 3 & 3 & 1 & 1/3 & 1 & 5 & 6 & 1 \\ 7 & 7 & 7 & 3 & 1 & 1/3 & 3 & 7 & 3 \\ 5 & 5 & 5 & 1 & 3 & 1 & 5 & 5 & 3 \\ 1/7 & 1/7 & 1/7 & 1/5 & 1/3 & 1/5 & 1 & 7 & 1 \\ 1/9 & 1/9 & 1/9 & 1/6 & 1/7 & 1/5 & 1/7 & 1 & 1/5 \\ 1/5 & 1/5 & 1/5 & 1 & 1/3 & 1/3 & 1 & 5 & 1 \end{bmatrix}$$

The matrix is proved for satisfying conformity. By the eigenvalue method we find the eigenvector of the judgment matrix, and the eigenvector corresponding to the maximum eigenvalue of the judgment matrix shall be the weight coefficient of each evaluation attribute relative to the overall jamming capability. Through computation, the weight coefficient of each attribute is (0.08, 0.08, 0.08, 0.15, 0.26, 0.25, 0.03, 0.02, and 0.05).

36.2.5 *Determining the Decision Model*

There are many multi-attribute decision models, and we may select the proper one according to a specific issue. As each attribute is playing a certain role in the comprehensive evaluation, the linear weighted model is therefore adopted, and the value of each evaluation scheme is as follows:

$$E = \sum_{i=1}^{9} E_i w_i \quad (36.9)$$

In the above equation, E_1 is the normalized value f each attribute factor, and w_1 is the weight of each attribute factor.

36.3 Case Study

Consider the US army AN/ALQ-99, AN/ALQ-131 and AN/ALQ-126B jammers for example in this case study and apply the aforementioned models to compute their jamming capacities. In addition, suppose the carrier's scattering area is 2 m^2, the jammer antenna gain is 0 dB, the equivalent attenuation F_A is 70 dB, and K_j is 0.3 [5].

We can compute the values of all attribute factors of each jammer as is shown in Table 36.3:

E_5 and E_7 need normalizing. After normalization, obtain the jamming threat value of evaluation scheme set $V = \{V_1, V_2, V_3\} = \{0.54, 0.5, 0.66\}$, thus the jamming sources can be sequenced as $V_3 > V_1 > V_2$, consistent with experts' qualitative understanding of those jammers, which means that the model is rational and feasible.

Table 36.3 Values of radar's attribute factors

Jammer	E_1	E_2	E_3	E_4	E_5	E_6	E_7	E_8	E_9
V_1	0	1	0	0.17	2	1	0.7	0.9	10
V_2	0.3	0.5	0.5	0.34	2	1	0.9	0.9	6
V_3	1	0	1	0.75	6	1	0.9	0.6	10

36.4 Conclusion

This paper applies the multi-attribute decision theory and gives an approach to the judgment on jamming threat by combining qualitative and quantitative methods, on the basis of a number of empirical data and experts' opinions, fairly accurate and true evaluation results are obtained, because sufficient consideration is given to various attributes that may influence the judgment on jamming threat, particularly to some qualitative attributes that are difficult to be quantified.

References

1. Shen Y, Chen Y (2006) Rating of radar radiation source threat in air raid resistance combat. Radar Sci Technol Hefei 12(6):328–331
2. Zhang Y, Tong N, Zhao G (2010) Principle of radar electronic war 8(8):132–134. National Defense Industry Press, Beijing
3. Zhou L (2001) Research on the applications of multi-attribute decision in radar jamming effect assessment. Radar Confront Xian 2(2):14–18
4. Zhang X, Xiao K, Gu J (2010) Introduction to new system radar confrontation. Beijing Polytechnic University Press, Beijing, pp 13–15
5. Li R (2002) Theory and applications of fuzzy multi-criterion decision. Science Press, Beijing, pp 138–147

Chapter 37
A Study on Teaching Environment of Military Practical Courses

Jiang Wu, Zhibing Pang, He Wei, Honglei Li and Meng Kang

Abstract This paper is purposed to analyze possible influences upon military practical courses posed by teaching environmental factors and their characteristics. By combining theoretical analysis method and practical teaching statistical method, it sums up and sorts out conditions of climate, conditions of the site and natural conditions, summarizes the merits and demerits and provide basis for scientific decision-making, and identifies various direct and indirect influences upon military practical courses by environmental factors. Through analyses of conditions of climate, conditions of the site and natural conditions in field training of military practical courses, this paper gives countermeasures on how to effectively utilize environmental factors and improve military practical courses.

Keywords Military practical courses Teaching · Environmental factors · Study

37.1 Introduction

Class instruction in the military academy of professional education is divided into theoretical lecturing and practical training. Theoretical lecturing is centered on fundamental and professional theories, while practical training focuses on operation and training administration, including weapon operation, tactical exercises, comprehensive drills, live firing, and so on. Practical training is an important mode to improve the quality of instruction, because it serves as a guarantee to motivate trainees' initiative and acts as a catalyst to internalize knowledge and convert it into capabilities [1]. As practical training involves man (trainee), machine (equipment), and environment, it has remained to be one of the most difficult issues to fall it in instructional methodology of tertiary education [2].

J. Wu (✉) · Z. Pang · H. Wei · H. Li · M. Kang
Air Defense Forces Academy, 450052 Zhengzhou, China
e-mail: wujiang629143@outlook.com

© Springer-Verlag Berlin Heidelberg 2015

S. Long and B.S. Dhillon (eds.), *Proceedings of the 15th International Conference on Man–Machine–Environment System Engineering*, Lecture Notes in Electrical Engineering 356, DOI 10.1007/978-3-662-48224-7_37

Military MME (Man-Machine-Environment) system engineering researches on military men, weapons and equipment, training, and combat environment [3]. Military practical training is conducted on the basis of MME system engineering conditions. Only in accordance with the direction and demand of MME system engineering theories we can realize the goal to bring the system into full display and improve the quality of instruction of military practical training. It is therefore of great significance to strengthen studies on environmental factors in military practical training so as to highlight the characteristics of field training, promote the effectiveness, and for trainees to acquire knowledge, master skills, and to increase their capabilities.

37.2 Environmental Factors that Influence Military Practical Training

Consider field exercises in military practical training as the object of study, and we deem that environmental factors include climatic conditions, conditions of the teaching site, and sociological conditions.

37.2.1 Climatic Conditions

Teaching and studying go along with certain objective environment, in which climatic conditions are influential and included. Hot summer makes a man sleepy and freezing winter brings him agony. Favorable climatic and weather conditions ensure high study-effectiveness. Climatic conditions are important for military building, strategic planning, combat and training, weapon-equipment development and testing as well as national defense project construction [4]. Climatic conditions can be categorized into natural climatic conditions and abrupt climatic conditions. Natural climatic conditions refer to the general climatic conditions in a certain area through years of observation and statistics; abrupt climatic conditions are understood as sudden climatic changes in military march or exercises.

37.2.2 Conditions of the Teaching Site

Conditions of the teaching site are understood as the conditions supporting practical training in a teaching site where trainees learn to operate, generate technical ability, and consolidate skills. Good conditions, advanced facilities, and reasonable disposition in a teaching site are positive for training administration, high training efficiency, as well as physical and psychological conditions of instructors and trainees. Conditions of teaching site consist of the size, safety, and suitability. The size of a teaching site should conform to the scale of the training subject; safety means elimination of threats to personnel and equipment; suitability is related with whether or not the site is suitable for training regarding evenness and stiffness.

37.2.3 Natural Conditions

In the field, the bright sunshine, green hills, clean air, and wide vision together provide favorable conditions for study on the one hand, but on the other hand, they may distract trainees from lectures or operation. A siren, a horn beep, a rumbling machine, or a dog's bark may make trainees absent-minded and affect the effect of instruction adversely. Concentration of trainees on lectures or operation is crucial for field instruction.

37.3 How to Improve Military Practical Training with Environmental Factors

Compared to indoor lecturing, military practical training is conducted in larger space and for wider influence. Instructors should therefore take all possible advantages of and optimize environmental factors by basing interests on practical training, paying attention to scientificalness, seeing from the angle of cultivation and education and creating favorable atmosphere by various means.

37.3.1 Carrying out Active Study to Have an Intimate Knowledge of Teaching and Training Environment

37.3.1.1 Having Initiative of Meteorological Conditions in Hand

Collect weather information before class and conduct training accordingly. On condition that it's windy or slightly rainy, ideological education such as developing fine traditions could be done so that the trainees should proceed with confidence to overcome difficulties and enhance study efficiency. Although much prone to be affected by meteorological conditions, field training is of its unique features which may result in good effect if properly used.

37.3.1.2 Doing Terrain Reconnaissance

Selection of teaching site is dependent upon demand of a specific subject. For facilitating field training, an unfamiliar terrain is a better choice, as new environment requires trainees to apply knowledge and skills more flexibly. Instructors should do on-the-spot terrain reconnaissance in preparation, including route of march, site for resting, lecturing and training, place to set equipment, and so on.

37.3.1.3 Studying Sociological Conditions

Arouse the enthusiasm of trainees according to various sociological conditions. If allowed, the instructor should select a site nearby a landscape and make arrangements for trainees to have a visit, which not only motivate trainees but also make them feel humanistic instruction.

37.3.1.4 Setting up EPPs (Emergency Pre-arranged Plan)

Emergencies may come up during field training like extreme weather condition, conflicts with local residents, injury or sudden ailment, in accordance to which EPPs should be set up for safe and successful training. The instructor's tackling with emergencies is a good example of education for trainees as well. In the meantime, instructors may develop some questions beforehand in order to bring the enthusiasm in trainees so that they solve problems by themselves and raise their initiative in field training.

37.3.2 Focusing on Needs to Create Good Environment for Study in Field Training

Field training is conducted in a *big* environment; therefore a standard teaching site is usually prioritized unless there should be any special requirement. In addition to the needs of the training subject, study efficiency of trainees should also be taken into consideration.

37.3.2.1 Good Exterior Environment

Take such points into account like less exterior-disturbance, high air quality, easy for entry and exit, convenient for vehicle parking and assembling, quick and safe access to encampment, and free from masses, in order not to be adversely influenced.

37.3.2.2 Small-Class Instruction

Small-class instruction is of small scale and easy to organize, convenient for personalized teaching and interaction between the instructor and trainees. In accordance with the characteristics of different subjects, split the class into squad or platoon in field lecturing, and form the class by corresponding levels (squad, platoon or battery) in light of different stages of comprehensive exercise.

37.3.2.3 Physical Factors of Environment

Temperature, light, sound, and smell are all influential factors on the instructor and trainees which should be dealt with properly. Get closer to trainees by forming them up in U shape, L shape or O shape; speak out loud and concisely so trainees to follow easily; turn their back to the sun or keep them away from cold wind and seat them on portable stools rather than on the ground, or seat them on flat and dry ground but not on uneven and muddy ground if possible; take measures for them to do warm-ups in cold season and avoid summer heat when hot.

37.3.3 Enlightening Awareness to Create Positive, Harmonious and Interactive Atmosphere

Bring the subjective initiative of trainees into full display and create harmonious instructional atmosphere, which significantly unifies teaching and learning, and improve the effectiveness of teaching activities.

37.3.3.1 To Respect Trainees

Give full cognition to the fact that trainees are the subjects of learning activity. Set up the notion to serve trainees and respect their personality. Use civilized words in instruction, respect the innovation of trainees, and encourage them to express themselves and listen to them, and keep in mind to treat trainees differently as their individual performances are varied.

37.3.3.2 To Prioritize Practice

Practice and theory should be well proportionated in instruction. Follow the generative law of techniques and skills by grasping the characteristics of such stages as imitation, consolidation, promotion and internalization, and inspire trainees to participate in training with all positive factors.

37.3.3.3 To Strengthen the Guidance of Instructional Environment

The guidance of instructional environment is understood as how the instructor makes full use of or adjusts factors of external environment to improve instructional effectiveness [5]. Trainees are influenced by the environment; meanwhile, they are building and reconstructing the environment as well. The instructor should arouse the enthusiasm of trainees so that they create harmonious, free and comfortable training atmosphere.

37.3.4 Putting Strictness into Effect to Build Normative Training Environment

Normative environment is the guarantee for orderly training. It is a must to adhere to strictness in training and discipline.

37.3.4.1 To Adhere to Normativeness in Proceedings

The instructor ought to overcome arbitrariness and carry out training in strict accordance with normative proceedings. Usually follow the preparation—implementation—review sequence, and in preparation stage, guide trainees to make preparations for equipment, personnel, and teaching site, in implementation stage, play a good role in guiding trainees to do exercises as per stipulated in teaching plan in good order, in review stage, pay attention to pertinence while commenting on full elements truly.

37.3.4.2 To Display Normativeness in Dressing

Neat formation is a good display of military image. Trainees are usually dressed in combat uniform with woven belt and necessary outfits. The instructor should be strict in training and discipline.

37.3.4.3 To Seek for Normativeness in Details

Strictness is the foundation of good teaching style. The instructor should follow the demand of strictness and create strict training atmosphere, make timely adjustment in minor aspects for strictness in light of teaching language, interaction, training and relaxation administration, disposal of equipment and so on.

37.3.5 Targeting Real Combat to Create Approximate Environment of Real Combat

To create approximate environment of real combat is favorable to forge brave, steadfast, and resolute traits for the enhancement of physical, technical, and psychological quality.

37.3.5.1 To Strengthen the Public Opinion Environment of Training for Combat

The instructor should seize all opportunities to mold public opinion and do ideological work by associating features of the subject, and improve the trainees' awareness of by education, inspiration, and publication.

37.3.5.2 To Be Adept at Using and Reforming Site Environment for Real Combat

Site, as the support of training, is the material element of training environment. Lay a solid foundation on general terrain and familiar sites prior to practicing commanding on complex terrain and unfamiliar sites [6]. In the meantime, make efforts to create approximate environment of real combat in line with practicability by means of situation induction and situation display.

37.3.5.3 To Create United and Tense Instructional Atmosphere

Carry out such activities like comparison, contest and assessment, make fair but competitive atmosphere and intensify awareness of competition; carry out such activities like mutual help, mutual teaching and mutual learning, make united- and cooperative-atmosphere, and cultivate consciousness of unity and cooperation.

37.4 Conclusions

Practical courses are now becoming very important in professional education in military institutions. In order to ensure better teaching and learning effect, instructors should make better use of field teaching environment and focus on professional demands to create better learning environment, enlighten awareness to create positive and harmonious environment of interaction, put the notion of strictness into effect to build normative training environment, and target on real combat to create approximate environment.

References

1. Feng X (2010) Building and implementation of assessment index system for practical courses. China Electr Power Educ 26(13):125–126
2. Zhang S, Wang G, Wang J et al (2014) Ergonomic thinking on practical instructional methodology based on target orientation. Ergonomics 47–48

3. Pang Z (1999) Air defense forces man-machine-environment system engineering. Air Defense Forces Academy, Zhengzhou, P5 (Ch)
4. CPLA Military Terminology Management Council (2011) CPLA military terminology. Military Science Press, Beijing, p 418
5. Lin M, Zhang D et al (1998) Dictionary of science of study. Xinhua Press, Beijing, p 547
6. Zhang X (2012) Thinking on optimization of field teaching environment. J Nanchang Army Acad (3):52

Chapter 38
Quantitative Estimation of Porosity of Solidified Sludge Based on GIS

Fei Xu

Abstract The microstructure of soil body is considered determining its macro-characters. Meanwhile, the gray level that the SEM images are present is regarded to have a close correlation with microscopic distribution of the soil. Thus, based on GIS, the SEM images of solidified sludge (SS) with different cement content are studied and quantitatively analyzed in this paper. 3D porosities are extracted as the micro-perspective research methods for SS, and the extracted results are compared with the soil test ones to analyze the errors. Conclusions are drawn despite the minor errors between estimated results and soil test ones, quantitatively estimating 3D porosities in a feasible way to offer technical parameters for soil mechanics research, as well as instructing for construction.

Keyword Solidified sludge GIS Microstructure porosity

38.1 Introduction

With the fast pace of urbanization in China, treatment rate of domestic sewage is increasingly sustainable. According to statistics, a sewage treatment plant can produce more than four million tons [1] of dewatered sludge per year. Discharging randomly or improper disposal may lead to serious environment problems [2]. Treating sludge through landfill is a proper way that suits the current situation domestically [3]. Since the soil mechanical property appears terrible, shear strength of sludge is poor [4] and permeability is low. It is unable to enhance the strength of sludge through sun-curing, which leads to landfill site safety accidents such as landslide [5]. Before landfill, sludge should be solidified to enhance the strength.

F. Xu (✉)
Hohai University, No.1 Xikang Road, Nanjing 210098, Jiangsu, China
e-mail: babyshaq@126.com

© Springer-Verlag Berlin Heidelberg 2015

S. Long and B.S. Dhillon (eds.), *Proceedings of the 15th International Conference on Man–Machine–Environment System Engineering*, Lecture Notes in Electrical Engineering 356, DOI 10.1007/978-3-662-48224-7_38

Relevant researches in China tend to focus on macro performance of clay and other traditional soil [6]. Studies from micro-perspective mainly focus on clay and polluted soil, rather than sludge. Analyzing the microstructure from micro-perspective would help reveal the variation regulations of sludge and SS. Currently, relevant researches have made it possible to estimate porosity and other structure parameters, through understanding images based on GIS [7]. This has provided the methods for estimating structure parameters of sludge and SS [8].

From micro-perspective, SEM images of sludge are studied in this paper. To examine reasonableness of estimated results, and to analyze the sources of errors, estimated porosity and soil test results of SS are compared. Conclusions appear that the variation trends between estimated values and test ones are similar and it is feasible to obtain porosity of sludge and SS based on GIS, to instruct construction.

38.2 Materials and Preparation

Dewatered sludge from a sewage treatment plant in a certain city in China is applied; parts of the basic indexes are listed in Table 38.1. $32.5^{\#}$ ordinary Portland cement is applied as solidified agent. Sludge are divided into five samples, all the samples are maintained for seven days till its the moisture contents reaches 100 %. Four of the samples are mixed with 5, 10, 15, and 20 % cement of their weight differently, with the last one being regarded as a comparison.

The SEM images, after solidification, are shown in Fig. 38.1. All the samples are treated using vacuum freeze-drying before taking SEM images. To ensure the pore water in the sample congeal to non-swelling crystalline ice, samples should be conserved in the fridge for over 8 h with the degree of −40 °C. Then, they are cut to get impact fracture surface, and are conserved in the freezer-dryer for vacuuming for over 8 h with the degree of −68 °C, to get the drying solidification body without damaging its structure. In this way, samples are sent to take SEM images. In this paper, SEM magnification is set to 2000 times, and it advices to choose pore distribution area as far as possible to avoid the large aperture caused by cutting off or breaking off.

Table 38.1 Basic indicators of sludge

Moisture content/%	TOC/mg	pH	Organic matter/%	Plastic limit/%	Liquid limit/%	Cd/ (mg kg)	Cu/ (mg kg)	Zn/ (mg kg)	Pb/ (mg kg)
136	14.87	8	49.76	31.3	286	27	199	510	110

Fig. 38.1 SEM images of SS with different cement content

38.3 Research Methods

DEM is often used to analyse the texture of the surface of earth and river in GIS. As to SEM images of soil particles, they describe the surface of soil particles using the depth of gray level. The depth of a certain pixel represents the height of this point on the particle. Therefore, the correlation between the gray level of images and the surface of particle is reliable. Considering that the three phrases of sludge are similar to that of clay and other soil, and that the microstructure of soil determines the mechanical property of them, it is feasible to evaluate the porosity of solidified sludge (SS) from micro-perspective based on GIS.

In an 8 bit image, for example, the gray level of a pixel, which reflects the height of soil particle, is between 0 and 255. In SEM images, the closer a pixel stays away from the light, the higher it reflects on gray value, and lighter on images. On the contrary, the farther a pixel stays away from the light, the darker it appears on images. In this paper, the software ArcGIS is used for analyzing SEM images. Through using the 3D analyst of the software, the 3D model of the image can be built, and the relevant parameters of the particle are able to be auto-calculated.

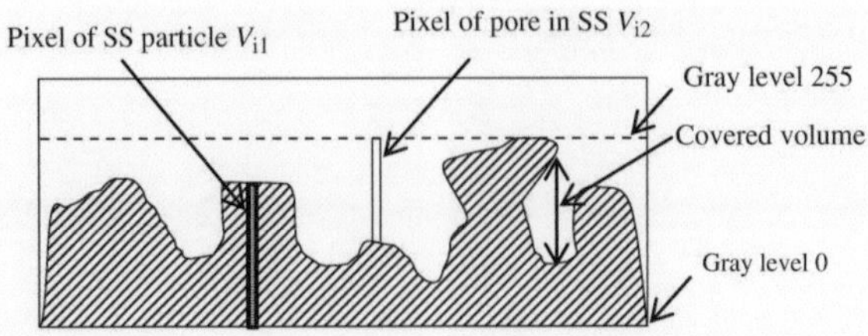

Fig. 38.2 The schematic diagram of evaluating porosity quantitatively

38.3.1 The Extraction of 3D Porosity of SS, based on GIS

The porosity of soil is the ratio of pore volume and total volume of soil. The equation to calculate the ration lies below.

$$n = \Sigma V_{i2} / \Sigma (V_{i2} + V_{i1})$$

In the equation, n stands for porosity, expressed in percentage. V_{i1} stands for volume of SS on a single pixel, and V_{i2} stands for pore volume of SS on a single pixel. Both of them can be obtained by using 3D analyst in ArcGIS. Before calculating, all selected particles are considered existing in a cuboid, which is regarded as the total volume of SS. The length and width of cuboid is the length and width of the image, whereas the height of cuboid is the maximum value of gray level. Picture units are adopted in equation, without the need to be transferred to the measurement ones. The real existing volume in the cuboid stands for the volume of SS particles, and the leftover volume stands for that of pore. In this way, it is relatively convenient to obtain the porosity by the two kinds of volume. The schematic diagram of estimating porosity quantitatively is shown below as in Fig. 38.2, and the empirical example is written in the following.

38.4 Applications and Analysis

38.4.1 Extraction of 3D Porosity of SS based on GIS

Four groups of SS are studied in this paper. In each group, four SEM images are analyzed quantitatively, and the operation platform is ArcGIS 10.0. The specific operating steps can be summarized as the following. First, the original image is transferred to grayscale picture, and is saved as the one with the extension of tiff. Second, the image with tiff extension is used to build 3D model in ArcScene module. The 3D model that built is shown in Fig. 38.3. Third, the total volume of particles and pore are calculated by 3D analyst module of the software. Finally,

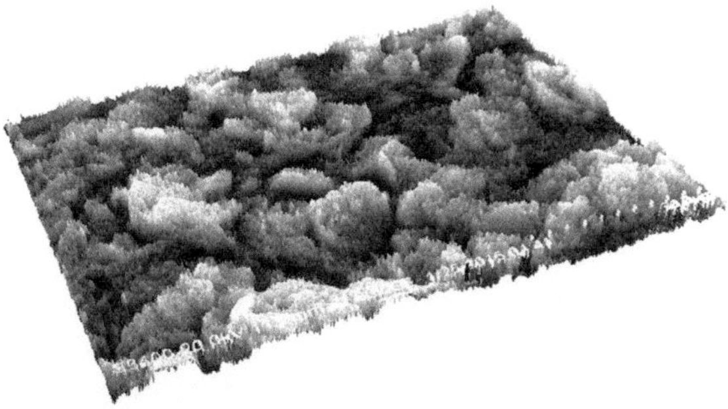

Fig. 38.3 Sketch map for 3D models of SS

Table 38.2 Porosity of SS in different cement content

Samples (%)	Particle volume	Total volume	Porosity	RMSE	Root-mean-square (RMS) (%)	Measured value (%)
10	69894641.8	244,085,639	0.7136	0.009837292	70.77	73.30
	71032522.3	251,034,475	0.717			
	73344003.6	251,443,609	0.7083			
	80266983.5	260,125,932	0.6914			
15	76357749.2	240,217,947	0.6821	0.019774208	69.34	70.75
	74975295.7	274,911,873	0.7273			
	88058589.3	274,827,761	0.6796			
	72870352.7	230,198,648	0.6834			
20	95230671.6	264,659,472	0.6402	0.005719744	63.38	67.92
	92183759.4	247,196,720	0.6271			
	93038546.1	250,895,310	0.6292			
	97571423.9	270,031,985	0.6387			
25	94168437.4	239,842,719	0.6074	0.037645752	62.44	67.30
	87370164.7	261,196,340	0.6655			
	90147155.2	2579,48,651	0.6505			
	107517348.1	249,756,230	0.5695			

it comes to the results (porosity, RMSE, and mean-root-square) and the specific calculated outcomes are listed in Table 38.2.

To ensure the accuracy of results, root-mean-square (RMS) and RMSE are calculated for the four results in each group. Recalculate the RMS when the dispersion degree of certain result in the group is relatively high, and the final result is considered negative if the RMSE of certain group is relatively high. In this way, the second result in sample 15 % is significantly higher than the measurement result, which is considered to be high dispersion degree and negative. As to sample with 25 % cement content, RMSE of the four results is relatively high, and the RMS is considered negative. Finally, the modified results are shown in Table 38.3.

Table 38.3 Comparisons between modified results and measurement results

Samples (%)	Estimated results (%)	Measurement results (%)
10	70.77	73.30
15	68.17	70.75
20	63.38	67.92

For those results of low RMSE, the porosities of original samples are measured by soil test. To make comparison, results of the measurement and estimation are listed in Fig. 38.4. Although the estimated results are relative lower than measured ones, the trends of two groups are similar. Therefore, it is feasible to instruct construction through estimating the porosity based on GIS, when conditions are inadequate for soil test.

38.4.2 Error Analysis

1. The reason for deleting the result of sample 15 % in Table 38.2 is that the RMSE is relatively high. From micro-respective, the reason may be due to the way of preparing sample 15 %. On the one hand, breaking off, or other ways may cause certain particles to be away from its origin fracture surface, leaving the pore where the particle originally exists that is relatively large. On the other hand, the complexity of composition of sludge may cause this situation. Plant roots or other substances that are uneasy to decompose may exist on the fracture surface, which also makes the pore of sample is relatively large.
2. Though no certain result in sample 25 % is relatively higher than the others, the RMS is still regarded as negative, due to the relatively high RMSE. This situation may be due to the uneven distribution of the surface of the sample. To reduce the RMSE of sample, it is recommended to take more SEM images of the surface for estimating, or to select another sample of SS 25 %.
3. The estimated results are relatively lower than the measured ones; this may be due to the fact that SS particles appear to cover each other, or sunk on the

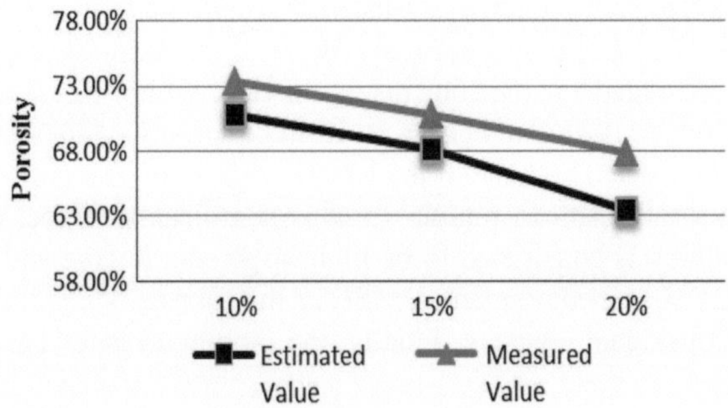

Fig. 38.4 Comparison chart of modified estimated value and measured value

fracture surface (shown in Fig. 38.2), leaving the real situation that is unable to reflect on SEM images. In this way, the volume of pore that the image shows is smaller than the real one, leaving the porosity relatively low.

4. As shown in Table 38.2, the gap between estimated and measured value appears larger with the increase of cement content of SS. This may be because, the hydration reaction between cement and sludge becomes more intense, when more cement is mixed with sludge, and the level of solidification is enhanced. The cluster of the reaction product makes the lump particles bigger than before, which boosts the cover up and stacking between particles.

38.5 Conclusions

Based on the application of GIS, methods for estimating porosity of SS quantitatively has been discussed and verified. The correlation and variation trend between estimated and measured results has been compared, and conclusions are listed below:

1. From micro-perspective, as shown in Fig. 38.1, particles cluster to bulk body after solidification, pore among the particles is fulfilled with massive hydration reaction production, reducing the porosity of SS, which is regarded as the key reason for enhancing the strength of SS.
2. The correlation and variation trend between estimated and measured results are favorable. Based on application of GIS, quantitative estimating is able to reflect the porosity of SS, which is simple and instructive to operation and construction.
3. The results of quantitative estimation appear lower than the measured ones, this is probably caused by the cover up and sunk among the particles, which are unable to be reflected on SEM images and to be calculated by software.
4. On condition that the real porosity values remain unknown, it is feasible to reduce the overall errors of estimation by studying different SS SEM images of one sample, and sifting through RMSE.

References

1. Shi J, Wang J (2012) Mechanical behaviour test study of sludge mixed with municipal solid waste. Rock Soil Mech 33(11):3208–3212 (in Chinese)
2. Russell H (1992) Race to clear sludge landslide[C], 20th. New Civil Engineer, Institution of Civil Engineers, London
3. Zhu W, Zhang CL, Abraham C, Chiu F (2007) Soil-water transfer mechanism for solidified dredged materials. J Geotech Geoenvironmental Eng ASCE 133(5):588–598
4. O'Kelly BC (2008) Geotechnical properties of a municipal water treatment sludge incorporating a coagulant. Can Geotech J 45(5):715–725

5. Li L, Zhu W, Ji S, Zheng X (2008) Influence of micro-organism on long-term strength of solidified/stabilized sludge. Chin J Geotech Eng 30(12):1778–1782 in Chinese
6. Cheng L, Wei Z, Jie H (2013) Strength and leachability of solidified sewage sludge with different additives. J Mater Civ Eng 25(11):1594–1601
7. Wang B, Shi B, Liu Z et al (2004) Fractal study on microstructure of clayey soil by GIS. Chin J Geotech Eng 26(2):244–247 (in Chinese)
8. Wang S, Guan J, Jiang X (1999) Study on microstructure and engineering property of saturated clayey soil. Port Waterw Eng 5:3–7

Chapter 39
Determination of Organochlorine Pesticides in Saposhnikovia Divaricata by Ultrasonic-Assisted Extraction Coupled with GC-ECD

Lang Lang, Bo Wang and Yubin Ji

Abstract A simple, rapid, and sensitive method for determination of organochlorine pesticides in Saposhnikovia divaricata by ultrasonic-assisted extraction coupled with GC-ECD was developed and validated. The extracting agent was a mixture solvent of acetone–methylene chloride (1 + 1). The extracts were cleaned up with concentrated sulfuric acid as sulfonating agent. The samples were separated by the capillary column HP-5. 6 organochlorine pesticides in Saposhnikovia divaricata were determined by gas chromatography with an electron capture detector (GC-ECD) simultaneously. 6 pesticides reference substances had good linear relationship in the concentration range of 8–150 ng/g. The average recoveries were, respectively, 83.5–97.3 %, 84.4–99.7 %, and 89.9–98.7 % at three added levels of 15, 30, and 60 ng/g. Relative standard deviation (RSD) was, respectively, 3.0–7.3 %, 4.6–7.4 %, and 3.9–8.7 %. The method was simple, rapid, and high selectivity. It was suitable for determination of organochlorine pesticides in Saposhnikovia divaricata.

Keywords Saposhnikovia divaricata · Gas chromatography · Organochlorine · Pesticide residues

39.1 Introduction

Traditional Chinese medicine has made an enormous contribution to the prevention and treatment of disease and nutrition health care for human beings. But pesticides are widely used in the planting process of Chinese herbal medicine against a variety of weeds and insects. These pesticides will pollute Chinese herbal medicine through

L. Lang (✉) · B. Wang · Y. Ji
Postdoctoral Station on Chinese Materia Medica, Harbin University of Commerce, Harbin 150076, China
e-mail: lang8968@163.com

© Springer-Verlag Berlin Heidelberg 2015

S. Long and B.S. Dhillon (eds.), *Proceedings of the 15th International Conference on Man–Machine–Environment System Engineering*, Lecture Notes in Electrical Engineering 356, DOI 10.1007/978-3-662-48224-7_39

direct or indirect modes [1]. At present, pesticide residues in Chinese herbal medicine are very serious. Pesticide residues not only seriously affect the medicinal cure effect and patient safety, seriously but also restrict the export of Chinese herbal medicine [2, 3]. It hinders the promotion of Chinese herbal medicine to the world. Therefore, strengthening the quality control of Chinese herbal medicine attracts more and more attention. Among them, determination of pesticide residues in Chinese herbal medicine is critical [4–7].

Saposhnikovia divaricata is the desiccative radix of umbelliferae Saposhnikovia divaricata (Turcz.)Schischek. It plays a part in defervescence, sweating, analgesia, and dispelling the wind. It is clinically used to the treatment of cold headache, rheumatic pains, tetanus, itching caused by rubella and so on. Due to its good quality and exact cure effect of Saposhnikovia divaricata, it draws more and more extensive attention. The demand is growing increasingly. Therefore, controlling the quality standard of Saposhnikovia divaricata strictly and determining pesticide residues of Saposhnikovia divaricata become very necessary.

Pesticides are a numerous and diverse group of chemical compounds. Because of their mobility, aggregation, and long-period effect on living organisms, pesticides are considered to be one of the most toxic and hazardous environmental contaminants [8]. Pesticides primarily contain organochlorine pesticides, organophosphorus pesticides, and pyrethroid pesticides. Among them, organochlorine pesticides are one of the most common pesticides. Because organochlorine pesticides are highly toxic and highly residual, they draw more and more extensive attention [9–12]. Although some high-toxic organochlorine pesticides had been banned for years, the excessive use of these pesticides in the early period has polluted the environment seriously. Pesticide residues remained in water and soil will continue to pollute Chinese herbal medicine, and at present, there is still unreasonable use of pesticides. These lead to inevitable residues of organochlorine pesticides in Chinese herbal medicine. Organochlorine pesticides are fat soluble and difficult to be metabolized in humans. Furthermore, this kind of pesticides is high toxic; therefore, it will pose significant risk to humans [13]. Hence, determination of organochlorine pesticides in Saposhnikovia divaricata becomes very necessary.

Based on consulting a large number of literatures at home and abroad, determination of organochlorine pesticides in Saposhnikovia divaricata was investigated systematically in the paper. Pesticide residues were extracted by ultrasonic-assisted extraction. The extracts were cleaned up with concentrated sulfuric acid as sulfonating agent. The samples were determined by gas chromatography with an electron capture detector (GC-ECD) [14, 15]. A method to determine 6 organochlorine pesticides in Saposhnikovia divaricata was established. The method was simple, rapid and had good selectivity. It was suitable for determination of organochlorine pesticides in Saposhnikovia divaricata.

39.2 Experimental Section

39.2.1 Chemicals and Reagents

Pesticide standards including γ-BHC, p.p′-DDT, hexachlorobenzene, aldrin, heptachlor, and trans-chlordane were purchased from National Research Center for Certified Reference Material (Beijing, China). Stock solutions of each pesticide were prepared in n-hexane at 100 μg/mL and standard working solutions at various concentrations were prepared by diluting the stock solutions in n-hexane. These solutions were stored at 4 °C. Petroleum ether (60–90 °C), n-hexane, acetone, and dichloromethane were of analytical grade and purchased from Tianjin Fuyu Fine Chemistry Limited Company (Tianjin, China). Petroleum ether 60–90 °C) and n-hexane were of HPLC grade and purchased from Tianjin Kemiou Chemical Reagent Limited Company (Tianjin, China). Concentrated sulfuric acid (98 %) was purchased from Beijing Chemical Works (Beijing, China). Anhydrous sodium sulfate was of analytical grade and purchased from Tianjin Fengchuan Chemical Reagent Limited Company (Tianjin, China). Sodium chloride was of analytical grade and purchased from Tianjin Basf Chemical Limited Company (Tianjin, China). They were stored in a desiccator before use. The Saposhnikovia divaricata sample was purchased from Jiangye Chinese Herbal Medicine Planting Professional Cooperatives (Daqing, China). It was ground using a grinder, ranked by mesh screen (0.425 mm), and then stored in a desiccator at room temperature.

39.2.2 Instrumentations

An Agilent 6890N gas chromatograph (Agilent Technologies, Santa Clara, USA) equipped with an Agilent 7683 autosampler (Agilent Technologies) and a HP-5 column (Agilent Technologies) was used for the separation of analytes. An electron capture detector (Agilent Technologies) coupled with Agilent Chem Station (Agilent Technologies) was used for determination of analytes. A JP-030ST ultrasonic cleaner (Shenzhen Jiemeng Cleaning Equipment Limited Company, Shenzhen, China) was used for accelerating extraction rate. A R201 rotary evaporator (Shanghai Shensheng Biotech Limited Company, Shanghai, China) coupled with a W201 numerical control constant temperature water bath (Shanghai Shensheng Biotech Limited Company), and an A-1000S aspirator (Shanghai Ailang Instruments Limited Company, Shanghai, China) was used for concentrating the extracting solution. A TDL80-2B bench top low speed centrifuge (Jinan Boxin Biological Technology Limited Company, Jinan, China) was used for the separation of analytes and impurities.

39.2.3 GC Analysis

GC analysis was carried out using an Agilent 6890 N gas chromatograph equipped with an Agilent 7683 autosampler and an electron capture detector. The analytes were separated via a HP-5 column (30 m × 0.32 mm i.d., 0.25 μm film thickness). Nitrogen (purity ≥ 99.999 %) was used as carrier gas at a constant flow of 2.5 mL/min. The column temperature program: set at 100 °C for 1 min; increased at a rate of 10 °C/min up to 200 °C; held for 2 min; increased at a rate of 8 °C/min up to 280 °C; held for 5 min. Injector temperature was maintained at 290 °C. Detector temperature was maintained at 300 °C. The injection volume was 1.0 μL in the splitless mode.

39.2.4 Extraction and Clean-Up

The Saposhnikovia divaricata sample was accurately weighed (2.0 g) and thoroughly mixed with anhydrous sodium sulfate (10.0 g) which was used as desiccating agent. The mixture was placed in a 100 mL conical flask with stopper. 60 mL acetone–methylene chloride (1 + 1) mixtures were added to the conical flask to soak the sample for 12 h. The sample was extracted by ultrasonic-assisted extraction for 20 min. The filtered extracting solution was collected. The residue was extracted by ultrasonic-assisted extraction with 30 mL acetone and then filtered. Repeat the operation three times. The filtrate was combined and then concentrated under reduced pressure to approximate dryness on a water bath at 40 °C. Following adding a little n-hexane, repeat the previous operation again and again until methylene chloride and acetone were removed. The analytes were dissolved with n-hexane and transferred to a 10 mL graduated centrifuge tube with stopper. Add n-hexane to 5 mL. Add 1 mL concentrated sulfuric acid into the centrifuge tube carefully and shake for 1 min. The mixture was centrifuged at 3000 rpm for 10 min. The supernatant was filtered through a 0.45 μm organic membrane. Appropriate filtrate was transferred to 2 mL autosampler vial and then placed in the autosampler for GC analysis.

39.3 Results and Discussion

39.3.1 Optimization of Extraction and Clean-up Conditions

39.3.1.1 Selection of Extraction Solvent

Based on researching the polarity of organochlorine pesticides and consulting a large number of literatures at home and abroad, acetone, methylene chloride, n-hexane, and petroleum ether have often been used for determination of multi-residues of organochlorine pesticides as the extraction solvent. Among them, acetone is the most common extraction solvent and mixed solvents have better extraction efficiency than a single one. Hence, three mixed solvents of acetone-n-hexane, acetone–petroleum ether, and acetone–methylene chloride were selected as the extraction solvent. For testing their extraction ability for Saposhnikovia divaricata, parallel experiments were carried out. Acetone–methylene chloride mixtures as the extraction solvent achieved a relatively higher detection value than other mixtures. The introduced impurities would not cause interference to determination of analytes. Moreover, the results of parallel experiments also demonstrated that the best recoveries of organochlorine pesticides were achieved with acetone–methylene chloride mixtures as the extraction solvent. Thus, acetone–methylene chloride mixtures were chosen as the extraction solvent to obtain more accurate detection results. Furthermore, the effect of the concentration of methylene chloride in the extraction solvent was researched. Parallel experiments were carried out with 60 mL acetone–methylene chloride mixtures of various concentration proportions. The results demonstrated that the detection value increased along with the proportion of methylene chloride raised, and then remained relatively constant when the proportion of methylene chloride was higher than 50 %. The introduced impurities also increased along with the proportion of methylene chloride raised. So acetone–methylene chloride (1:1) mixtures were chosen as the extraction solvent to obtain more accurate detection results, simplify the clean-up process and keep the chromatographic system from contamination.

39.3.1.2 Optimization of Clean-up Conditions

Some impurities in the extracting solution of Saposhnikovia divaricata, such as saccharides, fat, and pigments can interfere the determination of analytes. Therefore, the extracting solution need be cleaned up before GC analysis. A clean-up method was confirmed via parallel experiments. Concrete steps were shown as follows: 1 mL concentrated sulfuric acid as sulfonating agent was added to the extracting solution which was concentrated to 5 mL. And then, the solution was centrifuged at 3000 rpm for 10 min. If sulfonation was carried out before the extracting solution was concentrated, then the dosage and usage frequency of concentrated sulfuric acid could increase. It would lead to waste of concentrated sulfuric acid, increase of risk and loss of analytes.

39.3.2 Validation of the Method

Chromatogram of mixed reference substance of 6 organochlorine pesticides is shown in Fig. 39.1. The calibration curves all represent good linearity in the concentration range of 8–150 ng/g. The correlation coefficients ranging from

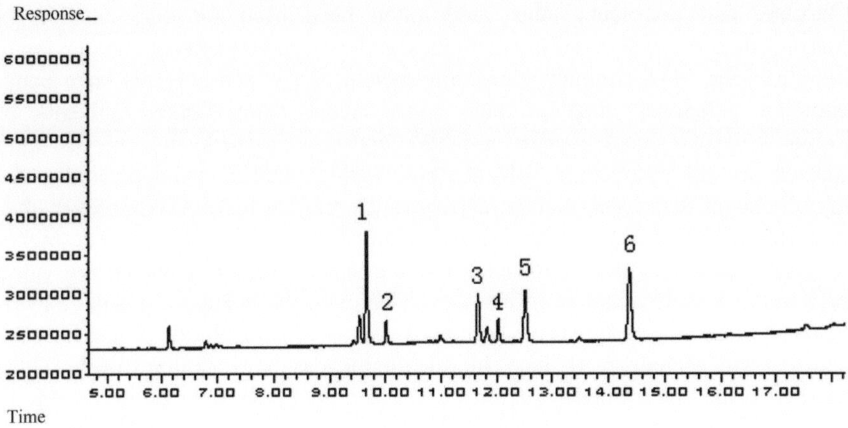

Fig. 39.1 Chromatogram of mixed reference substance of 6 organochlorine pesticides, 1-hexachlorobenzene; 2-γ-BHC; 3-heptachlor; 4- p.p′-DDT; 5-aldrin; 6-trans-chlordane

0.9991 to 1.0000 were obtained. The details of the calibration curves, the correlation coefficients, and the linear range are shown in Table 39.1. Method accuracy and reproducibility were evaluated via recovery experiments. The recoveries were conducted in matrix at three fortified levels (15, 30, and 60 ng/g). The average recoveries and relative standard deviations (RSDs) are given in Table 39.2. The average recoveries of organochlorine pesticides were in the range of 83.5–99.7 %, and the RSDs were in the range of 3.0–8.7 %, which can fulfill the requirements of pesticide residue analysis.

Table 39.1 The regression equation, correlation coefficient, and linear range of organochlorine pesticides

Pesticide	Retention time (min)	Regression equation	Correlation coefficient (r)	Linear range (ng/g)
γ-BHC	10.05	$y = 384,841 x - 627,974$	0.9996	8–350
p.p′-DDT	12.06	$y = 848,959 x - 151,3768$	0.9998	8–150
Hexachlorobenzene	9.67	$y = 7,394,751 x - 615,4316$	0.9994	3–150
Aldrin	12.53	$y = 1,285,014 x - 1,327,458$	0.9995	8–300
Heptachlor	11.68	$y = 1,901,233 x - 264,945$	0.9998	8–300
Trans-chlordane	14.39	$y = 1,592,395 x - 2,732,429$	0.9991	8–300

Table 39.2 Recoveries and RSD of organochlorine pesticides (n = 3)

Pesticide	15 ng/g		30 ng/g		60 ng/g	
	Recovery (%)	RSD (%)	Recovery (%)	RSD (%)	Recovery (%)	RSD (%)
γ-BHC	96.7	4.6	89.2	5.6	93.7	6.4
p.p′-DDT	92.5	3.5	93.4	3.8	97.6	5.2
Hexachlorobenzene	88.4	5.6	95.7	4.2	94.6	5.4
Aldrin	93.5	3.7	87.6	6.3	92.1	6.3
Heptachlor	93.7	6.4	94.6	5.4	96.7	4.6
Trans-chlordane	97.6	5.2	92.1	6.3	92.5	3.5

39.4 Conclusions

The work has employed ultrasonic-assisted extraction coupled with GC-ECD for determination of organochlorine pesticides in Saposhnikovia divaricata. The method improved the extraction efficiency, minimized the consumption of organic solvents, and simplified the clean-up process of the extracting solution. High efficiency, low cost, and acceptable recoveries were achieved. Organochlorine and pyrethroid pesticide residues in the test samples conformed to the national standard. Through the study of determination methods of pesticide residues in Saposhnikovia divaricata, it could provide reference for determination of pesticide residues in Chinese herbal medicine of similar properties and be conducive to controlling quality standard and promoting foreign trade of Chinese herbal medicine.

Acknowledgments This work was supported by China Postdoctoral Science Foundation (2013M531059), Heilongjiang Postdoctoral Science Foundation (LBH-Z12145) and Heilongjiang Province Science Foundation for Youths (QC2012C002).

References

1. De Wilde T, Spanoghe P, Sniegowksi K, Ryckeboer J, Jaeken P, Springael D (2010) Transport and degradation of metalaxyl and isoproturon in biopurification columns inoculated with pesticide-primed material. Chemosphere 78(1):56–60
2. Wan Zh et al (2003) Study on removing organochlorine pesticide in Angelica sinensis (Oliv.) Diels by supercritical fluid xtraction-determination. Shenyang Pharm Univ J 20(3):187
3. Zhao C, Bai L, Li H, Li F, Xin C (2007) Efficient removal of the organochlorine pesticide and heavy-metal residues in epimedium brevicornum maxim by supercritical fluid extraction. Arch Pharm Res 30(12):1558–1565
4. Zhao C, Bai L, Ou Y, Li D, Xin C (2009) Decontamination of organochlorine pesticides residue and heavy metal in Rehmannia glutinosa Libosch by SFE. J Chromatogr Sci 47 (10):919–924
5. Li X, Wang L, Wang Z, Zhang H, Su R, Xu X, Zhang H (2011) Ultrasonic nebulization extraction coupled with headspace hollow fiber microextraction of pesticides from root of Panax ginseng C.A. Mey. J Sep Sci 34(13):1582–1589

6. Wu J, Li L (2004) Determination of the residues of eleven organophosphorus insecticides in Job's-tears by gas chromatography with a nitrogen-phosphorus detector. J AOAC Int 87 (5):1260–1263
7. Wan Y, Mao X, Yan A (2010) Simultaneous determination of organophosphorus and organonitrogen pesticides residues in Angelica sinensis[J]. J Environ Sci Health B 45 (4):315–324
8. Stocka J, Tankiewicz M, Biziuk M, Namieśnik J (2011) Green aspects of techniques for the determination of currently used pesticides in environmental samples. Int J Mol Sci 12 (11):7785–7805
9. Mao X, Wan Y, Yan A (2012) Simultaneous determination of organophosphorus, organochlorine, pyrethriod and carbamate pesticides in Radix astragali by microwave-assisted extraction/dispersive-solid phase extraction coupled with GC-MS. Talanta 97:131–141
10. He M et al (2011) Determination of Organochlorine Pesticides Residues in RADIX RHIZOMA GLYCYRRHIZAE by GC. Chem Compon Pharmacol Action 2(5):25–27
11. Zheng G, Selvam A, Wong JW (2012) Enhanced solubilization and desorption of organochlorine pesticides (OCPs) from soil by oil-swollen micelles formed with a nonionic surfactant. Environ Sci Technol 46(21):12062–12068
12. Wang X, Xu J, Guo C, Zhang Y (2012) Distribution and sources of organochlorine pesticides in taihu lake, china. Bull Environ Contam Toxicol 89(6):1235–1239
13. Yang M, Wang L (2008) Advances in techniques on analysis and removal of pesticide residues in traditional chinese herbal medicines. World Sci Technol
14. Al Mahmud MN, Rahman M, Na TW, Park JH, Yang A, Park KH, Abd El-Aty AM, Nahar N, Shim JH (2012) A QuEChERS-based extraction method for the residual analysis of pyraclofos and tebufenpyrad in perilla leaves using gas chromatography: application to dissipation pattern. Biomed Chromatogr
15. Xu R, Wu J, Liu Y, Zhao R, Chen B, Yang M, Chen J (2011) Analysis of pesticide residues using the quick easy cheap effective rugged and safe (QuEChERS) pesticide multiresidue method in traditional Chinese medicine by gas chromatography with electron capture detection. Chemosphere 84(7):908–912

Chapter 40
Environmental Stress Selection Based on GRA and Preference

Guanqian Deng, Zhi Li, Ning Yan, Yongwu Yang, Xiaojian Li and Jun Zhan

Abstract Environmental stress (ES) is a major cause of faults, the different ES has various contributions to the fault, and someone ES usually plays a dominant role. In order to enhance the efficiency and accuracy of diagnosis, a novel approach to select the most-related ES with the fault based on gray relational analysis (GRA) and preference is present. First, the related ESs are selected preliminarily by analyzing the environmental condition when the fault occurring and the output of FMEA. Second, decision and preference matrix are constructed which denotes the relationship of the fault and ESs. Third, the sort of related ESs which express various contributions to the fault is given by calculating the gray relational degree. Finally, case study is presented to demonstrate the proposed approach. This approach can be used to select the most-related ES to study the failure mechanism, and, hence, help diagnosing, and prognosticating the faults which are environmentally sensitive.

Keywords Fault · Environmental stress · Gray relational analysis · Preference

40.1 Introduction

Statistical data and failure analysis show that environmental stress (ES) is a major cause of faults; the different ES has various contributions to the fault, and someone ES usually plays a dominant role. For example, multi-chip module (MCM) is sensitive to temperature, vibration, humidity, and so on [1, 2]. Considering taking all related ESs into account will influence the efficiency and accuracy of diagnosis, sometimes it is unnecessary and unpractical. Therefore, it is imperative important to analyze the relationship between fault and ESs, and the most-related ES should be selected to help diagnosing the fault.

G. Deng (✉) · Z. Li · N. Yan · Y. Yang · X. Li · J. Zhan
Institute of Reliability Engineering, Beihang University, Beijing 100191, China
e-mail: guanqiandeng@163.com

© Springer-Verlag Berlin Heidelberg 2015

S. Long and B.S. Dhillon (eds.), *Proceedings of the 15th International Conference on Man–Machine–Environment System Engineering*, Lecture Notes in Electrical Engineering 356, DOI 10.1007/978-3-662-48224-7_40

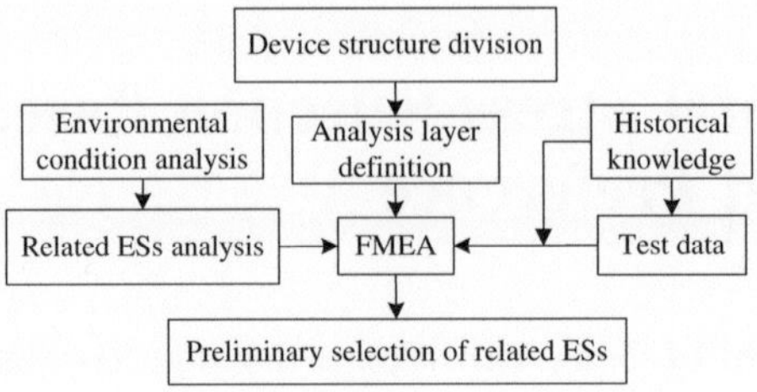

Fig. 40.1 The flow chart of preliminary selection of related ESs

Although it has long been recognized that ES has a great influence on fault, limited to technical reasons, it was until in the early 1980s that a device to measure and record ES, logistic history, and other information was developed. The emergence of microprocessors and non-volatile memory made time stress measurement device (TSMD) possible which is a device with the capability to measure and record ES. The original TSMD effort started in 1986 at Rome Air Development Center (RADC). From then on, the topic of TSMD has attracted many scholars and institution [3]. However, since fault cause is anfractuous. In addition, ES is interval number, instead fault is a state. It is difficult to establish the relationship between ES and the fault, and it is hard to obtain the most-related ES. To the best of ours knowledge, there is fewer study on the problem mentioned above at all.

40.2 Preliminary Selections of Environmental Stresses

FMEA is usually used to analyze fault mode, cause, and effect. However, it does not analyzing the environmental condition when the fault occurring. In this paper, we select the related ESs preliminarily by combining the analysis of environmental condition with the output of FMEA. The flow chart of preliminary selection of related ESs is illustrated in Fig. 40.1

In Fig. 40.1, the device structure is divided according to the operating principle and function at first. Second, the analysis layer is defined and FMEA of each module is carried out to get fault mode and attributes set. Then the environmental condition when the fault occurring and the related ESs are analyzed. Finally, with the help of prior information such as historical knowledge, and test data, the preliminary selection of related ESs can be obtained.

40.3 Relationship Analysis Based on GRA and Preference

The fault cause is anfractuous. In addition, ES value is usually interval number; instead the fault is a state. Therefore, it is difficult to establish quantitative relationship between ESs and the fault. Aiming at the problem of uncertain and incomplete relationship and system, gray relational analysis (GRA) adopted in this paper is reckoned as a very promising way [4]. Differing from the traditional mathematical analysis, GRA provides a simple scheme to analyze the series relationships or the system behaviors.

To calculate the relational degree, the compared and referenced series should be constructed first. In this paper, the fault is treated as referenced series. It is denoted by fault attributes such as fault location, modes, cause, and mechanism, $F = \{F_1, F_2, \ldots, F_m\}$. The related ESs are treated as compared series. It is denoted by $E = \{E_1, E_2, \ldots, E_n\}$.

In order to analyze the relativity between fault attributes and ESs, a decision matrix can be constructed. The elements of matrix are evaluation values which are given by the decision-maker. Considering the ambiguity of cognition, the evaluation values are given as internal numbers and denoted by $a_{ij}(\otimes)$, $a_{ij}(\otimes) \in [a^l_{ij}, a^u_{ij}]$, $i = 1, \ldots, n$, $j = 1, \ldots, m$, a^l_{ij} and a^u_{ij} are the left and right limits, respectively. Then the decision matrix $D = (\widetilde{a}_{ij}(\otimes))_{n\times m}$ can be established. Since the ESs have different contributions to the fault, the preference on the ESs can be expressed by the decision-maker according to his experience and knowledge. The preference is also given as internal number and denoted by $\widetilde{p}_{ij}$, $\widetilde{p}_{ij}(\otimes) \in [p^l_{ij}, p^u_{ij}]$, then the preference matrix $P = (p_{ij}(\otimes))_{m\times n}$ can be established.

There are benefits and costs criteria of attribute types in MADM [5]. Toward profit criteria, the larger the criterion value is, the better is the object. Toward profit criteria, the smaller the criterion value is, the better is the object. In order to avoid the difference of interval numbers in quantity, unit, and type, the interval numbers should be normalized first. Equations (40.1) and (40.2) are benefits and costs normalized equations, respectively.

$$\left.\begin{aligned} r^l_{ij} &= a^l_{ij} \Big/ \sqrt{\sum_{i=0}^{n} (a^u_{ij})^2} \\ r^u_{ij} &= a^u_{ij} \Big/ \sqrt{\sum_{i=0}^{n} (a^l_{ij})^2} \end{aligned}\right\}, i \in n, j \in m, \tag{40.1}$$

$$\left.\begin{aligned} r^l_{ij} &= (1/a^u_{ij}) \Big/ \sqrt{\sum_{i=0}^{n} (1/a^l_{ij})^2} \\ r^u_{ij} &= (1/a^l_{ij}) \Big/ \sqrt{\sum_{i=0}^{n} (1/a^u_{ij})^2} \end{aligned}\right\}, i \in n, j \in m. \tag{40.2}$$

The elements in matrix $D = (\widetilde{a}_{ij}(\otimes))_{n\times m}$ can be converted into corresponding ones in normalized decision matrix $\widetilde{D} = (\widetilde{r}_{ij}(\otimes))_{m\times n}$, $\widetilde{r}_{ij}(\otimes) \in [r^l_{ij}, r^u_{ij}]$ according to Eqs. (40.1) or (40.2). Similarly, the normalized preference matrix $\widetilde{P} = (v_{ij}(\otimes))_{m\times n}$, $\widetilde{v}_{ij}(\otimes) \in [v^l_{ij}, v^u_{ij}]$ can be obtained.

Definition 1 Let $r_i(\otimes) = (r_{i1}(\otimes), r_{i2}(\otimes), \ldots, r_{im}(\otimes))$ to be a normalized attributes vector, $r_{ij}(\otimes) \in [r_{ij}^l, r_{ij}^u]$, $0 \le r_{ij}^l < r_{ij}^u \le 1$. Note

$$r_{ij}^{l+} = \max_{1 \le i \le n}\left\{r_{ij}^l\right\}, r_{ij}^{u+} = \max_{1 \le i \le n}\left\{r_{ij}^u\right\}, j = 1, \mathrm{L}, \mathrm{m}. \tag{40.3}$$

Then,$r^+(\otimes) \in \left[r_{ij}^{l+}, r_{ij}^{u+}\right]$ is defined as ideal optimal attributes vector.

Definition 2 Let $x(\otimes) = [x^l, x^u]$ and $y(\otimes) = [y^l, y^u]$ are two internal numbers, then the distance $d(x(\otimes), y(\otimes))$ between $x(\otimes)$ and $y(\otimes)$ is defined as follow

$$d(x(\otimes), y(\otimes)) = \sqrt{(y^u - x^u)^2 + (y^l - x^l)^2} \tag{40.4}$$

Definition 3 The gray internal relational coefficient is defined as follows:

$$\xi_{ij} = \frac{\min\limits_i \min\limits_j d(r_{ij}(\otimes), r_j^+(\otimes)) + \rho \max\limits_i \max\limits_j d(r_{ij}(\otimes), r_j^+(\otimes))}{d(r_{ij}(\otimes), r_j^+(\otimes)) + \rho d(r_{ij}(\otimes), r_j^+(\otimes))}. \tag{40.5}$$

$$i = 1, 2, \ldots, n,\ j = 1, 2, \ldots, m, \rho \in [0, 1]$$

The relational matrix $R = (\xi_{ij}(\otimes))_{n \times m}$ between characteristic parameters and ideal optimal attributes can be established according to Eq. (40.5). Let $w = \{w_1, w_2, \ldots, w_m\}'$ be the weight of attributes. w is usually unknown and can be calculated as follow. The sum of difference between $r_{ij}(\otimes)$ and $v_{ij}(\otimes)$ is denoted by $D_i(w)$

$$D_i(w) = \sum_{j=1}^{m} d(r_{ij}(\otimes), v_{ij}(\otimes))w_j^2,\ i = 1, 2, \mathrm{L}, n. \tag{40.6}$$

Let $\sum_{i=1}^{n} D_i(w) = \sum_{i=1}^{n} \sum_{i=1}^{m} d(r_{ij}(\otimes), v_{ij}(\otimes))w_j^2$. Therefore, calculating the weight of attributes equivalents to solve the follow equations

$$\begin{cases} \min \sum\limits_{i=1}^{n} D_i(w) = \sum\limits_{i=1}^{n} \sum\limits_{j=1}^{m} d(r_{ij}(\otimes), v_{ij}(\otimes))w_j^2 \\ s.t. \sum\limits_{j=1}^{m} w_j = 1,\ w_j \ge 0, j = 1, 2, \mathrm{L}, m \end{cases}. \tag{40.7}$$

Then

$$w_j = \left(\sum_{i=1}^{n} d(r_{ij}(\otimes), v_{ij}(\otimes))\right)\left(\sum_{i=1}^{n}\left(\sum_{j=1}^{m} d(r_{ij}(\otimes), v_{ij}(\otimes))^{-1}\right)\right)^{-1}. \tag{40.8}$$

The gray relational degree between attributes vector and ideal optimal attributes vector is defined as follows:

$$Z(r^{+}(\otimes), r_{\mathrm{i}}(\otimes)) = \sum_{j=0}^{m} w_j \xi_{ij}, \ i = 1, \mathrm{L}, n. \tag{40.9}$$

The gray relational degree Z reflects the similarity between compared and referenced series. The bigger Z is, the greater the similarity is. If the weight of attributes known before, the gray relational degree Z can be calculated with Eq. (40.9) directly.

40.4 Case Study

We illustrate our approach with an example of aeronautic gyroscope which is a key component of helicopter attitude control system [6]. The analysis layer is shop replaceable unit (SRU). 13 kinds of main fault modes were got by the FMEA. The related ESs include temperature, humidity, vibration, electronic stress, strike, load, electromagnetism, sand, salt, and contamination are denoted by ES1, ES2, ES3, ES4, ES5, ES6, ES7, ES8, ES9, and ES10. Limited to the paper length, only electronic static convertor is listed in Table 40.1.

We take the fault mode of power tube breakdown for example. According to Table 40.1, ES1 (temperature) and ES3 (vibration) are the related ESs. The ES1 includes high temperature, low temperature, heating rate, and fall temperature rate and are denoted by $E_1 \sim E_4$. The ES3 includes acceleration peak and root-mean-square value and are denoted by $E_5 \sim E_6$. ES1 and ES3 are selected as scheme set and are denoted by $E_1 \sim E_6$. Fault modes, fault location, fault cause, and fault mechanism are treated as object set and are denoted by $F_1 \sim F_4$. Then the decision and preference matrix can be established in Tables 40.2 and 40.3.

Since the attributes are benefit type, the decision and preference matrix are normalized according to Eq. (40.1). The ideal optimal attributes vector can be calculated with Eq. (40.3)

$$r^{+}(\otimes) = ([0.4154, 0.6532], [0.4085, 0.6545], [0.4183, 0.6495], [0.4251, 0.6622]).$$

The relational coefficient matrix can be established according to Eq. (40.5)

Table 40.1 FMEA and the related ESs of electronic static convertor

Component	Mode	Location	Cause	Related ESs
Electronic static convertor	Breakdown	Power Tube	Overstress	ES1, ES3
	Open	Connector	Fatigue damage	ES1, ES4

Table 40.2 Decision matrix of power tube breakdown

	F_1	F_2	F_3	F_4
E_1	[0.92,0.95]	[0.9,0.92]	[0.91,0.93]	[0.94,0.97]
E_2	[0.86,0.89]	[0.9,0.95]	[0.91,0.93]	[0.9,0.92]
E_3	[0.85,0.9]	[0.91,0.94]	[0.94,0.96]	[0.9,0.91]
E_4	[0.88,0.91]	[0.91,0.94]	[0.92,0.94]	[0.86,0.89]
E_5	[0.93,0.96]	[0.85,0.88]	[0.87,0.88]	[0.87,0.9]
E_6	[0.83,0.87]	[0.8,0.82]	[0.82,0.86]	[0.81,0.82]

Table 40.3 Preference matrix of power tube breakdown

	F_1	F_2	F_3	F_4
P_1	[0.9,0.92]	[0.89,0.93]	[0.92,0.95]	[0.94,0.96]
P_2	[0.85,0.88]	[0.9,0.92]	[0.83,0.88]	[0.82,0.84]
P_3	[0.91,0.94]	[0.93,0.96]	[0.9,0.92]	[0.89,0.92]
P_4	[0.9,0.93]	[0.92,0.96]	[0.89,0.91]	[0.88,0.9]
P_5	[0.9,0.92]	[0.89,0.91]	[0.85,0.89]	[0.89,0.92]
P_6	[0.8,0.83]	[0.82,0.85]	[0.81,0.83]	[0.83,0.85]

$$R = \begin{bmatrix} 0.5798 & 0.6320 & 0.5992 & 1.0000 \\ 0.3333 & 0.8900 & 0.5992 & 0.4846 \\ 0.3450 & 0.8406 & 1.0000 & 0.4478 \\ 0.3895 & 0.8406 & 0.6916 & 0.3567 \\ 0.6525 & 0.3967 & 0.3677 & 0.3878 \\ 0.4484 & 0.3951 & 0.4046 & 0.3455 \end{bmatrix}.$$

The weight of attributes vector can be calculated through Eq. (40.8)

$$w = (0.2412,\ 0.2521,\ 0.2580,\ 0.2488).$$

The gray relational degree between attributes vector and ideal optimal attributes vector can be calculated according to Eq. (40.9)

$$Z(r^{+}(\otimes), r_{\mathrm{i}}(\otimes)) = (0.7025,\ 0.5799,\ 0.6645,\ 0.5730,\ 0.4487,\ 0.3981).$$

The gray relational degree orders according to descending sort as follow:

$$E_1 f\, E_3 f\, E_2 f\, E_4 f\, E_5 f\, E_6.$$

The above result show that the contribution of ESs to the fault mode of power tube breakdown according to descending sort are high temperature, heating rate, low temperature, fall temperature rate, acceleration peak, and root-mean-square value. Therefore, one can find that E_1 (high temperature) is the most-related ES to induce the power tube breakdown.

40.5 Conclusions

The approach present in this paper is not only utilizes the subjective information (the relativity between fault attributes and ESs), but also the attributes preference (preference to some ES) is considered based on the experience and knowledge of decision-maker. It is very suit for solving the problems of uncertain and incomplete relationship and system. This approach can be used to select the most-related ES to study the failure mechanism, and, hence, help diagnosing and prognosticating the faults which are environmentally sensitive.

References

1. Deng G, Qiu J, Liu G et al (2013) Environmental stress level evaluation approach based on physical model and interval grey association degree. Chin J Aeronaut 26(2):456–462
2. Deng G (2013) Research on research on key technology of intermittent faults diagnosis in extreme temperature environment [dissertation]. National University of Defense Technology, Changsha
3. Rouet V, Foucher B (2004) Development and use of a miniaturized health monitoring device. IEEE Annual 42nd International Reliability Physics Symposium, pp 645–646
4. Deng JL (2002) The primary theory of grey system. Huazhong University of Science and Technology Press, Wuhan
5. Zhang Q (2009) An objective approach to MADM with interval number. J Commun Comput 10 (5):37–51
6. Deng G, Qiu J, Liu G et al (2013) A novel fault diagnosis approach based on environmental stress level evaluation. J Aerosp Eng 227(5):816–826

Chapter 41
Peak Inhalation Air Flow and Minute Volume Measurements Used for Powered Air-Purifying Respirator Design

Ying Li, Xuezhi Zhang, Songtao Ding and Xiangxuan Liu

Abstract The supply air volume of powered air-purifying respirator (PAPR) determined by the peak inhalation air flow (PIAF) and minute volume (VE) of human body. Six healthy men participated in this study were pedaling a bicycle ergometer at step workload test. We measured their PIAF, VE, and maximal oxygen uptake (VO_{2max}) during the test. According to ISO 8990:2004 and data analysis, we got the averages of PIAF and VE under light (40 % VO_{2max}), moderate (60 % VO_{2max}), and heavy (80 % VO_{2max}) workloads. Moreover, the measured values were done to confirm the PIAF/VE ratio within the ISO16976-1:2007 range. Finally, we obtained the relation between PIAF and VE by the fitting analysis. The study offered some key data to PAPR design.

Keywords Peak inhalation air flow · Minute volume · Powered air-purifying respirator · Supply air volume

41.1 Introduction

The pressure in mask will reduce while inhaling [1], if the inhalation flow is higher than the supply ability, then there will form a negative pressure around the face; this will lead to mask leakage, and thus the respirator protection will reduce. In addition, the increase of negative pressure on the wearer face can cause discomfort [2–4]. The gas source of PAPR is a combination of power supply and gas filtration, which can supply the pure air to the user. Its advantage is can reduce respiratory resistance, at the same time, can form a certain positive pressure inside the mask, which

Y. Li (✉) · X. Liu
Six Department, Xi'an High Technology Research Institute, Xi'an 710025, Shaanxi, China
e-mail: li_ying_0318@126.com

Y. Li · X. Zhang · S. Ding
State Key Laboratory of NBC Protection for Civilian, Beijing 100191, China

© Springer-Verlag Berlin Heidelberg 2015
S. Long and B.S. Dhillon (eds.), *Proceedings of the 15th International Conference on Man–Machine–Environment System Engineering*, Lecture Notes in Electrical Engineering 356, DOI 10.1007/978-3-662-48224-7_41

can improve the safety and comfort of the mask. The minimum supply air volume in unit time of PAPR is the minimum threshold to guarantee adequate fresh air supply. Therefore, the minimum supply air volume is one of key research contents of PAPR.

Determine the size of the minimum supply air volume need to consider many factors, but the main fact is the PIAF of human body [5]. In the European and American countries, there are some high-value results on PIAF research [6]. Due to the different physical character from westerners, the PIAF of Chinese is different, so we cannot use westerner's data directly. At present, there is no comprehensive experiment on PIAF in China. And since 2007, there has no new study on this field. This paper researched on PIAF, VE, and VO_{2max} of human body during a step load test. The test results offer a scientific basis for the PAPR function design and the PAPR standard establishment.

41.2 Materials and Methods

This study was performed at State Key Laboratory of NBC Protection for Civilian in Beijing, China. The subjects participated in the test were voluntary and could withdraw from test at any stage.

41.2.1 Subjects

There were six male subjects participated in the test. General physical fitness was established with each subject before the test. Before the test, all subjects had trained how to operation the bicycle ergometer and familiar with the whole test procedure. Physiological characteristics of the subjects who participated in this test are summarized in Table 41.1.

41.2.2 Conditions in the Test Room

The temperature was 25 ± 1 °C, relative humidity was 65 ± 5 %, and wind velocity was 1 ± 0.3 m/s.

Table 41.1 Summary of physical characteristics of the subjects

Subject characteristics	Mean	Std. Dev.	Min	Max
Age (years)	21.4	1.3	18	22
Weight (kg)	63.6	3.8	59.2	65.1
Height (cm)	170.2	4.7	169	175
VO_{2max} (L/min)	2.60	0.23	2.3	2.8

41.2.3 Test Procedures

The subjects were dressed in cotton jersey. The test bicycle ergometer ("ER 500", Jaeger, Germany) was connected to a computer and calibrated in accordance with the manufacturers instructions. The test protocol was developed by means of the software supplied with the bicycle. The VE and VO_2 were measured by cardio-pulmonary exercise testing instrument ("Oxycon Pro" Jaeger, Germany). The heart rate was measured using a heart rate monitor ("Polar 430", Polar, Finland), downloading to Jaeger software. A Certifier FA Test System (TSI, USA) was used to measure PIAF.

Test starting power was 0 W, step was 25 W, and every step was keeping 4 min. The test was terminated by the test officer if the test subject felt exhaustion, or when 90 % of the theoretical max heart rate was reached, whichever occurred first.

41.2.4 Data Collection and Statistical Analyses Methods

Data were collected for every breath during the entire test, at 100 samples per second. The VO_{2max} is an important index to reflect the body's aerobic capacity. According to ISO 8996:2004, this study elicited 40, 60, and 80 % of each subject's VO_{2max}, these percentages of VO_{2max} were correspond with light, moderate, and heavy workloads [7].

Collected PIAF and VE data in 1 min under three workloads, and these parameters' mean expressed as $\overline{PIAF}$ and $\overline{VE}$.

41.3 Results

41.3.1 Parameters Statistics

First, we got the VO_{2max} mean, and then obtained the oxygen uptake value under variety workloads through numerical calculation (Table 41.2).

The oxygen uptake which corresponding to the workload should be an interval, therefore this study extracted all data in 30 s before and after the VO_{2max} percentage

Table 41.2 VO_{2max} and workload interval of the study

Parameters	VO_{2max} Mean	VO_{2max} Std. Dev.	Light workload 40 % VO_{2max}	Moderate workload 60 % VO_{2max}	Heavy workload 80 % VO_{2max}
VO_2 (L/min)	2.601	0.229	1.04	1.56	2.08

Table 41.3 Respiration parameters statistics of the study

Parameters	Light workload		Moderate workload		Heavy workload	
	Mean	Std. Dev.	Mean	Std. Dev.	Mean	Std. Dev.
VE (L/min)	27.67	2.78	46.55	3.52	62.08	3.56
PIAF (L/min)	90.17	8.24	**128.36**	8.23	159.34	17.66
$PIAF_{max}$ (L/min)	101.82	5.78	**140.12**	6.04	176.43	15.34

for respiration parameters statistics. The $PIAF_{max}$ is expressed as the maximum PIAF in the value interval. This study parameters statistics are described in Table 41.3.

41.3.2 The Relationship Between VE and PIAF

There are some research shows that the ratio of PIAF and VE is generally within (2.5 and 3.9) interval. Test data analysis (Table 41.4) of this study proves that the ratio of PIF and VE of Chinese people is also within (2.5 and 3.9) interval.

Due to the measurement method of VE is simpler than that of PIAF, if we get a linear relationship between PIAF and VE, we can get a relatively accurate PIAF value by numerical calculation. This study applied Excel software to data processing (Fig. 41.1) and obtained a linear Eq. (41.1) of PIAF and VE.

$$\text{PIAF} = (2.1102 \times \text{VE}) + 21.6391 \quad \left(R^2 = 0.9904\right). \tag{41.1}$$

Table 41.4 The ratio of PIAF and VE

Parameters	Light workload		Moderate workload		Heavy workload	
	Mean	Std. Dev.	Mean	Std. Dev.	Mean	Std. Dev.
$\overline{\text{PIAF}}/\overline{\text{VE}}$	3.25	0.38	2.76	0.31	2.57	0.24

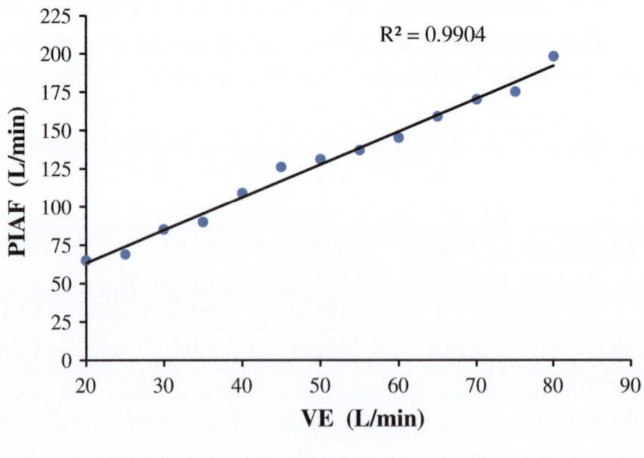

Fig. 41.1 Linear relationship between PIAF and VE

Table 41.5 Respiration parameters statistics of Anderson research

Parameters	Light workload		Moderate workload		Heavy workload	
	Mean	Std. Dev.	Mean	Std. Dev.	Mean	Std. Dev.
VE (L/min)	30.3	6.2	47.4	13.5	72.3	22.9
PIAF (L/min)	100.3	18.9	150.1	35.4	218.4	53.7
$PIAF_{max}$ (L/min)	125.6	29.2	185.3	37.1	254.7	57.8

41.4 Discussion

Research has shown that division of workload level by energy consumption index, oxygen consumption, heart rate, rectal temperature, sweat rate, the concentration of lactic acid, relative metabolic rate, and so on has the same meaning [8]. This study adopted ISO 8996:2004 method to grate workload by oxygen uptake percentage, the method is simple, and it is easy to be realized.

Anderson [9] researched the physiological characteristics of 13 people (8 male and 5 female) by cycle ergometer test, some results is shown in Table 41.5. Table 41.6 shows respiration parameters statistics of some PAPR standards. Figures in Tables 41.3, 41.5 and 41.6 indicate that the VE and PIAF of Chinese people are lower than West people under the same workload [10, 11]. There were male and female subjects in Anderson research, and it had a large age span, so the discrete of research figures were very highly. This study gave full consideration to the individual differences in order to reduce its impact on the reliability of the data.

The European and American countries generally choose PIAF mean under moderate workload as the minimum recommended flows of PAPR. According Table 41.6, the loose-fitting PAPR minimum recommended flows is 170 or 185 L/min, and tight-fitting PAPR minimum recommended flows is 115 L/min. we can see the PIAF mean under moderate workload of Chinese people (Table 41.3) is 128 L/min and the $PIAF_{max}$ mean 140 L/min. Usually the VO_{2max} and VE values of women and weak people are lower than that of young man. In this study all subjects were young man, so we recommend using negative double standard deviation plus

Table 41.6 Respiration parameters statistics of PAPR Standard

Standard		Parameters	Light workload	Moderate workload	Heavy workload
ISO 16976-1		VE (L/min)	22	36	50
		PIAF (L/min)	122	**185**	244
NIOSH 2007 PAPR	Tight-fitting	VE (L/min)	25	40	57
		PIAF (L/min)	/	**115**	170
	Loose-fitting	VE (L/min)	25	40	57
		PIAF (L/min)	115	**170**	235

the PIAF mean as the lower limit of minimum recommended flows value range, and using negative standard deviation plus the PIAF mean as the upper limit of its range. That is to say, the minimum recommended flows value should be in (112, 120) L/min range.

Because of the measurement method of VE is easier than that of PIAF; the numerical value of PIAF is obtained generally through the calculation of measured value of VE. There are some equations about the relationship between VE and PIAF, Lusa [12], Eq. (41.2) is a linear equation, and Caretti [13] Eq. (41.3) is a power function equation.

$$\text{PIAF} = (2.346 \times \text{VE}) + 20.828 \quad \left(R^2 = 0.9867\right), \tag{41.2}$$

$$\text{PIAF} = (5.605 \times \text{VE})^{-0.1675} \quad \left(R^2 = 0.5997\right). \tag{41.3}$$

Equations (41.1) and (41.2) are linear equation, and the R-square values of them are higher than that of Eq. (41.3). Therefore, the study firmly believes that Eq. (41.1) is practical and effective, and more suitable for Chinese people respiratory parameters study.

41.5 Conclusions

This research helps to modernize and establish baseline ventilation data for respirator design. Furthermore, this research would enrich relevant theories and practices about people respiratory characteristic, and it would provide a reference to similar researches in the future.

The present study has its own limitations and calls for improvement in future-related studies. First, the number of samples analyzed may be not large enough to be generalized to all population. Second, human movement method was simplistic. Future research need to base upon larger sample and more movement method.

41.6 Compliance with Ethical Standards

The study had obtained approval from the State Key Laboratory of NBC Protection for Civilian Ethics Committee.

All subjects participated in the experiment had signed the informed consent.

All relevant ethical safeguards had been met in relation to subject protection.

References

1. Clayton MP, Bancroft B, Rajan B (2002) A review of assigned protection factors of various types and classes of respiratory protective equipment with reference to their measured breathing resistance. Ann Occup Hyg 46:537–547
2. Bostock JG (1985) An investigation into the performance of positive pressure powered dust hoods and blouses at low flow rates. Ann Occup Hyg 29(3):415–420
3. Dahlbäck GO, Novak L (1983) Do pressure-demand breathing systems safeguard against inward leakage. Am Ind Hyg Assoc 44:336–340
4. Hinds WC, Bellin P (1987) Performance of dust respirators with facial seal leaks: I. Experimental. Am Ind Hyg Assoc J Oct 48(10):836–411
5. NIOSH Respiratory protective equipment manual[M] (1983)
6. Janssen LL, Anderson NJ, Cassidy PE et al. (2005) Interpretation of inhalation airflow measurements for respirator design and testing. J Int Soc Respir Prot 22
7. ISO 8996:2004 Ergonomics of the thermal environment—Determination of metabolic rate [S]
8. Zhiqiang C (2000) Labor hygiene and occupational medicine [M], People's medical publishing house, 06 2000. ISBN: 9787117039628
9. Anderson NJ, Cassidy PE, Janssen LL, Denger DR (2006) Peak inspiratory flows of adults exercising at light, moderate and heavy workloads. J Int Soc Respir. Prot 23:53–63
10. ISO 16976-1:2007(E) respiratory protective devices-human factors-part 1: metabolic rates and respiratory flow rates [S]
11. NIOSH Proposed Concept (2007) Powered air-purifying respirator (PAPR) standard Subpart P, 21 Dec 2007
12. Lusa S, Louhevaara A et al. (1993) physiological responses of firefighting students during simulated smoke-diving in the heat. Am Ind Hyg Assoc 54:228–231
13. Caretti D (2003) Work rates and anticipated ventiliation parameters. J

Chapter 42
Auto-Brightness Curve Model for a Night Vision Imaging System Compatible with White LED Integrally Illuminated Information Panels in Airplane Cockpits

Duanqin Xiong, Xiaochao Guo, Lili Zhang, Muzhe Zhang, Chunmei Gui, Yu Bai, Jian Du, Yanyan Wang and Wei Zheng

Abstract This chapter explores the optimal brightness of a night vision imaging system compatible with green LED integrally illuminated information panels when pilots are reading in different ambient illuminance, in an attempt to build an auto-brightness curve model. A total of 285 fighter pilots took part in the experiment. The equipment used in the experiment included an LED panel and a system to simulate light environments. In every grade of the simulated brightness, pilots adjusted the minimum brightness, the optimal brightness, and the maximum brightness when they read the LED panel. According to the data, the auto-brightness curve models of the minimum brightness, optimal brightness, and maximum brightness were found for the 5th, 50th, and 95th percentile of the 285 pilots. The engineering applications for these models need to be verified by the users and might be revised according to the verification.

Keywords Night vision white light · LED lighting control panel · Auto-brightness curve · Illuminance

42.1 Introduction

A light control panel in an airplane cockpit is a version of integrally illuminated information panels used in the home. It is a lighted signal device presenting painted or carved characters or phrases on the panel. When lights behind the panel are illuminated, information is presented to pilots through the transparent brightness of

D. Xiong (✉) · X. Guo · L. Zhang · M. Zhang · C. Gui · Y. Bai · J. Du · Y. Wang · W. Zheng
Institute of Aviation Medicine PLAAF, Beijing 100142, China
e-mail: xdqin123@163.com

© Springer-Verlag Berlin Heidelberg 2015

S. Long and B.S. Dhillon (eds.), *Proceedings of the 15th International Conference on Man–Machine–Environment System Engineering*, Lecture Notes in Electrical Engineering 356, DOI 10.1007/978-3-662-48224-7_42

the panel. A night vision compatible light control panel can be a comprehensive guided information panel when its wavelength, brightness, and uniformity do not interfere with the expected usage of night vision goggles, allowing pilots to view it with the naked eye [1]. According to U.S. standards, light sources that may be used by night vision compatible light control panels include incandescent lamps, electron tubes, and light-emitting diodes (LEDs). Newer aircraft use LEDs as a light source to varying degrees, and LED panel lighting has become a trend in aircraft cockpit design.

Night Vision Imaging System (NVIS)-compatible lighting technology is an important requirement for the safety of night vision [2]. LED night vision that is compatible with the light control panel is a new type of display device; the optical properties and the display quality will directly affect the visual observation and information read by the pilot, as well as flight ergonomics. Because the human eye processes changes in the light environment as the presence of dark and light, the light environment in the cockpit should not undergo significant changes. Light variations in the LED control panel cause a pilot to frequently adjust his or her eyes, leading to adverse cognitive effects and binocular visual fatigue. In the design of advanced aircraft, cockpit lighting uses automatic dimming technology that adapts to external environment changes, automatically adjusting the brightness of the lighting system in the cockpit to meet the visual demands of the pilot's need for observational information, reduce manual intervention and regulation, and ultimately reduce the pilot's workload [3]. For the cockpit lighting core technology of automatic light adjustment, we aimed to establish an automatic brightness curve model.

For night vision compatibility, green light and white light are the two character colors currently used in LED controls panel in aircraft cockpits. The light adjustment curves for pilots have been established with NVIS green LED characters [4]. This chapter examines pilots' automatic brightness demands for NVIS-compatible white-light control panels. Experiments that adjusted the light intensity of characters on the light control panel were carried out. Automatic brightness curves for white-light night vision compatible characters were built according to the experimental data, providing a scientific basis for the design of an automatic light adjusting system in the aircraft cockpit's guide panel lighting.

42.2 Method

42.2.1 Instruments and Equipment

42.2.1.1 LED Control Panel for Night Vision White Light

As shown in Fig. 42.1, the light control panel is arranged on test pieces of 32 military display characters [5] in italics, complying with the night vision white color coordinates of ($u' = 0.190$, $V' = 0.490$, $r = 0.040$) [6]. The panel uses character color

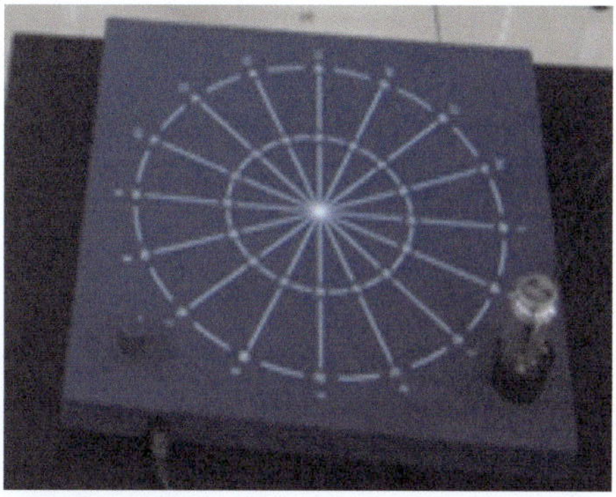

Fig. 42.1 LED panel in the experiment without electric power

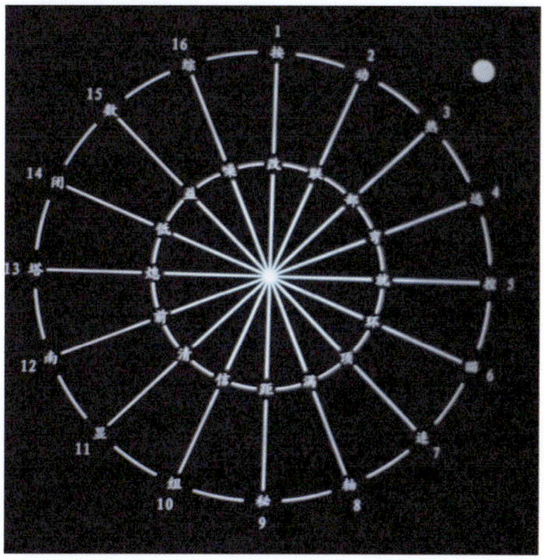

Fig. 42.2 LED panel in the experiment with electric power

of MIL-STD-3009, a character size of 5 mm × 5 mm, adjustable brightness, dark gray background, and a reflection coefficient of 0.8. There is a solid circle on the upper right corner of the light control panel, through which the brightness of a light source can be measured. The panel has an option to adjust the brightness of the knob, which can be manually adjusted in the range of 0–50 cd/m^2. In the bench experiment, the light control panel was placed right in front of the pilot's seat, 400 mm from the ground (Fig. 42.2).

42.2.1.2 Simulated Ambient Light

The simulated ambient light system (Fig. 42.3) has different ambient light intensities. Following standards for the lighting design of buildings [7], illuminance values of the light source for the control panel center are divided into 5 files, with a difference of 2× between illuminance values: 4lx, 8lx, 16lx, 32lx, and 64lx.

Fig. 42.3 Simulated system of ambient light

42.2.1.3 ST-86L Screen Lux Meter

By placing a lux meter probe supinely in the center of the light control panel, the simulated environmental illuminance can be obtained.

42.2.1.4 Brightness Meter

A ST-86LA screen brightness meter probe was placed in the solid circle of the light control panel, and light source brightness was measured.

42.2.1.5 Stopwatch

To determine the adaptation of the human eye to darkness and brightness, the pilots' manual regulation times were recorded.

42.2.2 Participants

A total of 285 fighter pilots took part in the experiment. Their age ranged from 22 to 42 years, and their flight time was in the range of 240–3500 h. Their naked visions were above 1.0, color senses were normal, and dark adaptation functions were normal.

42.2.3 Experimental Procedures

The entire experiment was conducted in an environment similar to a darkroom. After the pilots entered the laboratory, the experimenters first explained the experiment goal and the method. Two groups of pre-experiments were then conducted to familiarize the subjects with the experimental tasks and the procedures.

Before the experiment, 5 grades of illumination intensities (error < 5 %) with the simulated environment light were adjusted. After the environment illuminance changes, the pilots needed to adjust to the minimum brightness, the optimal brightness, and the maximum brightness as soon as possible, while simultaneously reading aloud two Chinese characters as assigned by the experimenter. Finally, the experimenter recorded pilots' reading results of the Chinese characters, the brightness adjusting time, and the light-source brightness of the control panel.

42.2.4 Data Processing

The raw brightness data were converted into compound brightness according to following formula (42.1) [8]:

$$\begin{aligned}\text{Compound brightness} &= \text{source brightness of the LED light control panel} + \text{ambient light illuminanance} \\ &\quad \times \text{reflection factors of the light control panel} \div \pi.\end{aligned} \tag{42.1}$$

The SPSS statistical analysis software package was used to carry out the statistics analysis of the empirical data. The pilots' adjustment time was used to determine whether the results surpassed the dark adaptation requested time limit for calculating the minimum brightness, optimal brightness, and the maximal brightness. The relationship between the environment illuminance and light-source brightness was analyzed and the pilots' brightness curve for the 5th (P_5), 50th (P_{50}), and 95th (P_{95}) percentiles were established.

42.3 Results and Analysis

42.3.1 Pilots' Light Adjustment Time

The pilots' light adjustment time data are given in Table 42.1. There is no statistically significant difference ($F = 0.094$, $P > 0.05$) between the adjustment time for the minimum brightness, the maximum brightness, (average 4.64 s, 4.81 s), and the optimal brightness (average 4.65 s). Of the 4275 light adjustment time points for 285 pilots, 99.96 % were no more than 20 s; the longest duration was 40 s. This

Table 42.1 Pilots' time to adjust LED brightness (n = 285, s)

Ambient illuminance (Lx)		4	8	16	32	64
Minimum brightness	$\bar{x}$	5.89	4.67	4.61	4.22	3.81
	$s_{\bar{x}}$	0.25	0.13	0.15	0.13	0.13
Optimal brightness	$\bar{x}$	5.38	4.85	4.51	4.35	4.15
	$s_{\bar{x}}$	0.17	0.16	0.15	0.15	0.13
Maximum brightness	$\bar{x}$	6.24	4.74	4.63	4.37	4.09
	$s_{\bar{x}}$	0.23	0.14	0.16	0.15	0.13

result complies with the regulation that the rapid dark adaptation time of fighter pilots cannot exceed 60 s. Therefore, the experimental data is valid and available.

42.3.2 *Pilots' Light Adjusting Curves*

Under the 5 grades of environment illumination conditions, 285 pilots were measured on their ability to correctly read displayed characters on the LED guide panel; also measured were the results for the minimum brightness, optimal brightness, and

Table 42.2 Luminance data with NVIS white LED light sources in different ambient illuminance (n = 285, cd/m^2)

Reading standard	Ambient illuminance (Lx)	$\bar{x}$	$s_{\bar{x}}$	P_5	P_{10}	P_{50}	P_{90}	P_{95}
Minimum brightness	4	0.98	0.06	0.10	0.10	0.70	2.50	2.90
	8	1.54	0.09	0.10	0.10	1.20	3.64	4.50
	16	2.10	0.12	0.10	0.10	1.60	4.94	6.34
	32	2.96	0.18	0.10	0.10	2.20	7.34	9.40
	64	3.64	0.24	0.10	0.10	2.50	9.60	12.28
Optimal brightness	4	4.06	0.19	0.80	1.10	3.30	8.52	10.97
	8	4.71	0.21	0.80	1.40	3.80	9.48	11.70
	16	6.22	0.24	1.13	1.96	5.20	11.24	14.44
	32	8.02	0.30	1.20	2.10	7.20	16.14	18.04
	64	10.47	0.39	0.50	1.62	9.80	20.26	22.40
Maximal brightness	4	10.63	0.37	2.70	3.60	9.10	21.00	22.50
	8	11.09	0.37	2.80	3.46	10.00	22.10	22.60
	16	12.75	0.39	3.30	4.46	12.10	22.50	22.80
	32	14.97	0.38	3.59	5.42	15.10	22.70	22.90
	64	17.05	0.37	3.70	7.86	19.70	22.90	23.00

Table 42.3 Luminance data with NVIS white LED characters in different ambient illuminance (n = 285, cd/m^2)

Reading standard	Ambient illuminance (Lx)	$\bar{x}$	$s_{\bar{x}}$	P_5	P_{10}	P_{50}	P_{90}	P_{95}
Minimum brightness	4	2.06	0.06	1.11	1.16	1.77	3.55	4.02
	8	3.66	0.10	2.14	2.19	3.24	5.81	6.59
	16	6.27	0.12	4.23	4.26	5.72	9.05	10.45
	32	11.21	0.18	8.28	8.38	10.48	15.66	17.68
	64	20.04	0.24	16.42	16.47	18.88	26.01	28.69
Optimal brightness	4	5.15	0.19	1.84	2.15	4.41	9.55	12.07
	8	6.82	0.21	2.84	3.52	5.83	11.66	13.85
	16	10.39	0.24	5.27	6.07	9.40	15.36	18.56
	32	16.26	0.30	9.38	10.35	15.48	24.37	26.25
	64	26.87	0.39	16.81	17.97	26.23	36.75	38.89
Maximal brightness	4	11.76	0.37	3.91	4.72	10.23	22.07	23.63
	8	13.23	0.37	4.95	5.75	12.14	24.18	24.81
	16	16.97	0.39	7.47	8.71	16.19	26.69	27.02
	32	23.26	0.38	12.19	13.69	23.39	30.98	31.17
	64	33.51	0.37	20.21	24.31	36.13	39.28	39.41

maximum brightness with a NVIS-compatible white LED control panel (see Table 42.2). The same data were also collected for a NVIS white LED guide panel (see Table 42.3).

To satisfy the P_5–P_{95} pilots' use (i.e., 90 % satisfaction) and determine a standard for light source brightness and character brightness of night-view white-light control panel, the pilots' brightness curves were calculated (Figs. 42.4 and 42.5).

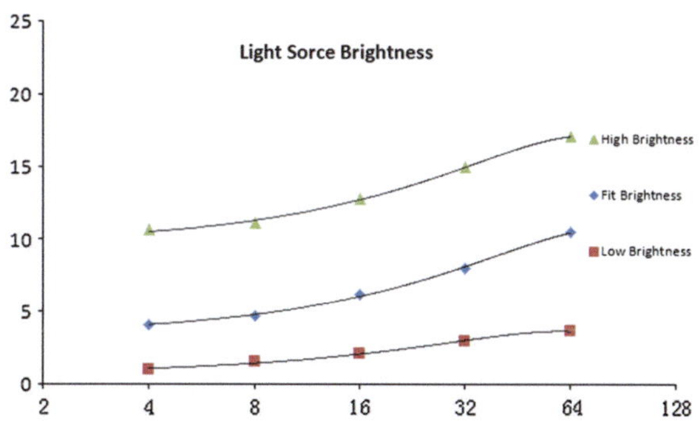

Fig. 42.4 The pilots' brightness curves for light source

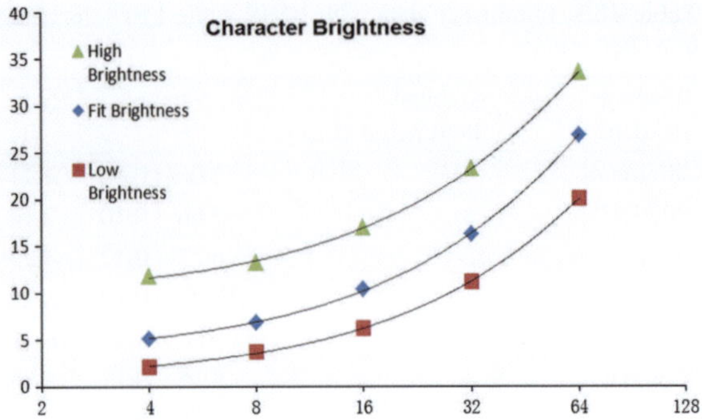

Fig. 42.5 The pilots' brightness curves for character reading

42.4 Conclusions

This chapter introduced experimental research on the visual ability of 285 pilots to read characters on a NVIS white guide panel. The minimum brightness, optimal brightness, and maximum brightness of the light source and characters were obtained for different ambient illuminance conditions as pilots read characters on the NVIS white guide panel. The automatic brightness curves of pilots were calculated using mathematical modeling. However, the engineering applications of the curve models still need user verification, and fine-tuning or corrections of curve models should be performed in accordance with these results.

42.5 Compliance with Ethical Standards

The study was approved by the State Key Laboratory of NBC Protection for Civilian Ethics Committee.

Each pilot was briefed on the aims of the study and consent was obtained. All pilots were paid ¥50/h for their participation.

All relevant ethical safeguards were met in relation to subject protection.

References

1. SAE ARP 4967 (2004) Night vision imaging systems (NVIS) integrally illuminated information panels
2. Men JF, Cheng HF, Cheng ZH et al (2008) Development of ANVIS compatible lighting technology. J Appl Opt 29(3):354–359
3. AC25–11A-2007 Electronic flight deck displays
4. Xiong DQ, Guo XC, Ye JB et al (2013) Study on auto brightness curve of night vision imaging system compatible green LED integrally illuminated information panels. Space Med Med Eng 26(2):92–95

5. GJB302-1987 Chinese characters and phraseology for electric/optic display in aircraft
6. MIL-STD-3009 Lighting, aircraft, night vision imaging system (NVIS) compatible
7. GB50034-2004 Standard for lighting design of buildings
8. Fudan University Department Of Optics, Electric Light Source Laboratory (1977) Principle of electric light source. Shanghai People's Publishing House, Shanghai, pp 22–33

Chapter 43
Research on Gravity Compensation Technology for Extravehicular Activity Training Facility

Shang Huan and Hua Deng

Abstract **Objective:** To improve the current extravehicular activity facility, new gravity compensation technology is researched. **Methods:** To assure the rapidity and precision of gravity compensation, high torque motor and low torque motor cooperation are applied to generate contragravity; to offload the maximum added mass, the system tracks the target position actively by servo control; to meet astronaut specific needs of extravehicular activity, the ingenious lift spreader of suspension system is designed. **Results:** Design analysis and model verification show that the position track error is less than 10 mm and the accuracy of the single suspended cable tension force is 99.5 % at the maximum limit of 0.1 m/s velocity and 0.1 m/s^2 acceleration. **Conclusion:** The facility that applied new gravity compensation technology is able to generate near-zero microgravity environment, which is more suitable for training of extravehicular activity.

Keywords Extravehicular activity · Gravity compensation · Track actively

43.1 Introduction

Extravehicular activity, a quite complex procedure, is an important symbol for humans in space. The astronaut should complete a series of works like wearing extravehicular spacesuit under weightlessness condition; so that onground training facility of extravehicular activity is absolutely indispensable. To complete SZ-7

Aided information: Ergonomics Engineering Fundamental Research Subject YJGF141202.

S. Huan (✉) · H. Deng
China Astronaut Research and Training Center, Beijing 100094, China
e-mail: huan_shang@126.com

© Springer-Verlag Berlin Heidelberg 2015

S. Long and B.S. Dhillon (eds.), *Proceedings of the 15th International Conference on Man–Machine–Environment System Engineering*, Lecture Notes in Electrical Engineering 356, DOI 10.1007/978-3-662-48224-7_43

Fig. 43.1 Procedure training simulator of extravehicular activity

mission, our country has researched and developed extravehicular activity procedure training simulator [1, 2] (hereinafter referred to as EVA simulator, Fig. 43.1).

EVA simulator mainly simulates extravehicular operation and procedure to the astronaut under weightlessness condition and plays an important role in SZ-7 mission. During simulation training process, the suspension system provides constant tension force to extravehicular spacesuit for gravity compensation in order to offset influence of load and own gravity on the astronaut. In the vertical direction, the astronaut is able to move up-and-down or hover; the arc and radial air-bearing guide rail support the astronaut's minimal damp horizontal motion. Restricted by expenditure and technical conditions, although EVA simulator meets the training demand for extravehicular activity, it faces the following problems in gravity compensation: first, the sector activity area restricts work space; second, influenced by inertia of bearing beam, it track possibly lagged or overshot, which causes suspension cable swing, poor rapidity and stability of system; third, when operating, although the astronaut's attitude is able to spin around Z axle (yaw), it has no roll and pitch motion and causes the astronaut's simulating extravehicular operation to be restricted and ordinary human−machine relationship to be experienced, which is unsuitable for complex task training such as on-orbital maintenance, repair, etc. This article proposes optimal design in three respects: the truss structure of suspension system, position tracking mode, and lift spreader structure to solve the above-mentioned problems.

43.2 Truss Structure

The truss bears constant tension force mechanism for gravity compensation, which decides the area of simulation operation of the astronaut. Although the semi-arc truss structure is simple, its motion area and characteristics are restricted. Even in

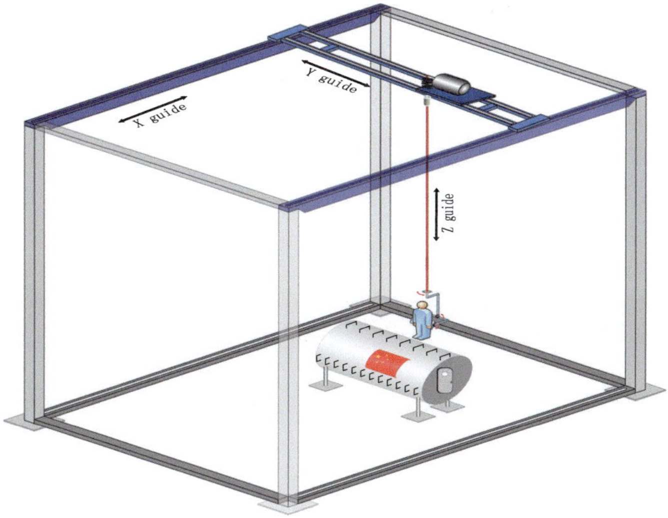

Fig. 43.2 H guide truss

single axle translation, the radial and arc directions also must be adjusted in linkage and the sector operation area is disadvantageous for large cabin placing. Relatively speaking, H-shaped truss structure is obviously advantageous. The structure is as listed in Fig. 43.2: axle X guide rail and axle Y guide rail are located on the top of the truss, which are able to provide rectangle motion space together with axle Z suspension cable.

43.3 Position Tracking

Gravity compensation mechanism of EVA simulator adopts passive position tracking method, applies component horizontal force generated by suspension cable offsetting vertical line to drive air-bearing cantilever beam, and constant tension force bracket to track position change of the target till the suspension cable holding vertical state and component force disappeared. Although it is minimal damp motion, the added mass is not offloaded and the inertia must cause lagging and overshooting, thus it will be overlapped on the spacesuit and form swing, causing weightless motion state distorted and satisfaction is down.

New type of simulator gravity compensation mechanism adopts servo control (Fig. 43.3) and actively tracks position change of target. Therein, axle X guide rail adopts gear rack, axle Y guide rail adopts ball screw, and takes suspension cable angular displacement as control system input. According to measured value of angular displacement, the controller calculate offset (Δx, Δy) of axle X and axle Y at the end of load, then generate control vector ($\vec{x}$, $\vec{y}$) to drive torque motor work of

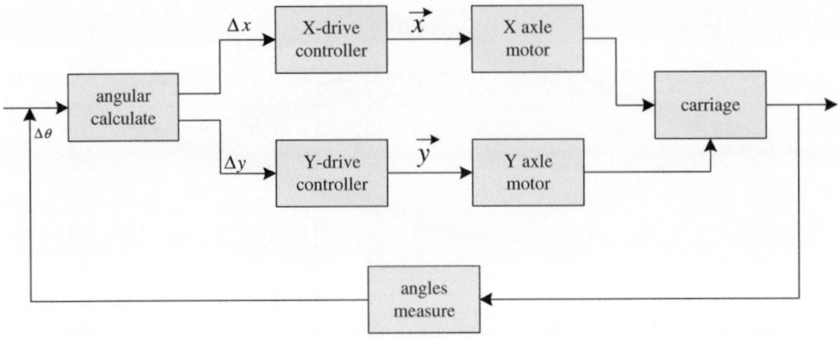

Fig. 43.3 Control block diagram

axle X and axle Y, respectively, rapidly track planar position change of target, and maintain suspension cable (axle Z) vertical always. Thus the new means completely offload the added mass containing cantilever beam, slider and torque motor, etc. [3].

43.3.1 Angular Displacement Measurement

As angular displacement is considered as input for two-dimensional tracking system, the data accuracy and instantaneity directly decide the system performance and the regular angular sensor is unable to meet the requirement; therefore, a kind of new method to measure angular displacement is needed.

As shown in Fig. 43.4, a horizontal annulus extends below the constant tension force subsystem bracket, H(m) away from the bracket; the suspension cable vertically penetrates through the circle center; four position sensors allocated symmetrically on the annulus shall in real-time monitor the displacement S of suspension cable deviating from circle center; in accordance with the ratio of S and

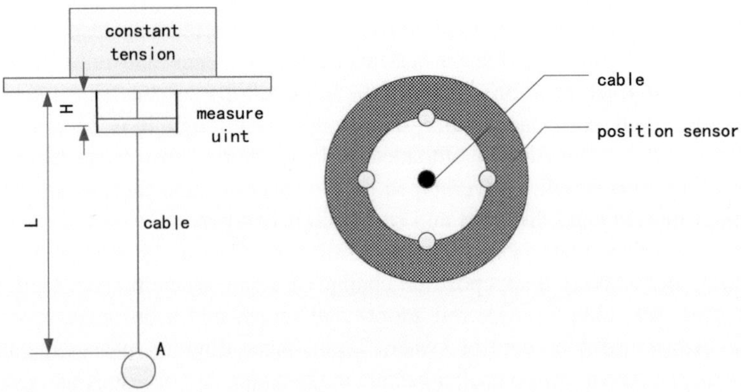

Fig. 43.4 Angle measure mechanism

H, angular displacement of suspension cable can be calculated. As length of L is measurable, displacement of target A can be inversely calculated with angular displacement. This method possesses advantages of high accuracy, rapid response, simple layout, sound anti-interference performance, and is better than angular sensor [3].

43.3.2 Constant Tension Force System

In gravity compensation, the direction of force-application-point is ensured by two-dimensional position tracking system while the amplitude of force is ensured by constant tension force system. The constant tension force subsystem structure is shown in Fig. 43.5, where the constant tension force is generated by cooperation between high torque motor and low torque motor [4]. High torque motor transit speed reducer and torsion spring coupling to roller to provide high torque and bear load weight; the low torque motor provides high frequency response to realize rapid adjustment; the tension force sensor feedback forms a closed-loop servo control. This gravity compensation mechanism maximizes both rapidity and accuracy. Besides, the spring buffer device set for force-application-point stabilizes constant tension force and absorbs cable impact caused by sudden change of load velocity [4].

43.4 Lift Spreader Structure

EVA simulator is able to hang the astronaut through balance beam and offer a yaw attitude. If it needs to simulate all operation attitudes (including roll and pitch), the corresponding lift spreader must be designed. The lift spreader strives to be light and handy, putting on and off conveniently, not interfering with the astronaut's motion.

Figure 43.6 shows the lift spreader structure, where the yaw, roll, and pitch bearing points keep load (extravehicular spacesuit) gravity center coinciding with

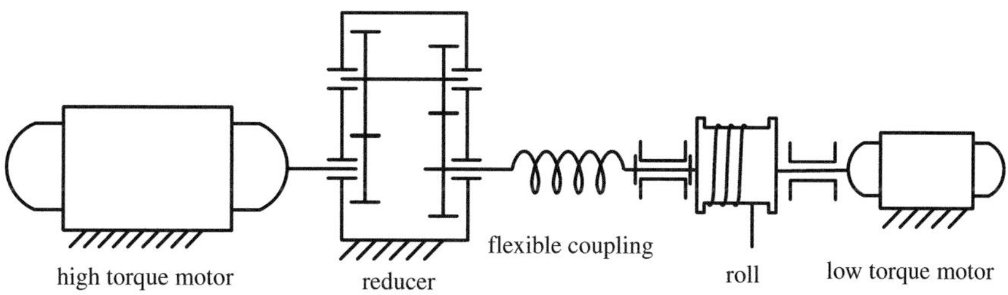

Fig. 43.5 Driving gear

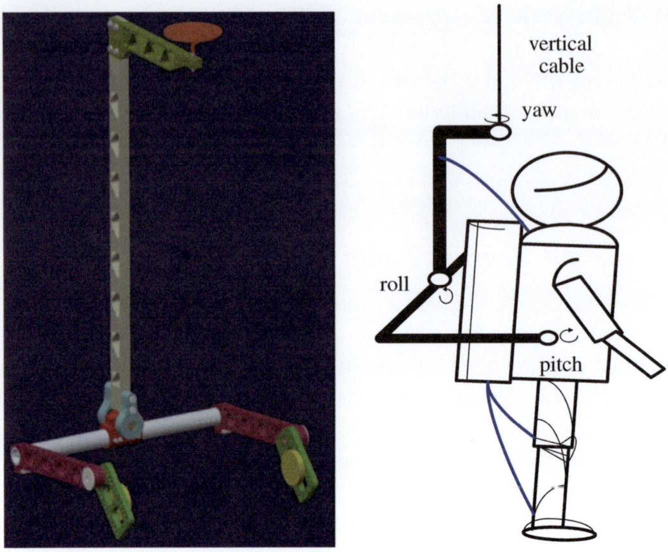

Fig. 43.6 Construct of lift spreader

suspension cable and the astronaut in vertical state. The attitudes of yaw 360° and roll ±20° are easy to realize. As roll demand scope is small, setting mechanical position limitation can prevent side turn from sudden excessive movement, etc. It is difficult to free pitch 90°.

To guarantee the spacesuit elbow activity scope without being influenced, the pitch bearing is on the lower part of two sides of the waist of spacesuit, which causes gravity center of spacesuit to deviate upward from midpoint of line between pitch bearings. When the astronaut pitches downward, the torque pulling down will be generated. To ensure the astronaut has good experience in weightlessness, the corresponding torque compensation must be conducted to ensure torque balance in pitching process. Besides, it also needs to prevent misoperation by the astronaut, which causes instantaneous turning and hanging upside down. If it compensates actively, the electromechanical device should be added. Inevitably increasing added mass and intervenes with upper limb movement, the compensation cost is also too much. Through force analysis of pitch, using suitable elastic hauling cable realizes perfect passive compensation.

Add an elastic hauling cable between the back shoulders of the spacesuit and the upper end of the lift spreader. In vertical state, the elastic hauling cable is in initial state and the compensation force is zero. With increase of pitch angle, the elastic hauling cable stretches longer and the compensation force is stronger, hence the torque is kept balanced. The compensation is real-time and effective.

In Fig. 43.7, P_0 is the waist supporting point for the astronaut; P_a is the connection point of the elastic hauling cable on the lift spreader; P_b is the connection

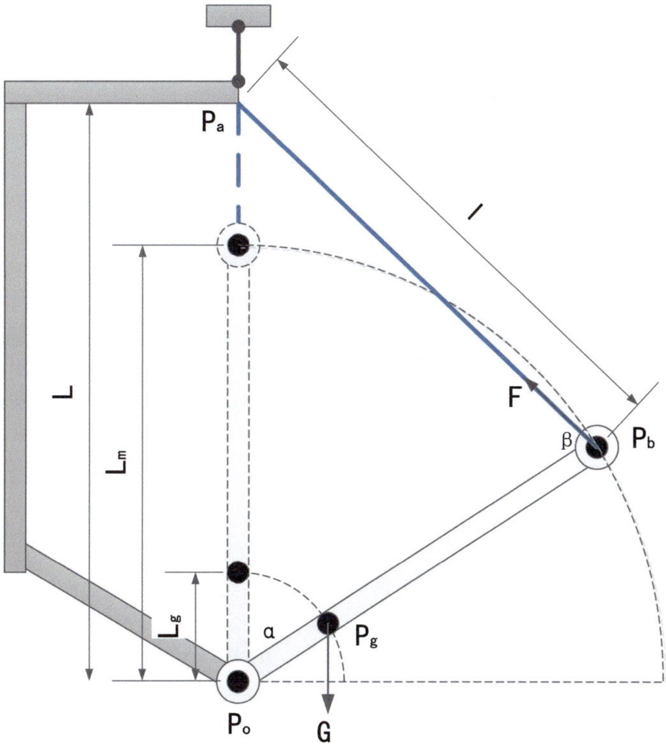

Fig. 43.7 Pitch analysis

point of the elastic hauling cable on the astronaut's shoulders; P_g is the gravity center location of the spacesuit. When the astronaut pitches at α angle, the included angle between the elastic hauling cable and the astronaut is β. At this time, the length l of the elastic hauling cable is

$$l = \sqrt{(L_m \sin \alpha)^2 + (L - L_m \cos \alpha)^2} \tag{43.1}$$

Sin function value of angle β is

$$\sin \beta = \frac{L \sin \alpha}{l} \tag{43.2}$$

Assume that torque is balanced:

$$G \sin \alpha L_g = F \sin \beta L_m \tag{43.3}$$

Substitute formulas (43.1) and (43.2) into formula (43.3):

$$F = \frac{GL_g}{L_m L} l \tag{43.4}$$

Formula 43.4 shows that the tension force amplitude and length provided by the elastic hauling cable is proportional. If G is given, $GL_g/L_m L$ are constant and accords with first-order elastic component characteristics; therefore, to meet the pitch motion demand, the elastic hauling cable for each astronaut is customized at elastic constant $GL_g/L_m L$.

Because of the flexibility of the leg part of the extravehicular spacesuit, when pitching downward, the legs will not be kept consistent with the body movement for reason of gravity action, so it needs to add nonshape transformation hauling cable on the leg's knee joints and ankle points to pull the legs' part of the spacesuit when pitching downward and compensate gravity for the astronaut's legs.

43.5 Conclusions and Prospect

Apply servo control technology to actively track target position change, keep compensation force direction, and offload added mass; adopt high torque motor and low torque motor cooperation method to ensure compensation force amplitude, which is able to conduct gravity compensation in real-time with accuracy. Through the design analysis and model verification, in work space of 12 m × 6 m × 6 m and under constrained motion condition of the maximum speed of 0.1 m/s and the maximum acceleration of 0.1 m/s^2, the accuracy of the single suspended cable tension force can reach 99.5 % and the error of position tracking is less than 10 mm. Compared with the present EVA simulator, the new microgravity environment on-ground is more realistic, which enables the astronaut to move in six degree of freedom and the training experience can be greatly improved. If it is combined with air-bearing platform, the microgravity environment for comprehensive collaborative operation can be built. At that time, the training of material transfer and on-orbital maintenance, repair in space, etc. will be included [5–7].

References

1. Chao J, Chen S (2008) Study on technology of space flight training simulator and Its engineering implementation. Space Med Med Eng 21(3):233–239
2. Huan S, Deng H (2011) Design and implementation of the interface system for a space flight training-simulator based on CAN bus. Space Med Med Eng 24(5):361–365
3. Qi N (2013) Project design on micro-gravity for astronaut training. Harbin Institute of Technology

4. Gao H (2013) Fundamental and foresight technology on gravity compensation. Harbin Institute of Technology
5. Xu W, Liang B, Xu Y (2011) Survey of modeling, planning, and ground verification of space robotic systems. Acta Astronaut 68:1629–1649
6. Moore SK, Gast MA (2010) 21st Century extravehicular activities: synergizing past and present training methods for future spacewalking success. Acta Astronaut 67:739–752
7. Zhu Z, Yuan J (2013) Construction on spacecraft operation micro-gravity environment. China aerospace press

Chapter 44
The Measurement and Analysis of Noise for a Military Vehicle

Yuping Luo, Longtang Xu, Ruiping Niu, Yonggang Sun, Yajuan Bai and Yaofeng He

Abstract Vehicle interior noise is one of the harmful environmental factors that affect the overall efficiency of the vehicle through the physiological and psychological influence on the occupant. In the noise reduction design, noise source identification is the primary work of noise control. In a vehicle standing idle or in low speed, the engine noise is the main noise source of the vehicle. The noise of the engine can come into the passenger compartment through the engine compartment bulkhead. In this paper, interior noise of a military vehicle is tested while the vehicle is standing idle and the spectrum characteristics of the noise are analyzed. The results show that the noise transferred to the passenger compartment is a steady and continuous sound when the engine is idle . The noise pressure level is increased with the increase in engine speed. The noise frequency is mainly concentrated in the low and intermediate frequency band.

Keywords Bound pressure level · A-weighted · Steady noise · 1/3 octave · Background noise · Noise spectrum

44.1 Introduction

The particularity of military vehicle structure and usage environment, the bad interior environmental conditions, and the vehicle interior noise that usually exceeded 90 dB(A) seriously affect the physiological and psychological health of the crew and the overall efficiency of the vehicle. In this paper, we have analyzed the distribution and spectrum characteristic of the vehicle interior noise through concentrating on the vehicle as research object and measuring the vehicle noise started in situ, which has provided a reference for the vehicle interior noise control design.

Y. Luo (✉) · L. Xu · R. Niu · Y. Sun · Y. Bai · Y. He
Beijing Special Vehicle Institute, Beijing 100072, China
e-mail: rx051610@163.com

© Springer-Verlag Berlin Heidelberg 2015

S. Long and B.S. Dhillon (eds.), *Proceedings of the 15th International Conference on Man–Machine–Environment System Engineering*, Lecture Notes in Electrical Engineering 356, DOI 10.1007/978-3-662-48224-7_44

44.2 The Testing Conditions and Methods

On the day of testing, the weather shall be sunny and calm, the temperature shall be 22.5 °C, and the relative humidity in the air shall be 48 %.

The testing site shall be flat and wide, its surface shall be covered with vegetation, and there shall be no big noise source within a circumference of 1 km.

The engine of the tested vehicle shall be located in the right front part of vehicle body.

The measuring point 1 shall be placed in the interior engine compartment bulkhead, where it is near the smoke outlet. The microphone shall be orientated with the smoke outlet.

The measuring point 2 shall be placed in the back part of vehicle body where it is near the right side of the deck.

The measuring point 3 shall be placed in the back part of the vehicle body where it is near the left side of the deck.

The measuring point 4 shall be placed in the middle part of the vehicle body.

Each measuring point shall be about 80 cm away from the face height of the seat. While measuring, all the doors and windows shall be closed.

In the course of testing, the vehicle interior background noise shall conduct a 30 s continuous testing with an average of 42.0 dB(A), in order to make sure the objectivity of datum; the vehicle interior noise, when starting, shall be a minimum of 75 dB(A), and if the background noise exceeds more than 30 dB(A), the measuring results do not need to be revised.

The measuring methods: When the vehicle brakes, in neutral, and the engine starts in a revolution speed of 800 r/min, after running 30 s, the revolution speed increases to 1500 r/min for 50 s through fast stepping on the accelerator, finally loosens the accelerator, and continuously measures the noise at every moment for three times after 30 s running in a revolution speed of 800 r/min.

44.3 The Testing Results and Analysis

The changing curve of each measuring point A-Weighted in the course of testing is shown in Fig. 44.1.

It can be seen from Fig. 44.1, that the vehicle interior noise transmitted from the working engine is a continuously steady noise. When the vehicle starts in situ, the vehicle interior noise will increase with the increase in the engine revolution speed. When the engine revolution speed is 800 r/min, the average noise near the engine compartment bulkhead is 79 dB(A); when the engine revolution speed is 1500 r/min, the average noise is 89 dB(A). The noise near the right side of the deck of the back part of vehicle has increased from 83 to 86 dB(A). The noise near the left side of the deck of the back part of vehicle has increased from 80 to 86 dB(A). The noise of the middle part of vehicle has increased from 75 to 83 dB(A).

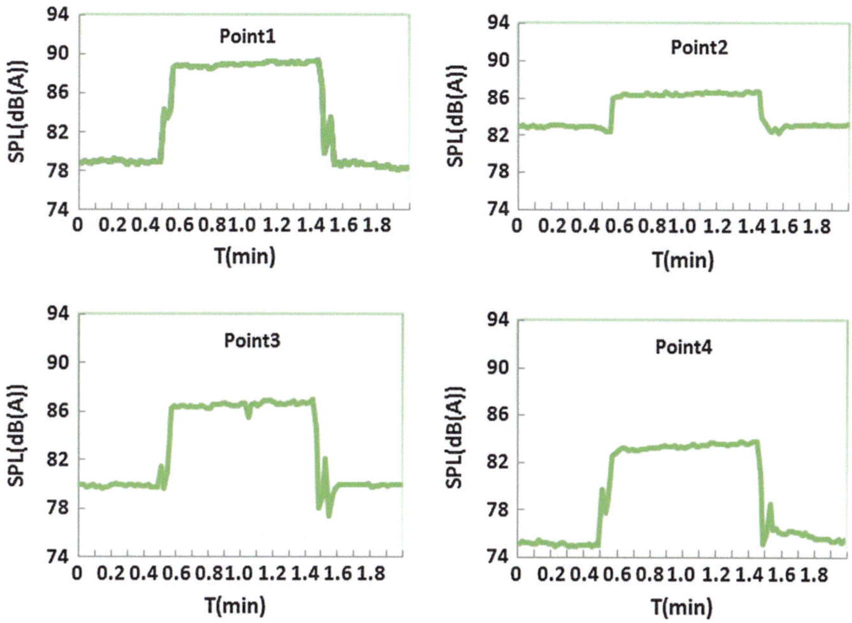

Fig. 44.1 A-weighted SPL

Seeing from the noise value of each measuring point, when the idle speed is 800 r/min, the order of noise from high to low is: "the right side of the deck- the left side of the deck-the engine compartment bulkhead-the middle part"; when the revolution increases to 1500 r/min, the order of noise from high to low is: "the engine compartment bulkhead-the right side of the deck-the left side of the deck-the middle part". The engine compartment bulkhead has a maximum of increasing amplitude which has reached 10 dB(A). This change indicates that the engine compartment bulkhead is the most important part for vehicle interior noise control when using the real vehicle. At the same time, the two sides of the deck have thin thickness, which can radiate larger noise when the engine compartment bulkhead is vibrating, especially when near one side of the deck; when the idle speed is 800 r/min, the noise of one side is obviously stronger than the other side; however, after increasing the rotating speed, the increasing amplitude is in a minimum of 3 dB(A). The two sides' noise level has reached a saturation value. Because the middle part of the vehicle locates in the high position, the noise is relatively lower; but as it is closer to the two sides of the deck than the engine compartment bulkhead, after increasing the rotating speed, its increasing amplitude will be larger.

Each testing stage will undergo a 1/3 octave power spectrum analysis, as shown in Fig. 44.2.

It can be seen from the spectrum curve that sound pressure in frequency band 20–160 Hz is higher (It mentions the center frequency of 1/3 octave, the same as below-mentioned), sound pressure in frequency band 160–800 Hz is lower than the

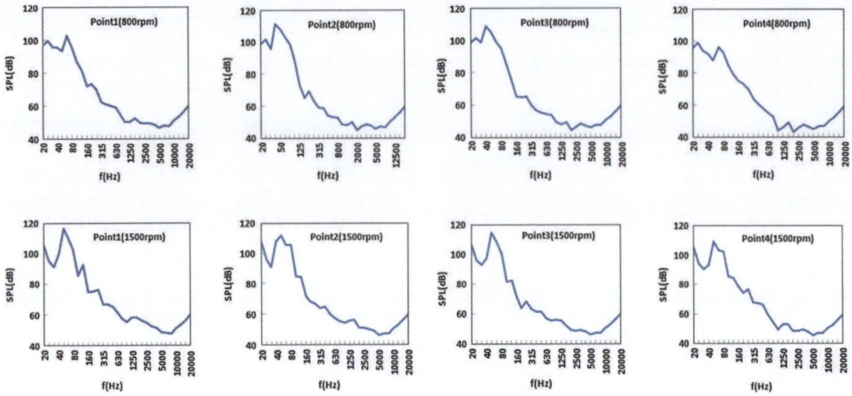

Fig. 44.2 1/3 octave SPL

20–160 Hz, and sound pressure in frequency band 800–8000 Hz is relatively lower. When the revolution speed is low in frequency band 40–63 Hz, sound pressure of the two sides of the deck is obviously higher than the other measuring points'.

Figure 44.3 shows the changing situation of sound pressure level (SPL) in each measuring point when the revolution speed increases from 800 to 1500 r/min.

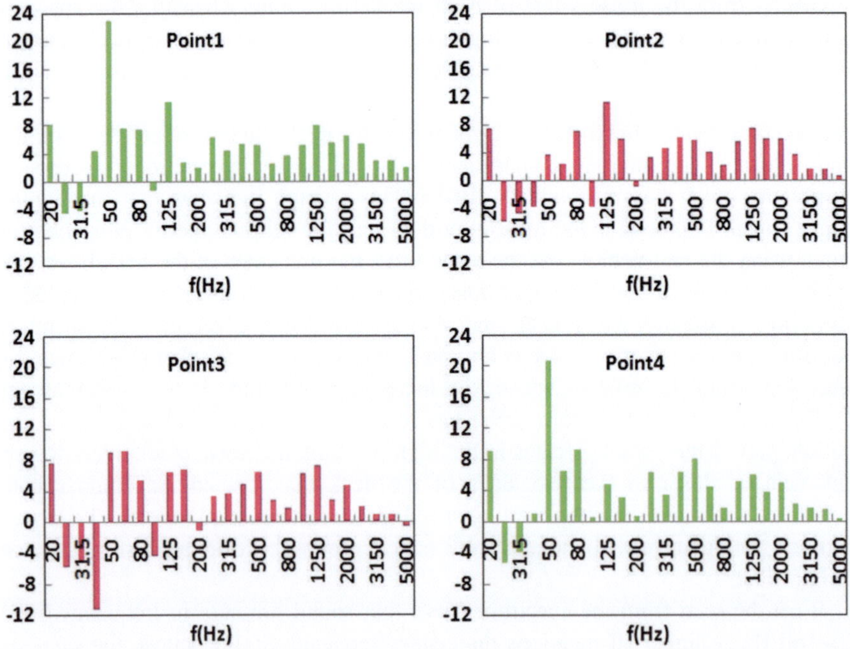

Fig. 44.3 1/3 octave SPL changes (from 800 to 1500 rpm)

As shown in Fig. 44.3, the strengthened frequency band of SPL in each measuring point divides into the low-frequency region (20–200 Hz), the intermediate frequency region (200–800 Hz), and the high frequency region (800–5000 Hz). After increasing the revolution speed, all the measuring points in the intermediate frequency region and the high frequency region have the same change, but have different reactions in each measuring point of the low-frequency region. Meanwhile, in the 11 frequency bands of the low-frequency region, SPL of the frequency band 25 and 31.5 Hz in each measuring point have decreased. Except the above-mentioned two frequency bands, SPL in the frequency band 40 and 100 Hz of the two sides of the deck have decreased. The SPL of the engine compartment bulkhead and the middle part, both in the frequency band 50 Hz, have reached 23 and 21 dB, respectively. The increasing amplification of noise from the right side of the deck is in the frequency band 125 Hz, and has reached 11.3 dB.

Under the action of noise, Signal-to-noise ratio (SNR) of the language and communications system input or output has been reduced, which has affected people's normal working. When the noise level comes close to the conversation sound level, the normal conversation shall be limited and disturbed. When sound pressure level (SPL) of this vehicle noise goes far beyond the level of the normal conversation, people inside the vehicle cannot change the information in language. If people stay in the noise environment more than 90 dB for a long time, it will cause organic disease of the auditory apparatus as well as lead to hearing loss [1].

The noise influences on human body is not only limited to impact on sense of hearing; the physiology and psychology reaction to noise from human body is related to the noise frequency; and the influences of the low-frequency noise on human body is especially significant. As for vehicle interior noise, 20–200 Hz low-frequency noise shall be paid more attention to, this low-frequency noise sounds like a so-called "booming noise", which can make people inside the vehicle feel uncomfortable intensively [2, 3]. As the inherent frequency of the low-frequency noise and the human organ is the same, the low-frequency noise is easily vibrated with the human body simultaneously, such that the inherent frequency of human eyeball is 30–80 Hz, the palm is 50–200 Hz [4], if appears the situation of vibrating simultaneously, it will seriously affect crew's observing, aiming, and operating equipment. According to relevant research, as for the operating personnel of some type of weapon equipment for air-defense, if the noise lowers by 15.5 %, the operating error rate will be reduced to 24 % [5]. The noise spectrum from the working engine mainly concentrates on the low-frequency part, and shall be taken steps to control.

44.4 Conclusion

The control of the engine noise shall be paid more attention in the design stage. When the engine of this vehicle is working, the noise transmitted to the vehicle interior is continuously steady noise. The vehicle interior noise increases with increase in the engine revolution speed. The noise frequency mainly concentrates on the intermediate frequency and the low-frequency bands.

References

1. Dayou M (2002) Noise and vibration control engineering handbook. China Machine Press, Beijing
2. Weiping D (2006) Quantitative relationship between low frequency interior noise and suspension characteristic parameters of a vehicle, Noise and Vibration Control
3. Jian P, Gang Z, Hua H (2006) Noise and vibration of vehicle, Beijing Institute of Technology Press, Beijing
4. Shuang D, Yuanqing Z (2004) Effects of environmental vibration on man. Noise and Vibration Control
5. Hongyan O (2009) Study on affection from environment factors to operating power. In: Proceedings of the 9th conference on man-machine- environment system engineering

Part IV
Research on the Man–Machine Relationship

Chapter 45
Experimental Study of Optimized Interface Displays of Navigation Information System

Chuan Wang, Jianguo Zhang, Hao Yu, Duming Wang, Xiaowei Yang, Qiao Wang and Fei Li

Abstract In the man–machine interface, display information is very important. How to display information to optimize the design is the key that is related to the ability to display information retrieved by the user effectively. Most existing studies carried out in this manner are shown in the model of the role of man−machine interface design assessment; optimization of the design requirements for specific studies is not in-depth. In this study, the design of the existing information system navigation information display location, manner, and color matching are studied experimentally. The results show navigation information system information display message box that is presented at the upper left point of the display mode which is the best choice. In a fixed position in the display, the message box is presented at the top right corner of the optimal display mode. In the case of a black background, the target display information is displayed in white for the best.

Keywords Navigation information system · Display interface · Optimal design

There are many factors that influence the effect of display in the design of display which includes resolution, scanning speed, brightness, contrast, color, etc. But for information display, it has a relation with symbols, words, figures, etc. [1]. There have been a lot of studies about information display in the monitoring homework display interface. Most of them are focused on the flight cabin, cockpit personnel

C. Wang
Naval University of Engineering, Wuhan 430033, China

C. Wang · J. Zhang · H. Yu (✉)
Naval Medical Research Institute, Shanghai 200433, China
e-mail: hg04381@163.com

D. Wang
ZheJiang SCI-TECH University, Hangzhou 310012, China

X. Yang
91656 PLA Troops, Shanghai 200433, China

Q. Wang · F. Li
92730 PLA Troops, Sanya 572000, China

© Springer-Verlag Berlin Heidelberg 2015

S. Long and B.S. Dhillon (eds.), *Proceedings of the 15th International Conference on Man–Machine–Environment System Engineering*, Lecture Notes in Electrical Engineering 356, DOI 10.1007/978-3-662-48224-7_45

work space, and other work space that is relatively small and display information is in a relatively centralized place [2, 3]. But it has no relevant studies and specific conclusions about appropriate matching of position and color of displayed information on the navigation information system display interface. The compulsive problems, such as the appropriate place to display the objective information, the mode of the fixed headed information and the variable parameter information, would appear in the interface of navigation information system.

By optimizing the display interface of navigation information system, it not only can make the information display on the navigation information system interface that is clearly communicated to the users, but also can reduce the work load and the probability of operation error for operation staff and enhance the operation quality of working staff. It plays an important significance to improve the information quality and guarantee the navigation information system in C4I system for its generality of important equipment.

45.1 Subjects and Methods

45.1.1 The Tested Subjects

The experienced navigation information system users are interfered by their practical experience in evaluation on a certain interface. In this experiment, it selects 30 university students who have no relevant experience in it. The male and female proportion is half-to-half, and average age is 22.1 ± 1.8.

45.1.2 Experimental Apparatus and Equipment

It is to prepare the simulated navigation information system display interface program, that uses DELL19 inch LCD monitor display, and use the similar Jupiter trackball of navigation information system controller as input control equipment. In the different points showing position experiment, it uses ordinary mouse as a control device.

45.1.3 Experimental Design

With reference to the existing navigation information system display mode and color, it is applied with double-factors repeated measurement experimental design within test, independent variables for the target information box displaying positions (four kinds: upper left, upper right, right middle, and that point to follow), and parameter information color (two kinds: green and white), when variable is the rate of reaction and correctness.

45.2 Results and Analysis

45.2.1 The Experiment of Information Displaying Position

According to the experimental target, it is to set experiment task; on the display interface that presents five targets, which requires the participants to check how many conditions these comply with for these five targets, and react by pressing the corresponding number keys.

The participant is first familiar to trackball operation for 5 min, after entering the practice stage and familiar to experiment process, it would stage into the formal experiment process. The first is to display the question conditions, such as "Please judge how many following objectives that speed is >50, and press the corresponding number key response". After pressing the corresponding number key, it would display the five objective points randomly distributed, and then it would perform the pressing reaction after participants check the objective information. After the completion of experiment, the participant performs the color preference selection between fourteen kinds of displayed location and parameters information.

45.2.1.1 The Reaction Time

The length of the reaction time is an important index to evaluate the different locations, and it is a two-factor analysis of variance of the repeated measurement on the data after removal of extreme value except three standard deviations, and the reaction time only counts the tested right reaction time length.

Figure 45.1 shows the different reaction time length for different displaying locations and different parameters colors. It can be seen from the figure, that the information windows are located at the upper right and right middle for the shortest of overall reaction time length, and the second is the points to follow; the longest is at the upper left. The two-factors repeated measurement deviation analysis shows that there is no obvious deviation on the different displayed locations ($F(3,87) = 0.53$, $P > 0.05$), different parameter information colors exist with different obvious deviation ($F(3,29) = 30.1$, $P < 0.01$), and the display place have no obvious interaction with parameters colors ($F(3,87) = 1.65$, $P > 0.05$).

45.2.1.2 Accuracy

Accuracy is an important index to display excellent modes or not.

Figure 45.2 shows the highest of overall accuracy in the mode of point to follow (95), the second is the information windows at the upper right (94 %), and the lowest is at the upper left (90 %). However, the two-factors repeated measurement deviation analysis shows different display locations and different parameters information colors have no obvious deviation ($F(3,93) = 2.1$, $P > 0.05$; $F(1, 31) = 0.13$, $P > 0.05$), and no obvious interaction ($F(3,93) = 0.64$, $P > 0.05$).

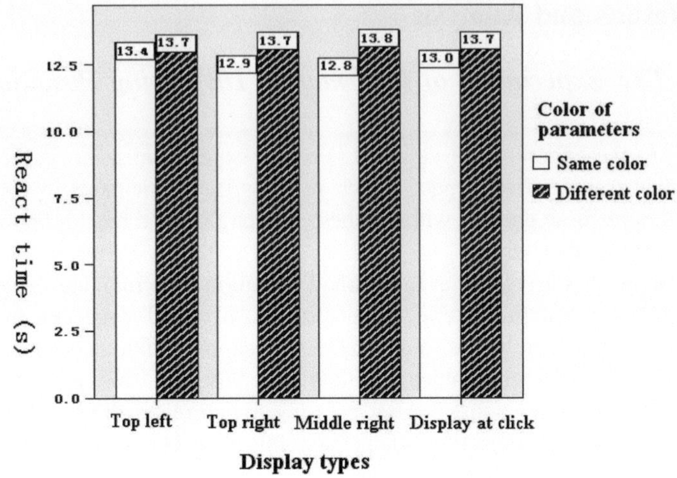

Fig. 45.1 Reaction time under different display positions and parameters of different colors conditions

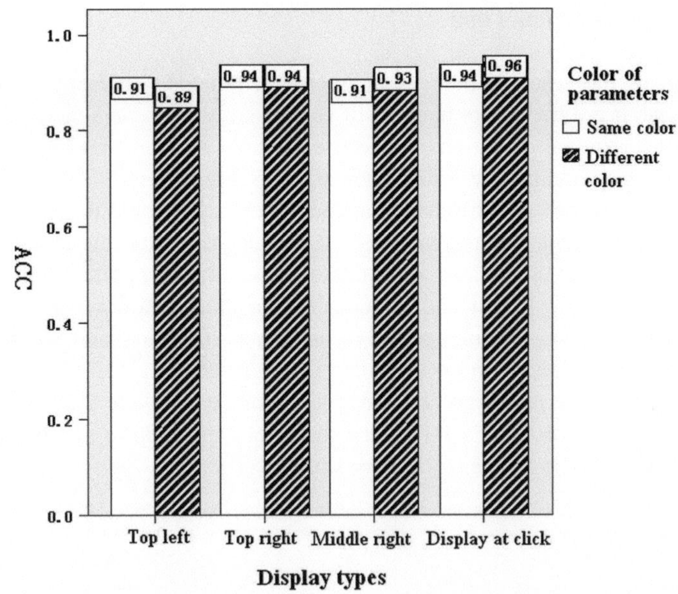

Fig. 45.2 Correct rate at different locations and different parameters of the display colors conditions

The experiment results after tidy reaction time length and accuracy in the tabular form can be clearly seen with deviation of interaction, according to the different modes, it also have participants to perform the subjective preference selection.

Table 45.1 The results showed that the reaction of different parameters and subjective preferences

Display mode	Choose	Reaction time (s)	Correct rate (%)	The proportion of subjective preferences (%)
Target position information presentation	The upper left	13.5	90	20
	Top right	13.3	94	6.7
	Right middle	13.3	92	33.3
	That point display	13.4	95	40
The color of the header information and parameter information	Same	13.0	92	20
	Different	13.7	93	80

As can be seen in Table 45.1, considering the overall performance, balancing reaction time, and result of accuracy, it is suggested that objective information display location are better to displace at the random-point-show and upper right, the title information and parameters information are displayed in different colors, namely, the parameter information is better to apply with color.

45.2.2 The Experiment of Different Random Points Display Position

Because the previous experiment shows that the random point effect is the best, and then the location of that point shows whether there are deviations, this experiment would be continued on the basis of previous experiment. In order to more clearly display, the analog display interface program is displayed on the 32 inch display. In order to allow the users to make quicker selection reactions, the ordinary optimal mouse is used as input control device.

The display interface presents five random objective points, which requests the participants to check how many points for these five objective points can meet the limited conditions, and press the corresponding number key for reaction.

In this study, it is designed by single factor repeated measure. Each participant completes seven kinds of information display positions for 84 mission's judgments, in order to increase the effects of order, seven kinds of display positions are displayed with order of random, recording subjects' reaction time, and accuracy.

Under conditions of seven kinds of position in use of ordinary mouse, the completion time of mission and error rate is as shown in Fig. 45.3:

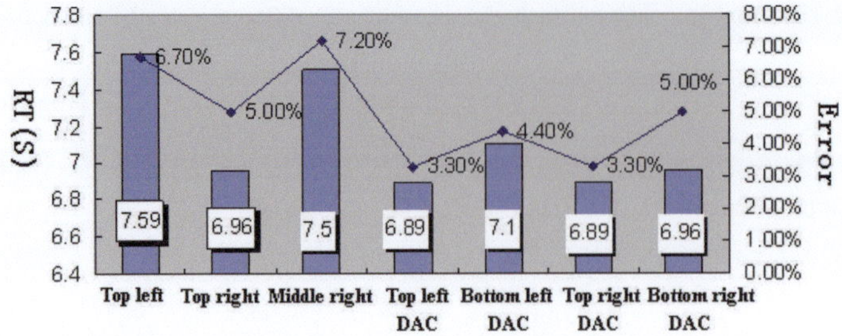

Fig. 45.3 Operating performance objectively different display modes

Figure 45.3 shows that, the fourth position following the upper left and the sixth position following the upper right reactions are the shortest, and error rate is the lowest. The reaction time from low to high by turn is 4, 6, 2, 7, 5, 3, and 1. The error rate from low to high is 4, 6, 5, 2, 7, 1, and 3. The result of repeated measurement deviation analysis to the experiment data shows the objective indicator of reaction time, seven kinds of location deviation reach obvious standards ($P = 0.001 < 0.05$).

Therefore, the experimental results show that: when using an ordinary mouse, there are deviations between the seven kinds of positions, and they reach a significant level, objective data show that the performance is the best for information window shown at the upper left and upper right in form of random point displaying.

In addition, the personnel subjective feeling to different displayed position is also a factor to influence the objective information display effect. Combined with experiment, it is to conduct the object evaluation at different positions, and results are shown in Fig. 45.4.

In the objective evaluation, it is most welcomed for the position of following the upper left.

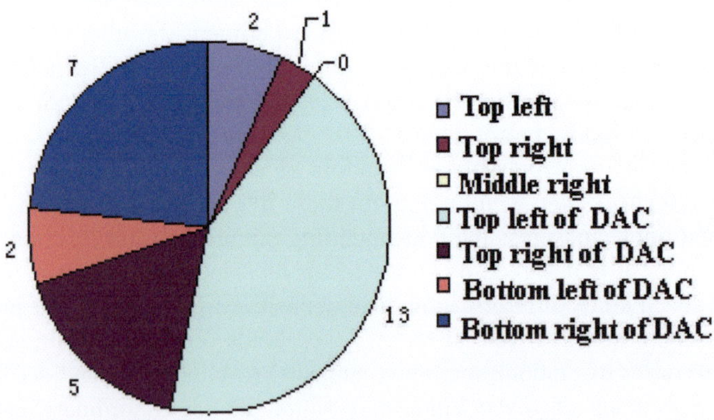

Fig. 45.4 Presents different location is selected as the optimal frequency

45.3 Conclusions

In the display interface, the interface design consistency and good information feedback is an important factor for good interface display [4, 5]. How to determine the exact requirements of these factors is the critical factors that influences whether the study result can be well applied accordingly.

Being integrated with both subjective and objective data, it is believed that the random point display mode of information windows at the upper left is the best in the information display of navigation information system. In the fixed position of display, it is the best to the display mode of information window displaying at the upper right corner. In the case of black background, the objective display information is the best to display the white color.

This work is supported by the opening foundation of the Science and Technology on Human Factors Engineering Laboratory, Chinese Astronaut Research and Training Center (grant no. SYFD140051810K).

45.4 Compliance with Ethical Standards

The study had obtained approval from Navy Logistics Department for Civilian Ethics Committee.

All subjects participated in the experiment had signed the informed consent.

All relevant ethical safeguards had been met in relation to subject protection.

References

1. Zhu Z (2003) Engineering psychology course, 1st edn. People's Education Press, pp 130–171
2. Kang W, Yuan X, Liu Z (2008) Mental workload optimization based on a visual display interface cockpit. Beijing Univ Aeronaut J 7(34):782–785
3. Wei K, Huo L (2003) Engineering psychology and human operations, 1st edn. East China Normal University Press, pp 82–144
4. Li L (2004) Color ergonomics and human-computer interface design color. Ergonomics 9 (10):54–57
5. Yu Z, Ding F (2004) The basic principles of human-computer interface design information systems. Netw IT 23(3):44–45

Chapter 46
Research on the Integrated Assessment Method for HCI Design of Digitized OM System in NPPs

Kun Yu, Rui Ma and Ying Zhang

Abstract The human−computer interface (HCI) of digitized operation and monitoring (OM) systems has been the most important interactive interfaces between operators and nuclear power plants (NNPs). The integrated assessment methods were researched for HCI design of the digitized OM systems. The assessment indexes structure of HCI was constructed based on the NNPs standard. A fuzzy theory-based multi-experts subjective assessment model was established to reduce the influences of ambiguity and uncertainty of the subjective assessment comments on the final results. Eye movement analysis method was researched and applied to HCI design during operation task. It was expected to optimize HCI design to reduce operation burden of operators and to improve the operation safety of NPPs.

Keywords HCI · Assessment · Set-valued statistics · Eye movements · Npps

46.1 Introduction

In traditional NNPs, the recorders, instruments, and switches were widely used in the OM system. For operators gaining more information about NNPs, normally a large variety of instruments and switches were arranged on control panels. With electricity demands growing rapidly, more advanced NNPs have been established in China. And the digitized OM systems were used widely in main control rooms of these advanced NPPs. The digitized OM system was the main control center of advanced NPPs, and video display terminals and large-scale screens have been widely used in the digitized OM system [1] as shown in Fig. 46.1. The computer-based HCI have become the main information interaction medium between operators and NPPs, and it has the following features: (1) more and more

K. Yu (✉) · R. Ma · Y. Zhang
China Ship Development and Design Center, Wuhan 430064, China
e-mail: kunyu2013@163.com

© Springer-Verlag Berlin Heidelberg 2015

S. Long and B.S. Dhillon (eds.), *Proceedings of the 15th International Conference on Man–Machine–Environment System Engineering*, Lecture Notes in Electrical Engineering 356, DOI 10.1007/978-3-662-48224-7_46

information needs to be display in HCI; (2) a wider display area for information; (3) more focused information display [2].

After the Three Mile Island accident and Chernobyl disaster, the International Electrotechnical Commission summarized that most of the NNPs accidents were related to human errors [3]. In fact, the HCI design of digitized OM systems has a direct impact on the operational efficiency and power plants operation safety. For new NPPs, the numbers display pages of HCI is more, typically in hundreds and thousands. To perform their functions and tasks, operators must access information in these display pages. Obviously, the appropriate HCI design can support information searching, state understanding, manual operation, and decision-making and can indirectly reduce the burden and mental stress of operators. The integrated assessment method for HCI design of digitized OM systems was researched. The subjective assessment method and objective eye movement data were used. The aim was to assess the human factor design quality of HCI design and optimize the HCI design.

46.2 The Assessment Indexes System

Digital HCI in advanced NPPs commonly use the mimic displays(e.g., process piping) to descript the whole or part of plant, and use the overview map, trends map, bar charts, and function charts to describe the NNPs system flow equipment status, operation parameters and gauge, and so on. The display formats contain continuous text displays, mimics, tables and lists, graphs, and so on, as shown in Fig. 46.1.

46.2.1 The Structure of Assessment Indexes System

The factors of the HCI design include: (1) the display pages design: display elements and forms, screen layout, mimic displays design, and critical parameters and variables; (2) user-interface interaction design: menu navigation, information input and feedback, and managing displays; (3) operation and monitoring-related system properties: system performance, user assistance, and error detection and correction. The basic principle of analytical hierarchy process was used to simulate the

Fig. 46.1 Case of the traditional OM system and the advanced digital OM system of AP1000

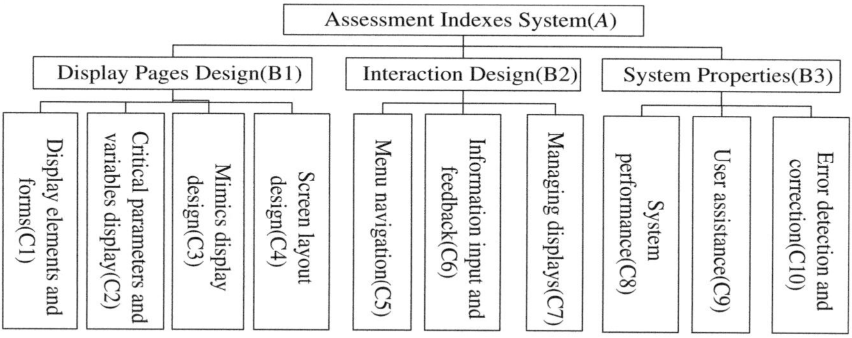

Fig. 46.2 The assessment indexes structure of HCI for digitized OM systems in NNPs

decomposition, judgement, and comprehensive process of human decision-making during the construction of assessment system (Fig. 46.2). The basic assessment indexes of HCI were extracted by considering the HCI design review guidelines (NUREG-0700(Revision2)) and relevant China ergonomics standards. The detailed basic assessment indexes of HCI are shown in Table 46.1.

46.2.2 The Weights of Assessment Indexes

The important degree of each basic assessment index for the upper items is different, and also for the final result. So, weights of basic indexes system are different for each level of assessment indexes. Analytical hierarchy process method is a commonly used subjective weighting method. It was used to fix the weight of each basic assessment index in assessment indexes system in this research. The steps for calculating the relative weights are as follows [4]:

(1) product of indexes in each sequence of pair-wise comparing judgement matrix can be calculated by (46.1).

$$M_i^k = \prod_{j=1}^{n} c_{ij}^k,\ i = 1, 2, \ldots, n \tag{46.1}$$

(2) the nth root of M_i and vectors normalization can be calculated by (46.2).

$$\bar{M}_i^k = \sqrt[n]{M_i^k},\ W_i^k = \bar{M}_i^k \Big/ \sum_{i=1}^{n} \bar{M}_i^k \tag{46.2}$$

$W^k = \left[W_1^k, W_2^k, \ldots, W_n^k\right]^T$ is eigenvector. The weights of other assessment indexes can also be determined by the same way.

Table 46.1 The detailed assessment indexes of HCI of advanced digital OM system in NNPs

Items	Basic Assessment Indexes	Items	Basic Assessment Indexes
Display elements and forms($C1$)	1. Display format consistent with user tasks (c_{11}) 2. Consistent design conventions for all display features (c_{12}) 3. Display information consistent with user conventions (c_{13}) 4. Display information consistent with control requirements (c_{14}) 5. Display elements and formats are clear and simple (c_{15}) 6. Display elements and formats are visually distinctive (c_{16}) 7. Colors is based on user conventions (c_{17}) 8. Color coding is conservative and consistent (c_{18}) 9. Display flexibility (c_{19})	Information input and feedback ($C6$)	1. User input actions are simple (c_{61}) 2. Entry procedures match to user skill (c_{62}) 3. Provide flexible means of entering information (c_{63}) 4. Data entry field are visually distinctive (c_{64}) 5. Form compatible for data entry and display (c_{65}) 6. Logical organization of data entry forms (c_{66}) 7. Feedback for any user entry (c_{67}) 8. Entry results is compatible with user expectations (c_{68}) 9. Interrupts and aborts can't modify or remove stored or entered data (c_{69})
Critical parameters and variables display($C2$)	1. Display in directly usable form (c_{21}) 2. Convenient and ready access to parameters and variables (c_{22}) 3. The representative meaning is clear (c_{23}) 4. Length of information is reasonable (c_{24}) 5. Display accuracy is sufficient for tasks (c_{25}) 6. Sensitivity of information display (c_{26}) 7. Group information in a display (c_{27}) 8. Rapid recognition of safety status change (c_{28}) 9. Freeze rapidly changing information (c_{29})	Managing display($C7$)	1. Overview of display network (c_{71}) 2. Cues to display network structure (c_{72}) 3. Location cues (c_{73}) 4. Display page titles and identification codes (c_{74}) 5. orientation coding (c_{75}) 6. flexibility in display system interaction (c_{76}) 7. evident direction of paging (c_{77}) 8. Anticipation of automated interface management actions (c_{78})

(continued)

Table 46.1 (continued)

Items	Basic Assessment Indexes	Items	Basic Assessment Indexes
Mimics display design($C3$)	1. Contain the minimum amount to yield a meaningful pictorial representation (c_{31}) 2. Graphic display enhancement with text (c_{32}) 3. Device icons and symbols are clear and simple (c_{33}) 4. Closed figures when possible (c_{34}) 5. Icons and symbols are easily discriminable (c_{35}) 6. Components represented on mimic lines are identified (c_{36}) 7. Flow paths are coded to indicate important information (c_{37}) 8. Text is understandable (c_{38})	System performance ($C8$)	1. Information update rate match to user ability (c_{81}) 2. Display time delay consistent with task requirements (c_{82}) 3. Reasonable response time to operation (c_{83}) 4. Flexibility of system control (c_{84}) 5. Display reliability (c_{85}) 6. Protection of destructive action (c_{86}) 7. System function meet user needs (c_{87}) 8. System stability (c_{88})
Screen layout design($C4$)	1. Display screen partitioning for functions (c_{41}) 2. Functional zones are visually distinctive (c_{42}) 3. Spatial proximity for related information (c_{43}) 4. Screen layout is consistent across and within displays (c_{44}) 5. Highlight the important information (c_{45}) 6. Different types of information are visually distinctive (c_{46}) 7. Colors coding is based on user conventions (c_{47}) 8. Color coding is conservative and consistent (c_{48}) 9. Use a uniform nondistracting background color (c_{49})	User assistance ($C9$)	1. Normal value reference index (c_{91}) 2. Guidance/help information is clear and simple (c_{92}) 3. On-line dictionary of display element definitions (c_{93}) 4. User control of automated guidance/help (c_{94}) 5. Explicit actions are required to access guidance/help (c_{95}) 6. Timeliness of guidance/help to the current operation (c_{96}) 7. Ease of access to guidance/help information (c_{97}) 8. The clarity of guidance/help information (c_{98}) 9. Guidance/help information is understandable (c_{99})

(continued)

Table 46.1 (continued)

Items	Basic Assessment Indexes	Items	Basic Assessment Indexes
Menu navigation ($C5$)	1. Function of the menu is evident to user (c_{51}) 2. Consistent display of menu options in different displays (c_{52}) 3. Screen locations of menus are consistent for all mode (c_{53}) 4. Visual grouping of menu options (c_{54}) 5. Breadth and depth of menus match to user (c_{55}) 6. Selection logic of hierarchic menus is consistent (c_{56}) 7. Menu structure match to mission (c_{57}) 8. Menu icons are understandable (c_{58}) 9. Menu icons are consistency (c_{59})	Error detection & correction ($C10$)	1. Indication of display failure (c_{101}) 2. Warning messages draw users' attention (c_{102}) 3. Users aids to error detection (c_{103}) 4. The difficulty to correct input errors personnel (c_{104}) 5. The ability to avoid misuse (c_{105}) 6. Ease of recovery the misuse (c_{106})

46.3 Set-Valued Statistics Based Integrated Assessment Model

The fuzzy set-valued statistics method [5, 6] was constructed to integrate assessment results of multi-specialists. The assessment result of an assessment index by a specialist is represented as: $\left[u_1^{(k)}, u_2^{(k)}\right]$. In which, k represents the kth specialist.

The assessment results of n specialists form a random interval set U: $\left[u_1^{(1)}, u_2^{(1)}\right], \ldots, \left[u_1^{(n)}, u_2^{(n)}\right]$. In set-valued statistics, the projection represents the statistical frequency that a random interval covering a fixed point. $\overline{X}(U)$ called as the sample projection function, representing the probability of all assessment result ranges containing a numerical value u. As shown in (46.3).

$$\overline{X}(U) = \frac{1}{n}\sum_{k=1}^{n} \chi\left[u_1^{(k)}, u_2^{(k)}\right](u), \quad \chi\left[u_1^{(k)}, u_2^{(k)}\right](u) = \begin{cases} 1, & \left[u_1^{(k)}, u_2^{(k)}\right] \\ 0, & \text{else} \end{cases} \tag{46.3}$$

According to the distribution of random interval set, the gravity center can be represented as shown in (46.4).

$$G(U) = \int_{u_{\min}}^{u_{\max}} \overline{X}(U)\text{gudu} \bigg/ \int_{u_{\min}}^{u_{\max}} \overline{X}(U)\text{du} \quad (46.4)$$

In which, $u_{\max}$ represents the max value of the random interval set; $u_{\min}$ represents the min value of the random interval set. Considering the differences of the n assessment specialists, the integrated weight of each specialist is integrated into the sample projection function, and $G(U)$ can be represented as shown in (46.5).

$$G(U) = 0.5\sum_{k=1}^{n} \omega_k \left[\left(u_2^k\right)^2 - \left(u_1^k\right)^2 \right] \bigg/ \sum_{k=1}^{n} \omega_k \left[\left(u_2^k\right) - \left(u_1^k\right) \right] \quad (46.5)$$

46.4 Integrated Assessment Case

46.4.1 Multi-experts Based Subjective Assessment

In this research, eight assessment specialists were selected. Age was between 30–55 years. As differences in the influence power, knowledge, understanding, and preference of assessment specialists, the assessment result of each assessment specialist have different weight and may not be fully reliable. The importance and reliability of each assessment specialist were considered for the integrated weights. The important weight set of eight specialists was determined by the analytical hierarchy process method, as shown below: $\lambda_k = \{0.120, 0.164, 0.111, 0.100, 0.062, 0.143, 0.214, 0.086\}$. The reliability coefficient: $w_k = \{0.920, 0.930, 0.900, 0.930, 0.950, 0.900, 0.900, 0.900\}$, and the integrated weights were calculated as: $\omega_k = \{0.1209, 0.1670, 0.1094, 0.1018, 0.0645, 0.1409, 0.2109, 0.0847\}$.

The assessment interval set of eight assessment specialists for the nine basic assessment indexes of "Display elements and forms($C1$)" were shown in Table 46.2. Using the set-valued statistics-based integrated assessment model, the

Table 46.2 The assessment interval set of eight assessment specialists for "Display elements and forms($C1$)"

	c_{11}	c_{12}	c_{13}	c_{14}	c_{15}	c_{16}	c_{17}	c_{18}	c_{19}
Specialist 1	[75,80]	[70,75]	[60,65]	[76,82]	[70,74]	[60,65]	[50,55]	[80,84]	[80,85]
Specialist 2	[80,83]	[66,70]	[66,70]	[76,80]	[70,74]	[58,60]	[54,60]	[79,83]	[78,84]
Specialist 3	[75,79]	[71,75]	[62,65]	[82,85]	[65,72]	[61,65]	[52,55]	[80,85]	[82,84]
Specialist 4	[78,80]	[68,70]	[61,65]	[80,85]	[68,72]	[55,63]	[51,55]	[80,83]	[79,85]
Specialist 5	[74,78]	[65,68]	[55,62]	[78,83]	[70,75]	[61,68]	[48,52]	[82,86]	[80,85]
Specialist 6	[75,80]	[70,70]	[64,70]	[85,85]	[70,75]	[60,60]	[50,55]	[75,80]	[85,90]
Specialist 7	[77,80]	[70,75]	[65,65]	[80,88]	[65,75]	[55,63]	[55,55]	[81,83]	[86,88]
Specialist 8	[75,78]	[72,76]	[60,65]	[78,82]	[68,74]	[50,55]	[50,55]	[80,84]	[80,84]

Table 46.3 The integrated assessment results of "Display elements and forms($C1$)"

	c_{11}	c_{12}	c_{13}	c_{14}	c_{15}	c_{16}	c_{17}	c_{18}	c_{19}
Results of basic assessment indexes	78.1	71.3	64.2	81.6	70.7	59.8	53.7	81.1	83.3
Weights	0.295	0.115	0.086	0.082	0.088	0.111	0.084	0.075	0.064
Integrated result	72.2								

assessment result of "Display elements and forms($C1$)" were shown in Table 46.2. The weights of $C1$ were $W_{C1} = [0.295, 0.115, 0.086, 0.082, 0.088, 0.111, 0.084, 0.075, 0.064]$. The final integrated assessment result of eight assessment specialists of $C1$ was shown in Table 46.3.

46.4.2 *Objective Assessment HCI Based Eye Movements Data*

Eye movement analysis method was to assess and validate the HCI using the task completion time, gaze path, gaze points data for the saccades, and smooth pursuit movement of eyes [7]. The procedure of the objective assessment HCI based eye movement data: for specific HCI, select the typical operator tasks; record the original eye movement data when experimenters are performing the typical operator tasks; statistics the total time, operation errors, saccade distance, average saccade speed indexes form the original eye movement data; analysis defects of HCI design based on the statistical data. Normally, the longer the total time the typical operator spend for tasks, it is described that the information searching and extracting of the HCI was more difficult. The saccade distance means the fixation moving distance of experimenters on the HCI during the typical operator tasks. The longer the saccade distance, it is described that the fixation track is not smooth and it is confusing. Average saccade speed means the average saccade distance of all the fixations on the HCI during the typical operator tasks. The longer the average saccade distance, it is described that the fixations and fixation track on the HCI during the typical operator tasks is less. Figure 46.3 shows the eye movement analysis case for HCI of circulating water system in NPPs for monitoring tasks. 20 specialists, who were nuclear power related graduate students, were selected as eye movement analysis data samples in the case. And final eight specialists' valid data were selected for HCI analysis. Table 46.4 shows the statistical eye movement data for HCI of circulating water system.

According to the analysis of the statistical eye movement data, the reason of higher cognitive load during the task process was confirmed. For example, during the task process, the average error for each experimenter was 1.25; the saccade distance was 77.5 pels; and the average saccade speed task was 260.3 °/s. It proves

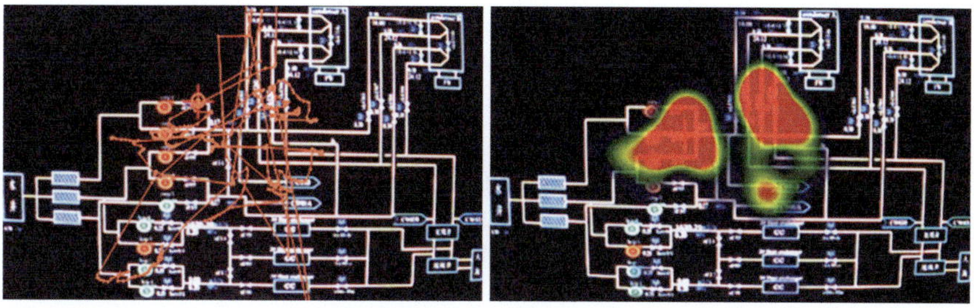

Fig. 46.3 Scanned chart and hotspot chart case of HCI of circulating water system in NPPs

Table 46.4 The statistical eye movements data for HCI of circulating water system

Specialists	1	2	3	4	5	6	7	8	Mean
total time (s)	222	300	286	278	269	284	216	281	265.4
operation errors	0	1	3	0	0	1	3	2	1.25
saccade distance (pels)	82	73	64	137	81	66	61	63	77.5
average saccade speed (°/s)	235	306	169	351	378	307	172	185	260.3

that the target information visual search target is not smooth, and there are many meaningless fixation points on the HCI of circulating water system.

46.5 Conclusions

The computer-based HCI design of digital OM system can obviously affect operator's efficiency and the safety operation of modern NNPs. The HCI design integrated assessment method was studied in this paper, the objective and subjective data were used in this integrated assessment. The main factors of interface ergonomics design were studied. According to analysis results, the multi-level structure assessment indexes system of HCI was constructed. Considering the related ergonomics design standards of HCI, the basis assessment indexes were extracted from the related standards. And the weight of each assessment indexes was determined. A fuzzy theory-based multi-experts subjective assessment model was established based on the set-valued statistics, to reduce the influence of ambiguity and uncertainty of the subjective assessment. The HCI assessment using eye movement analysis was researched. The integrated assessment method combines the objective and subjective method; it can reflect synthetically the HCI design level. Through applying this method in HCI design, the design defects of HCI can be discovered in time. It was expected to optimize HCI design to reduce operation burden of operators and to improve the operation safety of NNPs.

References

1. Zheng M, Lin Z (2002) The development of NPP advanced main control room and self-design research objectives. Chin J Nucl Sci Eng 20(4):297–303
2. Shinichiro H, Yujiro S (1999) Designing methods of human interface for supervisory control systems. Control Eng Pract 7:1413–1419
3. Yang M (2002) Development of the nuclear power human factors engineering field. Eng Sci 4 (8):13–19
4. Benítez J, Delgado-Galván X, Gutiérrez JA, Izquierdo J (2011) Balancing consistency and expert judgment in AHP. Math Comput Model 54:1785–1790
5. Tang H, Zhang J (2007) Study on fuzzy AHP group decision-making method based on set-valued statistics. In: Fourth international conference on fuzzy systems and knowledge discovery, vol 3. Haikou, China, pp 689–693
6. Foata D, Han GN (2009) New permutation coding and equidistribution of set-valued statistics. Theor Comput Sci 410(38–40):3743–3750
7. Jowers I, Prats M, McKay A, Garner Steve (2013) Evaluating an eye tracking interface for a two-dimensional sketch editor. Comput Aided Des 45(5):923–936

Chapter 47
Research on Subjective Evaluation Method in Human–Machine Interface Based on Grey Theory

Chuan Wang and Jianguo Zhang

Abstract Subjective evaluation is an important basis for judging and improving the usability of human–machine interface. Subjective evaluation theories and methods require further exploration. A subjective evaluation method based on grey system theory is put forward, based on human thinking having "grey characteristic". By establishing describable indicator information, the unknown or uncertain influence factors of human–machine interface become concrete and describable, making the comprehensive subjective evaluation questions solvable.

Keywords Grey theory · Human–machine interface · Subjective evaluation method

Evaluation of human–machine interface includes objective evaluation and subjective evaluation. Objective evaluation refers to the degree of difficulty for people to "see" or "touch" the human–machine interface display and control device. The evaluation divisions are based on the national standard about the human eye's visual field [1], palpable domain to the hands and feet [2] and other human body function scales; the subjective evaluation refers to easy comprehension of human–machine interface, easy operation, good appearance and other subjective feelings. The subjective evaluation is mainly to regard the objective feeling of the evaluation staff on evaluation basis. The objective evaluation and subjective evaluation are important bases to measure the human–machine interface usability [3], and both are indispensable [4]. Otherwise, it easily leads to misunderstanding, error in operation and induction of accidents.

Because the subjective feelings of a human has relationship with the ability to identify, process to recognize and read, comfort and system function and also has relationship with the evaluation of people's knowledge, experience, preference and other many known, unknown or uncertain knowledge factors it makes the

C. Wang (✉)
Naval University of Engineering, Wuhan 430033, China
e-mail: hg04381@163.com

C. Wang · J. Zhang
Naval Medical Research Institute, Shanghai 200433, China

© Springer-Verlag Berlin Heidelberg 2015

S. Long and B.S. Dhillon (eds.), *Proceedings of the 15th International Conference on Man–Machine–Environment System Engineering*, Lecture Notes in Electrical Engineering 356, DOI 10.1007/978-3-662-48224-7_47

subjective evaluation of human−machine interface not been satisfactorily resolved due to its complexity, and is so far still in exploration [4]. Therefore, it is of practical significance to seek a scientific and reasonable human−machine interface subjective evaluation theory and method.

Grey system theory [5] provides an effective way to solve the unknown or uncertain information system—information incomplete system evaluation. Grey theory about "the human mind has grey features" is the basic idea of the theory applied to human−machine interface based on subjective evaluation. Based on the pointer instrument taken as an example, this paper, on the basis of analyzing the pointer instrument display design factors, is applied with grey clustering analysis method [6] to perform a subjective evaluation on the design quality of pointer instrument.

47.1 The Grey System Theory Evaluation Method

Grey system is a system where part of the information is known and the other part is unknown; namely, it is an incomplete information system. The so-called "incomplete information", generally refers to: (1) system factors that are not entirely clear; (2) the factor relationship is not entirely clear; (3) the system structure is not completely clear; (4) system function principle is not fully understood.

Grey system theory views "Grey" as absolute and "white" as relative (grey nature immortal). Human mind has the character of grey, with periodic. In the premise of incomplete and uncertain information, grey system answer is not unique.

Grey theory does not consider the internal factors leading to the change in index, but only focuses on the change in subjective evaluation index. The unknown and uncertain affecting factors are changed into the exact, described representation information—evaluation index, in order to resolve the complicate problems.

47.2 Grey Clustering Analyses and Evaluation Model

Grey clustering is a method that is based on the number of grey whitening weight function to cluster some observation index or observation objects to several definable categories.

Assuming n clustering objects, m clustering index and s different grey classes; according to the $i(i = 1, 2, \ldots n)$ objects about $j(j = 1, 2, \ldots m)$ index samples values $x_{ij}(i = 1, 2, \ldots n; j = 1, 2, \ldots m)$, put the i object classified into the $k(k \in \{1, 2\ldots, s\})$ grey classes, called as grey clustering.

The n objects about index j values are classified correspondingly into S grey class, called j index subclasses. J index K subclass whitening weight function are recorded as $f_j^k(\tau)$.

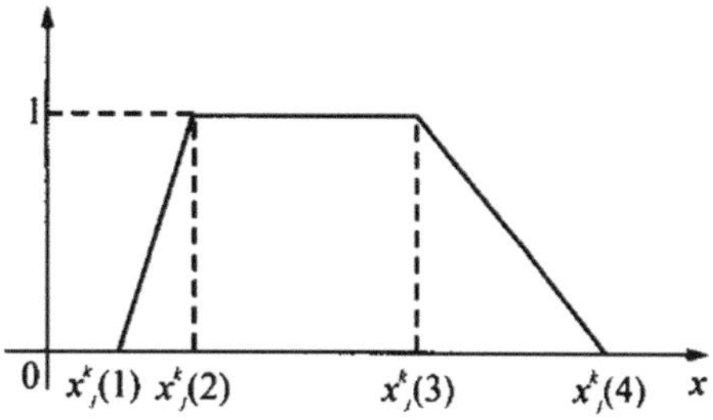

Fig. 47.1 Typical whitenization weight function

Assuming j index k subclass of whitening weight function (recorded as $f_j^k(\tau)$) as shown in Fig. 47.1 of the typical whitening weight function, and called as $x_j^k(1)$, $x_j^k(2)$, $x_j^k(3)$, $x_j^k(4)$ is $f_j^k(\tau)$ turning point. The typical whitening weight function is recorded as

$$f_j^k\left[x_j^k(1), x_j^k(2), x_j^k(3), x_j^k(4)\right] \tag{47.1}$$

As shown in Fig. 47.1 the typical whitening weight function are

$$f_j^k(x) = \begin{cases} 0, & x \notin \left[x_j^k(1), x_j^k(4)\right] \\ \frac{x - x_j^k(1)}{x_j^k(2) - x_j^k(1)}, & x \in \left[x_j^k(1), x_j^k(2)\right] \\ 1, & x \in \left[x_j^k(2), x_j^k(3)\right] \\ \frac{x_j^k(4) - x}{x_j^k(4) - x_j^k(3)}, & x \in \left[x_j^k(3), x_j^k(4)\right] \end{cases} \tag{47.2}$$

If the whitening weight function $f_j^k(\tau)$ of the first and the second turning points $x_j^k(1)$, $x_j^k(2)$ and then call $f_j^k(\tau)$ as the lower limit measure whitening weight function, recorded as $f_j^k\left[-, -, x_j^k(3), x_j^k(4)\right]$; the third and fourth turning $x_j^k(3)$, $x_j^k(4)$ and call $f_j^k(\tau)$ as the top limit measure whitening weight function, recorded as $f_j^k\left[x_j^k(1), x_j^k(2), -, -\right]$.

This paper is applied with grey fixed weight clustering. Grey fixed weight clustering can be performed according to the following steps:

(1) Provide j index k subclasses whitening weight function $f_j^k(\tau)(j = 1, 2 \ldots m; k = 1, 2 \ldots, s)$
(2) According to the qualitative analysis conclusion, it is to determine each indexes clustering weight $\eta_j(j = 1, 2 \ldots, m)$
(3) From (1) and (2) it can be concluded that the whitening function, clustering weight $\eta_j(j = 1, 2 \ldots, m)$, as well as the object i about j index sample value $x_{ij}(j = 1, 2 \ldots, m)$, to calculate the grey fixed weight clustering coefficient:

(4)
$$\sigma_i^k = \sum_{j=1}^{m} f_j^k(x_{ij})\tau\eta_j \quad (i = 1, 2, \ldots, s) \tag{47.3}$$

(5) If

$$\sigma_j k^* = \max_{1 \le k \le s} \{\sigma_i k\} \tag{47.4}$$

Then we call the object i as grey category of k^*. Typical whitening weight function, lower whitening weight function and top limit whitening weight function singlet and multiple values in a certain range that is in compliance with human thought characteristics and subjective evaluation of the uniqueness of solution; it can be used as mathematical model of subjective evaluation.

47.3 Subjective Evaluation Real Case

It is to perform the subjective assessment on the design of pointer instrument display for example. The design of pointer instrument design includes the dial, scale, pointer, character and colour lighting with other aspects' design [7]. Because the visual sense is characterized by integrity, selectivity, constancy, delusion and others, and has relations with many known, unknown or uncertain influence factors (system factors are not completely clear), pointer instrument design of comprehensive effect is not just dial, scale, pointers and others' sample overlay of single effect (factors relation is not entirely clear). Therefore, pointer instrument display design of subjective evaluation is a question of grey system.

47.3.1 The Selection of Pointer Instrument Subjective Evaluation Environment

The subjective evaluation environment of this paper is 135° column episodic virtual reality studio. First, it is to build-up and assemble pointer instrument digital model in the UG environment and then to change it to JT format. In the virtual reality environment, it is to open JT format pointed instrument model with disconcert three-dimensional visual software and then, the evaluators in stereoscopic glasses conduct the subjective evaluation on the instruments.

The selection of subjective evaluation on instrument in the virtual reality environment can find design flaws in the digital model phase, and to avoid the evaluation method after matters resulting in time delays and economic losses.

47.3.2 The Subjective Evaluation of Pointer Instrument

The subjective evaluation applied with familiar percentile of people is divided into five categories: excellent, good, fair, poor and very poor . Each category of whitening weight function turning point is divided as below:

$$\text{Excellent}: f_j^1[89, 90, -, -]; \tag{47.5}$$

$$\text{Good}: f_j^2[79, 80, 90, 91]; \tag{47.6}$$

$$\text{Fair}: f_j^3[69, 70, 80, 81]; \tag{47.7}$$

$$\text{Poor}: f_j^4[59, 60, 70, 71]; \tag{47.8}$$

$$\text{Very poor}: f_j^5[-, -, 60, 61]; \tag{47.9}$$

Formula (47.5)–(47.8) turning points $x_j^k(x)$ are substituted into Formula (47.2), which immediately acquires excellent, good, poor and very poor typical whitening weight function $f_j^k(\tau)$. The top limit whitening weight function is $f_j^1[89, 90, -, -]$ and the lower limit measure whitening weight function is $f_j^5[-, -, 60, 61]$; whitening weight function can be acquired by reference to the literature [7].

With regard to evaluation question of movements involved with people's consciousness, the comprehensive integration method from the qualitative to quantitative is the only effective methodology [8].

All subjective evaluation index of pointer instrument by survey questionnaire, clustering category weight η_j and three pointer instrument subjective evaluation indicators value are shown in Table 47.1:

From Formula (47.3), we obtain three instruments subjective evaluation grey fixed weight clustering coefficients:

$$\sigma_1^k = (\sigma_1^1, \sigma_1^2, \ldots \sigma_1^5) = (0.12, 0.363, 0.39, 0.113, 0.113)$$

Table 47.1 Clustering weight h_j and value of subjective evaluation indicators

Evaluation	Clustering weights	Evaluation scores		
	η_j	Metre 1	Metre 2	Metre 3
Scale division	0.07	85	80	70
Scale spacing	0.05	85	70	80
Tick mark	0.08	75	90	85
Scale standard number	0.07	65	70	85
Scale direction	0.11	85	95	95
Pointer	0.04	75	85	70
Pointer width	0.10	80	85	95
Dial size	0.12	95	85	90
Character size	0.07	75	70	75
Character shape	0.04	60	80	75
Stroke width	0.05	70	60	80
Colour Matching	0.10	75	70	80
Lighting effects	0.10	80	50	80

$$\sigma_2^k = (\sigma_2^1, \sigma_2^2, \ldots \sigma_2^5) = (0.163, 0.36, 0.23, 0.13, 0.117)$$

$$\sigma_3^k = (\sigma_3^1, \sigma_3^2, \ldots \sigma_3^5) = (0.29, 0.39, 0.283, 0.037, 0)$$

From Formula (47.4) it is known that, $\sigma_1^{k^*} = \sigma_1^3 = 0.39$, namely instrument 1 subjective evaluation is fair; $\sigma_2^{k^*} = \sigma_2^2 = 0.36$, instrument 2 subjective evaluation is good; $\sigma_3^{k^*} = \sigma_3^2 = 0.39$, and instrument 3 subjective evaluation is good.

47.4 Conclusions

As human thinking has grey characteristic, it is introduced with grey system theory evaluation method. The establishment of subjective evaluation indicator makes the unknown or uncertain factors of subjective evaluations become concrete and describable explanatory information, in order to make the comprehensive subjective evaluation questions solvable.

Typical whitening weight function, lower whitening weight function and top limit whitening weight function singlet and multiple values in a certain range are in compliance with human thought characteristics and subjective evaluation of the uniqueness of solution, which is applicable to people subjective assessment of mathematics. This method provides a subjective evaluation for a simple and effective way, and also makes the subjective evaluation results full of theoretical basis.

This work is supported by the opening foundation of the Science and Technology on Human Factors Engineering Laboratory, Chinese Astronaut Research and Training Center (grant no. SYFD140051810K).

References

1. GB/7269-87, Layout, type and size of the electronic device basic console. GB/7269-87 Arrangement, types and basic dimensions for electronic equipment consoles
2. GB/14775-93, Manipulator general ergonomics requirements. GB/14775-93. General ergonomics requirements for controller
3. Dutarte D, Martensson L (2001) Keeping-in-touch with the environment: an approach to ship and aircraft control. Design and evaluation of human-machine system. Kasel Germany
4. Yuan Q, Cao Y, Liu N et al (1998) Research of comprehensive evaluation theory and system in product design quality. China Mech Eng 9(10):8–10
5. Deng J (1992) Basic method of grey system. Huazhong Institute of Technology Press, Wuhan
6. Liu S, Guo T, Dang Y (1999) Grey system theory and application. Science Press, Beijing
7. Ding Y (1992) Man-machine engineering. Beijing Institute of Technology Press, Beijing
8. Chen X, Yuan X (2000) Human-machine-environment system. Beijing University of Aeronautics and Astronautics Press, Beijing

Chapter 48
Application Research of Gaze Tracking Technology in Battlefield Situation Visualization Assessment

Yanjun Zhang, Zhibing Pang, Chenhui Li, Hualiang Xu, Xiaofei Zhai and Shili Chu

Abstract Battlefield situation visualization provides imaginable, intuitive, plentiful battlefield information for command decision. The quality of visualization and the recognition of commanders will directly affect the war. For the situation information display, this paper introduces an evaluation method which can assess the effect of visualization by extracting the visual region of interest and calculate changes in visual attention on the time domain. According to analyzing changes in visual attention, the effect of battlefield environment rendering and fighting, target detection in battlefield situation visualization can be assessed. By paying attention to the important role of the human factor in the battlefield situation visualization system, this method can provide a scientific basis for improving the efficiency of situational information perception.

Keywords Gaze tracking technology · Battlefield situation visualization · Assess · Visual region of interest

48.1 Introduction

Because of the high-tech war characteristics, such as the possibility of a sudden war, greatly increases war process accelerates and opportunities which conducive the war is unpredicted, how to effectively and quickly establish the superiority over enemy information has become the primary task of battle, and the primary condition of success. By handling large amounts of information from different levels and uninterrupted time, and integrating the data that come from water

Y. Zhang (✉) · Z. Pang · C. Li · H. Xu · X. Zhai · S. Chu
Air Defense Forces Academy, Zhengzhou 450052, China
e-mail: fkb_mmes@163.com

© Springer-Verlag Berlin Heidelberg 2015

S. Long and B.S. Dhillon (eds.), *Proceedings of the 15th International Conference on Man–Machine–Environment System Engineering*, Lecture Notes in Electrical Engineering 356, DOI 10.1007/978-3-662-48224-7_48

surface, air, underwater, or tactical data links, the esthesia situation for commanders at all levels can satisfy the operational requirements under various battlefields strain conditions. By embedding the real-time information, such as combat units' movement and location, to the map and making it visualized, commanders can have a clearer grasp of the situation of both sides. The process of turning information caught by terminal equipments and low-level commanders on the battlefield into graphics and images is called battlefield situation information visualization. Battlefield situation information visualization can make sure that commanders get the information which covers all battlefields, and assign the sharing resources of battlefield information reasonably. The quality of visualization and the degree of recognition of commanders will directly affect the ultimate success or failure of the war [1]. This paper introduces an evaluation method which can assess the effect of visualization by using visual gaze tracking technology to extract the visual ROIs, which can provide a scientific basis for improving the efficiency of situational information perception.

48.2 Battlefield Situation Visualization

There are two display modes of battlefield situation visualization: 2D situation visualization and 3D situation visualization. 2D situation visualization can provide a global view point for commanders to monitor the entire process of battlefield, and make a view plan of the entire battlefield by configuring the required electronic scale map. 3D situation visualization, based on virtual reality technology, can provide real and comprehensive virtual natural environment, such as topography and surface features, of a particular battlefield area, and can clearly describe the situation of all forces. Commanders will get an immersive viewing experience by 3D situation visualization [2, 3].

Battlefield includes many elements, including troops which constitute the battlefield situation, environment, events, and predictions. These elements will change with operational objectives and operational phases. Different battlefield situations contain different elements. On the whole, battlefield situation is mainly constituted of two types of elements:

(1) Battlefield environment: the objective environment of combat space in addition to personnel and weaponry. From objective factors involved in the war to analyze, battlefield environment should contain its geographical, weather conditions, electromagnetic environment, and special environment.
(2) Combat situation: intelligence information which is real-time or instant has amount of data and changes frequently in the situation display. It should contain information of all battle units, including our forces, enemy forces, friendly forces, and neutral forces and the information should describe current location, status, deployment, and trajectories of these troops (Fig. 48.1).

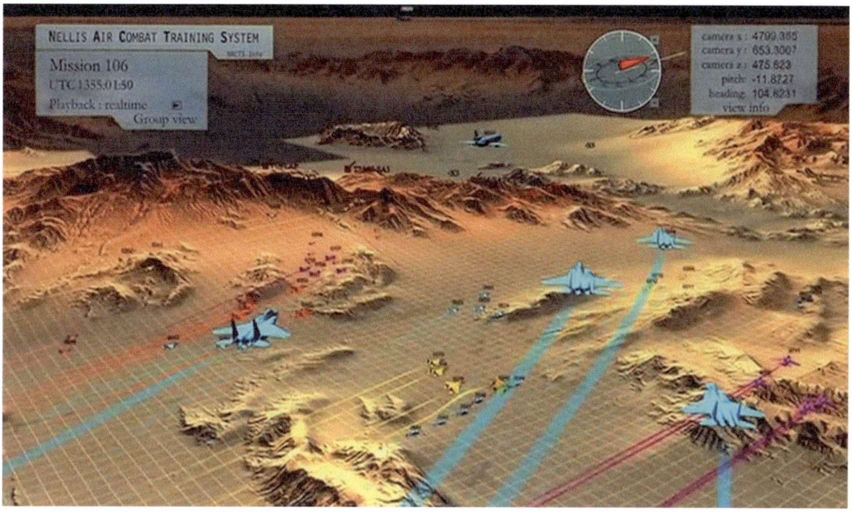

Fig. 48.1 Battlefield situation map generated by Nellis air combat training system

48.3 Displays of Battlefield Situation Information

According to the division of goals in the battlefield situation information visualization, there are two main displays: display of battlefield environment and display of military standard.

48.3.1 Display of Battlefield Environment

Researches on battlefield visualization are focused on the modes to build an electronic battlefield environment. There are three main modes to build an electronic battlefield environment [4, 5].

(1) 2D digital map environment: 2D digital map environment is developed with advances in information visualization technology. We can get a digital map by simulating a general map on the computer. Digital map with higher data integration tools and faster data processing speed can satisfy the storage and query the geographic data requirements for engineers.
(2) 2D digital map with local elevation information environment: With the development of geographic information visualization, the requirement of information visualization is far more than only query location on computers. Digital maps should satisfy a number of needs such as distance estimates, elevation calculating, and spatial analysis. Researchers put local elevation information into 2D digital map to satisfy some simple spatial analysis needs.

This map is called 2.5-dimensional digital map environment. It has a quite high computing speed.

(3) 3D virtual battlefield environment. 3D virtual battlefield environment is a real 3D simulation of comprehensive battle space. It needs technologies for generating 3D terrain, processing texture of image space, and integrating massive data. The realistic that the 3D space can construct, will let commanders grasp the battlefield situation intuitively and provide a 3D spatial analysis function to commanders. Currently, 3D virtual battlefield environment is not good at real-time rendering and the 3D environment may cause commanders to lose sense of battlefield situation, and let them get wrong decisions.

48.3.2 Display of Military Standard

The researches of combat forces visualization are focused on the visualization means of military standard information. Military standard is the main symbolic mean to edit and draw situations in all kinds of battlefield command system. There are two main modes to display visualization military standards:

(1) 2D symbols: 2D military symbols can display all kinds of military units professionally. Generally, the symbols can be obtained by symbolically simulating military standards used on traditional topographical sandbox. By using this kind of symbols, operations' officers can plot situations professionally and the commanders can get familiar battlefield awareness on situation maps.

(2) 3D solid model: With the development of 3D structure of the battlefield environment, using 3D solid model to plot 3D battlefield situation has become a hot spot of situation information visualization. Solid model military standard can be obtained by real graphics drawing military units fighting in 3D environment. This military standard can help commanders to get awareness about the integrated battlefield.

(3) 2D military standard with 3D attributes: Although the 3D solid model can descript the details of military units more really, it can lose commanders' sense of battlefield situation, and lead to wrong decisions. 3D military standards have confusion function, which means 3D entities simulation will lead people erroneously understand entities of real attribute. By implementing a variety of different psychology experiments, researchers compare 2D and 3D military standards awareness degree at all application levels. Researchers obtained a 2D military standard with 3D attribute. It can meet the needs of computing speed on real-time events demo. At the same time, in 3D environment, it can support commanders to make better decision, who will face massive data of the integrated battlefield.

48.4 Battlefield Situation Visualization Assessment Method Based on Gaze Tracking Technology

Battlefield situation visualization technology has been developed. Commanders can use this technology to clearly grasp situations of troops and battlefield. But the human factor is not fully considered by designers in research and development stage. Are the combat units and information displayed reasonable? Can the commanders get information successfully? These problems are more and more acute. Gaze tracking technology has a strong practical edge on the evaluation works about interface elements and interface layout. It can provide objective, comparable, and quantifiable measurement standards. Therefore, in this paper, the gaze tracking technology will be used to assess the battlefield situation visualization. By extracting the visual region of interest, and analyzing changes in visual attention, we can assess the effect of battlefield environment rendering and fighting target detection in battlefield situation visualization.

48.4.1 Eye Gaze Tracking Technology [6]

Eye gaze tracking technology is referred to as eye tracking,which is a technology to study how to track human visual accurate and without interference. It was mainly used on reading research in psychology from early twentieth century. With the development of imaging equipment and computer, real high-precision interference-free eye tracker hardware products are invented until this century. Eye tracking has been an active branch of cutting-edge technology after that and been widely used in psychology, HCI, aerospace, traffic safety, human factors engineering, behavior research, pattern recognition, market research, medical research, highway engineering, AI, etc.

There are three basic forms of eye movements:

(1) Fixations: In normal observation, human vision will rapidly beat between points on the observed targets. When vision stays on one target over 100 ms, it is called fixation. Most information only can be processed by fixations.
(2) Saccade: The sudden change of fixation point or direction is called saccade. It is a kind of high-speed eye movement between fixation points. Saccade means that the eye moves in high speed and large range; last time between 30 and 120 ms. No information will be obtained when eye movement is saccade.
(3) Smooth Pursuit: When there is relative motion between a target and eyes, eyes will move to follow the target. This movement is called smooth pursuit.

The current eye trackers are based on the principle that the infrared can capture the reflection of cornea and retina. Eye trackers can record data like tracks of the user's eye movements, the number of fixations, and fixation time. According to recently published studies and information, the main test parameters of eye tracker are:

(1) Eye movement frequency, a parameter related to searching efficiency. When the frequency is high, the efficiency is low. That could be because the layout elements are not reasonable.
(2) Size of pupil: By measuring the size of pupils, researchers can study examinees' psychological and physiological state with other detection indexes.
(3) Average fixation time: Fixation time means the difficulty degree of extracting information. The longer the time spent on fixing, the harder the examinee can get information from the display area.
(4) Fixation point sequence: Conversion of fixation point between interest areas can be used to test if the layout of the user interface is reasonable.
(5) The moment the vision first time reaches the target interest area.

48.4.2 Assessment Methods of Battlefield Situation Visualization's Effect

When facing a complex scene, human visual system HVS, will be able to quickly focus on a few significant visual objects and process priority. This is called visual attention and significant visual object is called region of interest ROI. By the function of the mechanism, HVS can assign limited information processing resources reasonably, given the ability to choose the visual perception process.

The emergence of the eye tracker provides a new effective tool to explore human visual information processing mechanism and observe the direct or indirect relationship between it and the mental activity, using gaze tracking technology in a variety of conditions. There are three main kinds of visualization methods to extract static image ROIs based on eye tracking experiment data [7]:

(1) Direct circle region of interest based on the coordinates and the fixation point of the image;
(2) Draw color temperature map on the original object to be observed, in which the warm colors express a higher degree of concern and cool colors represents relatively lower level of concern;
(3) By changing the brightness of the original image indicates the degree of interest in different parts, which is to maintain the original brightness of the area of interest and reduce the brightness of the area of unconcern.

But the battlefield situation is usually dynamically changing and the above-described extraction method is limited to static images. Therefore, it is important to apply the eye movement test data to dynamic battlefield situation. To facilitate data processing, the displayed window of confrontational exercise according to combat screenplay is recorded as a video file, with the help of which the eye tracking experimental data is got.

The information of fixation (the center coordinates of ROIs) obtained by eye tracking experiment consists of five parts: starting time t_{start}, ending time t_{end}, duration t, and coordinates x, y. There are two steps to extract video ROIs weight matrix from the experimental data obtained by eye tracking:

(1) The synchronization problem of video image frames and coordinates is solved according to the starting time and the ending time. By judgment condition of $t_{\text{start}} \leq nT \leq t_{\text{end}}$, the fixation of each frame is obtained. In the formula n is the video frames and T is duration of the video image.
(2) After obtaining the fixation point of each frame, it is the problem of how to get weight matrix of every frame ROIs and visualize eye tracking experiment data. First, generate an all zeros matrix $\mathbf{M}_{H \times W}$ of the same size as the original image (H is the height of the image, W is the width of the image). According to the experimental data of eye tracking, the coordinates(x, y) of fixation is obtained. By the following formula, the bitmap of ROIs is obtained and preprocessed by Gaussian filter [8].

$$f(x, y, t) = (\alpha t + (1 - \alpha)) \exp\left(-\frac{x^2 + y^2}{2\sigma^2}\right) \tag{48.1}$$

After normalizing weight matrix of ROIs in [a, b] interval, the weight matrix of ROIs representative of significantly characteristic image is obtained. In Formula (48.1), σ is the standard difference of analog foveal Gaussian function; $\alpha \in (0, 1)$ is the rights of duration in the weight matrix of ROIs.

Vision research shows that a large contrast, slender shape, larger size, central location, color highlight areas, and areas with the trend of the movement are prone to visual attention. For example, in the deduction of battlefield situation, large moving goal in screen should be a significant object. In this case, human eye can track a moving target in the battle driven by their interests and viewing purpose. However, if some of the color or shape of the abnormal areas appear in the background of the target area, the human eye will be drawn to these exceptions, and divert attention from the moving target to these areas, which affects the perceiving efficiency of situational information of the commanders.

Experimental video is time-series images, but its effect is not simply additive images. Instead, there is unique effect in the time domain, which brings great visual impact to the viewer, such as the sudden occurrence of some unusual events, (e.g., flame, smoke, explosions, and other effects in order to enhance the realism of war in battlefield situation visualization). The information on a single image does not attract attention. Therefore, the dynamic changeover time is an important difference between the battlefield situation map and the static map. Since the human eye is likely to be attracted by the salient partial region, the original desired line of sight changes. The magnitude of these changes in the time domain can reflect the effect of battlefield situation visualization. By calculating the standard deviation of the weight matrix of ROIs to measure saliency temporal variation (STV) [9] of battlefield situation map:

$$\mathrm{STV} = \mathrm{STV}_t(\mathbf{M}_{H\times W}) \tag{48.2}$$

In the formula, $\mathrm{STV}_t()$ calculates the standard deviation. The higher STV value of battlefield situation visualization system, the results displayed are better. The assessment result of battlefield situation visualization is obtained by comparing the STV value.

48.5 Conclusion

For the battlefield situation visualization system, this paper introduces an evaluation method which can assess the effect of visualization by using visual gaze tracking technology. Using the data measured by eye tracker, the ROIs and the weight matrix of the ROIs are obtained. By calculating the standard deviation of weight matrix, the significant changes in the time domain battlefield situation map are measured. According to the changes, the assessment of visualization is obtained. This method focuses on the important role of human factor in the battlefield situation visualization. The method provides scientific guidance for improving information perception efficiency of battlefield situation visualization systems. Because of the enormous amount of data measured by eye tracker, this paper merely extracts and computes the visual ROIs from reducing the amount of computation. For more objective and comprehensive evaluation process, the further work is done using more data measured by eye tracker.

References

1. Hayes PM, Simcox WJ (2002) Battlespace, visualization and the integrated meteorological system (IMETS). Military Intelligence Professional Bulletin
2. Cai ZH, Peng XY (2003) Research and implementation of GEO graphic information system-based virtual battlefield situation display system. J Syst Simul 15(7)
3. You X (2003) The battlefield environment simulation based on VR. Acta Geodaet et Cartographica Sinica 31(1)
4. Zhao Z, Chao M, Tang X, Chen M, Yang Q (2007) Research on implementing dynamic symbol.Comput Eng Des 28(12)
5. Yang Q, Chen M, Tang X, Zhao Z (2007) Real-time creating and protracting of 3D static unit symbols. Comput Eng Des 28(14)
6. Jacob RJK, Karn KS (2003) Eye tracking in human computer interaction and usability research: ready to deliver the promises. Comput Vis Image Underst 24:682–701
7. Epakov O, Miniotas D (2007) Visualization of eye gaze data using heat maps. Electron Electr Eng 74(2):55–58
8. Backer G, Mertsching B, Bollmann M (2007) Data and model-driven gaze control for an active-vision system. IEEE Trans Pattern Anal Mach Intell 23(12):1415–1429
9. Feng X, Yang D, Zhang L (2011) Saliency variation based (duality Assessment for Packet-loss-impaired Videos. Acta Automatica Sinica 11(37)

Chapter 49
Construction of Human–Machine Interface Evaluation with Multi-index Weights Allocation Method

Chuan Wang and Jianguo Zhang

Abstract Based on the analysis of traditional AHP, statutory right of the main problems exist; the evaluation of human–machine interface is based on the characteristics of the index weight allocation method and the multi-index weights allocation is based on the reliability coefficient of correlation based on grey-level analysis. Use HMI index weight allocation method, reliability coefficients-based grey theory of AHP correction, reliability coefficient from the perspective of the overall similarity of structure expert judgment to achieve a quantitative description of the cognitive characteristics to solve the multi-expert judgment gathering information. Case studies show that the weight distribution method is not only a clear concept, but is simple and easy to operate which reduces the influence of subjective factors on weight, so that the weight distribution becomes more clear and reasonable. The results are more scientific and reliable. Multi-level index weights based on grey relational analysis method for redistribution to the grey from the perspective of the overall analysis of the structure and type II and I judgment matrix to solve excessive number of indicators to determine the number of pairwise comparisons too many problems. Case studies show that the judgment matrix I is only simple and convenient, easy to operate and high resolution, suitable indicators when excessive weight distribution.

Keywords Human–machine interface evaluation · Multi-index weights · Allocation method

AHP analysis can make the mind judgment in quantification, and it plays important roles for the effective determination and evaluation on the weight of weight function. However, the human–machine interface evaluation system is a huge system, the index data volume is large and the indicators are interrelated. At the same time, many evaluation indicators also have a large of grey, at some level; the indicator

C. Wang (✉)
Naval University of Engineering, Wuhan 430033, China
e-mail: hg04381@163.com

C. Wang · J. Zhang
Naval Medical Research Institute, Shanghai 200433, China

© Springer-Verlag Berlin Heidelberg 2015

S. Long and B.S. Dhillon (eds.), *Proceedings of the 15th International Conference on Man–Machine–Environment System Engineering*, Lecture Notes in Electrical Engineering 356, DOI 10.1007/978-3-662-48224-7_49

number is relatively large, at this moment, if the pair comparison approaches are applied, the number of judgment is relatively large, which is probably out of people's judgment ability and makes people's thinking more confused, even influences the accuracy of weight distribution results. Therefore, in the case of a large number of indicators, how to apply the appropriate approach to constitute judgment of matrix scientific nature and reliability to weight function distribution result plays an important function.

49.1 Grey Relational Hierarchy Process Analysis Proposal

In light of the conditions above, taking into consideration that grey correlation analysis is an effective method to dispose the system that is characterized of grey system, and in the establishment of grey system judgment matrix, the number of judgment the evaluator makes is same with the number of indicators evaluated, all are n times. The number of analytic hierarchy process (AHP) pair judgment determination is $(N^2 - N)$. Obviously, the larger the evaluation indicator number n, the more times the pairs comparison judgment is, even much more larger than the times of judgment to establish grey judgment matrix. Therefore, the AHP has some problems in disposal of multiple indicators weight functions. The grey theory of grey theory is introduced to the AHP. The grey correlation analysis and hierarchical analysis are organically integrated. It utilizes the grey correlation analysis to constitute pairs comparison judgment matrix, to resolve the problems to judge that the matrix is hard to constitute in the case of multi-indicators, in order to realize the multi-indicators weight functions distribution. Grey correlation hierarchy implementation process is as shown in Fig. 49.1.

49.2 Grey Correlation AHP Basic Principle

49.2.1 *The Establishment of Grey Evaluation Matrix*

Assume $B = \{B_1, B_2, \ldots B_m\}$ as evaluated group sets, m as the number of evaluation experts, $A = \{A_1, A_2, \ldots A_n\}$ as evaluated indicator sets. Different evaluation experts B_j to different evaluation indicators A_i offered evaluation indicators' importance attribute value $x_{ij}, i = 1, 2, \ldots n; j = 1, 2 \ldots m$, and then judgment matrix is given as

$$X = (x_{ij})_{n \times m} = \begin{bmatrix} x_{11} & x_{12} & \cdots & x_{1m} \\ x_{21} & x_{22} & \cdots & x_{2m} \\ \vdots & \vdots & & \vdots \\ x_{n1} & x_{n2} & \cdots & x_{nm} \end{bmatrix} \tag{49.1}$$

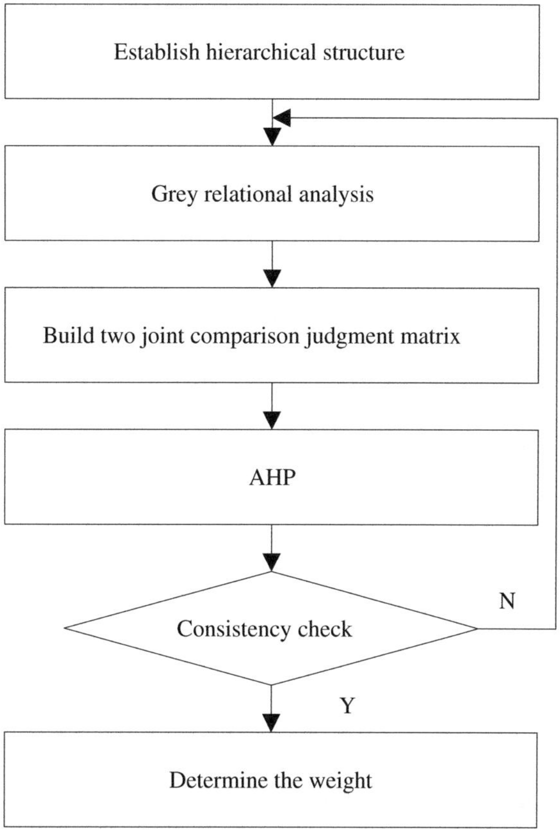

Fig. 49.1 Process of grey relation analytic hierarchy process

49.2.2 *The Determination of Reference Sequence*

Among $X_1, X_2, \ldots X_n$, it selects the largest important attribute value as "public" reference value, each expert reference value is given with this value, thus it constitutes reference data column $X_0, X_0 = (x_{01}, x_{02}, \ldots x_{0m})$. Among them, $x_{0j} = \max\limits_{1 \le i \le n} \max\limits_{1 \le j \le m} x_{ij}$. In order to increase the comparability of the sequence, it can apply the method of average method to conduct the normalized processing.

49.2.3 *The Calculation of Grey Correlation Degree*

According to the grey system theory, it applies formulas (49.2) and (49.3) to calculate each experts' experience judgment value to each indicator's importance grey correlation coefficient with "public" reference value and correlation matrix, and get the grey correlation degree for each indicator and reference sequence: $r_{0i} = \frac{1}{m}\sum_{j=i}^{m} \varepsilon_{ij}$

$$\varepsilon_{ki} = \frac{\min\limits_{1 \le k \le m} \min\limits_{1 < i < n} |y_{0i} - y_{ki}| + \rho \max\limits_{1 \le k \le m} \max\limits_{1 < i < n} |y_{0i} - y_{ki}|}{|y_{0i} - y_{ki}| + \rho \max\limits_{1 \le k \le m} \max\limits_{1 < i < n} |y_{0i} - y_{ki}|} \tag{49.2}$$

$$\varepsilon = (\varepsilon_{ki})_{m \times n} = \begin{bmatrix} \varepsilon_{11} & \varepsilon_{12} & \cdots & \varepsilon_{1n} \\ \varepsilon_{21} & \varepsilon_{22} & \cdots & \varepsilon_{2n} \\ \vdots & \vdots & \vdots & \vdots \\ \varepsilon_{m1} & \varepsilon_{m2} & \cdots & \varepsilon_{mn} \end{bmatrix} \tag{49.3}$$

The correlation degree of each indicator directly reflects relative correlative degree corresponding to reference sequence for each evaluation indicator. Because of each indicator is relative to the same degree of related reference sequence, it can use this correlation degree to constitute comparative judgment matrix according to this point.

49.2.4 Structure and Weight Function Calculation of Type I Judgment Matrix

Definition 1.1 $\beta^{I}_{ki} = \frac{r_{0k}}{r_{0i}}$ refers to indicator a_k which is compared with indicator a_i, under the same rule, the degree gets close to the ideal sequence, is called as degree of dominance I which reflects its precedence relation between indicators, namely its relatively importance degree.

If $\beta^{I}_{ki} = 1$, then the indicator a_k is of the same importance as indicator a_i. If $\beta^{I}_{ki} > 1$, then the indicator a_k is more important than indicator a_i. At this moment, if β^{I}_{ki} is larger, then it refers to the comparison of indicator a_k and indicator a_i, the laterbecoming more important; if $\beta^{I}_{ki} < 1$, then the indicator a_k is less important than the indicator a_i. At this moment, if β^{I}_{ki} is smaller, then it refers to the comparison of indicator a_k and indicator a_i, the later becoming less important.

Definition 1.2 The judgment matrix composed of dominance I is called as Type I judgment matrix E. Type I judgment matrix E can be expressed as

$$E = (\beta^{I}_{ki})_{n \times n} = \begin{bmatrix} \beta^{I}_{11} & \beta^{I}_{12} & \cdots & \beta^{I}_{1n} \\ \beta^{I}_{21} & \beta^{I}_{22} & \cdots & \beta^{I}_{1n} \\ \vdots & \vdots & \vdots & \vdots \\ \beta^{I}_{n1} & \beta^{I}_{n2} & \cdots & \beta^{I}_{nn} \end{bmatrix} \tag{49.4}$$

Theorem 1.1 *it can be easily to attain by Definition 1.1, type I judgment matrix has the following features:*

(1) $\beta^{\mathrm{I}}_{kk} = 1$
(2) $\beta^{\mathrm{I}}_{ki} = \frac{1}{\beta^{\mathrm{I}}_{ik}}$

For $n \times n$ degree matrix, if for all $k, i = 1, 2, \ldots, n$ meet the two attributes above, the matrix is positive reciprocal matrix; therefore, the constructed type I judgment matrix is the straight reciprocal matrix.
According to definition 1.1, it can be known that each vector in the type I judgment matrix meet: $\beta^{\mathrm{I}}_{ki} = \frac{r_{0k}}{r_{0i}}$, $\beta^{\mathrm{I}}_{kj} = \frac{r_{0k}}{r_{0j}}$, $\beta^{\mathrm{I}}_{ij} = \frac{r_{0i}}{r_{0j}}$, so there is $\beta^{\mathrm{I}}_{ki} = \frac{\beta^{\mathrm{I}}_{kj}}{\beta^{\mathrm{I}}_{ij}}$. By the related theory of matrix theory, the matrix in line with conditions above is consistent matrix.

It can be seen that the constituted type I judgment matrix is positive reciprocal consistency judgment matrix, it has all the attributes of positive reciprocal consistency judgment matrix, therefore by this judgment matrix it can directly obtain the weight function without consistency test. If judgment matrix has no consistency, although it can meet consistency requirement by many times adjustment, no doubt, it can increase the computing workload, and affect operability of AHP. When judgment matrix meets consistency requirement, judge gives judgment information that is effective, and explain; ns this constitute judgment matrix keeps judgment mind consistency.

On the basis of the use of this method to construct the judgment matrix, it uses AHP square root method to calculate the weight function of each evaluation indicators. Because the constituted type I judgment matrix naturally meets the consistency requirement, it can save consistency test, and make the calculation process of weight function more simple and convenient.

49.2.5 Type II Judgment Matrix Constitute and Weight Calculation

Definition 1.3 Orders $\beta^{\mathrm{II}}_{ki} = \frac{r_{0k}}{r_{0k}+r_{0i}}$ explains that indicator a_k is compared with indicator a_i, the degree that it gets close to the ideal sequence under the same rule, call it as dominance II, it reflects the precedence relation between indicators, namely relative importance degree.

If $\beta^{\mathrm{II}}_{ki} = 0.5$, the indicator a_k and indicator a_i are equally important; If $\beta^{\mathrm{II}}_{ki} \in (0.5, 1]$, the indicator a_k is more important than indicator a_i; when β^{II}_{ki} is more, it means the indicator a_k and indicator a_i become more important; if $\beta^{\mathrm{II}}_{ki} \in [0, 0.5)$, the indicator a_k is less important than indicator a_i; when β^{II}_{ki} is smaller, it expresses the comparison for indicator a_k and indicator a_i becomes less important.

Definition 1.4 The judgment matrix composed of dominance II is called as type II judgment matrix F. Type II judgment matrix F can be expressed as

$$E = (\beta^{\text{II}}_{ki})_{n\times n} = \begin{bmatrix} \beta^{\text{II}}_{11} & \beta^{\text{II}}_{12} & \cdots & \beta^{\text{II}}_{1n} \\ \beta^{\text{II}}_{21} & \beta^{\text{II}}_{22} & \cdots & \beta^{\text{II}}_{2n} \\ \vdots & \vdots & \vdots & \vdots \\ \beta^{\text{II}}_{n1} & \beta^{\text{II}}_{n2} & \cdots & \beta^{\text{II}}_{nn} \end{bmatrix} \tag{49.5}$$

From Definition 1.2, and Definition 1.3, it can be easily attained that the type II judgment matrix has some features as below:

(1) $0 < \beta^{\text{II}}_{ki} < 1$
(2) $\beta^{\text{II}}_{ki} = 0.5$
(3) $\beta^{\text{II}}_{ki} + \beta^{\text{II}}_{ik} = 1$

According to the matrix theory, to $n \times n$ degree matrix, if for all $k, i = 1, 2, \ldots n$ satisfies the attributes above, the matrix is called fuzzy complementary matrix; thus, the structure of a type II judgment matrix is a fuzzy complementary judgment matrix. For the fuzzy complementary matrix, when it satisfies $\beta^{\text{II}}_{ki} = \beta^{\text{II}}_{kj} - \beta^{\text{II}}_{ji} + 0.5$, it has consistency. But according to the above definition, it does not prove that type II judgment matrix is a fuzzy consistency matrix; therefore, it must conduct a consistency test if it uses the method mentioned above to obtain answer about weight function, or use the answer acquisition method proposed by Zhang Jijun for the fuzzy complementary judgment matrix method. At first, the matrix is transformed into a fuzzy complementary consistency matrix, and then the indicator weight function is determined; the exact method is as follows:

At first, it is to summation for all the rows in the fuzzy complementary judgment matrix $F = (\beta^{\text{II}}_{ki})_{n\times n}$, are recorded as $r_k = \sum_{i=1}^{n} \beta^{\text{II}}_{ki}$, k = 1, 2…, n, and transformed as:

$$r_{\text{ki}} = \frac{r_k - r_i}{2(n-1)} + 0.5 \tag{49.6}$$

Then, all acquired indicators weight functions are:

$$w_k = \frac{1}{n} - \frac{n}{4\alpha(n-1)} + \frac{n}{2\alpha(n-1)} \sum_{i=1}^{n} \beta^{\text{II}}_{ki} \tag{49.7}$$

In formula, α satisfies: $\alpha \geq \frac{n-1}{2}$, generally when n is larger, it is to apply $\alpha = \frac{n-1}{2}$. The corresponding weight function vector is recorded as $W = (w_1, w_2, \ldots, w_n)^T$.

49.3 Case Studies

In order to verify the feasibility of the above two proposed methods to determine the indicator weight function by constructing judgment matrix, for now it explains the proposed method by the example that in the computer monitor screen displaying screen evaluation indicator weigh functions affirmation process.

In order to determine each indicator weight function in the computer screen interface, eight experts are engaged at scoring for the above-mentioned 11 sub-evaluation indicators relative total indicators 'relative degree of influence, and establish the related grey evaluation matrix as below:

$$X = \begin{bmatrix} 7 & 6 & 7.5 & 6.5 & 8 & 7.5 & 6 & 6.5 \\ 4 & 5 & 4.5 & 6 & 7 & 5 & 5.5 & 4 \\ 8 & 8.5 & 7.5 & 6.5 & 8 & 8.5 & 8 & 9 \\ 6.5 & 7.5 & 6 & 7 & 8 & 7.5 & 6.5 & 8 \\ 7.5 & 8 & 6 & 7.5 & 6 & 7.5 & 8 & 6.5 \\ 8 & 8.5 & 8 & 7.5 & 6 & 6.5 & 7 & 8.5 \\ 7 & 7.5 & 8 & 6.5 & 7 & 5 & 6 & 6.5 \\ 8.5 & 8 & 7.5 & 8 & 8.5 & 7 & 7.5 & 8 \\ 7 & 6.5 & 7 & 6 & 7.5 & 6.5 & 6 & 7.5 \\ 8 & 8.5 & 9 & 7.5 & 8 & 8.5 & 7.5 & 8.5 \\ 4.5 & 5 & 6 & 5.5 & 6.5 & 6 & 6.5 & 5 \end{bmatrix}$$

Then, it is to perform the normalized processing, and acquired the normalized matrix:

$$X = \begin{bmatrix} 0.092 & 0.076 & 0.114 & 0.087 & 0.099 & 0.099 & 0.081 & 0.083 \\ 0.053 & 0.063 & 0.068 & 0.081 & 0.087 & 0.066 & 0.074 & 0.051 \\ 0.105 & 0.108 & 0.114 & 0.087 & 0.099 & 0.113 & 0.108 & 0.115 \\ 0.086 & 0.095 & 0.091 & 0.094 & 0.099 & 0.099 & 0.087 & 0.103 \\ 0.099 & 0.101 & 0.091 & 0.01 & 0.075 & 0.099 & 0.105 & 0.083 \\ 0.105 & 0.108 & 0.121 & 0.01 & 0.075 & 0.086 & 0.094 & 0.109 \\ 0.092 & 0.095 & 0.121 & 0.087 & 0.087 & 0.066 & 0.081 & 0.083 \\ 0.112 & 0.101 & 0.114 & 0.108 & 0.106 & 0.093 & 0.101 & 0.103 \\ 0.092 & 0.082 & 0.106 & 0.081 & 0.093 & 0.086 & 0.081 & 0.096 \\ 0.105 & 0.108 & 0.136 & 0.01 & 0.099 & 0.113 & 0.101 & 0.109 \\ 0.059 & 0.063 & 0.091 & 0.074 & 0.081 & 0.079 & 0.087 & 0.064 \end{bmatrix}$$

The corresponding reference sequence is as below

$$X_0 = (0.136, 0.136, 0.136, 0.136, 0.136, 0.136, 0.136, 0.136)$$

Then, according to the formula $r_{0i} = \frac{1}{m}\sum_{j=i}^{m} \varepsilon_{ij}$, the correlation degree of each indicator grey is calculated and the constituted type I matrix is as below:

$$E = \begin{bmatrix} 1.000 & 1.243 & 0.868 & 1.008 & 1.068 & 0.975 & 1.014 & 0.886 & 1.024 & 0.878 & 1.196 \\ 0.804 & 1.000 & 0.698 & 0.811 & 0.859 & 0.785 & 0.815 & 0.713 & 0.823 & 0.707 & 0.962 \\ 1.152 & 1.432 & 1.000 & 1.162 & 1.231 & 1.123 & 1.167 & 1.021 & 1.179 & 1.011 & 1.377 \\ 0.992 & 1.232 & 0.861 & 1.000 & 1.059 & 0.967 & 1.005 & 0.879 & 1.016 & 0.871 & 1.185 \\ 0.936 & 1.163 & 0.813 & 0.944 & 1.000 & 0.912 & 0.948 & 0.829 & 0.958 & 0.822 & 1.118 \\ 1.025 & 1.275 & 0.890 & 1.034 & 1.095 & 1.000 & 1.039 & 0.909 & 1.050 & 0.901 & 1.225 \\ 0.986 & 1.226 & 0.856 & 0.995 & 1.054 & 0.962 & 1.000 & 0.874 & 1.010 & 0.867 & 1.179 \\ 1.128 & 1.402 & 0.979 & 1.138 & 1.205 & 1.100 & 1.144 & 1.000 & 1.155 & 0.991 & 1.348 \\ 0.976 & 1.214 & 0.843 & 0.984 & 1.043 & 0.952 & 0.989 & 0.865 & 1.000 & 0.858 & 1.167 \\ 1.138 & 1.415 & 0.988 & 1.148 & 1.216 & 1.110 & 1.154 & 1.009 & 1.166 & 1.000 & 1.361 \\ 0.836 & 1.039 & 0.726 & 0.844 & 0.894 & 0.894 & 0.816 & 0.847 & 0.741 & 0.857 & 1.000 \end{bmatrix}$$

Based on the constituted type I judgment matrix, the characteristic root method is applied to obtain each indicator weight functions: $w_1 = 0.091$, $w_2 = 0.073$, $w_3 = 0.104$, $w_4 = 0.090$, $w_5 = 0.085$, $w_6 = 0.093$, $w_7 = 0.089$, $w_8 = 0.102$, $w_9 = 0.089$, $w_{10} = 0.103$, $w_{11} = 0.076$.

After calculation, the largest eigenvalue of constituted type I judgment matrix $\lambda_{\max} = n = 11$, is obtained which clearly specifies that the judgment matrix composed by this method is characteristic of consistency, and there is no need to conduct the consistency test; at the same time, it also appears that the constituted judgment matrix is in compliance with the characteristics of people's thinking.

At the same time, type II judgment matrix constituted by grey correlation degree can be expressed as below:

$$F = \begin{bmatrix} 0.500 & 0.554 & 0.465 & 0.502 & 0.516 & 0.494 & 0.503 & 0.469 & 0.506 & 0.468 & 0.545 \\ 0.446 & 0.500 & 0.411 & 0.448 & 0.462 & 0.439 & 0.449 & 0.416 & 0.452 & 0.414 & 0.490 \\ 0.535 & 0.589 & 0.500 & 0.537 & 0.552 & 0.529 & 0.538 & 0.505 & 0.541 & 0.503 & 0.579 \\ 0.498 & 0.552 & 0.463 & 0.500 & 0.514 & 0.492 & 0.501 & 0.468 & 0.504 & 0.466 & 0.542 \\ 0.483 & 0.538 & 0.448 & 0.486 & 0.500 & 0.477 & 0.487 & 0.453 & 0.489 & 0.451 & 0.528 \\ 0.506 & 0.560 & 0.471 & 0.508 & 0.523 & 0.500 & 0.519 & 0.476 & 0.512 & 1.474 & 0.551 \\ 0.497 & 0.551 & 0.462 & 0.499 & 0.513 & 0.490 & 0.500 & 0.467 & 0.503 & 0.464 & 0.541 \\ 0.531 & 0.584 & 0.495 & 0.532 & 0.547 & 0.524 & 0.533 & 0.500 & 0.536 & 0.498 & 0.574 \\ 0.494 & 0.548 & 0.459 & 0.496 & 0.511 & 0.488 & 0.497 & 0.464 & 0.500 & 0.462 & 0.539 \\ 0.455 & 0.510 & 0.421 & 0.457 & 0.472 & 0.449 & 0.459 & 0.426 & 0.461 & 0.424 & 0.500 \end{bmatrix}$$

On this basis, Eq. (49.7) is applied to calculate the weight function of each indicator as: $w_1 = 0.091$, $w_2 = 0.0878$, $w_3 = 0.0932$, $w_4 = 0.091$, $w_5 = 0.0901$, $w_6 = 0.092$, $w_7 = 0.0909$, $w_8 = 0.09294$, $w_9 = 0.0907$, $w_{10} = 0.09295$, $w_{11} = 0.0884$.

Thus, it can be seen that the application of grey correlation is to constitute two categories of judgment matrices, acquisition of weight functions of each indicator, and for realization of organic combination of grey correlation analysis and analytic hierarchy process, and resolution of weight function distribution problem of evaluation indicators when the evaluation indicators are large.

49.4 Result Analyses

(1) The Comparison and Analysis of Two Structure Methods

In order to compare the proposed two categories of judgment matrix structure method, right now the weight functions results that are calculated by different judgment matrices for the above case is conducted with tidy and sorting according to the importance of the indicators; the sorting results are shown in Table 49.1.

From Table 49.1, it can be clearly seen that each index important sequence results are complete same by calculation of the constructed type I judgment matrix and type II judgment matrix, it illustrates that it is workable to judge the matrix by grey correlation analysis method, both construction methods are valid and reasonable.

By observing the corresponding weight function of each indicator, it can be found weight function results' resolution ration acquired by type I judgment matrix is far more than the weight function results' resolution ration acquired by type II judgment matrix, and more clearly reflect the importance difference of indicators, and both theory and practice have proved the constructed type I judgment matrix can naturally meet the requirement of consistency, and save the tedious consistency test, enhance AHP operability, it is visibility, type I judgment matrix has more advantages than type II judgment matrix, By use of type I judgment matrix, it is not only more simple, convenient, and higher operability and resolution ration by used of type I judgment matrix to acquire the answer of weight function, but also has better the practical value. Therefore, when the number of indicators are more, it should be applied with type I judgment matrix to conduct evaluation of weight function distribution.

Table 49.1 Results compare of two structure method

Importance No.	I type judgment matrix		II type judgment matrix	
	Indicators No.	Weights	Indicators No.	Weights
1	3	0.104	3	0.09320
2	10	0.103	10	0.09295
3	8	0.102	8	0.09294
4	6	0.093	6	0.09200
5	1	0.091	1	0.09120
6	4	0.090	4	0.09100
7	7	0.089	7	0.09090
8	9	0.089	9	0.09070
9	5	0.085	5	0.09010
10	11	0.076	11	0.08840
11	2	0.073	2	0.08780

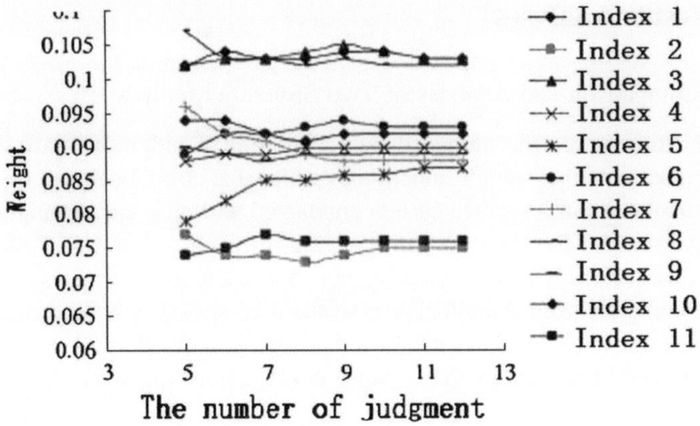

Fig. 49.2 Relation curves between index weight and amount of evaluation expert

(2) The Stability Analysis of Weight Allocation Results

When the subjective weight function is performed, the weight function distribution accuracy depends on the number of experts. The more amount the experts are, the more credibility the result is accordingly. In general, according to experience, the amount of selected experts is between the numbers of 5–11 [1, 2]. To analyze the stability of weights distribution outcomes, this paper refers to conduct the calculation, comparison, and analysis, it obtained when the type I judgment matrix is used to evaluate the weight functions, when the judgment experts amount is changed in 5–12, the corresponding weighing functions change conditions is acquired. As shown in Fig. 49.2.

As it can be seen from Fig. 49.2, with the amount of judgment expert increases, the change of each index weight change is generally stabilized. When evaluation experts amount reaches to ten people, each index weight has already basically change little, and weight distribution outcome becomes stable. This result is basically in line with the conclusions of the literature [3, 4]. Therefore, in the judgment, expert amounts can be selected with ten people.

49.5 Conclusions

This paper, based on the main problems of legal weights in analysis of traditional hierarchy, according to human–machine interface characteristics, proposes the index weight distribution method based on reliability coefficient and multiple index weight distribution method based on the grey correlation analytic hierarchy process. Based on the human–machine interface index weight distribution method of reliability coefficient, it is applied with grey theory to conduct revision on AHP, from the perspective of overall similarity structure expert judgment reliability coefficient;

it achieves the quantitative description towards cognitive characteristics; and solves the gatherings of multi-experts judgment information. The cases studies show that, this weight distribution method is not only clear and simple for their concept, but easy to operate; reducing the subjective factors influence on weight, and making the weight distribution more clear, and result more scientific and reliable. Based on multiple index weight distribution methods of grey correlation layer analysis, it is applied with angel structure I and II judgment matrix from the entire analysis viewpoint, it is to solve that pairwise judgment times is too much when index number is too much. The case studies show that type I judgment matrix is simple, convenient, strongly operable and high resolution; therefore, it is suitable for weight distribution problem when index amount is too much.

This work is supported by the opening foundation of the Science and Technology on Human Factors Engineering Laboratory, Chinese Astronaut Research and Training Center (grant no. SYFD140051810K).

References

1. Song G, Yang H (2000) Group decision making behavior analysis. Acad Explor 57(3):48–49
2. Dennis AR et al (1990) An experimental investigation of small, medium, and large groups in an EMS environment. IEEE Trans SMC 20(5):1049–1057
3. Zhang J (2005) A new method for ranking fuzzy complementary judgment matrix. Oper Res Manage 14(2):59–63
4. Jing H, Zhang L, Wen B (2009) Reallocation based on the product design quality right analytic hierarchy process. Northeast Univ J 30(5):712–715

Chapter 50
Basic Principles of Information System UI Design

Ruijie Yin, Bingjun Zhang, Meng Kang, Tao Li, Kang Chen and Yong Kang

Abstract Basic principles of information system UI design are user participation, user control, interface design, information feedback, rapid system response, help system, fault tolerance and security, and interface modification. Quality of user–system interaction interface is immediately related with computer input and output, user's operation and control, as well as the overall performance of the information system. Properly following the aforementioned principles including UIs are designed for users, user control, consistency in interface design, good information feedback, rapid system response, applicable help system, good error tolerance and security, repeated interface modification, in information system UI design, shall effectively improve user–system interaction.

Keywords Information system · UI · Principle of design

50.1 Introduction

Information system is a man–machine system. As the medium of user–system communication, information system UI plays a very important role in user–system interaction. Quality of user–system interaction interface is immediately related with computer input and output, user's operation and control, as well as the overall performance of the information system. In man–machine interface designing, system developer shall thoroughly analyze the demands of user's interactive property

R. Yin (✉) · B. Zhang · M. Kang · K. Chen · Y. Kang
Air Defence Forces Academy, Zhengzhou, China
e-mail: jerryin7908@163.com

T. Li
The Reserve Division of Antiaircraft Artillery, Shanghai, China

© Springer-Verlag Berlin Heidelberg 2015

S. Long and B.S. Dhillon (eds.), *Proceedings of the 15th International Conference on Man–Machine–Environment System Engineering*, Lecture Notes in Electrical Engineering 356, DOI 10.1007/978-3-662-48224-7_50

and task, select reasonable interaction pattern, distribute tasks between user and system scientifically, bring the advantages of both man and computer into full display, and develop optimal system UIs that are of high functionality and usability.

50.2 Basic Principles of Man–Machine Interface Design

50.2.1 User Participation

UIs are designed for users [1]. As UI color assortment, widget layout, semantic expression are fairly subjective and user-oriented, user participation in UI design shall effectively enhance its usability, and make man–machine interfaces more suitable for user's mind habit and interaction experiences. There are two ways of user participation: one is that a certain amount of users collaborate with system developers to complete interactive property and task analysis and make good preparations for man–machine interface prototypes in the stage of system design; the other is that a representative group of users are selected to test and assess man–machine interface prototypes, find out flaws or errors, and make necessary modifications if needed after prototypes are designed.

50.2.2 User Control

It is understood so as to make users believe that they are controlling the system other than being controlled by it.

1. Users should play an active role in man–machine interaction, and may be allowed to select ways and procedures to interact.
2. Different users may have different demands of man–machine interface, therefore, options should be available for users to change such elements as color and font according to their personal preferences.
3. Distribute interaction tasks and manage processes reasonably so that a certain operation can be done without loading the entire application module.

50.2.3 Consistency in Interface Design

Consistent interface allows users to effectively pass existing knowledge and experiences on to new tasks and easily gets oriented, and better concentrate on interactive tasks such as judgement, analysis, retrieval, classification, and so on. Consistency is primarily reflected in color, hotkey, operation mode, and widget.

50.2.3.1 Consistency in Color

Color is an auxiliary display of information, which arouses attention and incurs relevance. To avoid visual cognition difficulty, man–machine interface should include limited colors which meet the demand of interaction and accord with users' psychological cognition laws. Assortment of colors should be good-looking and fitting, and good for users to focus on.

50.2.3.2 Consistency in Hotkey

Hotkey is a specific key or a combination of particular keys which is corresponding to a menu command and functions equal to it. Scheme of hotkeys should be uniformly designed and allocated. Hotkeys should be allocated only for important and common functions; use CTRL+key combination or SHIFT+key combination, avoid using combinations of three or more than three keys, and keep away from system's existing hotkey combinations; do not allocate one hotkey to many an operation.

50.2.3.3 Consistency in Operation Mode

Consistency in operation mode helps users to get familiar with interfaces and predict the situation of the system, allows users to concentrate on specific interaction other than spending time and energy recognizing dissimilarities and being adapted to them. Consistency in operation mode is reflected in the consistency in data input and object operation [2]. Consistency in data input is understood as to abide by certain principles in data input, while object consistency means that there should be identical or similar property and operation mode for identical and similar objects in the system.

50.2.3.4 Consistency in Widget

Widget is a fundamental component of interface. There should be detailed regulations for the combination, layout, arrangement, and size of widgets. First, determine widget combination and man–machine interface widget package in accordance with system's interaction tasks. Second, arrangement and layout among widgets should accord with users' operation habits and interaction needs; interval should be arranged among widgets and between widgets and window to avoid covering and overlapping.

50.2.4 Good Information Feedback

System man–machine interface must give feedback to users' operation. Good feedback helps users confirm proceeded or proceeding operations [3]. Types of feedback are: morphological, grammatical, and semantic.

50.2.5 *Rapid System Response*

System man–machine interface should be quick in response. User may lose his head if no response comes 3 s after one operation.

50.2.6 *Applicable Help System*

A scientific and reasonable help system should be designed. Merely having a user's manual is not enough for an information system. It is hard for beginners to read through the manual; unexpected problems might be encountered during interaction while time does not allow users to consult the manual. Hence, there should be various helps such as context help, procedure help, and so on.

50.2.7 *Good Error Tolerance and Security*

User security measures should manage authority well and stop unauthorized access or hackers. Error tolerance allows users' mistaken operation and give audio or visual signals as reminders. Avoid system breakdown or other disastrous results due to users' mistaken operation. A certain data backup or recover function should be available. Man–machine interaction design should have protection function, be capable of preventing errors, and prevent system operation and information storage from being destroyed [4]. There should be necessary error information which is clear, easy to understand, and corresponding with the principle of consistency. The error information should include error location, error level, reason, and suggestions for correction.

50.2.8 *Repeated Interface Modification*

1. Interface modification is throughout UI development. Before designing, developers should modify and perfect the user property and man–machine interactive tasks analysis with the help from users; during interface programming, programmers should adopt users' suggestions and make necessary modifications according to interface demand analysis and design result; after completing interface prototypes, designers should have users' assess and verify the interface functionality and usability, find out possible problems and make pointed modifications by integrating software platform maintenance and modification [5].

2. In interface developing and testing, designers should not rely on users completely, but verify and modify by combining quantitative analysis and users' qualitative assessment.
3. Attention should be paid to the differentiation of user interface verification and system interface testing. User interface verification focuses on how to better meet users' demand, namely, how to improve interface usability.

50.3 Conclusion

Interactive interface is a bridge between users and system. Many users seldom consider the specific structure and information flow, but concentrate on user interface, namely, how to operate the system and complete man–machine interaction. The above-mentioned principles are helpful for information system development, particularly for aided decision-making system and information management system.

References

1. Ding Y (2000) Man-machine engineering science. Beijing Science and Technology University Press, Beijing
2. Microsoft Press (2001) Windows users experiences. Beijing University Press, Beijing
3. Microsoft Press (2000) Windows users interface design. University Press, Beijing
4. Han C (2002) Introduction to communication confrontation equipment interface operation mode. Electronic Engineering Academy, Hefei
5. Cheng J, Ni Y (1994) Man-machine interface design and develop tools. Electronic Industry Press, Beijing

Chapter 51
Human–Machine Interface Evaluation with Multiexpert Weighted Aggregation Methods

Chuan Wang, Jianguo Zhang, Wenhao Zhan, Xiaowei Yang, Qiao Wang and Fei Li

Abstract Cronbach given the information itself from digging out the size of the reliability of the expert, fully reflects the credibility of the information given to achieve a quantitative description of the characteristics of human cognition, human cognitive characteristics to solve the problem difficult to quantify; Meanwhile this paper Cronbach determine the index weight fully consider the human perception of gray and reduce the interference of subjective factors, the effective implementation of the multi-expert weight information gathered, to avoid the existence of the analysis of traditional AHP (analytic hierarchy process) to determine the weights of defects. Therefore, the use of indicators based on weight assignment of reliability coefficients determine the weight, weight assignment can become more clear and reasonable, the results of more scientific and reliable.

Keywords Human–machine interface evaluation · Multiexpert weights · Information gathering methods

To determine the analytic hierarchy process (AHP) multiple experts have gathered and weighted information to determine a method for human–machine interface

C. Wang (✉)
Naval University of Engineering, Wuhan 430033, China
e-mail: hg04381@163.com

C. Wang · J. Zhang
Naval Medical Research Institute, Shanghai 200433, China

W. Zhan
National Key Laboratory of Human Factors Engineering, China Astronauts Research and Training Center, Beijing 100094, China

X. Yang
91656 PLA Troops, Shanghai 200433, China

Q. Wang · F. Li
92730 PLA Troops, Sanya 572000, China

© Springer-Verlag Berlin Heidelberg 2015

S. Long and B.S. Dhillon (eds.), *Proceedings of the 15th International Conference on Man–Machine–Environment System Engineering*, Lecture Notes in Electrical Engineering 356, DOI 10.1007/978-3-662-48224-7_51

evaluation index weight assignments and ensure the science and objectivity of weight assignment outcomes. This paper aims to improve the AHP.

51.1 Index Weight Assignment Method for Reliability Coefficients

When there are <10 evaluation indicators, a group judgment method is conductive to reducing the influence of subjective factors. Ensuring the objectivity of weight assignment results in the application of the AHP method. There are generally two types of multiexpert information gathering methods. The first method is to equalize the experts' weights, directly performing hard multiplication or addition using geometric or arithmetic averages. Another method is to artificially and subjectively determine the weight of an expert's opinion according to the expert's cognition ability, influence, professional level and familiarity with evaluation problems. To some extent, the accuracy of evaluation results are influenced by these methods.

From the evaluation information itself, the credibility of information provided by the experts can be made more objective by considering the functions of each expert in the judging process and becoming more compliant with objective reality. Liu et al. [1] reflect experts' weight according to the degree of information deviation; they acquire the relatively stable synthesis weights by expert weighting and attributing the weight to an adaptive adjustment process. Liang [2] and Jing et al. [3] determine the credibility of expert weighting using a transfer matrix for comparison group decision-making, which is a relatively objective method for determining the experts' weight. Both methods, to a certain extent, solve the collective problems of multiexpert opinions. They essentially calculate experts' individual decisions and the group's decision deviation degree by distance and information incompleteness. Grey nature may affect the reliability of the results. In this case, the self-principle nature of judgment information and the comprehensive consideration given to various factors would play a basic role for the experts' opinions on collected information, in order to specifically reflect the all evaluation index weights.

In light of the problems outlined above and the vagueness and complexity of human–machine interface evaluation [4], this chapter puts forward a human–machine interface evaluation index weight assignment method based on credibility coefficient modification. This method, which is based on AHP for multiexpert opinion, fully considers the grey cognition influence of the evaluation results. It is applied with grey correlation analysis to handle incomplete information using an expert opinion credibility coefficient from the perspective of overall similarity, to determine the cognitive characteristics of quantitation, and to acquire a weight assignment for each component in compliance with the cognitive characteristics of a human–machine interface.

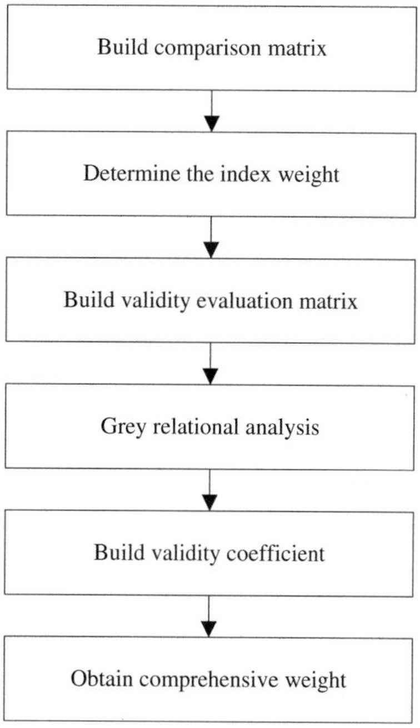

Fig. 51.1 Process of the weight assignment method based on reliability coefficients

51.2 Basic Principles of the Credibility Coefficient Index Weight Assignment Method

The AHP method can effectively integrate expert opinions and determine quantitative descriptions for an evaluation index weight. However, grey system theory focuses on information incompleteness in complicated systems to effectively deal with problems with grey information. The method of grey correlation analysis using grey system theory can determine the correlation degree between all comparison sequences from the perspective of overall similarity [5]. Based on this idea, grey correlation analysis can be used for judgment information to reflect experts' overall cognition degree, to determine the credibility matrix of information provided by experts, and to calculate an expert's credibility coefficient. In addition, it can be integrated in AHP to acquire an index weight from the modification and application of AHP. Based on the credibility coefficient, the process for the assignment of human–machine interface evaluation index weights is shown in Fig. 51.1.

51.2.1 Paired Comparison Judgment Matrix

Here, we assumed that m experts participate in n indicators for human–machine interface evaluation indicator weight assignment. The paired comparison judgment matrix provided by the K experts is $A_k(k = 1, 2 \ldots m)$.

$$A_k = (a_{ij}^k)_{n \times n} = \begin{bmatrix} a_{11}^k & a_{12}^k & \cdots & a_{1n}^k \\ a_{21}^k & a_{22}^k & \cdots & a_{2n}^k \\ \vdots & \vdots & & \vdots \\ a_{n1}^k & a_{n2}^k & \cdots & a_{nn}^k \end{bmatrix} \tag{51.1}$$

In this equation, a_{ij}^k refers to the relatively important degree judgment value of the i element and the j element at the same level.

51.2.2 Determination of Index Weight

The determination of index weight can be calculated by the maximum character root corresponding to eigenvectors in Eq. 51.1. For each evaluation, experts constructed a paired comparison judgment matrix A_k, which has a characteristic root λ_k and corresponding eigenvectors P_k:

$$A_k P_k = \lambda_k P_k \tag{51.2}$$

By AHP calculation of the weight's square root, it can be acquired with the largest characteristic root corresponding eigenvector P_{km} of judgment matrix A_k. The result is acquired by the normalization processing. That is to say that all index weight vector values are provided by the k expert, recorded as $W_k' = \left(w_{k1}', w_{k2}', \ldots, w_{kn}'\right)^T$. By doing so, the weight vector calculation must be performed with a consistency test. If the judgment matrix cannot meet the consistency requirements, the acquired weight vector is invalid. The process to determine the judgment matrix must continue until the requirement for consistency is met.

51.2.3 Structure of Credibility Evaluation Matrix

Assume that $B = \{B_1, B_2, \ldots B_m\}$ as an evaluation group set, where m is the number of evaluation experts, and $C = \{C_1, C_2, \ldots C_n\}$ is the evaluation index set. Different evaluation experts B_k for different evaluation indexes C_i give the index attributes of

$x_{ki} = w_{ki}, i = 1, 2, \ldots, n; k = 1, 2, \ldots, m$. Therefore, the credibility evaluation matrix is

$$X = \begin{bmatrix} x_{11} & x_{12} & \cdots & x_{1n} \\ x_{21} & x_{22} & \cdots & x_{2n} \\ \vdots & \vdots & & \vdots \\ x_{m1} & x_{m2} & \cdots & x_{mm} \end{bmatrix} \tag{51.3}$$

51.2.4 Determination of Reference Sequence

To better reflect the discrete degree of judgment information given by the experts, the experts' average value of evaluation information is chosen for each judgment index as a reference sequence X_0, which is recorded as follows:

$$X_0 = (x_{01}, x_{02}, \ldots, x_{0n}) \tag{51.4}$$

Here, $x_{0i} = \frac{1}{m}\sum_{k=1}^{m} x_{ki}$

At the same time, normative processing of the evaluation information is required to reduce interference from random factors and solve the comparability problem between the judgment matrixes. The average value method is as follows:

$$y_{ki} = x_{ki} / \sum_{k=1}^{m} x_{ki} \tag{51.5}$$

51.2.5 Calculation of Grey Correlation Degree

According to grey system theory, each index for important experience judgment values and reference value for grey correlation coefficients by each expert can be shown as follows:

$$\varepsilon_{ki} = \frac{\min\limits_{1 \le k \le m} \min\limits_{1 < i < n} |y_{0i} - y_{ki}| + \rho \max\limits_{1 \le k \le m} \max\limits_{1 < i < n} |y_{0i} - y_{ki}|}{|y_{0i} - y_{ki}| + \rho \max\limits_{1 \le k \le m} \max\limits_{1 < i < n} |y_{0i} - y_{ki}|} \tag{51.6}$$

As a formula, $\rho \in [0, 1]$ is the distinguishing coefficient, generally taken as 0.5, and y_{0i} is the i index reference value after normalization. The correlation coefficient matrix is therefore as follows:

$$\varepsilon = (\varepsilon_{ki})_{m\times n} = \begin{bmatrix} \varepsilon_{11} & \varepsilon_{12} & \cdots & \varepsilon_{1n} \\ \varepsilon_{21} & \varepsilon_{22} & \cdots & \varepsilon_{2n} \\ \vdots & \vdots & \vdots & \vdots \\ \varepsilon_{m1} & \varepsilon_{m2} & \cdots & \varepsilon_{mn} \end{bmatrix} \tag{51.7}$$

To easily compare information, we can describe the degree of correlation between the comparison sequence and reference sequence:

$$r_{0k} = \frac{1}{n}\sum_{i=1}^{n} \varepsilon_{ki} \tag{51.8}$$

Here, the correlation degree is between the information proposed by different experts and reference information. In short, it is the average value of the correlation degree for the information: the larger the value is, the closer is the expert opinion to a reference average value, and the discrete degree of information is smaller.

51.2.6 *Structure of Credibility Coefficient*

The information credibility degree provided by experts is referred to as the expert's credibility. The expert's credibility reflects the expert's function in the judgment process. Credibility coefficients are a quantitative display of the information credibility degree, referred to as δ.

As is known from the meaning of the correlation degree, the size of the correlation degree reflects the discrete degree of evaluation information, and the information discrete degree provided by each expert reflects the experts' recognition of information. The stronger this correlation is, the smaller is the information discrete degree, the stronger is the experts' recognition consistency and information credibility, and the greater is the credibility degree of evaluation information provided by experts. Therefore, the correlation can display the information credibility degree given by experts. The information credibility coefficient provided and the k judges' credibility coefficient can be displayed as follows:

$$\delta_k = r_{0k} / \sum_{k=1}^{m} r_{0k} \tag{51.9}$$

The credibility coefficient is acquired by calculation of evaluation information and reference information correlation degree of all experts: the greater the correlation is, the greater are the information effective degree proposed by the experts, the credibility, and the corresponding credibility coefficient. The size of credibility coefficient reflects the experts' functions and cognition consistency degree in the process of evaluation process, and quantitatively displays the people's recognition.

The credibility coefficient matrix for the m experts' credibility coefficient can be shown as follows:

$$d = (\delta_k)_{1\times m} = [\delta_1, \delta_2, \ldots, \delta_m]. \tag{51.10}$$

51.2.7 Acquisition of Comprehensive Weights

The experts' function is not fully considered by the index weight of AHP in the evaluation process, and the grey area of people's cognition is also overlooked. Therefore, the acquired indicator weight is characterized by a certain one-sidedness. However, the application of grey system theory structure to the credibility coefficient not only reflects the validity and credibility of expert information, but it also reflects each expert's function in the process of evaluation. Even more, it can solve such problems as insufficiency and incompleteness. Therefore, it is feasible to acquire an indicator weight for use in the credibility coefficient analytical hierarchy process. The revised comprehensive weight is as follows:

$$W = d * X \tag{51.11}$$

Each index of a comprehensive weight vector is $W = (w_1, w_2, \ldots, w_n)$.

51.3 Experiment Verification

To verify the feasibility evaluation indicator weight assignment method for the credibility coefficient, it can be applied to the method provided in this chapter to calculate rotation for a chosen operator's objective weight of evaluation indexes. The rotating chosen operator includes knob's height, width, length, and turning angle displacement—a total of four objective evaluation indexes.

First, the four experts are invited to develop a paired comparison judgment matrix for the four indicators:

$$A_1 = \begin{bmatrix} 1 & 5 & 3 & 3 \\ 1/5 & 1 & 1/3 & 1/3 \\ 1/3 & 3 & 1 & 1/3 \\ 1/3 & 3 & 3 & 1 \end{bmatrix} \quad A_2 = \begin{bmatrix} 1 & 5 & 4 & 2 \\ 1/5 & 1 & 1/3 & 1/4 \\ 1/4 & 3 & 1 & 1/4 \\ 1/2 & 4 & 4 & 1 \end{bmatrix}$$

$$A_3 = \begin{bmatrix} 1 & 4 & 3 & 2 \\ 1/4 & 1 & 1/3 & 1/3 \\ 1/3 & 3 & 1 & 1/4 \\ 1/2 & 3 & 4 & 1 \end{bmatrix} \quad A_4 = \begin{bmatrix} 1 & 1/5 & 1/4 & 1/2 \\ 5 & 1 & 3 & 3 \\ 4 & 1/3 & 1 & 3 \\ 2 & 1/3 & 1/3 & 1 \end{bmatrix}$$

According to Formula (51.2), the indicator weight vectors can be obtained respectively by four different experts:

$$W_1' = (0.513, 0.075, 0.150, 0.260)^T \quad W_2' = (0.484, 0.067, 0.125, 0.323)^T$$
$$W_3' = (0.452, 0.083, 0.144, 0.319)^T \quad W_4' = (0.078, 0.509, 0.278, 0.135)^T$$

The consistency verification can acquire $\mathrm{CR}_1 = 0.087 < 0.1$, $\mathrm{CR}_2 = 0.047 < 0.1$, $\mathrm{CR}_3 = 0.087 < 0.1$, and $\mathrm{CR}_4 = 0.035 < 0.1$. They all satisfy the requirement of consistency, and the weights acquired above are displayed as valid.

Then, the structure credibility evaluation matrix is as follows:

$$X = \begin{bmatrix} 0.513 & 0.075 & 0.150 & 0.260 \\ 0.484 & 0.067 & 0.125 & 0.323 \\ 0.452 & 0.083 & 0.144 & 0.319 \\ 0.078 & 0.509 & 0.278 & 0.135 \end{bmatrix}$$

The credibility evaluation matrix is applied. The average value generation method is performed with normalization and acquired as follows:

$$Y = \begin{bmatrix} 0.336 & 0.102 & 0.216 & 0.251 \\ 0.317 & 0.019 & 0.180 & 0.311 \\ 0.296 & 0.113 & 0.202 & 0.308 \\ 0.051 & 0.693 & 0.401 & 0.130 \end{bmatrix}$$

The acquired reference sequence is $Y_0 = (0.250, 0.250, 0.250, 0.250)$. Grey correlation analysis is applied to obtain the following grey correlation coefficient matrix:

$$\varepsilon = \begin{bmatrix} 0.724 & 0.602 & 0.871 & 1 \\ 0.771 & 0.585 & 0.763 & 0.787 \\ 0.832 & 0.621 & 0.826 & 0.796 \\ 0.529 & 0.335 & 0.597 & 0.652 \end{bmatrix}$$

The expert opinion and reference information correlation degree proposed by the four experts are, respectively, $r_{01} = 0.799, r_{02} = 0.727, r_{03} = 0.769, r_{04} = 0.528$. By Formula (51.9), it can be determined that each expert's credibility coefficients are, respectively, $\delta_1 = 0.283, \delta_2 = 0.258, \delta_3 = 0.272, \delta_4 = 0.187$. The index weight of traditional AHP is modified by the use of the credibility efficiency. It is determined with the revised comprehensive weight vector to be $W = (0.306, 0.156, 0.165, 0.269)$.

It is evident that the evaluation index weight assignment method, on the basis of the credibility system, is clear in physical meaning, simple and easy to understand, and a very workable solution to solve the compound problem of multiexpert weighted information.

51.4 Results Analysis

The calculated weight vector above is applied to the results after credibility coefficient modification. If the modification is not performed, the comprehensive weight vector using a conventional method of geometric average integration is $W = (0.292, 0.184, 0.174, 0.259)$. The weight assignment results determined by the two methods are contrasted in Fig. 51.2.

As can be seen from Fig. 51.2, the results are evidently different for the weights acquired from credibility coefficient modification and the average value method. After a comparison of weights determined by the conventional method and modification, each index weight deviation was calculated to be 4.57, 17.9, 5.45, and 3.71 %. The main reason for the occurrence of this phenomenon is that the conventional method ignores the experts' function and influence on the weight assignment results in its evaluation. Therefore, when experts make different determinations according to their backgrounds, cognition level, thought mode, and others, the method has to make modification to the weight. Otherwise, the results would appear to deviate: the greater the deviation degree proposed by each expert, the larger the deviation becomes.

When the information provided by judges tends toward consistency, the credibility coefficient of all experts should be consistent. At this moment, the weight value acquired by the application of credibility coefficient modification has to be consistent with the weight acquired by the regular geometric average assembly method. To illustrate this point, it is found in the analysis that the first three experts' judgments are basically consistent. For this purpose, in view of the first three experts' evaluation results, the methods proposed by this paper can be used to recalculate the weight, for which the results are respectively $W = (0.484, 0.074, 0.139, 0.297)$, $W_{常} = (0.483, 0.075, 0.140, 0.301)$ Both results contrasting conditions, as shown in Fig. 51.3.

As can be seen in Fig. 51.3, the weight results calculated by the two methods are basically similar; then, relative to the weight assignment method based on the credibility coefficient, all calculated index weight deviations using the regular method are, respectively. 0.5, 1.35, 0.72, and 1.36 %. It is evident that both deviation values are small. This just illustrates that the credibility coefficient value

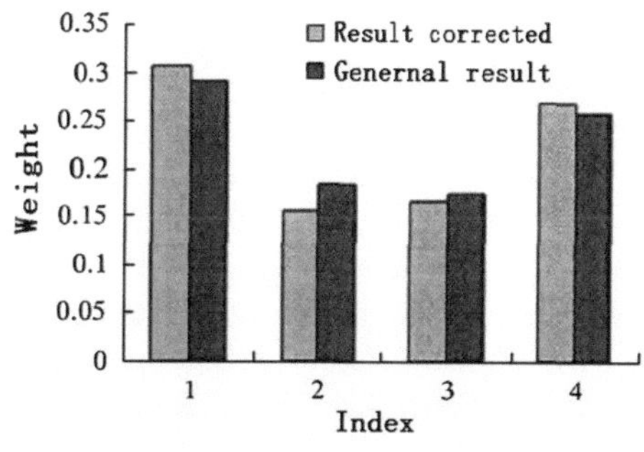

Fig. 51.2 Weight assignment results using two methods

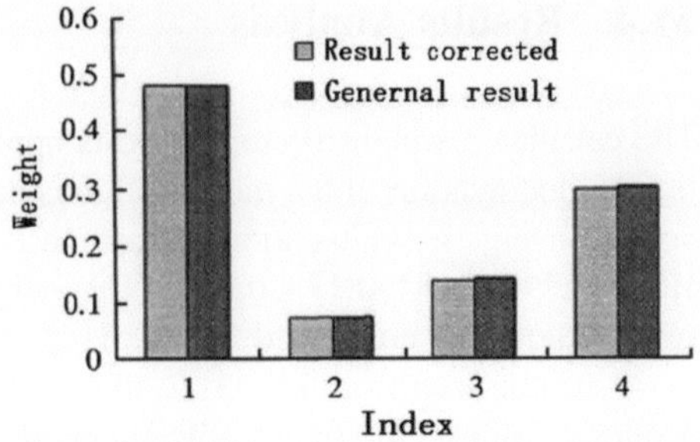

Fig. 51.3 Weight assignment results using two methods

depends on the discrete degree of information provided by experts. When the information provided by experts is highly consistent, each expert's knowledge of the problem tends to be consistent, all credibility coefficients are basically equivalent, and they are naturally acquired with results from the regular method.

However, when the real judgment is performed, because of differences in the judge's thinking, professional background, knowledge, and others, it is normal to obtain different judgments and very hard to provide completely consistent judgments. Therefore, weighting the opinions of many experts, in most cases, especially when the experts' opinions have a large divergence, is indispensable to performing the weight modifications.

Also, the above examples indicate that the extent of credibility coefficient would make the modification when it is different for the information provided by the evaluation experts, and the information itself would reflect its extent of credibility. Thus, the result would be characterized by strong objectivity. The advantage of credibility coefficient modified index weight is the introduction of grey theory. It reflects the credibility of information provided by experts and provides a quantitative description of information credibility. As the information changes, the adjustment also has change accordingly; furthermore, it ensures that the weight assignment results are correct and objective. Thus, when the judges' determinations are discrepant, the credibility coefficient has to be modified to acquire the correct weight assignment results.

51.5 Conclusion

In summary, the credibility coefficient aims to determine the extent of credibility of experts' opinions from the information itself, sufficiently reflect all information credibility, determine a quantitative description of experts' cognitive characteristics, and provides a solution to the puzzle of quantitation of cognition. When the index weight is ascertained using credibility coefficient, it should fully consider the human's grey cognition, reduce the interference of objective factors, effectively achieve the gathering of weight information from the different exports and prevent the conventional AHP defect from affecting weight determination. Therefore, the application of index weight assignment method is based on a credibility coefficient to determine weight; it can make weight assignment more clear and reasonable, as well as make the results more scientific and reliable.

This work is supported by the opening foundation of the Science and Technology on Human Factors Engineering Laboratory, Chinese Astronaut Research and Training Center (grant no. SYFD140051810K).

References

1. Liu Y, Xu D, Jiang Y (2007) Multi-attribute group decision making adaptive adjustment weights. Syst Eng Electron 29(1):45–48
2. Liang L, Xiong L, Wang G (2005) A new group decision-making to determine the objective weight method. Syst Eng Electron 27(4):652–655
3. Jing H, Zhang L, Wen W (2009) Reallocation based on the product design quality right Analytic Hierarchy Process. Northeast Univ J 30(5):712–715
4. Zhang J (2005) A new method for ranking fuzzy complementary judgment matrix. Oper Res Manag 14(2):59–63
5. Song G, Yang H (2000) Group decision making behavior analysis. Acad Explor 57(3):48–49

Chapter 52
An Analysis of Man–Machine Interface Design Principles of Military Hardware

Zhibing Pang, Honglei Li, Liang Yu, Hong He and Mingming Li

Abstract This paper aims to study the principles of man–machine interface design which are expected to give significant theoretical guidance on humanized man–machine interface of the military hardware. It is an important constituent of the military hardware design and an essential aspect for improvement of service life of the military hardware. It adopts man–machine–environment system engineering theory and practical experience in combination of definition, classification and composition of man–machine interface to make this research a creative approach by integration of theory and practice. In accordance with the general requirements of humanized design, the man–machine interface design should follow man-oriented principle, safety principle, high effectiveness principle, economic efficiency principle, comfortableness principle and unity principle. Thus, the concrete design principles proposed in this paper are of great significance to 'two adaptation approaches' and the whole service life of the military hardware, and further promote the development of the military hardware.

Keywords The military hardware · Man–machine interface · Design principles

52.1 Introduction

The military hardware are the important material bases of combat power of armed forces. Recently, with the further development of new RMA worldwide, a higher demand is imposed on the design and manufacture of the military hardware and the status of the man–machine interface design becomes much highlighted. Any man–machine system has an interface where man and machine act upon each other,

Z. Pang (✉) · H. Li · H. He · M. Li
Air Defence Forces Academy, Zhengzhou 450052, China
e-mail: fkb_mmes@163.com

L. Yu
Naval Aeronautical Engineering Institute, Yantai 264001, China

© Springer-Verlag Berlin Heidelberg 2015

S. Long and B.S. Dhillon (eds.), *Proceedings of the 15th International Conference on Man–Machine–Environment System Engineering*, Lecture Notes in Electrical Engineering 356, DOI 10.1007/978-3-662-48224-7_52

which is termed man–machine interface [1]. In terms of a friendly man–machine interface of the military hardware, four aspects should be taken into consideration: (1) coincide with the physiological need of human being to the maximum; (2) satisfy the psychological need of users to the full extent; (3) give sufficient consideration to the need of battlefield environment; (4) accord with the need of fierce combat [2]. The friendly interface may facilitate operation by overcoming the terrible external environment and decreasing the wrong operation so that the weapons and equipment may display maximum combat efficiency and improve the combat power. With the deepening of man–machine–environment system engineering theory, the man–machine interface of the military hardware is studied in a more scientific way.

52.2 Man-Machine Interface and Composition

52.2.1 *Definition of Man-Machine Interface*

Since the invention and use of tools, the interface has become a part of human culture. In man–machine system, the man–machine interface refers to the space where man and machine act upon each other. Simply, all the domains where man and machine may conduct information exchange are termed man–machine interfaces.

52.2.2 *Classification of Man-Machine Interface Design*

The man–machine interface design can be classified into three kinds: functional design interface, emotional design interface and environmental design interface. The design of an interface is an organic combination system based on functional interface, pivoted on emotional interface, and premised on environmental interface [3].

52.2.3 *Composition of Man-Machine Interface*

The man–machine interface of the military hardware is the convergence point between man and military hardware. It generally includes a system terminal (display), an operation controller and a user with initiative [4].

Display presents to the user all information about the military hardware and the target so that the user may fully acquire the status of the military hardware and the target. The information is presented naturally or manually.

Controller is manipulated to transmit the processed information to the military hardware to implement the user's intention. The controller is the device used by the user to change the state of the military hardware. It is an important part of man–machine interface which transfers the output command of the user into the input signal of the military hardware.

User plays the initiative in the man–machine interface of the military hardware and combines display and controller in an organic way. The user acquires information from three channels: eye, ear and sense organ of displacement, i.e. sense of touch, sense of hearing and sense of kinesthesis. The sense of kinesthesis is used the most, and the sense of hearing takes the second place. Ergonomics demands a correct distribution of the information to proper channels of senses of human being and then ensures the information flow of the man–machine passing through the man–machine interface successfully.

Simply speaking, information of the military hardware acts upon the user and the machine–man transmission is accomplished. Moreover, the user receives the information about the military hardware through sense of sight, sense of hearing and sense of touch, and processes it and makes decision in the brain and then gives a response. Thus man–machine transmission is accomplished.

52.3 Design Principles of Man-Machine Interface of the Military Hardware

The man–machine interface may affect the user's physical and psychological and ideological activities as well as his working efficiency. The complexity of man–machine interface design lies in the user. That is the interface is assessed by the user. In order to attain the perfect integration of man and machine, the design of man–machine interface should abide by the following six principles [5].

52.3.1 Man-Oriented Principle

Man-oriented principle means the man–machine interface is designed in consideration of human physiology and psychology. The design is oriented towards human beings. The human being is taken as the starting point of design as well as its destination. The man–machine interface of the military hardware interacting with the user should even more fully conform to the man-oriented principle.

The man–machine interface of the military hardware is the interactive hub of information between the user and the military hardware. It may directly affect the working performance of the user. Only in consideration of human factor and man–machine interactive principle in design, the concept of machine adapting to man and environment can be realized, and maximum coordination between user and military hardware can be achieved. The user may thus give full play to his working capability and take advantage of the military hardware. Furthermore, no matter how unreasonable the interface is designed in a pressed, rotated, held or pinched way, it may possibly cause occupational disease or other physical hurt. Therefore, the man–machine interface should be designed in accordance with anatomy, physiological structure of the user, ergonomics and man–machine–environment system engineering theory so that the user can be brought into full play.

52.3.2 Safety Principle

Safety principle means the man–machine interface design should satisfy users' safety requirements in the course of operation. The safety of the military hardware is the fundamental requirement of operation as well as the important goal of man–machine–environment system engineering theoretical research. The factors which affect safety are very complicated. It may be of the external environment, the military hardware or the user. The safety principle of man–machine interface of the military hardware can be studied in the following four aspects. First, the man–machine interface should not cause physiological hurt to the user. The improper man-machine interface design may cause physiological hurt to the user which should be avoided by the designer. Second, the man–machine interface should have strong anti-error measures. The user is not a machine. It is unavoidable to make mistakes in his judgement and operation. Thus, it should have a fault-tolerant and fault-preventive properties. Third, the man–machine interface should overcome the harm caused by the complicated environment. The environment may exert influence on the safety of the military hardware. In order to make the military hardware work normally in the complicated environment, the man-machine interface should overcome the effects caused by low visibility, high temperature, low air quality, vibration, etc. Fourth, the man–machine interface is capable of keeping the malfunctioning hardware working continuously. The trouble for the military hardware comes inevitably. The designer should endeavour to improve the reliability of the military hardware and reduce the troubles from perspective of design and technology.

52.3.3 High Effectiveness Principle

High effectiveness principle means the information between user and military hardware should be transmitted in a concise and a rapid way. As far as the man–machine interface is concerned, the high effectiveness principle is reflected in the following three aspects: first, the display subsystem should show the most needed information to the user. The traditional display subsystem can only provide limited information which gives high working load to the user. How the designer makes effective use of the limited space of the display subsystem and offers more information to the user is the primary problem to be considered. The display subsystem may provide the most needed information to the user according to the different working stages. Second, the display subsystem should provide a friendly display interface. The display subsystem should provide the most direct and vivid information such as standardized image, simulated digital information and Chinesized characters so that the user may obtain the information rapidly and accurately. Third, the user should effectively control individual subsystem of the military hardware. The military hardware has a lot of switches, buttons and handles which are used frequently. In order to make them to be used conveniently, the designer should give a reasonable layout according to their functions.

52.3.4 Economic Efficiency Principle

Economic efficiency principle means the design of the man-machine interface should comply with the reality. That is the design is made in consideration of the present technical and economic strength. The man–machine interface should abide by the economic efficiency principle. It is no need to overemphasize the high efficiency of the man–machine interface. The reasonable distribution of functions of man and machine is made on a condition that the investment is permitted. At the same time, the function cost must be analysed if function distribution is conducted so that the perfect systematic function is attained with the minimum investment. The design of the friendly man–machine interface is intended for its application. Thus, based on the actual level of science and technology and weapons and facilities, man and machine are made to match each other in function so as to achieve optimization, high-efficiency, reliability and safety in unity.

52.3.5 Comfortableness Principle

The comfortableness principle means the man–machine interface should provide a comfortable operating environment to the maximum. The modern technology tends to be man-oriented. The design of modern military hardware takes the human feeling into consideration to an increasing extent. The comfortableness of man–machine interface is not merely a problem of sensation. It also affects the user's psychological and physiological states. It relates directly to the full capability of the user. The comfortableness principle can be analysed in the following three aspects. The first is visual factor. We have a kind of experience that the harmonious colour and image may bring about a peaceful and pleasant feeling. The gloomy and disorderly environment leaves people an anxious and uneasy feeling. The second is audio factor. The strong noise makes people irritable and fatigued. The noise of the military hardware in some cases is inevitable. How to reduce the noise suffered by the user is a problem. The corresponding approaches have to be taken towards noises from different sources. The third is touch. The touch sensation given to the user by the hardware interface is felt by the direct contact between user and machine. The material and touch of switches, buttons and handwheels, etc. may directly affect the user's feeling, working efficiency and anti-fatigue duration.

52.3.6 *Unity Principle*

The unity principle means a full and thorough consideration should be taken in design of man–machine interface so as to achieve the optimum match between man and machine. The design of man–machine interface is on purpose to attain such effect that one may play to one's strong points and can supplement to the other's weak points in a functional distribution between user and hardware. Thus, the whole system is made safe, high-efficient, economic, comfortable and reliable. The high-efficient interface may reduce the working load of the user. The user will be released from fatigue and works in a pleasant state with low psychological pressure, where he naturally feels comfortable. Therefore, the designer of man–machine interface should keep the overall situation in mind and take the whole into consideration.

52.4 Conclusion

The study of the principles of man–machine interface design may realize 'three kinds of serving', that is, serving the scientific demonstration for industrial department, serving the standardized instruction of military institutions and serving the scientific training of troops [6]. The man–machine interface has undergone several stages of machine adapting to man and environment, man adapting to machine and environment, and mutual adaptation of man and machine in the process of man–machine–environment system research [7]. Now, it has moved to the stage of coordination among man, machine and environment. Thus, only by conforming to man-oriented principle, safety principle, high effectiveness principle, economic efficiency principle, comfortableness principle and unity principle, a perfect and harmonious man–machine interface can be obtained.

References

1. Pang Z (1999) Man-machine-environment system engineering of Air Defense Forces. Air Defense Forces Academy, Zhengzhou, pp 243
2. Long Sh (1997) A study on the relationship between staff's quality and his reliability. Space Med Med Eng (1)

3. Ye L (1997) Mathematical approach to man-machine interface. In: The 3rd man-machine-environment system engineering research proceedings. Beijing Science and Technology Press, Beijing
4. Luo Sh (2002) Design of man-machine interface. China Machine Press, Beijing
5. Yu L (2009) Assessment of man-machine interface design of operators' cabinet of TOP-M1 combat vehicle. Air Defense Forces Command Academy, Zhengzhou, pp 60
6. Li Ch, Pang Z, Ji Z (2014) On the eye movement technique-based HMI work efficiency evaluation system. In: Proceedings of the 14th international conference on man-machine-environment system engineering. Springer, Germany, pp 231–239
7. Pang Z, Li H, Niu J (2014) Practice of "two adaptation approaches" in the whole service life of the military hardware. In: Proceedings of the 5th academic conference on PLA military hardware theories, National Defense Industry Press, Beijing. pp 826–829

Chapter 53
Analysis and Construction of Evaluation Index Screening Methods Based on Comprehensive Multifactor Methods

Chuan Wang, Jianguo Zhang, Chunhui Wang, Xiaowei Yang and Fei Li

Abstract Based on a review and analysis of evaluation screening methods to determine the relevance, importance, and effectiveness of indicators, evaluation screening algorithms based on a comprehensive multifactor methods were constructed using gray relational clustering, vague analysis, and gray correlation analysis. The algorithm takes multiple factors into account, not only retaining the index system of the more important indicators but also allowing each index to be independent, so that both can be used in the overall index system without redundancy. On this basis, further analysis by validation ensures that the screening results are scientific and effective, providing new methods and ideas for the screening evaluation of complex systems. In addition, gray correlation analysis can be applied to evaluate the effectiveness of the overall approach, the degree of the curve in similar geometry to reflect the size of the effectiveness indicators, and the effectiveness of the inspection conducted. Exploratory research can broaden the scope of application of gray correlation analysis.

Keywords Human–machine interface · Evaluation indexes · Screening methods

C. Wang (✉)
Naval University of Engineering, Wuhan 430033, China
e-mail: hg04381@163.com

C. Wang · J. Zhang
Naval Medical Research Institute, Shanghai 200433, China

C. Wang
National Key Laboratory of Human Factors Engineering,
China Astronauts Research and Training Center, Beijing 100094, China

X. Yang
91656 PLA Troops, Shanghai 200433, China

F. Li
92730 PLA Troops, Sanya 572000, China

© Springer-Verlag Berlin Heidelberg 2015

S. Long and B.S. Dhillon (eds.), *Proceedings of the 15th International Conference on Man–Machine–Environment System Engineering*, Lecture Notes in Electrical Engineering 356, DOI 10.1007/978-3-662-48224-7_53

A scientific and reasonable evaluation index system is a prerequisite and basis for an evaluation. It can obtain accurate and objective evaluation results. At present, many studies aimed to examine evaluation index systems [1–3], but most are focused on a specific research area. Evaluation index systems constitute a variety of documents and resources. Analysis and integration often find different problems in various aspects, such as a lack of scientific theory and method as guidance, a too-large correlation between indicators, an unreasonable index system structure, lower effectiveness, poor maneuverability, and other problems. Therefore, a scientific and effective index screening method has significant importance in the structure of an evaluation index system [4].

53.1 Evaluation Index Screening Principle

A scientific, reasonable, and workable evaluation index system for evaluation index screening should be based on scientific theory. In addition, the correct guiding principles must be in place. Usually, the screening of an evaluation index should be in compliance with the following principles [5–7]:

(1) Comprehensive and representative
(2) Having a clear objective
(3) Scientific and objective
(4) Hierarchical and Independent
(5) Using a combination of qualitative and quantitative Indexes
(6) Maneuverable

53.1.1 Overview and Analysis

At present, the commonly used quantitative analysis methods for evaluation index screening are mainly classified as follows:

(1) Analytic hierarchy process (AHP)
(2) Information entropy method
(3) Principal component analysis (PCA)
(4) Clustering method
(5) Grey correlation analysis method
(6) Neural network method
(7) Rough set attribute reduction method

Many scholars have performed a lot work on evaluation index screening methods, obtaining some practical methods and solving some practical problems. However, some problems exist, as embodied in the following aspects:

(1) Most evaluation index are screened based on a combination of qualitative analysis and quantitative analysis. However, it is hard to consider the correlation between and importance of different indexes using quantitative analysis screening, which is often partially focused on the aspects of a single factor. Only when the screening of an evaluation index is fully considered with regard to all factors of the overall aspects can it become a reasonable and effective evaluation index system.
(2) The studies focus less on the effective testing of an evaluation index system. The same evaluation can be established with many evaluation index systems from different viewpoints. Which system truly reflects the evaluation target? How does it find a quantitative description of the effectiveness of the establishment of an index system? How does it perform an effective test of the evaluation index? These problems have important significance in establishing a reasonable index system. However, at present, relatively few have studied the effectiveness tests for an evaluation index system. The existing literature [8] adopts a statistical method to perform an effective analysis of the qualitative index system; thus, scientific and applicable methods are required for further discussion.

In summary, the establishment of an evaluation index system must be based on the specific evaluation problem, with overall consideration of many factors and discussion of appropriate evaluation index selection methods.

53.2 Construction of an Evaluation Index Screening Method

53.2.1 Multifactor Comprehensive Algorithm

As discussed, evaluation accuracy and objectivity should be developed on the basis of a scientific and reasonable evaluation index system. However, many problems and shortcomings remain in the establishment of an evaluation index system, especially for large complex systems such as human–machine interfaces, because its internal factors have a more complex relationship between factors; however, it also increases the complexity and difficulty of the construction of an evaluation system. Thus, in order to ensure that the established evaluation index system is more comprehensive instead of redundant, as well as truly reflecting the evaluation target, factors such as index correlation, importance, effectiveness, and other aspects have to be comprehensively considered when developing a scientific evaluation index system, as described in the following definitions:

- *Index correlation* refers to the correlation degree between various indicators, and the indicators' relative extent can be described by their relevancy. The greater the relevancy is among the indexes, the worse the independence of

indexes is and the greater the degree of redundancy is among indexes. On the contrary, the better the independence between the indexes is, the less the redundancy between indicators is; the index system would be better if it truly reflected the evaluation targets.
- *Index importance* refers to the extent of the important between all indexes. The greater the index importance is, the more influence the index exerts on the target as indicated; on the contrary, it would have little impact on the evaluation target.
- *Index effectiveness* refers to the extent of index effectiveness when the judge adopts an index to evaluate the evaluation target. The index effectiveness extent is measured by the validity. The greater the validity is, the higher its valuation effectiveness is when using the index to evaluate the target as indicated. On the contrary, the effectiveness is lower. In the evaluation index system, the sum of the validity of all indexes reflects the validity of the overall index system. In accordance with the index system validity, it can test the validity of the establishment of an evaluation index system.

According to the definitions above, the development of a reasonable and scientific evaluation index system has to minimize the relevancy between the indexes as much as possible. In addition, it should perform an analysis and determination to accept or reject it according to its specific meaning for an index with a large correlation. The single index has a less relative importance, but is still required to be screened out in order to guarantee the simplicity of index system. Finally, validity should be considered when conducting an effectiveness analysis to determine if an index system reasonable or not according to the validation results. Based on this viewpoint for a comprehensive consideration of the index's relevance, importance, and validity, we present a multifactor comprehensive algorithm for evaluation index screening.

53.2.2 The Basic Principle of the Multifactor Comprehensive Algorithm

The basic principle of multi-factor algorithm is developed on the basis of domestic and overseas standards, and evaluation index of the earlier literature establishment. Meanwhile, the earlier developed evaluation index system is also applied to the evaluation of target information. According to evaluation data, the correlation is calculated between the indexes using a grey correlation clustering method. At the same time, a vague analysis is applied to calculate each index's importance, according to all index correlations. The index importance acquired from the calculation is used to conduct the clustering and acceptance (or rejection). Finally, grey correlation analysis is applied to perform the index validity test. The algorithm

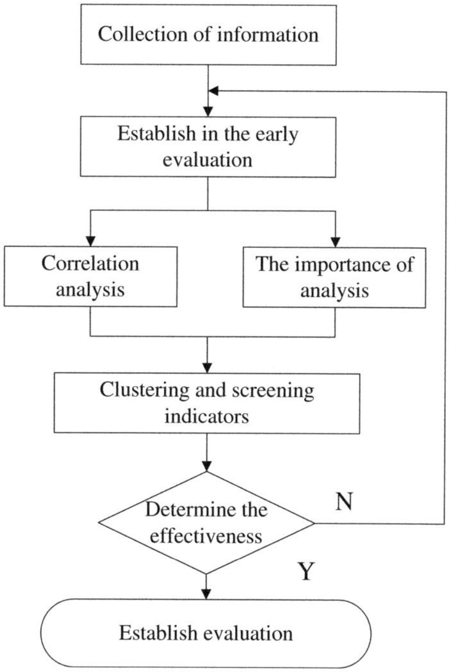

Fig. 53.1 flow chart of a multifactor synthetic algorithm

process is shown in Fig. 53.1. Based on the multifactor comprehensive selection algorithm, the acquired evaluation system is guaranteed be a comprehensive, instead of redundant, system. We highlight the key points and their independence from each other.

(1) Analysis of Index Correlation

There are a number of methods for index correlation analysis. Most are applied using a mathematical statistics method or clustering analysis and other methods, which require a lot of supporting statistical data. Thus, the calculation volume is huge, and some methods require a data sample in accordance with a certain linear relation. Grey correlation analysis makes up for the insufficiency that is caused by using the above-mentioned method to complete the relevant analysis. It has no strict requirements for data sample volume and characters, with easy calculations, and it is also applied for quantitative and qualitative indicators.

Grey correlation analysis reflects the close degree of all curves' association according to the degree of similarity in the geometric shape of comparative sequence curves. The more similar each comparative sequence curve's geometric shape is, the closer the degree of correlation among the relevant sequences becomes, and vice versa. Based on this thought, it can analyze the relevancy between all indexes according to the degree of correlation for all indexes. The more similar the shape is among the curves, the greater is the degree of correlation among the relevant indexes and the higher the correlation for these two indexes becomes.

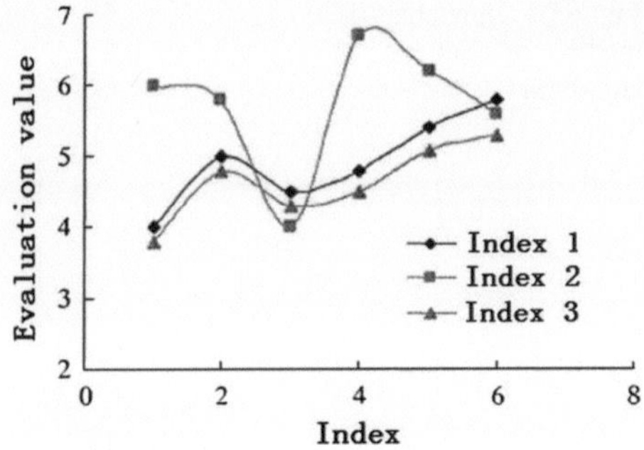

Fig. 53.2 Comparison of relation degrees between different indices

As shown in Fig. 53.2, the geometric shape of the curve between index 1 and index 3 has a much higher degree of similarity, which indicates that the correlation degree between index 1 and index 3 is larger and it should be considered for clustering. It is evident that grey correlation analysis principles can be adopted to conduct the index correlation analysis from the degree of curve geometry similarity. In addition, the screening of a complicated system evaluation index is affected by people's subjective factors. The screening would refer to a lot of unascertained and vague information, and the analysis process would have great grey character. The grey system analysis method acts as an effective method to solve such a problem. Therefore, it is feasible and reasonable to apply the grey correlation clustering method to conduct correlation analysis of an evaluation index.

The basic procedure of the grey correlation clustering method to calculate index correlation is as follows: For the index to be analyzed, assuming that each awaited analysis index contains m characteristic data, the relevant judgment matrix is as below:

$$X = \begin{bmatrix} x_1(1) & x_1(2) & \cdots & x_1(m) \\ x_2(1) & x_2(2) & \cdots & x_2(m) \\ \vdots & \vdots & & \vdots \\ x_n(1) & x_n(2) & & x_n(m) \end{bmatrix} \tag{53.1}$$

Because each index has different dimensions, the information comparability is lower. Therefore, standardized processing has to be performed on index information. The standardized processing rules are as follows:

$$y_i(k) = x_i(k) / \frac{1}{m} \sum_{k=1}^{m} x_k(k) \tag{53.2}$$

$i = 1, 2, \ldots n; k = 1, 2, \ldots, m$; parameter X_i and index X_j grey correlation coefficient ξ_{ij}, recorded as:

$$\xi_{ij}(k) = \frac{\min_j \min_k \left|\Delta_{ij}(k)\right| + \rho \max_j \max_k \left|\Delta_{ij}(k)\right|}{\left|\Delta_{ij}(k)\right| + \rho \max_j \max_k \left|\Delta_{ij}(k)\right|} \tag{53.3}$$

The formula $\Delta_{ij}(k) = \left|y_i(k) - y_j(k)\right|$ indicates the absolute values of corresponding points between two indexes, $i \neq j$

ρ为分辨系数, $\rho \in [0,1]$, 通常取$\rho = 0.5$. In this case, *Xi* and *Xj* grey correlations are obtained as follows:

$$\mathrm{r}_{ij} = \frac{1}{m} \sum_{k=1}^{n} \varepsilon_{ij}(k) \tag{53.4}$$

The triangular matrix under the corresponding grey correlations is as follows:

$$R = \begin{bmatrix} r_{11} & & & \\ r_{21} & r_{22} & & \\ \vdots & \vdots & \ddots & \\ r_{1m} & r_{2m} & \cdots & r_{mm} \end{bmatrix} \tag{53.5}$$

Definition 53.4 refers to matrix R as the correlation matrix between all indexes; the relevancy between the different indexes r_{ij} refers to index relevancy.

Definition 53.5 refers to relevancy *r*, which is the critical value of index correlation. Its value is set on the basis of the specific condition of the index system.

When $r_{ij} > r$, the index *i* and index *j* should conduct the clustering. According to the previously mentioned calculation method, the relevancy degree is analyzed between the indexes, the classification to indexes is performed with greater relevancy, and the purpose of optimization can be achieved for evaluation index systems.

(2) The Screening of Index Importance

For the evaluation index system, the relatively important indexes should remain and the unimportant indexes should be screened out. A guarantee of the evaluation index system's basic function can further simplify the structure of the system and facilitate analysis of the system. At the same time, one category of index clustering through correlation analysis, according to the index's importance degree, can be accepted or not, in order to ensure scientific and objective index selection. Because the screening evaluation of complex systems has great ambiguity, the importance of indicators should be analyzed using a vague analysis method and the unimportant indexes should be eliminated.

The detailed procedure is as follows:

We assume factors are set as $U = \{u_1, u_2, \ldots u_n\}$, among *i* = 1, 2, …, where *n* is the number of experts for importance judgment. For index *t*, the importance evaluation model is as follows:

$$B_t = A_t \circ R_t = (a_1, a_2, \ldots, a_n)\begin{bmatrix} \mu_{11} & \mu_{12} & \cdots & \mu_{1m} \\ \mu_{21} & \mu_{22} & \cdots & \mu_{2m} \\ \vdots & \vdots & \vdots & \vdots \\ \mu_{n1} & \mu_{n1} & \cdots & \mu_{nm} \end{bmatrix} \tag{53.6}$$

In the formula: $\circ$ is the vague operator. In order to retain all useful information, $M(\circ, +)$ is selected as the weighted average operator. A_t is the weight set composed of each expert's assignment $A = (a_1, a_2, \ldots, a_n)$; here, it can be applied with equal-weighted processing to embody the fairness principle, namely $a_i = 1/n$. R_t is the vague evaluation matrix, $R_t = (\mu_{ij})_{n \times m}$, in which μ_{ij} is the i element affiliated to the evaluation set and the j element is affiliated with the membership, $i = 1, 2, \ldots, n; j = 1, 2, \ldots, m$.

Typically, the judgment level can generally be divided for grades of 5–9. According to the characteristics of people with an m judgment grade, the evaluation set can be shown as $V = \{v_1, v_2, \ldots v_m\}$. To reflect the index importance grades, the judgment grades can be divided into five grades, namely $V = \{$not important, slightly important, important, more important, very important$\}$. The theoretical domain for the corresponding value V can be indicated as $W = \{0-2, 2-4, 4-6, 6-8, 8-10\}$.

According to the theory of vague mathematics and the value range in the theoretical domain defined above, it can constitute a subordinating degree function. Its expressed formula is as below:

$$\mu_{i1} = \begin{cases} 1 & x < 1 \\ \frac{3-x}{2} & 1 \le x < 3 \\ 0 & x \ge 3 \end{cases} \tag{53.7}$$

$$\mu_{i2} = \begin{cases} \dfrac{x-1}{2} & 1 \le x < 3 \\ \dfrac{5-x}{2} & 3 \le x < 5 \\ 0 & 其他 \end{cases} \tag{53.8}$$

$$\mu_{i3} = \begin{cases} \frac{x-3}{2} & 3 \le x < 5 \\ \frac{7-x}{2} & 5 \le x < 7 \\ 0 & \text{other} \end{cases} \tag{53.9}$$

$$\mu_{i4} = \begin{cases} \frac{x-5}{2} & 5 \le x < 7 \\ \frac{9-x}{2} & 7 \le x < 9 \\ 0 & \text{other} \end{cases} \tag{53.10}$$

$$\mu_{i5} = \begin{cases} 0 & x<7 \\ \frac{x-7}{2} & 7\leq x<9 \\ 1 & x\geq 9 \end{cases} \quad (53.11)$$

In the formula: $\mu_1 \cdots \mu_5$ respectively refer to index evaluation scores affiliated with judgment grade values obtained for the theoretical domain 0–2, 2–4, 4–6, 6–8, and 8–10 subordinate functions; x is the evaluation score.

For convenience, to contrast the evaluation results, the following formula can be adopted to exchange this for the score in the form of a centesimal system:

$$C_t = B_t \bullet E^T \quad (53.12)$$

In this formula, E refers to the transition matrix, E = [60, 70, 80, 90, 100].

Formula 53.6 refers to C_t as an importance degree in the evaluation index: the larger the value is, the more important the index is as indicated. According to the size of C_t, the screening can be performed for the index and the unimportant indexes can be eliminated, thus simplifying the index system structure.

(3) Effectiveness Test of Index

From different perspectives, the same evaluation may be jointly concluded from the different evaluation index systems. With them, it becomes more complete and effective to reflect the evaluation target essence and more effective to screen out the evaluation index, which would play an important role to affect the evaluation result. Therefore, it is necessary to conduct an effectiveness test for the evaluation index system.

At present, it is less for the effectiveness testing studies of evaluation index system. The literature [9] performs the effectiveness study of qualitative evaluation index system through information discrepancy provided by experts and based on the principle of statistics, but the main defect of statistics method is that the accuracy of result depends on quantity of data sample; furthermore, the samples are required to be in compliance with a certain typical probability distribution, thus it would bring about some difficulties to the analysis of effectiveness. To compensate for the defects of the statistical method, we applied grey correlation analysis in the evaluation index effectiveness test.

Suppose that different judges adopt a certain evaluation index system to perform an evaluation on the same evaluation target. If the evaluation results appear with a large discrepancy, it indicates this evaluation index system's failure to accurately reflect the evaluation target; thus, its effectiveness is lower. The basic purpose of grey correlation analysis in conducting an effectiveness test is to analyze the index sequence correlation degree with a corresponding reference sequence through the curves' shape similarity. For a certain index with a high degree of similarity for each index's comparative sequence curve and reference sequence curve, the experts would have a high consistency on the basis of this index judgment, thus indicating the better effectiveness of this index. It is evident that the theory of application of grey correlation in effectiveness analysis is also realized through analysis of the degree of the curves' geometric shape similarity. However, the key problem is to determine the reference sequence, because the essence is different for selecting the different reference sequences for reflection of the problem.

The basic principles and procedures for the application of grey correlation analysis in an effectiveness test is as follows. Assume that the evaluation index sets are $A = \{A_i | i \in N,\ N = (1,\ 2,\ \ldots n)\}$, where n is the evaluation index number. Evaluation experts are set as $B = \{B_k | k \in K,\ K = (1,\ 2,\ \ldots m)\}$, and then each index sequence can be shown as follows:

$$\begin{array}{llllll} X_1 = & (x_1(1), & x_1(1), & \ldots, & x_1(m)) \\ X_2 = & (x_2(1), & x_2(2) & \ldots, & x_2(m)) \\ \ldots & \ldots & \ldots & \ldots & \ldots \\ X_n = & (x_n(1), & x_n(2), & \ldots, & x_n(m)) \end{array} \quad (53.13)$$

Normalization are as follows:

$$y_i(k) = x_i(k) / \frac{1}{m} \sum_{k=1}^{m} x_k(k) \quad (53.14)$$

Formula 53.7 refers to $y_{oi}(k)$ as the reference sequence value of each index. According to a different index, its reference sequence value is different: $y_{oi}(k) = \frac{1}{m} \sum_{k=1}^{m} y_i(k)$. The i index reference is $Y_{oi} = (y_{oi}(1), y_{oi}(2) \cdots y_{oi}(m))$.

Formula 53.8 refers to ε_i as the validity of each index; it describes the size of the index's effectiveness. All indexes' effectiveness sum and validity of the evaluation index system consist of these indicators, recorded as ε.

According to each indictor, the correlation coefficient to calculate information and reference information given by all experts is the correlation coefficient as below:

$$\varepsilon_{io}(k) = \frac{\min_i \min_k |y_i(k) - y_{oi}(k)| + \rho \max_i \max_k |y_i(k) - y_{oi}(k)|}{|y_i(k) - y_{oi}(k)| + \rho \max_i \max_k |y_i(k) - y_{oi}(k)|} \quad (53.15)$$

The size of the correlation coefficient reflects the correlation extent of evaluation information provided by experts and reference information. The greater the correlation is, the less information is provided as indicated, the higher the consistency of expert recognition is, and the higher the index effectiveness is. Each index's effectiveness can be shown as follows:

$$\varepsilon_i = \frac{1}{m} \sum_{k=1}^{m} \varepsilon_{io}(k) \quad (53.16)$$

The larger the index's validity for a value is, which shows when this index is used to evaluate, the higher the experts' cognition consistency is and the larger the index effectiveness is. The overall corresponding index system effectiveness can be shown as follows:

$$\varepsilon = \frac{1}{n}\sum_{i=1}^{m} \varepsilon_i \tag{53.17}$$

A larger index system validity shows that this index system more truly reflects the nature of evaluation objects and that its effectiveness is higher.

It is evident that the key to conducting an effectiveness test in the application of grey correlation analysis is the determination of the reference sequence. The reasonable determination of a reference sequence can have grey correlation analysis play different functions and realize multiple specific functions. Through an effectiveness test, it not only can test a selected evaluation index and the effectiveness of the evaluation index system constituted by it, but it also can compare an indicator system's effectiveness change condition before and after selection and reasonability of test indicator selection.

53.3 Conclusions

This chapter presents an evaluation index screening method on the basis of a multifactor comprehensive algorithm in terms of index correlation, importance, and effectiveness in applications of grey correlation clustering, vague analysis, and grey correlation analysis after the evaluation index screening method is reviewed and analyzed. This algorithm, on the basis of multiple factors considered at the same time, not only remains an important index in the index system but also makes each index mutually independent and index system complete, not redundant. On this basis, we aim to further analyze the index system through an effectiveness test, so as to ensure the selected results' scientific nature and effectiveness, as well as to provide a new method for the selection of a complex system evaluation index. Meanwhile, grey correlation analysis can be tentatively applied for an evaluation index effectiveness test from an overall perspective; it uses the degree of similarity of the curves' geometric shape to reflect the extent of effectiveness. This exploratory study of the evaluation index effectiveness test therefore widens the application scope for grey correlation analysis.

This work is supported by the opening foundation of the Science and Technology on Human Factors Engineering Laboratory, Chinese Astronaut Research and Training Center (grant no. SYFD140051810K).

References

1. Cervantes O, Espejel I (2008) Design of an integrated evaluation index for recreational beaches. Ocean Coast Manag 51(2):410–419
2. Jim6ne AR, Ceres ZR, Pons JL (2001) A new adaptive filter and quality evaluation index for image restoration. Pattern Recognition 34(2):457–467
3. Liu H, Li J, Ge Y (2006) Design of customer satisfaction measurement index system of EMS service. J Chin Univ Posts Telecommun 13(1):109–113
4. Liu Y (2008) Index system based discrimination and the right measure of entropy theory redesign. Nanjing Univ Aeronaut Astronaut Master Thesis 13–14
5. Zhang YJ, Chen X, Ren XF (2007) Establishment of an index system for evaluating outstanding biomedical scientists for science foundation of Shanghai. J Med Coll PLA 22(6):191–194
6. Guo Y, Yu Z (2002) Comprehensive Evaluation of the reasonableness of the problem. Northeast Univ (Nat Sci) 23(9):844–847
7. Su W (2000) Research questions for comprehensive evaluation theory and methods. Xiamen University doctoral dissertation 17–20
8. Li S, Chen J, Zhao H (2001) Qualitative evaluation index system of decision-making. Syst Eng Sci Pract Perish 2001(9):22–28
9. Pawlak Z (1991) Rough sets-theoretical aspects of reasoning about data. KluwerAeademic Publishers, Dordrecht, pp 68–162

Chapter 54
Research on Display-Control Layout of Haptic Interaction Interface Based on Character Input Task

Lifen Tan, Shihua Zhou, Yu Tian and Chunhui Wang

Abstract This study investigated the influence of the layout of control and display areas of the haptic interface on the performance of word/number input task. Based on QWERTY/numeric keyboards, we designed several schemes of the haptic interface, and developed the haptic interaction software. Nineteen subjects participated in the test, and finished word/number input tasks on the different haptic interfaces; we recorded their task completion time and accuracy, and subjective ratings are recording and analyzed. Results show that the input speed and accuracy of the subjects on the several haptic interfaces do not have significant difference. However, from the subject ratings, we found the subjects favor the haptic interface with the keyboards on the right side of the interface. Chinese character input method can influence task performance significantly. The current study can provide support for the development of relevant ergonomic requirement of haptic interface in the space station.

Keywords Haptic interaction · Display-control layout · Layout of numeric keyboards · Task performance

54.1 Introduction

With the development of manned space technology, manned spacecraft human–machine interface has transformed from graphic–user interface to natural human–machine interacting interface direction. Human–machine interactional modes like electroencephalogram, audio, gesture, and haptic interaction has gradually become

Supported by Equipment beforehand Question for Study 9140A26070214KG57431/9140A26070213KG57387.

L. Tan (✉) · S. Zhou · Y. Tian · C. Wang
National Key Laboratory of Human Factors Engineering, China Astronaut Research and Training Center, Beijing 100094, China
e-mail: tanlifen163@163.com

© Springer-Verlag Berlin Heidelberg 2015

S. Long and B.S. Dhillon (eds.), *Proceedings of the 15th International Conference on Man–Machine–Environment System Engineering*, Lecture Notes in Electrical Engineering 356, DOI 10.1007/978-3-662-48224-7_54

the research highlights. Haptic interaction control technology has been successfully and widely applied in appliances, numeric machine tools, and smart mobile phones, and gradually been attached great importance to. In the future, in the fields of mechanic arm control, materials management and live entertainment of space station tasks, haptic interaction may be used. NASA STD-3000 illustrates the related standard of human–machine functional allocation, spacesuit fitness, and human–machine interface factors in the space environment. The related standard description in haptic interaction and touch screen is not complete and only involves in qualitative description basically. Therefore, it is of great realistic importance to develop the haptic interaction ergonomics based on touch screen controller.

In the manned space flight, it is required the interactional control interface which meets the cognition features and esthetic demand so as to promote astronauts to quickly and accurately compete the touch-control operation. Interactional interface, as the important communication channel with spacecraft, requires to research from the haptic interactional interface lay out from human–machine interface interactive designing angle [1]. Haptic interaction interface synthesizes the display area and control area at the touch screen. The layout of display control area will directly influence the control performance and operating experience [2]. With the application of a large number of electronic equipment, there are more and more equipments requiring astronauts for numeric input at the orbit. The current numeric keyboard layout includes computer layout and telephone layout means. The early research about the advantages and disadvantages refers to papers in 1960s [3, 4]. There are very fewer related researches in China [5]. Applying ergonomics principle, this paper aims at the haptic interaction interface layout design: display-control layout and layout of numeric keyboards to guide easy operational friendly haptic interaction interface design. Aiming for different display-control working conditions, two numeric keyboards layout working conditions design Chinese editing input, numeric input operating tasks, respectively, and develops the testing research.

54.2 Research Methods

54.2.1 Platform of Software and Hardware

Adopt VS2008 development environment C# advanced programming language research testing software. Software provides display-control functional layout, numeric keyboard layout functional modules. Adopt Access 2003 development testing database software to realize data collection, storage and simple processing. The database file can be derived out as Excel file.

Adopt the tablet PC of 12.1 in. Asus EeeSlate Win7 64-bit operating system as the testing touchscreen controller. It is required to be installed with Microsoft, NET Framework 4.0, and above-version complier.

54.2.2 Subjects

According to the men and women astronaut of Chinese space station task period, select 19 right-handed subjects, 4 of which are tested in the preliminary experiment and 15 of which are tested in formal experiment (10 males and 5 females). During the test, they shall settled down in sitting gesture and operate by a single hand (basically as right-hand index finger operation).

54.2.3 Experimental Flow

Before testing, the subjects shall be simply trained and introduced the basic flow of tests and module tasks of each test. Thereafter, subjects start to operate software drills until they think that the operating requirement of each module is very familiar. After practicing, each subject starts formal test. This test shall adopt tested internal testing design. That is to say, each subject shall complete different kinds of tests with different working conditions under each module. The testing order among different modules is randomly decided by subjects.

According to the requirement of tests, the subjects input variant Chinese–English characters based on virtual QWERTY/numeric keyboards or variant groups random numbers based on numeric keyboard editing. The input standard Chinese characters are provided by paper files. To reduce the learning effect, each character is randomly determined from dictionary. There are no meaningful connections among the neighboring characters. The input standard numeric groups shall be generated from the testing software randomly. The QWERTY keyboard working conditions in display-control layout module shall be input with 100 Chinese and English characters. The numeric keyboard working conditions require input 1 group of 30 digits randomly produced. Numeric keyboard layout modules test requires numeric editing and inputting 20 groups of 8 digits randomly produced based on calculator layout/telephone layout numeric keyboard.

During the testing process, adopt testing software to automatically collect the related testing data. After finishing each test, the subjects are required to finish the test electronic questionnaires provided by the testing software to collect the subjects' subjective feelings, opinions, and suggestion on the testing design and process.

54.3 Research Result and Conclusions

Obtain the related testing data of all modules through testing research, including the input time, input correct ratio, input subjective feelings, opinions, and suggestion. Based on testing the real-tested data, this paper researches and analyze the suitable display-control layout under QWERTY keyboard/numeric keyboard working conditions and the main factors influencing display-control layout. Moreover, it takes the number inputting speed, correct ratio, and subjective bias of subjects, analyzes and compares the two numeric keyboard layouts of calculator layout, telephone layout.

54.3.1 Results

54.3.1.1 Input Speed and Correct Ratio of Display-Control Module

As for different applying conditions, the control area of display-control layout may adopt QWERTY keyboard with large area and variant keys or small area with less keys. Therefore, this paper researches on the working conditions of the display-control layout based on QWERTY keyboard and numeric keyboards. According to the current published literary [2] summarized 4-type display-control layout, this paper designs 5 types of QWERTY keyboard display-control layout and 4 types of numeric keyboard display-control layout, referring to Figs. 54.1 and 54.2.

The analysis of testing data can obtain the input time of different display-control layout, referring to Fig. 54.3. The input correct rate refers to Fig. 54.4. The input time and correct rate at different display-control layout matches the "t" testing results refer to Tables 54.1 and 54.2, respectively.

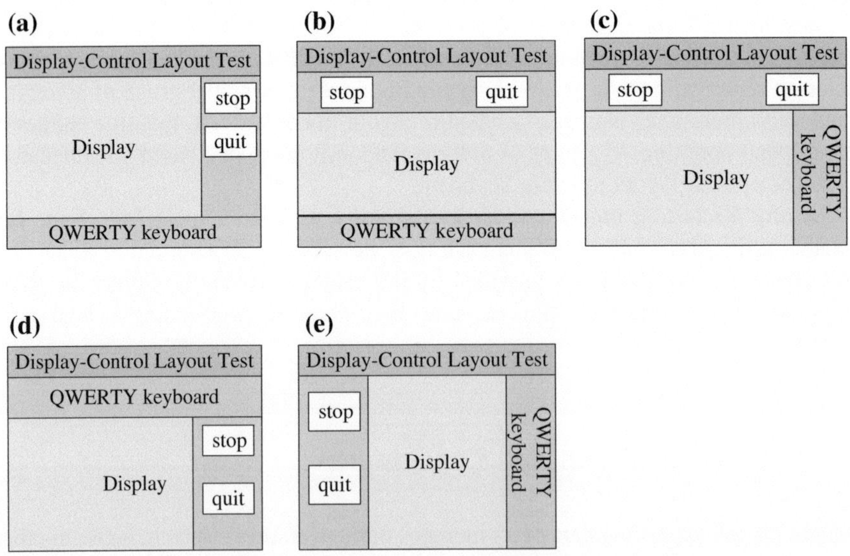

Fig. 54.1 Display-control interface with QWERTY keyboard. **a** Layout 1. **b** Layout 2. **c** Layout 3. **d** Layout 4. **e** Layout 5

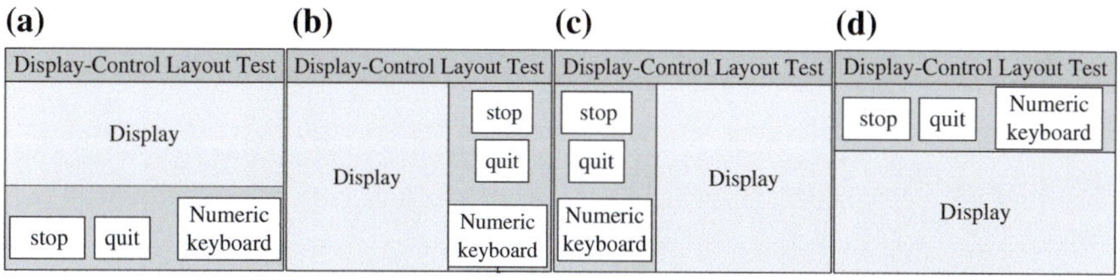

Fig. 54.2 Display-control interface with numeric keyboard. **a** Layout 1. **b** Layout 2. **c** Layout 3. **d** Layout 4

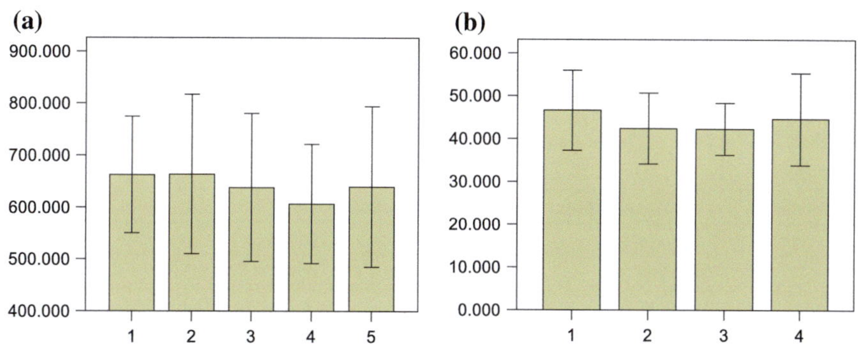

Fig. 54.3 Mean task completion time for different display-control interfaces. **a** QWERTY keyboard. **b** Numeric keyboard

Figure 54.3 describes the Chinese characters/numeric input time with different display-control distribution under the two different working conditions. As for the display-control layout working conditions with QWERTY keyboard, layout 4 has the shortest input time, then as layout 5, layout 3, and layout 1 with the increase of time. Layout 2 is the worst. The analysis result of Table 54.1 paired t-test shows that, the input speed, operating time/correct rate indicator of layout 4 are evidently superior to layout 1. The indicators of input speed and correct rate among other layouts do not show significant differences. As for the numeric keyboard display-control layout working conditions, Fig. 54.3 indicates that layout 3 has the shortest operating time, then successively as layout 2 and layout 4. Layout 1 has the longest layout operational time. Table 54.2 shows that the input time of 4 layouts do not show significant differences.

Figure 54.4 describes the Chinese characters/numeric input correct ratio under two working conditions. As for the working conditions of QWERTY display-control layout, layout 3 has the highest correct ratio, then successively as layout 2, layout 5, and layout 4. Layout 1 has the worst layout. The analysis result of Table 54.2 paired t-test shows that the correct ratio of layout 5 does not have significant difference. There are no differences among the correct ratios of the working conditions of 4 numeric keyboard display-control layouts.

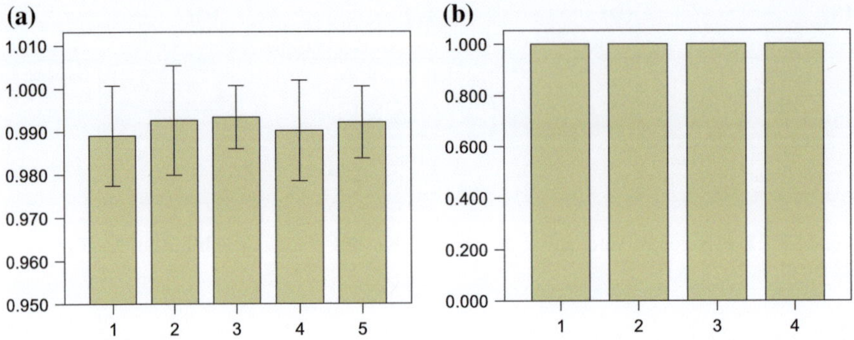

Fig. 54.4 Mean accuracy for different display-control interfaces. **a** QWERTY keyboard. **b** Numeric keyboard

Table 54.1 *P* values of Paired t-test between different display-control interfaces with QWERTY keyboard

Parameter	Work condition	Layout 1	Layout 2	Layout 3	Layout 4	Layout 5
Test of time	Layout 1	–	0.978	0.346	0.047	0.488
	Layout 2		–	0.434	0.310	0.414
	Layout 3			–	0.365	0.956
	Layout 4				–	0.547
Test accuracy	Layout 1	–	0.354	0.191	0.820	0.485
	Layout 2		–	0.864	0.447	0.898
	Layout 3			–	0.456	0.692
	Layout 4				–	0.410
Time/accuracy	Layout 1	–	0.970	0.287	0.037	0.431
	Layout 2			0.436	0.326	0.418
	Layout 3				0.377	0.939
	Layout 4					0.560

Table 54.2 *P* values of Paired t-test between different display-control interfaces with numeric keyboard

Work condition	Layout 1	Layout 2	Layout 3	Layout 4
Layout 1	–	0.168	0.089	0.668
Layout 2		–	0.690	0.334
Layout 3			–	0.334

54.3.1.2 Input Speed and Correct Ratio of Numeric Keyboard Layout Modules

Targeting the two kinds of layout working conditions of calculator layout and telephones (see Fig. 54.5), through the experimental research and statistical analysis, the average value of operation time of telephone layout is 146.260 s with the

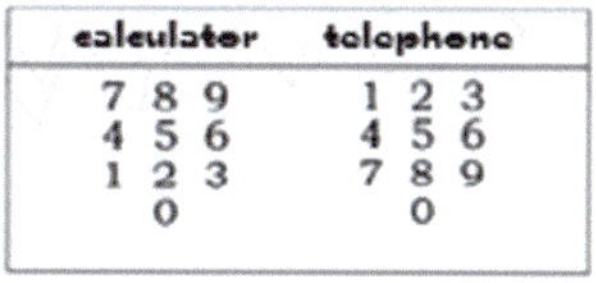

Fig. 54.5 Two kinds of numeric keyboards

standard differences of 27.672 s. The average value of operation time of calculator is 141.268 s with the standard differences of 21.443 s. Layout of calculator has shorter input time than telephone layout. The input time paired time t-test of two layouts is $*p = 0.376$, without significant differences. Calculator layout has lower input correct ratio than telephone layout. The average values of the two layouts are more than 0.995.

54.3.1.3 Subjective Comments Display-Control Display Modules

After testing, subjects shall select an item as their comment on the display-control layout from five options, that is, poor, discrepancy, general, reasonable, and good. The statistical results of the two working conditions of QWERTY keyboards and numeric keyboards refer to Fig. 54.6. Figure 54.6 shows that, as for the working conditions of QWETY keyboard, the subjects have the best subjective comment on layout 2. The one with worst subject comment is layout 4. As for the numeric keyboard, the subjects have the best subjective comment on layout 2. The one with worst subject comment is layout 4.

54.3.1.4 Subjective Comments Numeric Keyboard Layout Modules

After testing, subjects shall make subjective selection among the numeric keyboards with different layout at the inputting speed of "slow, a little slow, general,

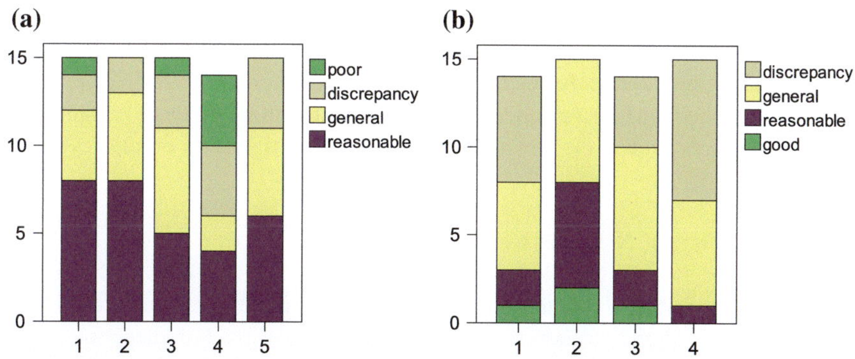

Fig. 54.6 Subjective ratings of the different display-control interfaces. **a** QWERTY keyboard. **b** Numeric keyboard

faster, fast" and correct ratio from "low, lower, general, higher and high". The statistical results show that, subjects have approximate different inputting speed comment of two layouts. The comments on two layout and correct ratio are different. The correct rate of calculator layout is a little higher than that of telephone.

54.3.2 Discussion

54.3.2.1 Display-Control Layout

The testing results of display-control layout QWERTY keyboard show that the operating performance of QWERTY keyboard under the below layout 1 and 2 has the worst operating performance, the lowest layout of input correct ratio and slowest input speed, respectively. The layout of QWERTY keyboard at the middle and upper part has the fastest inputting speed, but with low correct ratio. The input correct ratio and the comprehensive performance of inputting correct ratio and speed on the left layout 3 and right layout 5 of QWERTY keyboard are relatively good. The correct ratio of layout 3 is the best. The QWERTY keyboard of layout 4 is on the right above of display area. The distance between hand operation and eye movement is short. QWERTY keyboard of layout 1 is located at the right down part of display area. It is the vital reason that the input speed of layout 4 is faster than layout 1 evidently. QWERTY keyboards have the same value. That is, excluding the factor of Chines input methods, the correct ratio of layout 2 is higher than layout 1 and the correct ratio of layout 3 is higher than layout 5. It demonstrates that special functional keys control area locating at the above of screen is better than the two sides of screen.

As for the working conditions of display-control layout numeric keyboards, the operation performance of numeric keyboards layout at the left and right sides is better than the layout at the down position and down-above layout. The speed at the left down side of numeric keyboard is fastest. The result is basically equivalent to the subjective comment result. The subjects' subjective comparison and bias is superior to layout 2.

To sum up, display-control layout method and Chinese input methods will influence the hand and eye movement distance of touchscreen interactional control, then influence the performance. In the future, the Chinese input method using touch control shall be researched. Adopt the testing design method of QWERTY keyboard to input English characters (avoid the disturbance of inputting methods). Implement further testing research at the down, medium upper, and right area on the keyboard.

54.3.2.2 Layout of Numeric Keyboard

Bell laboratory research pointed out that the average inputting time of calculator real keyboards is slightly slower than the real numeric keyboard of telephone layout [4]. Literary [5] research on numeric real keyboard points out that the input correct ratio of calculator layout is high and the inputting speed of telephone layout is fast. Literary [6] pointed out that telephone layout is superior to telephone layout in terms of ergonomics with less errors and time. It is different from the testing results of touchscreen numeric keyboards. This testing research demonstrates that the input speed of the two kinds of distribution and the correct operating performance do not have significant differences. The input speed of virtual numeric keyboard of calculator layout is faster and the correct ratio of telephone layout virtue numeric keyboard is high. Through analysis, the main reasons of the difference lie in the following two aspects. First, the operating features are different between the real keyboard and virtual keyboard: touchscreen display and control are integrated on display screen; touchscreen virtual keyboards do not have blind operation. They have different features with PC keyboard and display area separation. Second, most subjects are technicians using PC usually, who are very familiar with the numeric keyboard of calculator layout.

54.4 Conclusions

This paper is based on the ergonomics testing software of research taking Chinese and English characters or numeric input time, correct ratio, and subjective feelings as indicators. It deeply researches the ergonomics of display-control layout and numeric keyboard layout. The research results show that display area/control area layout, numeric keyboard layout, Chinese input way, hand and eye moving distance are the key influencing factors of touchscreen interacting interface. This paper also gives the suggestions of display-control layout and numeric keyboard layout way meeting different indicator requirement. According to the requirements of different indicators, the designing scheme and results of touchscreen interacting layout of the test may provide the basic data for the related touchscreen interacting interface design in space station.

References

1. Sanders MS, McCoimick EJ (2002) Human factors in engineering and design. Tsinghua University Press, Beijing
2. Xu Y, Yan S (2004) The design principle study of human-machine interface of numerical control machine. Mach Tool Hydraul 12:118
3. Gordon LE, Winwood M (1973) Searching through letter arrays. Ergonomics 16:177–188
4. Deininger RL (1960) Human factors engineering studies of the design and use of pushbutton telephone sets. The Bell Syst Tech J 34:995–1012
5. Shen H, Yang J, Tang B (2005) Study of ergonomics of numeric keyboard layout. Pack Eng 26 (4):129–130
6. Tong S (2011) The human-machine engineering design and application of the manual. China Standards Press, Beijing, p 310

Chapter 55
Optimization Design of Mechanical System Human–Machine Interface

Siyuan He, Chuan Wang, Xiaowei Yang, Qiao Wang and Fei Li

Abstract The human–machine interface of mechanical system includes lots of contents such as display, manipulation, geometric position, lighting, sound, vibration, temperature, etc. It is very difficult to comprehensively study their impact on the operators. All along, the evaluation and optimization of mechanical system human–machine interface geometric position matching relation is one of the inevitable questions of designing human–machine system. This paper elaborates the research situation of mechanical system human–machine interface, and overviews the traditional mechanical system human–machine interface design, and proposes mechanical system human–machine interface optimization design methods based on evaluation and research.

Keywords Mechanical system · Human–machine interface · Optimization design

Human–machine system means a system composed of human and machine, and the size of the system is changeable; human and spacecraft, human and car, human and seat, human and indoor environment, and human and outdoor environment can all constitute human–machine system [1]. Geometric position of human–machine interface occupies the main position in mechanical system design, and has the direct bearing on matching goodness of mechanical system human–machine interface, and reflects that matching relation of mechanical system human–machine interface affects

---

S. He · C. Wang (✉)
The Naval Medical Research Institute, Shanghai 200433, China
e-mail: hg04381@163.com

C. Wang
Nuclear Science and Engineering Department of Naval Engineering University, Wuhan 430033, China

X. Yang
91656 PLA Troops, Shanghai 200433, China

Q. Wang · F. Li
92730 PLA Troops, Sanya 572000, China

© Springer-Verlag Berlin Heidelberg 2015

S. Long and B.S. Dhillon (eds.), *Proceedings of the 15th International Conference on Man–Machine–Environment System Engineering*, Lecture Notes in Electrical Engineering 356, DOI 10.1007/978-3-662-48224-7_55

the overall efficiency of mechanical system to a large extent, and it is a factor which must be considered when overall design of most mechanical system. So it is necessary to do optimization design of mechanical system human–machine interface.

55.1 Overview

Human–machine interface of mechanical system means interaction area between operators and machines, and it is the medium of transferred information between human and machines. The interface contains three parts: interface for information channel between machine's display and human; interface for locomotion organ between machine's manipulator and human; interface between human–machine system and environment. The structure of machine has its own rules, and operating environment or living environment could be limited on space and time because of various factors, such as economic feasibility, technological feasibility, conditions required by performance of machine, external environment conditions when using machines (e.g., high-temperature operation), etc. In order to adapt these situations, requiring limitation and training of human factors, and giving play to human factors with certain "plasticity" characteristic to the greatest extent, making human adapt requirements of machines, so as to make sure optimum efficiency of human–machine system. Actually, except the above hardware equipment of human–machine interface, there should contain softwares such as operating instruction, maintenance manual, etc. Some people also name the above human–machine interface as hardware human–machine interface, and name human–machine interface of human and computers as software human–machine interface [2].

There are many contents in human–machine interface, and studying human–machine interface has a certain difficulty of system operating efficiency and influence for operators' health. So far, the evaluation method of domestics and overseas human–machine interface optimization matching is still not perfect. The arrangement of components' geometric position is very important in human–machine interface design, so optimization design of mechanical system matters much.

55.2 Elements

Mechanical system human–machine interface design contains multiple factors. Based on basic principles of ergonomics and design principles of mechanical system human–machine interface, and referring general mechanical system human–machine interface's characteristics and design requirements, the elements of mechanical system human–machine interface mainly include as shown in Fig. 55.1.

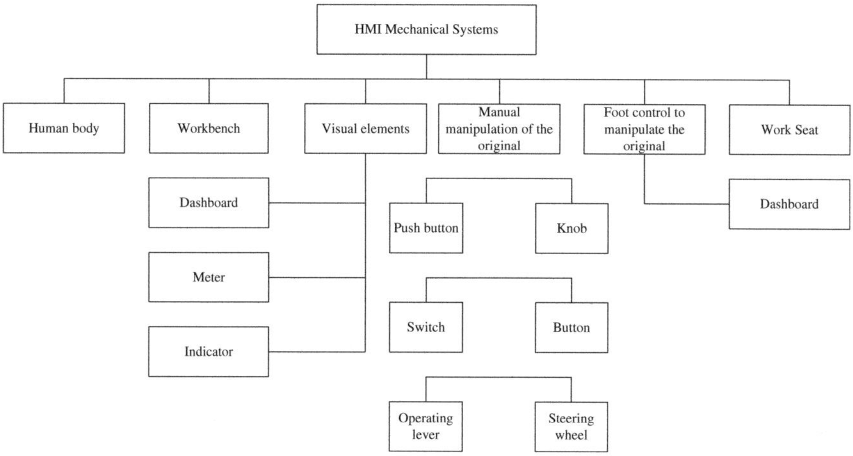

Fig. 55.1 Components of the mechanical system human–machine interface

Operating platform: table-board height of operating platform is T1; knee space height of operating platform is T2; depth of operating platform is T3; width of operating platform is T4; foot space height of operating platform is T5; width of operating platform is T6; depth of operating platform is T7; table-board width of operating platform is T8; installation site of operating platform in coordinates' X axis, Y axis, and Z axis is, respectively, T9, T10, and T11.

Visual component: installation site of dash board on operating platform in coordinates' X axis, Y axis, and Z axis is, respectively, P1x, P1y, and P1z; included angle of dash board and table-board is P2; view distance between arrangement position of instrument and operators' eyes is VI1; arrangement area of instrument on dash board's vertical plane and horizontal plane is VI2 and VI3; distance between arrangement position of indicator light and operators' eyes is VL1; arrangement area of indicator light on dash board's vertical plane and horizontal plane is VL2 and VL3.

Manual control manipulation component: diameter of component is H1; (The components' hand stretch and degree is H2) degree of component is H2; installation site of component in coordinates' X axis is H3; height of component is H4.

Foot control manipulation component: length of foot plate is F1; width is F2; dip angle between board and horizontal plane is F3; installation site of foot plate in coordinates' X axis is F4; distance of body SIP point is F5; installation site of foot plate in coordinates' Z axis is F6.

Work seat: height of seat surface is S1; depth is S2; width is S3; inclination angle of seat back is S4; installation site of seat in coordinates' X axis is S5; installation site in Y axis is S6; installation site in Z axis is S7.

55.3 Optimization Design Objective Function and Constraint Conditions

Overall evaluation index of mechanical system human–machine interface—system matching goodness, is overall objective function, and matching goodness of components in human–machine interface is single objective function, optimization in multiple targets by different level.

Matching goodness of mechanical system human–machine interface call matching goodness, meaning matching reasonableness among display of mechanical system, operating interface and operators, and it is a number between 0 and 1, the closer to 1, the better matching goodness of geometric interface between human and machine. The expression of matching goodness, that is to say the objective function of mechanical system human–machine interface optimization design is

$$\langle Es \rangle i = \frac{E_{si}}{\sum_{i=1}^{n} E_{si}}$$

As seen from design variables and objective function, geometric position optimum value of components in human–machine interface relates to body model selection of this human–machine interface. So if the body model selection varies, then constraint conditions of components in human–machine interface also vary and constraint conditions are constructing with body dimensions as variables. Body dimension variables related to human–machine interface optimization design is shown in Table 55.1.

First of all, body in coordinates' X axis, Y axis, Z axis is Bx, By, and Bz, respectively; Second, constraint conditions of each component are put in same constraint equation and conduct computer processing. Constraint conditions of design variables are as follows (number unit: mm).

Table 55.1 Body size and variable names

Human size	Variable names	Human size	Variable names
Long arm	A1	Foot wide	A9
Forearm length	A2	Sitting Shoulder	A10
Chest width	A3	Seated elbow height	A11
Sitting hip width	A4	Sitting knee high	A12
Thigh length	A5	From hip knee	A13
Leg length	A6	Waist Width	A14
Sitting lower limb	A7	Thigh diameter	A15
Foot length	A8	IPD	A16

Operating platform. Installation site of operating platform relates to its attribute parameter optimization results, after confirming geometric parameter of operating platform, system shall conduct optimization of operating platform's installation site based on position of body and work seat and body dimensions.

Constraint condition of operating platform's other attribute parameter is

$$\begin{cases} T_0 = (2 \times (A_{11} + S_1) + A_{10} - 0.5 \times A_1 + S_1)/3.0 \\ T_1 \geq T_0 \\ T_1 \leq T_0 + (A_{10} - 0.5 \times A_1 - A_{11})/4.5 \\ T_1 \geq T_2 \geq A_{12} + 100 \\ T_3 \geq A_{13} - A_{14} \\ T_8 \geq T_4 \geq A_{42} \\ A_8 \leq T_5 \leq T_8 + 20 \\ 2 \times A_9 + 20 \leq T_6 \leq T_8 \\ T_7 \geq A_8 \\ T_8 \geq A_3 \end{cases}$$

Visual component. First optimize dash board. Installation site of dash board is P1x, P1y, P1z, and restrain by installation site of operation platform. Here-into, P1y is sum of height of operating platform (T1) and installation parameter of operating platform in Y axis (T10), and no need to optimize.

Instrument and indicator light arranged on dash board. Reading time displayed on instrument is longer than indicator light's reading time, so as to arrange instrument in the central part of dash board and extend arrangement area of indicator light. Restraint condition of other dash board parameter is

$$\begin{cases} T_9 - 200 \leq P_{1x} \leq T_9 + 500 \\ T_{11} - T_8/2 \leq P_{1x} \leq T_{11} + T_8/2 \\ 60^\circ \leq P_2 \leq 70^\circ \end{cases}$$

Arrangement area of visual component in vertical plane is confirmed by vertical direction's included angle of visual component's central ligature and horizontal ligature, and arrangement area in horizontal plane is confirmed by horizontal direction's included angle of eyes and visual component's central ligature and horizontal ligature [3].

Manual control manipulation components contain many contents and components. There are select key, knob, switch, button, steering wheel, and operating lever as typical components of manual control manipulation component [4, 5]. Because these typical components vary much from each other, when conducting optimization design, setting constraint conditions for above typical components, respectively. Foot plate is selected as typical components of foot control manipulation component and conducting optimization design. There are foot plates on the both sides of human body, but their constraint conditions are different. Constraint condition of the left foot plate of human body is

$$\begin{cases} 150 \le F_1 \le 180 \\ 95 \le F_2 \le 115 \\ 45^\circ \le F_3 \le 53^\circ \\ B_x - 2 \times (f_1 - f_2)/21 \le F_4 \le B_x - (3 \times f_1 + 4 \times f_2)/7 \\ 0.6 \times f_3 \le F_5 \le 0.68 \times f_3 \end{cases}$$

55.4 Traditional Design Method

In terms of design method and design means of human–machine interface, currently the design of mechanical system human–machine interface is mainly in the traditional product development stage based on prototype. Throughout domestics and overseas research methods and research results in related fields, traditional design methods of mechanical system human–machine interface mainly refer to predecessors' research theories, results and data, and analyzing characteristics and requirements of human–machine interface that are to be designed, and conducting human–machine interface design based on designers' experience and analysis results of human–machine interface. After finishing the design, model machine is made and experimented. If experiment results meet design requirements, then stop it; otherwise, redesign it and repeat the above processes until the model machine meets the requirements. Moreover, in the traditional mechanical system human–machine interface design, mechanical system human–machine interface involves many factors, and designers need to inquire huge number of standard data and refer to other researchers' related research results, so designers are hard to weigh and get satisfied design effect. Furthermore, design advantages and disadvantages of mechanical system human–machine interface with traditional human–machine interface design methods depend on designers' individual factors such as knowledge level, aesthetic standards, etc., to a large extent. As a result, human–machine interface design could not meet realistic requirements and the design fails. The reasons why mechanical system human–machine interface could only adopt empirical design are as follows: there are no appropriate, operable evaluation methods and optimization design methods, designers could not conduct reasonable evaluation and optimization design of human–machine interface, designers could not distinguish advantages and disadvantages of different design, the design cycle is too long, funds are wasted much, and it is helpless for more complicated human–machine interface. Therefore, designers basically shall not adopt this design method anymore.

55.5 Optimization Design Method

The purpose of conducting human–machine interface optimization design is to build best human–machine interface matching relation, on the one hand, making "machines suit human", on the other hand, making "human adapt machines". So-called "machines suit human", that is to say in human–machine system, machines' function design suit human, and designing machines based on human characteristics parameter, and making machine system meet users' condition requirements such as physique, physiology, psychology, esthetics standard, social values, etc., to the largest extent.

Under the drive of electronic computers, optimization theory and method get a wide range of applications in economic planning, engineering design, production management, transportation, etc., and becomes an active subject. In scope of applying optimization design theory and method to mechanical product structure design, there are three optimization parameters namely, structural parameter optimization, shape optimization, and topology optimization. Product design shall first settle parameter optimization, and it is also one of the earliest introducing contents of mechanical design. Good design is directly obtained through searching optimum parameter, and completing optimization of parameter design under fixed structure concept, component structural shape, and selecting material. Shape optimization means under the fixed conditions of structure type, material and layout, conducting optimization of structural geometric shape, including two-dimensional and three-dimensional shape optimization and parameter optimization related to structure, and this is deepening of optimization design. Topology optimization means conducting optimization of structural component layout and node's connection relation, that is to say first searching from structural concept, type, layout under requirements of external design; this is a higher level optimization, and also a more innovative conceptual design. As seen from current research and application maturity, although shape optimization has been put forward nearly 20 years, it is still in the research and application exploratory stage, but topology optimization is in the inoculation phase. Specifically, when conducting optimization design of mechanical system human–machine interface shall be accordance with following two procedures: first, making components' matching goodness of human–machine interface as objective function, confirming constraint conditions of each component's geometric parameter based on human–machine design principles, conducting optimization design of components' geometric dimensions, getting dimension parameter optimum solution of components, and making each component in the optimum situation of matching goodness; then, making matching goodness of human–machine interface system as objective function, confirming constraint conditions of components' arrangement position based on human–machine interface design principles and predecessors' research results, conducting optimization design of components' geometric arrangement positions, getting geometric arrangement position optimum solution of components, and making overall human–machine interface achieve system matching goodness maximum, as shown in Fig. 55.2.

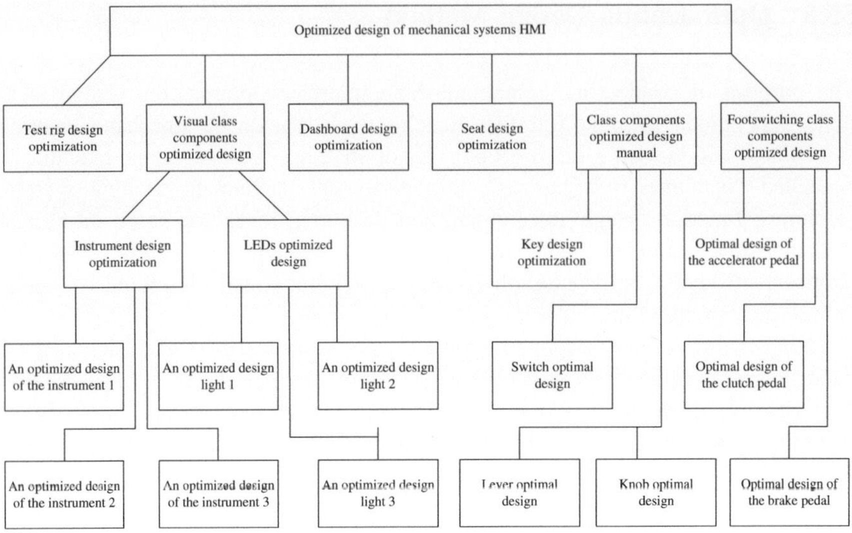

Fig. 55.2 HMI mechanical systems hierarchical design optimization

55.6 Conclusion

Optimization design method of mechanical system human–machine interface, that is put forward, can absolutely realize optimization design of mechanical system human–machine interface, improving design quality and design efficiency of mechanical system human–machine interface, shortening design cycle of related designers, and reducing design cost.

This work is supported by the opening foundation of the Science and Technology on Human Factors Engineering Laboratory, Chinese Astronaut Research and Training Center (grant no. SYFD140051810K).

References

1. Xu Z (2007) Thermal power plant control room display and control system research and application of human factors evaluation. Harbin Engineering University, Harbin
2. Li P (2008) Fuzzy comprehensive evaluation parcel sorters man-machine interface design. Packag Eng 07
3. Lin J (2001) Evaluation of the human-machine interface in mechanical systems. J Fujian Agric Univ 30(1):119–122
4. Cifrek M, Medved V, Tonkovic S et al (2009) Surface EMG based muscle fatigue evaluation in biomechanics. Clin Biomech 24(4):327–340
5. Al-Mulla MR, Sepulveda F, Colley M (2011) A review of non-invasive techniques to detect and predict localised muscle fatigue. Sensors 11:3545–3594

Chapter 56
Design and Evaluation of Human–Machine Interface of Simulation Training System

Meng Li

Abstract Human–machine interface is an important component of the simulation training system. First, the general principle of the human–machine interface design is briefly introduced. Based on the principles, considering the specialty of the users of simulation training system and the feature of war field application, the tactical requirement of the interface design is discussed, as well as some typical problem in the design process. Finally, the evaluation index system for the human–machine interface of simulation training system is put forward, and some advices for the uniform design of the human–machine interface are provided.

Keywords Simulation training · Human–machine interface · Human–machine interaction · Design principles · Evaluation index

Human–machine interface of simulation training system is the section that conducting information exchange between human and simulation training system and realizing interaction interface [1]. Human–machine interface is an important component of technology system structure of whole simulation training system. Successful human–machine interface shall greatly motivate training enthusiasm of participating trainees, and make simulation training more efficient, that is to say an excellent user interface is an important factor for success of application system.

Human–machine interface is a medium for information delivery of human and machines, also the interface for communication of human and machines [2]. The definition of human–machine interface could be divided into two types, broad sense and narrow sense; human–machine interface in narrow sense mainly aims at computer system model [3]. Human–machine interface in broad sense means medium or level for human–machine interaction of human–machine system, and that is interface. Processes of human–machine interaction are shown in Fig. 56.1.

M. Li (✉)
Naval Medical Research Institute, Shanghai 200433, China
e-mail: hg04381@163.com

© Springer-Verlag Berlin Heidelberg 2015
S. Long and B.S. Dhillon (eds.), *Proceedings of the 15th International Conference on Man–Machine–Environment System Engineering*, Lecture Notes in Electrical Engineering 356, DOI 10.1007/978-3-662-48224-7_56

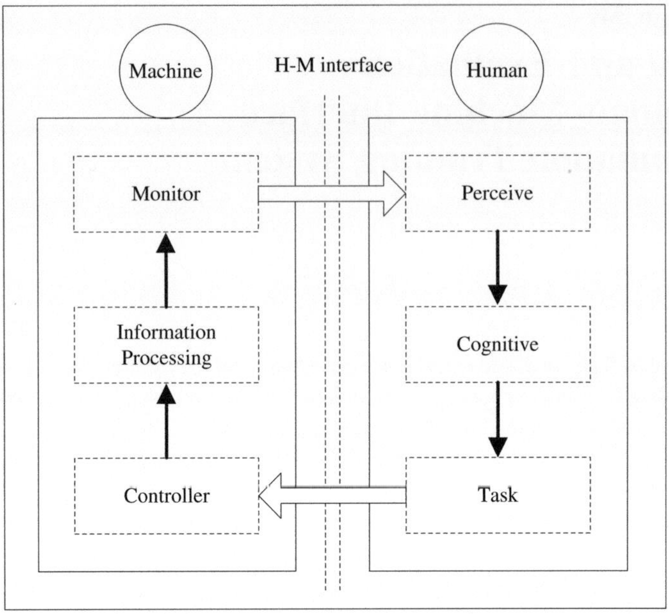

Fig. 56.1 Human–computer interaction

56.1 Design Requirements of Simulation Training System Human–Machine Interface

56.1.1 General Principles of Simulation Training System Human–Machine Interface Design

Generally speaking, a good human–machine interaction interface must follow some principles, and these principles constitute some kind of design standards, based on the precondition that following these principles, conducting appropriate improvements against system characteristics, one can get a good interaction interface.

(1) Uniformity of interface operation

Commonality of simulation training system gets better and better, and interface with good uniformity could extend operating experience of participating trainees to new operating tasks, in the meantime, completing training tasks in a better way and reducing memory work of participating trainees. Uniformity of interface operation usually presents the following aspects:

(a) Uniformity of style: for example, color matching, training terms, battlefield icon, and operational symbol;
(b) Uniformity of controls: conduct grouping of controls based on operation habits and interaction needs [4]; and

(c) Uniformity of operation: As for those common or important functions, keep uniformity of operating mode and pay attention to accord with operation habits of most participating trainees at the same time [5].

(2) Aesthetics of interface layout

Screen layout of human–machine interface should take users as principal thing; reasonable layout could not only improve comfort level of participating trainees during operation process, but also enable trainees to participate in training more prompt and efficient, such as follows:

(a) Screen color matching: color matching is an important design for screen display; colors have aesthetic value except being an efficient strength technology, constraining colors that are simultaneously displaying, usually they should not exceed four or five kinds of colors in the same picture, and coordinating colors with different levels and shapes to improve changes [6].
(b) Screen's information display: there is a huge amount of information in simulation training, and information need to simultaneously display might be huge, so reasonable choices are needed during design development based on timeliness and significance of information. Emphasizing on considering whether amount of information is too huge, whether important information are highlighted, and whether different kinds of information are distinguished obviously.
(c) Screen's area arrangement: try to keep balance of UDLR of screen. No pile-up data, too crowded display, would make participating trainees have visual fatigue and reception mistakes, emphasizing on considering numbers of area arrangement and reasonability of area arrangement.

(3) Friendliness of human–machine interaction

As for training system, friendliness of human–machine interaction is an important aspect, and this is fundamental difference between computers and other display systems [7, 8]. People passively accept information from TV, but in the human–machine interaction, there is two-way information between human and machines, and participating trainees could actively get needed information, so friendliness of human–machine appears very important.

(4) Diversity of interaction methods

Currently, human–machine interaction methods contain key, mouse, operating lever, and touch screen; with the development of voice recognition and compound, character recognition, and image recognition technology, human–machine interaction methods keep growing, and choosing appropriate interaction methods allows convenience for operation of participating trainees and improves training efficiency.

(5) Fault tolerance of interface interaction

A good interface enables users not to make severe mistakes. If mistakes have already appeared, then it should detect mistakes and provide simple and comprehensive handling methods of mistakes, and should not appear system crash [9], conducting error information tips for users and allowing users re-operate disoperation.

56.1.2 *Tactics Requirements of Simulation Training System Interface Design*

In view of user characteristics of simulation training system, users (trainers and trainees) shall also participate in design except designers. Based on battlefield application characteristics of systems, the design process must conform to specific tactics requirements except following general design requirements of human–machine interface.

(1) Flexible training conditions convenience for trainers

During operational process of simulation training system, trainers could input various combat conditions through human–machine interface conveniently based on requirements of training contents, and set different difficulty degrees. It shows terminals in the form of trainees' training work conditions, reaching the purpose of different tactics command training.

(2) Flexible control combat process convenience for trainers

During operational process of system, trainers provide with different kinds of means for the control and influence of combat process, enable to suspend or terminate operation, and could reappear operational command and results of specific trainees, in order to organize and implement teaching work.

(3) Convenience for trainers conducting training evaluation

Display battlefield situation of operational training simulation and operation of trainees through screen monitoring system, audio and video switching matrix, projector, camera, etc. It is convenient for trainers to know battlefield comprehensive situation in time and evaluate training condition scientifically and efficiently.

(4) Convenience for trainees accepting combat condition information

During practical training, trainees could conveniently, easily, and flexibly get combat condition information in system by various means at any time, and standard ability, commonality, and friendliness of operational interface should be better.

(5) Convenience for trainees handling combat condition

After trainees get combat condition information through various means, they enable to achieve tactics intention by friendly operation interface and conduct operation in the briefest means, and enable to conduct modification of previous mistaking operation or non-ideal operation.

(6) Convenience for trainees observing and affecting combat process

The system shall provide various status displays of combat process for trainees, including text and data report, sound effects, voice prompt, dynamic condition display, etc., with three-dimensional battlefield view, strengthening intuitive feelings of trainees and making simulation training more realistic. In the meantime, they also need to provide various means that may affect combat process in actual combat for trainees and trainees can choose, in order to foster overall awareness of trainees, and form information quality of modern combat and improve command ability of trainees.

56.1.3 Considering Typical Questions in Simulation Training System Interface Design

General principle of simulation training system interface design is furthest reducing operating tasks and mental stress of participating trainees, simplifying operating process, and need to consider following questions in detailed design.

(1) Considering expertise level, psychological quality, and memory restriction of participating trainees

User model in battlefield environment should include factors such as expertise level, knowledge structure, thinking ability, reaction capacity, physiology ability, and memory ability. When realizing it, adopting advanced computer technology and multimedia technology creates various visual or audio output interfaces, trying to adopt intuitive and simple input means, providing various operating approaches, which are convenient for the use of practiced operators and also meeting requirements of unskilled users.

(2) Information preprocessing and integrating

Various sensors, detectors, and weaponry from outside world and different information in network enter into simulation training system in a scattered and non-systematic way, before they reach output interface, and screening and integrating work should be done, and complete functions such as detection, classification, search target, logical reasoning, combat analysis, and aid decision making try to provide participating trainees with intuitive and succinct information.

(3) Better efficiency of human–machine interaction

Combat opportunity of modern combat is fleeting; high efficiency of human–machine interaction is the pursuant important performance of simulation training system. High efficiency mainly shows in two aspects. One is usefulness: interface content shall give prominence to the key points, have distinct gradation, display valuable information according to high–low priority level, and prevent gibberish occupying screen space. The other one is response time: as for an interface, if it cannot respond quickly, then no matter how powerful it is, and finally it will be eliminative.

56.2 Evaluation Index System of Simulation Training System Human–Machine Interface

There is no unified standard of human–machine interface evaluation; evaluation index system of simulation training system could put forward evaluation index system of simulation training system human–machine interface against particularity of simulation training system on the basis of general evaluation standards, which is shown in Fig. 56.2.

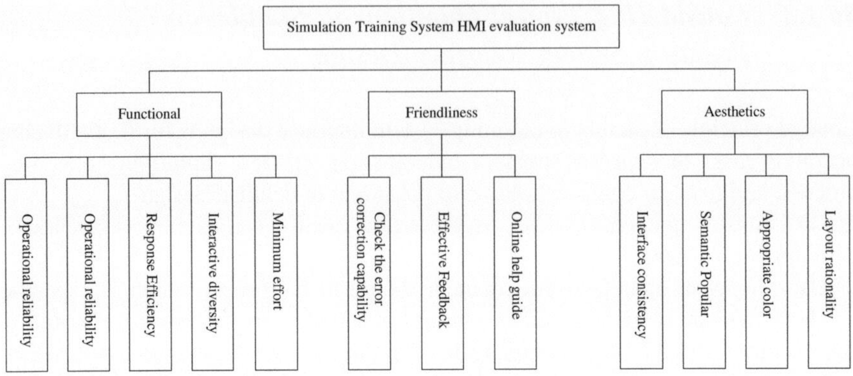

Fig. 56.2 Simulation training system HMI evaluation system

56.2.1 Functionality

Emphasis of functionality index is used to evaluate whether realized functions of system meet requirements of participating trainees, including (1) operation comfort level: inner feelings of users through operation steps; (2) operation reliability: the capability of system could operate within set time and no occurrence of malfunction; (3) response efficiency: fast or slow response process degree of system for users to submit events; (4) interaction diversity: users could conduct human–machine interaction through various ways; and (5) minimum workload: participating trainees reach purpose through minimum interaction steps and interaction operation.

56.2.2 Friendliness

Friendliness index emphasizes on evaluating whether human–machine interface design is humanization, including (1) error correcting capability: warning false operation in time and putting forward solutions; (2) efficient feedback: providing feedback mechanism and structure for interface problems and users' problems; (3) online help guidance: as escalation of training system performance, providing help timely is very important; as for participating trainees, providing help, guiding demonstration, monitoring, and giving advice could make users learn operation of system fast, and it is important for the success of failure of system.

56.2.3 *Aesthetics*

Aesthetics index emphasizes on evaluating appearance effect of interface, making operators feel invigoration in the operation process, including (1) interface uniformity: mainly including uniformity of style, controls, and operation; (2) semantic commonality: having coherence, no cognitive disorder, and easy to learn and memorize; (3) appropriate color: reasonable color matching, having aesthetic feeling; and (4) layout reasonability: content displaying on screen giving prominence to the key points and having distinct gradation.

56.3 Conclusion

As an acceleration of military information construction process, simulation training system will appear on training ground in a huge number. Simulation training system human–machine interface should integrate design philosophy of “people first” into development process of system, and improve cognition, aesthetics, fault tolerance, and learn ability of interface. As the development of artificial intelligence technology, virtual reality technology and other related technologies, human–machine interface, will develop toward better adaptive capacity of multimedia and intelligent zed way.

This work is supported by the opening foundation of the Science and Technology on Human Factors Engineering Laboratory, Chinese Astronaut Research and Training Center (grant no. SYFD140051810K).

References

1. Chen M, Zhang D (2007) Research and application based HMI development model. Packag Eng 28 (4):113–115
2. Hou J (2002) Talk about the man-machine interface design. Weifang Univ J 2(2):69–71
3. Yi X (2007) From the man-machine interface to the user interface. Art Des 1:8–9
4. Chen Q (2005) Optimal design of man-machine interface. Softw World 12:64–65
5. Luo S (2009) Human-machine interface design. Mechanical Industry Press, Beijing, pp 39–43
6. Jin C, Zhang N (2005) Human-machine interface design principles based on standardized cognitive theory. Ergonomics 11(3)
7. Li H (2005) Airborne information systems HMI software design. Modern Electron Eng (2):59–63
8. Li R, Huang Y (2006) User model study of human-machine interface design. Changchun Univ Technol 29(2):59–61
9. Shen X, Tao Jin, Wei D (2008) Thinking of CNC machining equipment design psychology from the perspective of human-computer interface design. Ind Des 35(6):66–67

Chapter 57
The Man–Machine Interface Research on Manned Lunar Rover

Baogui Qiu, Chunhui Wang, Fan Li, Zheng Wang and Shoupeng Huang

Abstract A study of manned lunar rovers' ergonomic design is presented. Based on summarization and characteristic analysis of NASA's and other nations' manned lunar rovers, design status and development trend are resulted. Lessons learned from the ergonomic design of lunar roving vehicles are discussed, when the astronauts are on the lunar EVA suit. Then we put forward Chinese manned lunar rovers' conceptual design. By studying the fundamental configuration of man–machine interface, we carry out five conceptual design categories of man–machine interface for Chinese manned lunar rover. Finally, we list some key technical indexes related to man–machine interface of manned lunar rovers. As many ergonomists work extensively on the development of lunar rovers, we know that it will be a footstone for our manned lunar rovers to explore the moon and other planets in the near future.

Keywords Manned lunar rover · Lunar EVA suit · Man–machine interface · Ergonomics

---

C. Wang—Contributes as equal as the first author and is the co-first author.

The research is funded by pre-research project of the general armament department "Research on Concept and General Technology of Manned Lunar Rover" and the project number is 030102.

B. Qiu
Shanghai Key Laboratory of Spacecraft Mechanism, Aerospace System Engineering
Shanghai, Shanghai, China

C. Wang · F. Li (✉) · Z. Wang · S. Huang
National Laboratory of Human Factors Engineering, China Astronaut Research and Training Center, Beijing, China
e-mail: 19650769@qq.com

© Springer-Verlag Berlin Heidelberg 2015

S. Long and B.S. Dhillon (eds.), *Proceedings of the 15th International Conference on Man–Machine–Environment System Engineering*, Lecture Notes in Electrical Engineering 356, DOI 10.1007/978-3-662-48224-7_57

57.1 Research Background

In the task of lunar exploration and planet development, human cognition, decision-making ability in a complex environment, and body coordination and flexibility cannot be replaced by other human intelligent agent and mechanical tools. The expected missions to complete cannot be achieved without astronaut's participation and control. In order to effectively expand the range of astronauts' activity and assist astronauts to fulfill various operation tasks on lunar surface, manned lunar rover is the key media and exploration tool, which recently has been one of the hot issues in the research field of lunar exploration tasks [1]. At present, countries and organizations with the implementation plan of lunar exploration invest heavily one after another in developing the lunar rover applicable for various tasks and environments. Applied as a kind of special electrically driven vehicle running on lunar surface, on the one hand, consideration shall be taken into the influence of low gravity on lunar surface and other special environments; and on the other hand, great changes have been taken place in astronaut's operation ability on lunar surface with lunar EVA suit, which also leave various special requirements on ergonomics interface design for manned lunar rover. This article studies on the ergonomics design of man–suit–vehicle system for prevention and improvement measures under the special working condition, puts forward the system of ergonomics design indicator for man–suit–vehicle interface with lunar EVA suit in task of manned lunar exploration, and provides references for the design of manned lunar rover to ensure that astronauts can fulfill tasks safely, healthily, and efficiently.

57.2 Development Trend of Manned Lunar Rover

United States holds a leading place in research of manned lunar rover as early as the 1970s. United States has the successful experience in applying manned lunar rover to the exploration tasks of the moon crafts Apollo 15, Apollo 16, and Apollo 17 [2], which is the only three-time record in moon landing with manned lunar rover in human history. In the 1990s, various countries developed many manned detectors one after another and of which MOBEVR1B, LOTRAN, LUR, DMLRV, and Chariot were most famous. Chinese manned lunar rover co-developed by Joint Research Center for deep space exploration organized by Ministry of Education consisting of 27 major universities of 985 Project, which Chongqing University assumes main research work, was displayed on the eleventh Chongqing Hi-tech Fair and the Seventh International Civil-military Integration Exposition on April 10, 2014. It is expected to complete conceptual pre-research of the manned lunar rover by the end of 2015 [3]. By sorting out the latest researches above on manned lunar rover and manned mars rover at home and abroad, manned lunar rover can be divided into five parts according to the demand and influence of operation capacity of human–suit system to manned lunar rover including [4]:

1. Navigation bridge

Navigation bridge consists of two folding seats, belts, folding foot rests, handles, polings, splasher, operating control, and console. In case of the lunar rover to be folded, first fold the seats and foot rests and then rotate the operating control and console backward for about 30°. Handles and polings are provided for astronaut's convenience for getting on and off and splasher can be used to prevent lunar dust or rock debris from hitting astronauts, scientific instrument, or the lunar rover itself.

2. Navigation subsystem

Navigation subsystem mainly consists of three parts including directional gyro, millimeter (to display speed and distance), and solid-state minicomputer (used to save the speed, direction, distance, and other information of each flight, and figure out current position of lunar rover and distance direction to the lander according to such information which shall be shown on the display).

3. Control and display board

Control and display system is modeled after instrument panel. The upper half part includes display information of navigation and relevant control switches. Lower half of the part is used for displaying electric quantity and driving information, and also relevant control switches are included. Abnormality and alarm indicator are equipped on the roof.

4. Driving subsystem (manual controller)

Typically, driving subsystem is mounted between two seats for astronauts in the mode of one-hand operation so that two astronauts can operate. Of which, the controller applies T-shaped design recommended by the astronauts (who consider that the T-shaped controller is easier to operate with pressure suit on). Controller is designed for direction control, speed control, and braking.

5. Deployment subsystem

Deployment subsystem adopts automatic and manual modes (of which manual mode for backup) to fold and unfold the lunar rover.

57.3 Influence on Ergonomics Design for Man–Machine Interface of Lunar EVA Suit and Manned Lunar Rover

Lunar EVA suit is the outbound life support equipment designed to guarantee astronaut's life safety and work ability as they carries out extravehicular activities on lunar surface. As astronauts with lunar EVA suit on, their flexibility, visible range, reachable range, and so on will change and may have a significant influence on design of man–suit–vehicle interface.

1. Maximum motion range

NASA employs lots of maximum motion range tests in EVA spacesuits (EMU) with portable foot limiter for man–suit system. Research shows that [5] astronaut's motion range will be smaller in lunar EVA suit, so the layout of operating equipments shall be more compact to make sure that astronaut operates in a comfortable position.

2. Maximum vision range

In lunar EVA suit, helmet shall influence the astronaut's maximum vision range. In design of human–computer interaction interface of the manned lunar rover, it is required to put important information within the key vision area for astronauts to complete various outbound observations and operation tasks. In addition, NASA also did some research on the key vision area of side window. In this case, it can not only make sure all visual tasks done within the maximum field of vision, but also can make sure layout of position of key equipments within key vision area of side window [5].

3. Flexibility of hand joints

Range of joint motion and joint torque in lunar EVA suit is the main factor in affecting astronaut's mobility and operational capability, especially the flexibility of hand joints. In the operation tasks on lunar surface, perceptive flexibility and mobility of finger joints such as comfort, temperature, strength, and shape of gloves of lunar EVA suit are the key factors in affecting the operational capability of man–EVA suit which may directly influence astronaut's ability to use tools and operate equipment. Control button of manned lunar rover and operating rod shall be in accordance with the requirements under the working conditions for gloves of lunar EVA suit with regard to shape and size [6].

4. Mass distribution

Mass distribution of lunar EVA suit may also affect the mobility performance of man–lunar EVA suit. In design of lunar EVA suit, make sure that the center of mass of lunar EVA suit shall be near to that of human body as possible or both on the same straight line, so as to make the astronauts only bear the weight vertically other than additional load from other directions, and consequently they can keep different still and balanced postures for easier walking to prevent from falling down. Under the working condition of lunar EVA suit, astronaut's balance is going down and the standing posture changes which requires auxiliary fixing device to be added to design of manned lunar rover (such as safety belt, handrail foot limiter, and so on), especially during driving in vertical position, astronauts are more prone to losing balance [7].

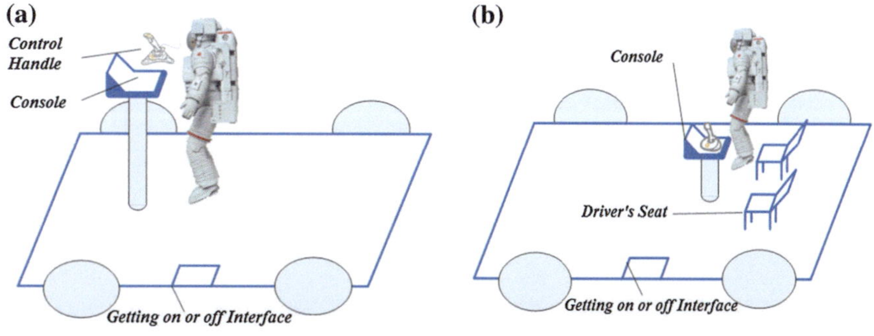

Fig. 57.1 **a** A concept design of driving manned lunar rover by standing. **b** A concept design of driving manned lunar rover by sitting

57.4 Analysis of Man–Machine Interface Ergonomics Design of Man–Suit System and Manned Lunar Rover

In order to put forward specific ergonomics interface design specification, from earlier research on manned lunar rover at home and abroad [8], simplified and universal concept design of manned lunar rover is raised and Fig. 57.1a, b shows separately the concept design drawings for driving in standing position and driving in sitting position.

Concept design of universal ergonomics between manned lunar rover and man–suit system can be divided into five categories according to typical operating tasks by manned lunar rover:

- Interface design of deployment lunar rover;
- Interface design to get on and off the manned lunar rover;
- Interface design to drive and take the lunar rover;
- Interface design for maintenance/repair; and
- Interface design for man–vehicle joint detection.

57.4.1 *Interface Analysis of Deployment Lunar Rover*

At the beginning of lunar exploration, in consideration of smaller volume of the lander, usually the manner lunar rover shall be packed to mount on the lander. At the same time, it is required to minimize installation operations due to great difficulty in installation of lunar rover under the working condition on lunar surface in lunar EVA suit. Therefore, in principle, the main body frame of lunar rover shall be foldable which shall automatically (or by astronaut) unload and unfold after landing on lunar surface, antenna, and seats, and other components shall be installed by

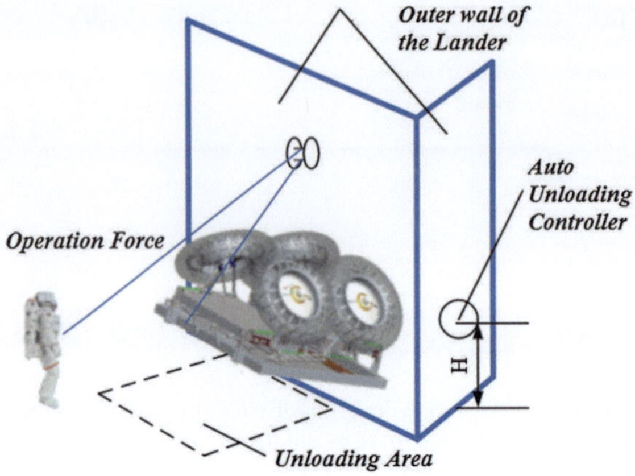

Fig. 57.2 Sketch map of lunar vertical rover reconfigure

astronauts on lunar surfacc aftcr unfolding under the lunar rover. In addition, the lunar rover shall be packed to be fixed outside of the lander in which way it can save inner space of the lander for deployment. In future, tasks of human lunar exploration, how to deploy the manned lunar rover, is an importance operation and usually the whole operation can be divided into two key steps which are unloading and unfolding of main body and other components installation. Unloading and unfolding of main body can be divided into automatic and manual modes, of which the latter is for backup. Here, it emphatically analyzes ergonomics interface design during unloading and unfolding by hand and the manned lunar rover is packed to mount on the outer wall of lander which is H from the ground (including support height). How to unload the manned lunar rover is shown in Fig. 57.2, and the operations to unload the manned lunar rover can also be divided into automatic and manual modes:

- Automatic unloading: Use mechanical device or electric device to assist astronaut to unload the lunar rover. Astronaut unloads the lunar rover on lunar surface by operating the automatic unloading controller (it could be a button, an operating rod or pull rod, and so on) on outer wall of the lunar lander; if the lunar rover is folded, then the lunar rover shall automatically unfold during unloading.
- Manual unloading: It is the backup mode for automatic unloading. Astronauts use simple auxiliary device to unload the lunar rover and pulling rope, and auxiliary unloading device of pulley was used during Apollo Project. Unloading weight of Apollo manned lunar rover is 208 kg which also requires bigger operating force even if under the 1/6 gravity environment. In the future, the weight of manned lunar rover may increase, so manual unloading can be designed as cooperative mode to ensure that astronauts can unload the lunar rover successfully. Furthermore, attention shall be paid to operating force and unloading area during manual unloading.

57.4.2 Interface Design to Get on and Off the Manned Lunar Rover

With lunar EVA suit on, astronaut's mass distribution changes and balance weakens, and body flexibility also greatly reduces. In this case, astronauts are prone to fall down due to unbalanced gravity. Therefore, auxiliary device shall be designed to assist astronaut to get on and off the lunar rover. Now there are three proposals for assisting astronauts to get on and off the lunar rover:

- According to height of car body and mobility of astronaut, astronauts can get on the lunar rover step by step using other tools as the step shown in Fig. 57.3a. Step height shall be dependent on the mobility range of astronaut's knee and hip joints with lunar EVA suit on. In addition, lunar surface environment shall be the main factor to be taken into consideration for design of vehicle height, while lunar surface environments at different locations are quite different which requires the lunar rover that can be able enough to climb slope and run on gravel pavement.
- The second proposal is to lift astronauts who are standing on the lift table to on and off the lunar rover by electric device or rocker arm, as shown in Fig. 57.3b. Handrail is required in reasonable location to make sure that astronauts can keep steady during lifting.
- The third proposal is to make further improvements based on the second proposal by installing guide rail on the pedestal, in which the seats can move horizontally or vertically on the guide and rail/astronauts can control movement of seats by controller or rocking bar mounted on seats so as to get on and off the lunar rover, and moreover, the proposal is only applicable to the manned lunar rover driven in sitting position.

Getting on and off the lunar rover has been the focus of ergonomics for the existing lunar EVA suit that may affect astronaut's performance in many aspects (such as stability of gravity, joint flexibility and tactile sensitivity, etc.), which makes getting on and off lunar rover an operation with risks. To avoid risks or

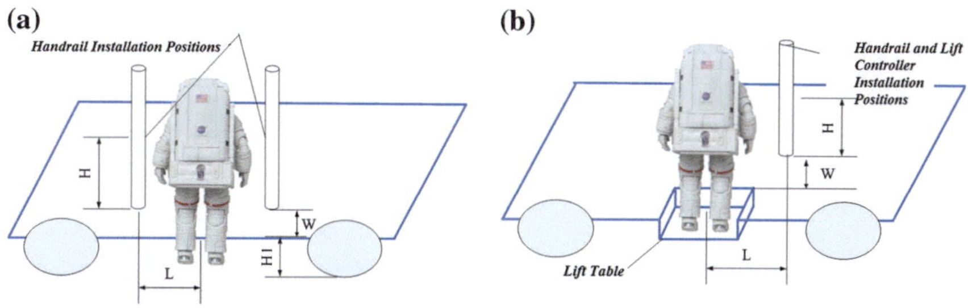

Fig. 57.3 **a** Astronaut landing the vehicle without lifting devices. **b** Astronaut landing the vehicle with lifting devices

improve comfort for operations, manned lunar rover is required to be provided with auxiliary equipment to assist astronaut to fulfill operations. In the future, lunar EVA suit must be lighter and fitter so the issues may be weakened accordingly.

57.4.3 Interface Analysis of Driving and Taking the Manned Lunar Rover

According to driving postures, manned lunar rovers can be categorized as manned lunar rovers controlled in sitting posture and manned lunar rovers controlled in standing postures, which are presented in Fig. 57.4a, b.

Operations in the process of driving are starting, driving, braking, and operating console. Relevant equipment includes a seat, a control rod, console, brake, auxiliary equipment, and so on. At present, the equipment can be divided, by control modes, into two types: hand controlled and foot controlled, and the analysis of their ergonomic interface is as follows:

1. Seat: It is only possessed by manned lunar rovers controlled in sitting posture. There is an interface between a seat and the life support system of lunar EVA suit. The seat supports the constraints of the life support system and reduces the load of the astronauts. The seat shall be equipped with safety belts in case of the instability of man–suit system caused by sudden stops, jerks, sharp turns, or bumpy roads. The position of the seat shall be adjustable so that astronauts can adjust their driving postures. After the technology of life support and lunar EVA suit become more mature, gas, cooling, and power sources can be installed in manned lunar rovers. Giving priority to the use of carried resources during the travel will help to realize farther exploration by manned lunar rovers. Therefore, an interface exists between the seat and the life support system of lunar EVA suit.

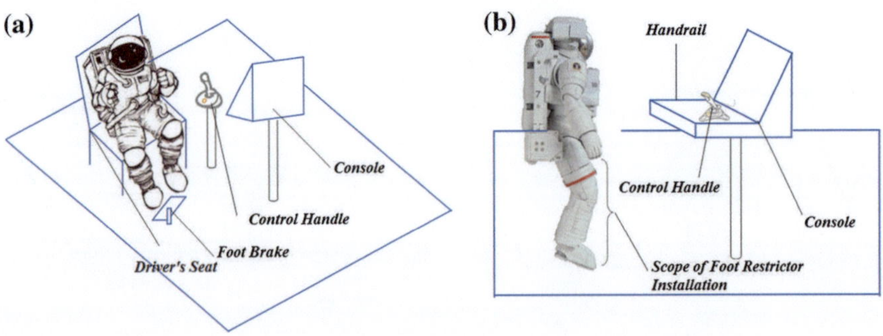

Fig. 57.4 **a** Controller module layout in sitting posture. **b** Controller module layout in standing posture

2. Control rod: Conventional controls of driving include controls performed by steering wheel and by control rod. The operation of the steering wheel needs great arm movement, which thus is not suitable to be used on the lunar surface. Control rods are mainly used for large engineering machines, which are relatively simple to operate. In Project Apollo, NASA found out that T-shaped control rod is the most convenient one. At present, the research of other control methods is being carried out at home and abroad, such as voice, brain, and other controls. Of course, these control messages need to be conveyed to the lunar rover after converted into electrical or radio signals. In the future, these advanced control methods can be used as auxiliary driving methods.
3. Brake: It can be divided into two types: hand controlled and foot controlled. Pedal brake has an interface with astronauts' legs, whose major ergonomic parameters are distance from the astronaut's coronal plane, operating force, inclination angle of the worktop, and so on. Major ergonomic parameters of hand brake are operating force, space capacity for hands, thickness (of the brake rod), and so on.
4. Console: It includes monitoring and control equipment. Monitoring information includes navigation display information, battery, battery temperature, speed, mileage, rotate speed of the wheels, body posture, abnormal alarm indicator, and so on. On the premise that the monitoring interface is within astronauts' visible range, identifications of all kinds of instrument must be clear and installation position must be not only reasonable but also consistent with human cognitive habits. Console switches, button sizes, and intervals need to meet the requirements of ergonomics.
5. Auxiliary devices: It includes handrails, foot limiters, and lighting tools. Driving in sitting posture, astronauts need handrails to be installed in order to help them stabilize their bodies. Shadows of other objects on the console are easily formed due to the direct light from the lunar. Therefore, auxiliary lighting means like fluorescent display are required to illuminate the instruments, meters, witches, buttons, and so on.

57.4.4 Interface Analysis of Maintenance/Repair of the Manned Lunar Rover

As the flexibility of hand joints and operation capability are lowered, which are vital factors for the design for maintenance/repair, simplified structure and shape design shall be adopted. The standard fittings must be given priority to be selected so as to improve the compatibility and the universal degree. Modular design is supposed to be applied to reduce the types and quantity of components, thus to simplify the maintenance/repair procedure. Once modular design is applied for vulnerable parts in manned lunar rovers, the rest can be re-used even if some damage has been caused to somewhere of a part. That is a method to minimize the number of spare parts. Interfaces of vulnerable parts shall adopt pluggable standard ones so that other interfaces can be replaced when a failure pops up to some interface. Backup module size and weight shall be moderate enough to be directly manipulated by the astronauts, mechanical arms, or other simple maintenance tools.

57.4.5 Interface Analysis of Man–Vehicle United Detection

With the help of the manned lunar rovers, lunar robots, and other auxiliary tools, astronauts can also carry out the discovery and exploitation of mineral resources and set up the complex lunar base on the moon. As operating articulated manipulator in the sites is representative, the man–machine united detection is still on the experimental stage. It has not been applied yet on the surface of the lunar or Mars. However, this kind of detection method can fully use the power of machines, reduce the consumption of manpower, and improve work efficiency, which is the development trend in the future. Therefore, in future tasks of man–vehicle united detection, the operation of manipulators will be more reasonable for which it is mainly done by terminal control. An important design principle of terminal controller of manipulator is that the operation direction of controller and the movement direction of the object are in the same direction. The control rod and handle (handle is controlled mainly through the lever handle or control keys) in Fig. 57.5a, b meet the need. The former should be fixed on the operating table, while the latter is movable. In addition, as astronauts are in lunar EVA suits all the time in the process of operating the manipulators, the visible range and the flexibility of hand joints of astronauts are reduced. Therefore, those factors shall be considered in the design of the interface of manipulator operation.

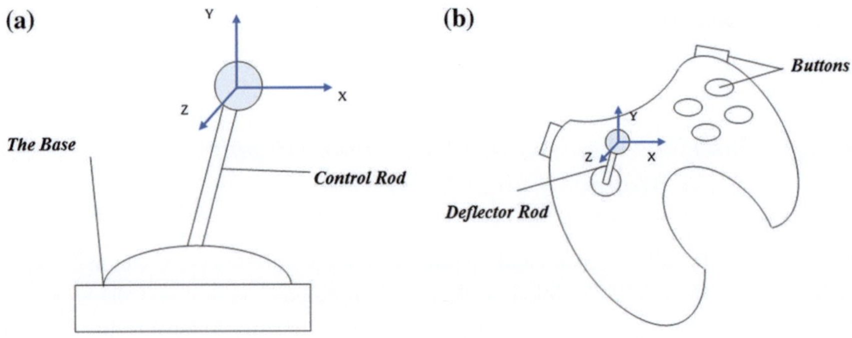

Fig. 57.5 **a** Control rod design. **b** Control pad design

57.5 Conclusion

There is no doubt that manned lunar landing has become the global focus of manned space development in the twenty-first century. For landing on the moon with the help of manned lunar module, astronauts' participation in the investigation, and experimental research of lunar science by manned lunar rover and other auxiliary tools, as well as the development and utilization of lunar resources, have great significance not only in politics, economy, and science, but also in the promoting the development of all kinds of space technologies [9]. Based on the induction and summary of current research status and development trend of lunar rovers and lunar suits abroad, and combined with the analysis of astronauts' ability changes after dressed in lunar EVA suits, this thesis has mapped key ergonomic problems of the interface between lunar EVA suits and manned lunar rovers, and also has drawn the design framework and implementation principles for ergonomic design of man–machine interface of Chinese manned lunar rover. Although China has started to follow the lunar exploration trends and results in 1960s, both the hardware configuration and software design are on the initial stage. Manned lunar landing is such a big idea and a huge project that today's exploration and research on manned lunar rover are just one small step in the long March of "Human oriented" manned space engineering and constitute a little technical reserves for the study of man–suit system and ergonomic research on man–machine interface of manned lunar rovers. With the gradual deepening of the study of manned lunar landing, more new problems and challenges will pop up, so we should carry out steady and solid work step by step.

References

1. Deng Z, Fan X et al (2012) Review and key techniques for locomotive system of manned lunar rovers. J Astronaut 33(6):676–689
2. Song G (2008) Study on control system of lunar rover based on CAN bus. Master thesis, Shanghai Jiao Tong University
3. http://www.cmse.gov.cn/know/show.php?itemid=79
4. Harrison DA, Ambrose R et al (2014) Next generation rover for lunar exploration. Technical report, NASA Johnson Space Center
5. Rochlis J, Delgado F, Graham J (2006) Science crew operations and utility testbed. Ind Robot: Int J 33(6):443–450
6. Scheuringa RA, Jonesa JA et al (2008) The Apollo medical operations project: recommendations to improve crew health and performance for future exploration missions and lunar surface operations. Acta Astronautica 63:980–987
7. Olofinboba O, Hamblin C et al (2008) The design of controls for NASA's orion crew exploration vehicle. In: Proceedings of the European association for aviation psychology 2008 conference. Valencia, Spain, 27–31 Oct
8. Steven C, Gregg M (2010) Roving Mars: using function analysis to develop a concept system. Ergon Des: Q Human Factors Appl. doi:10.1518/106480410X12887326202951
9. Chen S (2013) Medical challenges of manned lunar mission. Astornaut 7:33–39

Chapter 58
Research on Assessment of Eye Movement Sensitivity Index Through Aircraft Cockpit Man-Machine Interface Based on Eye Movement Tracking Technology

Yanyan Wang, Qingfeng Liu, Duanqin Xiong, Xiaozhao Guo, Yu Bai, Jian Du and Huajun Xiao

Abstract It aims at finding sensitivity index of the eye tracker in the evaluation of aircraft cabin static human-machine interface usability, and sets up the model of static eye movement test data. Method: based on the typical aircraft cockpit display interface, two groups of subjects have been set up with different difficulty, participants are required to find the corresponding position in the original image according to the given target image in the shortest time. Eye movement data of 40 participants, who take part in the tests with different degree of difficulty, has been used for paired T test; factor analysis has been done to eye movement indicators of statistical significance to T inspection results. Results are in the commonly used 25 eye movement indicators, there are 17 indexes with significant difference and $P < 0.05$. Results of factor analysis show there are five indicators with single factor contribution greater than 1; 5 indicators' cumulative contribution rate is 87.75 %. The first main components have been extracted, namely the index directly related to performance of the eye movement, including seven factors: Fixation points, total saccade time, spent time in AOI, total fixation duration, number of fixation points before arrival AOI, duration before first fixation arrival, total saccade distance. The sensitivity indicators: Number of fixation points before arrival AOI. The second main components of the pupil-related indicators include four factors: Average pupil area, average pupil diameter, average pupil diameter in AOI, average pupil area in AOI. The sensitivity indicators: Average pupil area in AOI. The third main components of the first eye movement indicators include three factors: First saccade distance, first saccade speed, and first fixation duration. The sensitivity indicators: First saccade distance. The fourth main components of the AOI eye movement index include two factors. The fifth main components include one factor: Average fixation duration in AOI.

Y. Wang · Q. Liu · D. Xiong · X. Guo · Y. Bai · J. Du · H. Xiao (✉)
Beijing University of Aeronautics and Astronautics, Beijing 100191, China
e-mail: 46999743@qq.com

© Springer-Verlag Berlin Heidelberg 2015

S. Long and B.S. Dhillon (eds.), *Proceedings of the 15th International Conference on Man–Machine–Environment System Engineering*, Lecture Notes in Electrical Engineering 356, DOI 10.1007/978-3-662-48224-7_58

Keywords Eye movement tracking technology · Human-machine interface · Ergonomics · Aircraft cabin

58.1 Introduction

As a direct media interface pilot control of the aircraft cockpit display, its ergonomic directly affects the level of aircraft control effect, especially for aircraft, the man-machine interface performance will directly determine the level of operational effectiveness. Eye tracking technology is applied in the human-machine interface evaluation has been nearly a hundred years of history [1], the eye tracker is the early invasive, with the development of technology of Dodge and Cline (1901) according to the principle of corneal light reflection developed the first non-intrusive eye movement instrument [2]. In 1947, Fitts and his colleagues began using a film camera on the pilot's cockpit eye movement control device and instrument landing aircraft, which is first used in aviation field eye tracker [3]. In twentieth Century 60 time, study on the pilots' scan mode to researchers, many models are all based on the eye movement data and eye movement data support for. Using eye movement data to evaluate the pilot instrument scanning behavior, to understand the pilot to control the aircraft attitude, position and velocity in three-dimensional space and obtain information from the instrument [4]. From the beginning of the 1970s at the end of the Anliker (1976) as the representative of a team of researchers has not satisfied to clarify the subjects in the "what", but to pay attention to the meaning behind the [5] index. 1990 years after the eye tracking system has been widely used in usability evaluation by eye tracking technology, put forward a lot of opinions on the improvement of interface design, product.

The interface is always in the front of the aircraft cockpit display interface, with "human centered design" gradually into the aircraft design, various stages of development; eye tracking technology efficiency of man-machine interface has become one of the most important means to evaluate the [6, 7] evaluation. Get the eye tracker on the market participants of the eye movement data is less than 20 and more than 10 more, how to use these eye movement data, the data processing speed faster, more play an important role in the human-machine interface evaluation function has become an urgent problem. Therefore, the author hopes through methodology test efficiency, found the eye tracker in the sensitivity index of the aircraft cabin static interface usability evaluation, and the establishment of the static test of eye movement data model. Already provide useful support for aircraft cockpit man-machine interface design.

58.2 Method

58.2.1 Test Object

Test subjects were a total of 40 people between the ages of 26–49 years old (an average of 36, with a standard deviation of 6), the total time of flight is 221–7455.20 h (an average of 2142, with a standard deviation of 1723).

58.2.2 Equipment

The test instrument used for America Chamber Test Applied Science (Applied Science Laboratory, ASL) ASL—D6 eye tracker (Model D6 Eye View produced by Monitoring Systems). The sampling rate of 60 Hz. Subjects were recorded: eye movement trajectory, eye movement time (including the fixation time, regression time, saccade time, saccade distance, pupil diameter, etc. data).

58.2.3 Experimental Design

Experiment with complete randomized block experimental paradigm. The test procedures for the program, participants in the fastest time based on the target image to find the corresponding position in the original image. Elements of design program for cockpit typical picture of a total of 10, part of the picture from the picture capture meaningful subjects need to search the "target," set target map editor according to the color, size, position three, 30 total system problems, then the 30 question intention increase the difficulty, such as the size is reduced to the original 1/2. A total of 60 such pairs: 30 groups of problems, the efficiency of the A segment, B segment efficiency is not good. The set itself two groups of problems obviously ergonomics differences, through the test of 25 commonly used indexes of eye movement instrument selection, in order to find the sensitive indexes of cockpit ergonomics evaluation of man machine interface.

58.2.4 Test Procedure

Each participant agrees to participate in the test, first adjust the posture, the main test calibration. The end of the program presents calibration instructions, subjects into practice in a clear mode of operation after the test, a total of three exercises, which are the correct answer into the formal test. The formal test consists of 60 questions, in order to avoid the practice effect, the questions were presented randomly. For each question first appeared “target,” and “target” disappeared, with “target” emergence, subjects with the fastest speed in the big picture is found in the “target” by the space bar to confirm, automatic calibration, then move on to the next question until all of the questions were completed. Each picture presentation time was 200 ms. Background program recording eye movement data, the correct rate, reaction time.

58.2.5 Statistical Data

Twenty-five commonly used indicators of eye movement (including: AOI, AOI back depending on the number of focus, AOI number in average fixation duration, average pupil diameter of AOI, AOI, AOI average pupil area in fixation duration, AOI, AOI for the first time at the end of time, processing time first fixation duration, AOI the average fixation duration, average saccade time, average saccade distance, average saccade speed, average pupil diameter, average pupil area, the total fixation time, total time, total fixation saccade distance, count, before entering the AOI fixation duration, fixation count before entering the AOI, first fixation duration, first saccade time, first saccade distance, first saccade velocity) are used for paired T test. Factor analysis is done to eye movement indicators, which have statistical significance to T test results.

58.3 Results

58.3.1 Analysis Results of Eye Movement to Aircraft Cockpit Display Images with Different Ergonomics Difference

As the eye tracker is found as different efficiency show differences between the tools we developed two sets of test questions, design elements for the active screen picture of typical aircraft a total of 10, part of the picture from the picture capture meaningful subjects need to search the “target,” set target according to the color map editor the location, size, a total of three ways, preparation of Title 30, then the 30 question deliberately increase the difficulty, such as the size is reduced to the original 1/2. A total of 60 such pairs: 30 groups of problems, the efficiency of the A segment, B segment efficiency is not good. Paired T test was used for data processing of the 25 indicators of eye movement in common use, and homogeneity of variance test. The results are given in Table 58.1:

Table 58.1 Paired t test results of commonly used eye movement test parameters

Index	df	F	Sig.
Fixation points A–B	1.00	97.50	0.00**
First fixation duration A–B	1.00	12.38	0.00**
Average fixation duration A–B	1.00	0.01	0.95
Total fixation duration A–B	1.00	94.87	0.00**
First saccade distance A–B	1.00	12.93	0.00**
Average saccade distance A–B	1.00	0.43	0.51
Total saccade distance A–B	1.00	91.83	0.00**
First saccade time A–B	1.00	2.76	0.10
Average saccade time A–B	1.00	2.35	0.13
Total saccade time A–B	1.00	94.04	0.00**
First saccade speed A–B	1.00	5.96	0.02*
Average saccade speed A–B	1.00	2.53	0.11
Average pupil diameter A–B	1.00	14.87	0.00**
Average pupil area A–B	1.00	13.03	0.00**
Fixation points in AOI A–B	1.00	3.21	0.07
Number of fixation points before arrival AOI A–B	1.00	115.09	0.00**
Number of times regression in AOI A–B	1.00	16.35	0.00**
First fixation duration in AOI A–B	1.00	3.03	0.08
Spent time in AOI A–B	1.00	98.31	0.00**
First pass time in AOI A–B	1.00	0.03	0.87
Average fixation duration in AOI A–B	1.00	6.28	0.01*
Fixation duration in AOI A–B	1.00	14.54	0.00**
Duration before first fixation arrival A–B	1.00	107.83	0.00**
Average pupil diameter in AOI A–B	1.00	43.96	0.00**
Average pupil area in AOI A–B	1.00	43.93	0.00**

*$P < 0.05$, **$P < 0.01$

58.3.2 Analysis of Aircraft Cockpit HMI Eye Movement Sensitivity Index

According to statistics, in the 25 eye movement indicators commonly used, there are 17 indexes have significant difference, and $P < 0.05$ or $P < 0.01$, it can be shown that the eye tracker can be used to determine the difference between the tool of ergonomics. Further to the 17 indexes by factor analysis. The results are as follows:

The KMO statistic was 0.69, the partial correlation is very weak, suitable for factor analysis, Bertlettball test, reject the original hypothesis of unit correlation matrix, $P < 0.001$.

It is known from Table 58.2 that five principal components of eigenvalue are greater than 1, their cumulative contribution to total variance is 87.75 %. Main axis

Table 58.2 Total Variance Explained

Component	Initial Eigenvalues			Extraction sums of squared loadings		
	Total	% of Variance	Cumulative %	Total	% of Variance	Cumulative %
1.00	6.36	37.42	37.42	6.36	37.42	37.42
2.00	3.88	22.81	60.23	3.88	22.81	60.23
3.00	1.96	11.53	71.76	1.96	11.53	71.76
4.00	1.61	9.47	81.23	1.61	9.47	81.23
5.00	1.11	6.52	87.75	1.11	6.52	87.75
6.00	0.84	4.95	92.69			
7.00	0.47	2.79	95.48			
8.00	0.27	1.59	97.08			
9.00	0.18	1.04	98.12			
10.00	0.13	0.75	98.87			
11.00	0.09	0.51	99.37			
12.00	0.04	0.25	99.62			
13.00	0.04	0.22	99.84			
14.00	0.02	0.09	99.93			
15.00	0.01	0.05	99.97			
16.00	0.00	0.02	99.99			
17.00	0.00	0.01	100.00			

Extraction Method Principal Component Analysis

factor extraction method is used with maximum variance rotation, the results after five iterations are given in Table 58.3:

Table t test results of 1, the first main components have been extracted, namely the index directly related to performance of the eye movement, including seven factors: Fixation points, total saccade time, spent time in AOI, total fixation duration, number of fixation points before arrival AOI, duration before first fixation arrival, total saccade distance. The sensitivity indicators: Number of fixation points before arrival AOI. The second main components of the pupil-related indicators include four factors: Average pupil area, average pupil diameter, average pupil diameter in AOI, average pupil area in AOI. The sensitivity indicators: Average pupil area in AOI. The third main components of the first eye movement indicators include three factors: First saccade distance, first saccade speed, and first fixation duration. The sensitivity indicators: First saccade distance. The fourth main components of the AOI eye movement index include two factors. The fifth main components include one factor: Average fixation duration in AOI.

Table 58.3 Rotated component matrix[a]

Index	Component				
	1	2	3	4	5
Fixation points	0.95				
Total saccade time	0.94				
Spent time in AOI	0.93				
Total fixation duration	0.93				
Number of fixation points before arrival AOI	0.89				
Duration before first fixation arrival	0.89				
Total saccade distance	0.89				
Average pupil area		0.98			
Average pupil diameter		0.98			
Average pupil diameter in AOI		0.98			
Average pupil area in AOI		0.98			
First saccade distance			0.93		
First saccade speed			0.92		
First fixation duration					
Number of times regression in AOI				0.86	
Fixation duration in AOI				0.78	
Average fixation duration in AOI					0.93

Extraction Method Principal Component Analysis
Rotation Method Varimax with Kaiser Normalization
[a]Rotation converged in five iterations

58.4 Conclusion

This study is based on a typical aircraft cockpit display interface, the establishment of two groups of subjects with different levels of difficulty, participants in the fastest time based on the target image to find the corresponding position in the original image, to participate in the trial of 44 participants completed the eye movement data of different degree of difficulty of the paired T test, eye movement indicators and statistical significance of T test results of the factor analysis. Results in the 25 eye movement indicators commonly used, there are 17, indexes have significant difference, $P < 0.05$. The results of factor analysis show there are six principal components with eigenvalue greater than 1, their cumulative contribution to total variance is 87.75 %. Result is obtained by using the main factor extraction method and after the maximum variance rotation and 5 times of iteration, combined with the average value of all factors and the result of T test, The first main components have been extracted, namely the index directly related to performance of the eye movement, including seven factors: Fixation points, total saccade time, spent time in AOI, total fixation duration, number of fixation points before arrival AOI, duration before first fixation arrival, total saccade distance. The sensitivity indicators: Number of fixation points before arrival AOI. The second main components of the pupil related indicators include four factors: Average pupil area, average pupil diameter, average pupil diameter in AOI, average pupil area in AOI. The sensitivity indicators: Average pupil area in AOI. The third main components of the first eye movement indicators include three factors: First saccade distance, first saccade speed, and first fixation duration. The sensitivity indicators: First saccade distance. The fourth main components of the AOI eye movement index include two factors. The fifth main components include one factor: Average fixation duration in AOI.

58.5 Compliance with Ethical Standards

The study had obtained approval from the State Key Laboratory of NBC Protection for Civilian Ethics Committee.

Each pilot was briefed on the aims of the study and informed consent was obtained. All pilots were paid $32/h. for their participation.

All relevant ethical safeguards had been met in relation to subject protection.

References

1. Javal E (1878) Essaisur la physiologie de la lecture. Annalesd'Oculistique 79, 97–117, 155–167, 240–274; 80 (1879), 61–73, 72–81, 157–162, 159–170, 242–253
2. Dodge and Cline (1901) The angle velocity of eye movements. Psychol Rev 8:145–157
3. Fitts PM, Jones RE, Milton JL (1950) Eye movements of aircraft pilots during instrument-landing approaches. Aeronaut Eng Rev 9(2):24–29
4. Anliker J (1976) Eye movements: on-line measurement, analysis, and control. In: Monty RS, Senders JW (eds) Eye movements and psychological processes. Lawrence Erlbaum Associates, Hillsdale, pp 185–199
5. Fisher DF, Monty RA, Senders JW (eds) (1981) Eye movements: cognition and visual perception. Lawrence Erlbaum, Hillsdale
6. Manhartsberger M, Zellhofer N (2005) Eye tracking in usability research: what users really see? Usability Symp OCG Publication 198:141–152
7. Hu F, Han J, Ge L (2005) Summary of eye tracking and usability testing research. Human Ergon 11(2):52–54

Part V
Research on the Man–Environment Relationship

Chapter 59
The Protection Measure Research into the Medical Diagnostic X-ray of "Peace Ark" Hospital Ship

Hangtao Cheng, Chuan Wang and Jianguo Zhang

Abstract In order to understand the radiology injury to the working personnel and subjects it is detected the dermal dose of subjects receiving radiology diagnosis on the "Peace Ark" Hospital Ship, to statistically calculate the average dose volume and weighted dose equivalent of each X-ray diagnosis posed on the subjects, and carry out comparison analysis for the risk degree of subjects of different ages and genders. Results demonstrated, for each diagnosis, the subjects to radiology diagnosis would all have different radiology dose on skin and other corresponding organs; with the increase of age, the radiology risk degree of subjects would decrease gradually, the male subjects have different risk degree in different radiology dose and risk degree compared with female subjects in different tissues and organs. It is suggested that radiology diagnosis working personnel and subjects should be careful to apply radiology diagnosis, to justify the radiology diagnosis, at the same time, reduce the exposure, to realize the optimization of radiology protection.

Keywords Radiology diagnosis · X-ray · Protection measure

Medical Radiology Diagnosis is the field in which X-ray has been most widely applied by human race, as the artificial radiation source with the most dose contribution to all the people as a collection, therefore, the protection for radiology diagnosis X-ray is of special significance [1].

Radiology Diagnosis could often find out the crucial reason of the patients' symptoms, enabling them with symptomatic treatment [2, 3], however, there also

H. Cheng
Military Agent's Room in Guangzhou Ship Company of Navy Equipment, Guangzhou 510382, China

C. Wang (✉)
Nuclear Science and Engineering Department of Naval Engineering University, Wuhan 430033, China
e-mail: hg04381@163.com

C. Wang · J. Zhang
Naval Medical Research Institute, Shanghai 200433, China

© Springer-Verlag Berlin Heidelberg 2015

S. Long and B.S. Dhillon (eds.), *Proceedings of the 15th International Conference on Man–Machine–Environment System Engineering*, Lecture Notes in Electrical Engineering 356, DOI 10.1007/978-3-662-48224-7_59

exists the risk of exposure to radiation damage. Therefore, when carrying out radiology diagnosis, it should first be justified for the irradiation. It should be carefully considered of the possible applicable information provided in the current radiology diagnosis, as well as the risk to be undertaken. In comparison, it should be considered whether the former is greater than the latter, in order to judge whether or not it should carry out radiology diagnosis. During radiology diagnosis, for the subjects themselves, the benefits of diagnosis for the diseases and the risk of radiology injury are embodied in the same person; therefore, it is unnecessary to formulate the upper limits of the dose for the subjects, in stead, it is only necessary to carry out balance between benefits and harms. However, when carrying out protection for radiology personnel, subjects and the public in the neighborhood, the optimization principle should be followed, in order to maximize the social benefit with the minimum collective dose.

There are two types of factors influencing the dose for working personnel and subjects, the former is controlled by the working personnel, such as which kind of X-ray should be applied, as well as what the radiation scope and exposure volume should be administered, and where the working personnel should stand, etc.; and the latter is related to machines, equipments, materials and architectural conditions, and other conditions applied.

59.1 Principle of Radiation Diagnosis

59.1.1 Justification for the Radiology Diagnosis to Be Applied

Physicians and technical personnel putting forward the application and implementing radiology diagnosis should all be fully awarded of the indications of radiology diagnosis as well as the basic knowledge of exposure injury. It should be understood in particular, under what conditions, certain radiology diagnosis is indispensable, under what stage of the pathogenesis, it would occur the physical sign that could be discovered by the radiology diagnosis, as well as how long it may take to perceive the changes occurred, therefore, it is necessary to carry out radiology diagnosis review [4]. In one certain radiology diagnosis, the subjects would receive how much dermal dose, and how much the organ absorption dose would be (referred to Tables 59.1 and 59.2), as well as how much radiology injury risk degree it may result in (referred to Table 59.3). In this way, the physician could be able to balance between benefits and risks, to decide whether this radiology diagnosis is justified or not.

In order to avoid or reduce the irradiation injury, first it should reduce the unnecessary irradiation. For example, if it is decided that one patient is in need of radiology diagnosis, it should be first inquired of the patient whether or not he/she has undergone radiology diagnosis of the same category, if the necessary data is

Table 59.1 Radiological diagnosis of the subject's skin dose on "Peace Ark" hospital ship

Species examined	The number of cases	Changes in skin dose range (mGy)
Chest fluoroscopy	300	2.4–50.4
Upper gastrointestinal contrast	300	106–1105
Esophagus angiography	300	11–498
Barium enema	300	74–514
Head photography	300	1.1–37.0
Cervical photography	300	0.5–12.0
Abdominal plain film	300	1.1–29.1

Table 59.2 The absorbed dose for each diagnostic X-ray examination in the subjects on "Peace Ark" hospital ship

Species examined	The number of cases	Absorbed dose to each organ or tissue (mGy)			
		Red bone marrow	Bone surface	Thyroid	Lung
Outpatient chest	300	0.26	0.58	0.03	0.72
Esophagus angiography	300	1.90	4.30	0.99	6.20
Upper gastrointestinal contrast	300	6.00	13.2	1.20	10.6
Head	300	0.47	1.04	4.84	0.01
Cervical vertebra	300	0.29	0.64	2.33	0.10
Abdominal plain film	300	0.43	0.95	0.02	0.11

Table 59.3 Risk of different age and sex by irradiation (The dimensionless, Unit 1) [6]

Male								
Tissue or organ	Age range/Year old							
	0–9	10–19	20–29	30–39	40–49	50–59	60–69	70
Red bone marrow	20	20	20	20	20	15	6	0
Lung	21	21	19	17	12	7	2	0
Thyroid	5	5	5	5	3	2	0	0
Bone surface	5	5	5	5	3	2	0	0
Skin	1	1	1	1	1	0	0	0
Others	51	50	47	40	30	17	5	
Female								
Tissue or organ	Age range/Year old							
	0–9	10–19	20–29	30–39	40–49	50–59	60–69	70
Red bone marrow	20	20	20	20	20	18	10	4
Lung	21	21	19	17	12	7	2	0
Thyroid	5	5	5	5	3	2	0	0
Bone surface	5	5	5	5	3	2	0	0
Skin	1	1	1	1	1	0	0	0
Others	51	50	47	40	30	17	5	0

available, it should not carry out this radiology diagnosis again (those who underwent similar inspection in other hospital recently should not repeat the inspection).

Second, it should substitute inspection of higher dose with inspection of as lower dose as possible, and replace radiology inspection with non-radiology inspection. For example, the average dermal dose of chest X-ray is 13 mGy, while one chest Photography has a dermal dose of 1/24 of the chest X-ray. In our country, chest X-ray cover 70 % of overall radiology diagnosis, if most of them are changed to photography, or shifted to X-ray machine with photography enhancer for chest X-ray, then the total irradiation dose of all people in our country could be greatly reduced [5].

When there is X-ray general survey for certain diseases, if it is discovered that there are enough early phase patients, enabling them to be treated in time without the risks of too greater irradiation injury, then such general survey should be encouraged. If there is any doubt for the value of such general survey or danger, then it should be studied earnestly. For example, epidemiological investigation in Japan demonstrated that during stomach cancer and chest diseases radiology diagnosis general survey, in population aged over 30 years old, the newly discovered cancer patients exceed the expected irradiation-induced cancer number, namely, the benefit-risk ratio is over 1. However, in population aged below 30 years old, the benefit-risk ratio is below 1 [6].

59.1.2 Reduce the Irradiation Volume to Optimize the Exposure Protection

Radiation biology research demonstrated, the severity degree of determination effect of radiation injury increase along with the dose, while the random effect incidence increase along with the dose. In order to obtain the maximum diagnosis effect with risk degree as minor as possible, to realize the optimized principle for radiology diagnosis, it should reduce the irradiation dose and irradiation position as much as possible. Therefore, it should commence in the following several aspects:

59.1.2.1 Improve the Record and Reduce Dose

The purpose of radiology diagnosis is to obtain clear X-ray image. The dose volume of subjects receiving irradiation, as to X-ray, depending on the irradiation rate necessary for the display system to show clear X-ray images, X-ray radiation time and the penetration rate of X-ray to subjects; as to photography, it depends on the radiation volume necessary for the imaging of film as well as the penetration rate of X-ray to subjects [7].

As to X-ray photography, the method for improving the record system sensitivity is to select high sensitivity intensifying screen and film combination, as well as high sensitivity development kit developing technique. Application of high speed intensifying screen would often reduce the image quality, which would require the balance between reducing dose of the subjects and maintaining image quality. However, applying rare-earth intensifying screen would often improve the sensitivity and maintain the image quality at the same time. When applying intensifying screen, attention must be paid to select the film it matches, since different intensifying screen would send out fluorescence of different color (blue or green), when applying intensifying screen, it is not suitable to use films commonly applied for X-ray photography.

59.1.2.2 Improving the Penetrating Coefficient of X-ray

The higher the tube voltage of X-ray machine, and the thicker the filtration lens it applied, the stronger the penetration strength of X-ray irradiation, when carrying out radiology diagnosis, the higher the ratio between average irradiation volume of X-ray sent out from the body surface of the subjects and the average irradiation volume shooting into the body namely the penetrating coefficient. In this way, it could obtain the same exit irradiation volume with smaller incidence irradiation volume, to form the image available for diagnosis [8]. For example, a set of X-ray machine with automatic luminance control when carrying out upper gastrointestinal tract barium meal, if 65 kVX radiation is applied for X-ray, the tube current is 4 mA. For X-ray of the same position, if it is shifted to 80 kVX radiation, then the tube current is 0.5 mA. Since the tube voltage is increased, X-ray output of the same tube current increases by 50 %, while the tube current is reduced to one-eighth of the original tube current here. Therefore, the dermal irradiation volume of subject would reduce to one-fifth of the original volume, so, it should apply high tube voltage as much as possible [9].

59.1.2.3 Reduce the Irradiation Area as Much as Possible

It is very important to reduce the irradiation fields for the protection of the working staff. The larger the exposure fields, the higher the odds for the working staff to enter the effective irradiation area. In order to obtain the exposure fields area as small as possible, the X-ray machine for photography should be equipped with adjustable effective irradiation rectangle irradiation field's device, and intuitively indicate the size and position of the exposure fields with light, in case it does not comply, it should be rectified in time. If the exposure fields is larger, it would increase the unnecessary exposure scope, however, if the exposure fields is smaller, it is liable to result in the missing shoot giving rise to the necessity for reshoot. All these cases would influence the quality of radiology diagnosis, making the subjects be exposed to larger dose of irradiation. In some cases, when there is no irradiation adjustment device, such as fluorescent X-ray photography and all chest X-ray photography, it should be set up with proper rectangle fixed irradiation device.

59.1.2.4 Novel Materials Should Be Applied, to Reduce the Unnecessary Attenuation

During X-ray photography, the X-ray penetrating the body of the subjects, would penetrate the diagnosis bed board of the bed, further more the irradiation filtering device and film cartridge would also absorb part of the X-ray. In order to obtain the irradiation volume necessary to form the image available for diagnosis, it should increase the X-ray irradiation volume penetrating the subjects' body. In order to reduce these unnecessary attenuation as much as possible, somebody use carbon fiber board with smaller absorption coefficient to manufacture the diagnosis bed board, film cartridge, and the connection strap between the lead strips in the radiation filtering device. Experiment indicate, when the peak voltage of X-ray machine is 70 kV, using carbon fiber board to manufacture film cartridge, would enable the subjects to receive 10 % to 30 % less of exposure dose than using plastic film cartridge, while using carbon fiber board to manufacture the diagnosis bed board could reduce 14 % of exposure.

59.1.2.5 Reduce the Exposure Time, to Control the Irradiation Volume of Subjects

In order to reduce the dose of subjects, on the one hand, it should enhance the sensitivity of image display system to reduce the body surface irradiation rate entering the subjects' body, on the other hand, it should reduce the exposure time as much as possible. When using ordinary screen to carry out X-ray inspection, the image is very faint. In order to enable the physician of the radiology department to see clearly of the image as soon as possible to carry out diagnosis, the X-ray room should avoid any external light exposure, even a faint light beam would sometimes spoil the contrast of the image. Apply the image intensifying device, it could not only greatly improve the sensitivity of display, therefore reduce the X-ray irradiation rate needed, but also obtain clear image to greatly reduce the X-ray time needed. In recent years, it emerges the image storage devices and pulse pressure X-ray machine, with all these devices, it could effectively reduce the exposure time, and reduce the subjects' irradiation dose.

59.1.2.6 Away from the Exposure Fields

When the subjects are receiving X-ray inspection, apart from the inspected position, it should make the other position of the body far away from the effective irradiation and exposure position as much as possible. The working staff carrying out specific imaging for gastrointestinal and cardiac catheter, cerebrovascular inspection, it is often necessary to carry out operations close to the working bench, at this time, the position of the working staff is standing with distance related to the exposure fields has very great influences on the exposure dose the working staff received. For example, working staff standing beside the subjects to inject the contrast agent, if they move away from the position 10 cm away from the exposure fields to 30 cm away from the exposure fields, their exposure volume would reduce by nearly tenfolds.

59.1.2.7 Carrying Out Necessary Screening

In order to reduce the exposure to working staff and the public, it is necessary to carry out screening for scattered radiation as needed. First, it must capture the direct irradiation not absorbed by the subjects. Therefore, the exposure fields must be smaller than the X-ray screen or image intensifying device, and the screening thickness for the maximum tube voltage for 100 kV X-ray machine should be 2 mm lead equivalent, for X-ray machine with maximum tube voltage of 150 kV, it should be 2.5 mm lead equivalent. For X-ray machine used for photography, it should be equipped with lead screening plate with a thickness of 2 mm as the direct irradiation capturer lead board to be as close as possible to the film cartridge, in this way, the capturer needed is the smallest in size and lightest in weight. In order to screening the scattered irradiation form the subjects, X-ray machine must have boat-shaped plate on both sides of the bed, under the screen and beside the screen, it should be equipped with lead rubber curtain of sufficient size of at least 45 cm × 45 cm and of sufficient thickness of at least 0.5 mm lead equivalent.

59.1.2.8 Ensure the Irradiation Photography Quality and Reduce the Reshooting Rate

For various reasons, for X-ray films in poor quality or failed to meet the clinical requirement, it is inevitable to carry out reshooting some times. However, it should reduce the reshoot times as much as possible. The rate of reshooting depends not only on the technical level of the radiology photographers, their working experience and earnest spirits, but also on the rigorous quality control system of a unit. According to the investigation in some countries [10], the re-shooting rate is approximately between 3 and 12 %. The main reasons lead to the reshooting are poor location, error in light exposure volume as well as the darkroom technical problems. Therefore, it should be strictly controlled of the light exposure conditions, quality of the developing agents as well as developing conditions, to ensure the optimal status of exposure, developing, and fixation procedures, so as to apply the minimum irradiation dose to achieve the images of the best quality, at the same time it could be avoided of the reshooting caused by problems during exposure and developing procedures.

59.2 Conclusion

As with all external irradiation source protections, the basic principle of medical X-ray radiology diagnosis protection is to reduce the exposure time, away from the X-ray sources and necessary screening. However, X-ray machines are controlled with width of the available beam, irradiation rate of 1 m from the focal spot, and the semi-value layer thickness of the beam, and with the changes of available beam, there would also be changes of leakage irradiation and scattered irradiation. All these conditions bring about some specific requirements for X-ray protection. The useless X-ray harmful in the instrument and equipment, should be sufficiently screened, to make its leakage volume would not exceed corresponding standard. The width of available beam for various X-ray machines should be reduced as much as possible. Because the irradiation rate of available beam is much larger than the leakage irradiation and scattered irradiation, it could reduce the odds and position of personnel receiving direct exposure, to reduce the width of available beam. At the same time, it is also advantageous for reducing scattered irradiation. Under most conditions, it should also enhance the sensitivity of the detector or the display system, to reduce all the necessary available beam irradiation.

This work is supported by the opening foundation of the Science and Technology on Human Factors Engineering Laboratory, Chinese Astronaut Research and Training Center (grant no. SYFD140051810K).

References

1. Shao W (2011) Talking about the importance of X-ray radiation on the body harm and protection. China Pract Med 11(6):72–73
2. Liang D (2003) Benefits X-ray diagnosis of clinical inquiry. Jinggangshan Med Coll 10 (6):65–66
3. Yang K (2009) The value of X-ray radiation in the secondary diagnosis. Environ Occup Med 26(6):606
4. Zhang X (2010) X-ray is a double-edged sword. Sino-Jpn Friendship Hosp 24(3):131–132
5. Duan Y, Ju Z (2008) Indexing and treatment of acute radiation sickness. X-ray spectra of the Tenth National
6. Li L, Yu X, Xia C (2004) Radiation exposure to medical diagnostic study guide dose levels. China Med Equip 4(2):28–30
7. Chen X, Li Y (2011) Medical rescue nuclear and radiation emergencies. Ninth army radiation medicine society of professional committee meeting, 2011, p 5
8. Liu Q (2011) According to the principle of X-ray machines, radiographic techniques on improving methods. Med Inf 1(4):1653–1654
9. Tang J, Jiang L, Deng S (2008) Advances in digital medical X-ray imaging system. China Med Imaging Technol 24 (Suppl):4–7
10. China Institute for Radiation Protection (1994) Nuclear and radiation protection. Atomic Energy Press, Beijing, p 79

Chapter 60
The Establish of Evaluation Index System for Equipment Personnel Effectiveness in High Altitude Area

Heping Wang, Xuegong Liu and Fang Zhang

Abstract The plateau environments make a great impact on soldier's physiological, biochemical, physical, and other personnel factors, which affect the soldier's battle effectiveness. In this paper, the soldier's SaO2, heart rate, physical endurance explosive force are analyzed in plateau and plain area. The test plan is designed for soldier's comprehensive ability based on ergonomics, and then the effects of plateau on the staff's operational efficiency are studied. Finally, the index evaluation system is established; it provides the basis for the evaluation of equipment personnel changes in plateau combat.

Keywords Plateau · Effectiveness · Index system

60.1 Introduction

The plateau area has special natural environment, which is characterized by low pressure, low oxygen content in the air, ultraviolet radiation, wind speed higher. Flexibility, stability of poor high altitude environment will affect the equipment operational personnel action, and the physical ability greatly decreased, greatly reduced the reliability of personnel system [1, 2], need to make the field test of the effectiveness. And the evaluation, puts forward the effective improvement measures, improve the effectiveness.

H. Wang (✉) · X. Liu · F. Zhang
Beijing Special Vehicle Research Institute, Beijing 100072, China
e-mail: whpzhjh@sina.com

© Springer-Verlag Berlin Heidelberg 2015

S. Long and B.S. Dhillon (eds.), *Proceedings of the 15th International Conference on Man–Machine–Environment System Engineering*, Lecture Notes in Electrical Engineering 356, DOI 10.1007/978-3-662-48224-7_60

60.2 Plateau Influences on the Capability of Body Parameters

Evaluation personnel ability is a complex problem; it is evaluated in many aspects. Ergonomics indexes affecting operational capability in plateau area include blood oxygen saturation, heart rate, and blood pressure. Work ability Index properties include physical endurance, speed, power, etc. These indicators reflect the plateau adaptation time, endurance, speed, etc.

60.2.1 Plateau Influences on the Physiological Indexes

Plateau influences on staff capacity are mainly caused by the decrease of oxygen. The oxygen partial pressure is reduced, so the blood oxygen saturation decreased. Oxygen saturation is an important index to reflect the degree of the supply to the body, it is often used as the individual labor ability and labor intensity index, when SaO2 is below 85 %, it can lead to impaired brain concentration and muscle ability reduced [3]. Heart rate variation is an important index to reflect myocardial hypoxia and antifatigue ability.

In order to assess the impact on personnel oxygen saturation, about 80 people aged 20–28 years are selected. Their SaO2, heart rate and blood pressure are measured after staying for 7, 15, and 30 days in altitude of 4200 m . And the data are analyzed. Some test results are shown in tables and figure (Tables 60.1 and 60.2).

It can be seen from the above chart staff, at the beginning of the oxygen saturation decreased rate is at around 13 %, the heart change rate is at around 25 %, it

Table 60.1 SaO2 change in plain and plateau (%)

Tester	1	2	3	4	5	6	7	Average
Plain area	96	98	99	98	97	95	99	97.4
Plateau(7)	85	82	86	84	85	83	86	84.4
Plateau(15)	88	86	87	85	89	87	87	87
Plateau(30)	93	92	94	91	90	90	91	91.6

Table 60.2 Heart rate change in plain and plateau (times/min)

Tester	1	2	3	4	5	6	7	Average
Plain area	63	65	72	71	66	75	69	68.7
Plateau(7)	82	85	90	95	88	93	94	89.6
Plateau(15)	75	78	85	82	80	78	81	79.9
Plateau(30)	68	70	79	78	75	76	73	74.1

proves that the supply oxygen of the body ability and antifatigue ability are decreased obviously. After 30 days of adjustment, the blood oxygen saturation increases, compared with plain change rate reduced to about 6 %, heart rate slows, compared with plain change rate at around 7 %, it proves the staff of antifatigue ability are increased.

60.2.2 Plateau Influence on the Personnel Manual Index

The PWC_{150} method is often used to test and evaluation personnel physical endurance changes. The PWC_{150} method refers to the subjects of heart rate in 150/min, works in the unit time, it is an objective index to evaluate labor ability and endurance. The value is greater so that the work ability and endurance is stronger. Eighty young people are tested by the bicycle ergometer method, wearing the heart rate zone, starting with 50 W, keeping 60 r/min continuous pedaling, every 2 min increments 50 W, pedal to the subjects' heart rate for 150 times/min. Each tester is tested both in plain and plateau area, each plain and plateau subjects to test for 3–5 time, the results are compared both in plain and plateau area.

Participants in the study of PWC_{150} test are shown Fig. 60.1. The test results of PWC_{150} in plain and plateau are shown in Table 60.3.

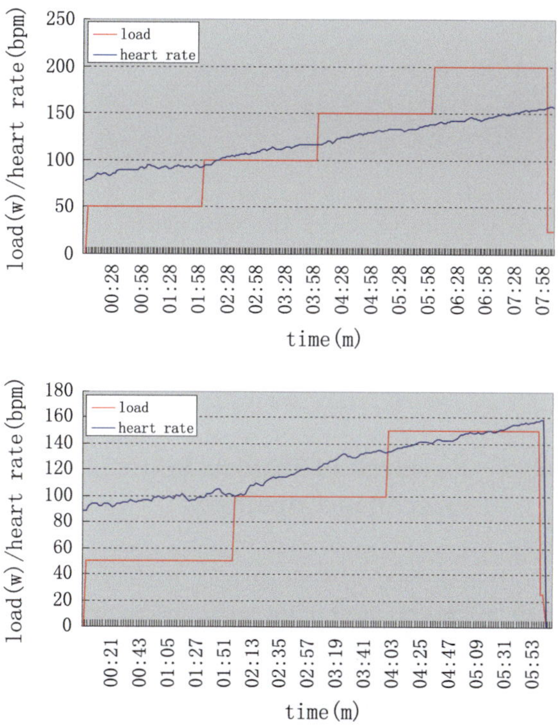

Fig. 60.1 The compared PWC_{150} results in plain and plateau

Table 60.3 The PWC_{150} test results in plain and plateau (W)

Tester	1	2	3	4	5	6	7
Plain	164	136	216	190	139	169	189
Plateau	135	115	156	149	116	140	159
Change%	17	15	27	21	16	17	16

Table 60.4 The back strength test results in plain and plateau (kg)

Tester	1	2	3	4	5	6	7
Plain	181	161	183	196	142	149	178
Plateau	173	155	169	185	133	140	162
Change%	4	3.7	7	5.6	6.3	6	8.9

Table 60.5 The grip strength test results in plain and plateau (kg)

Tester	1	2	3	4	5	6	7
Plain	63.9	53.6	44.7	55.5	50.7	62.9	72.1
Plateau	60.1	50.4	42.1	50.8	46.9	57.8	68.3
Change%	5.9	6	5.8	8.4	7.4	8.1	5.3

It can be seen from the above table and figure. In plateau the participant's labor ability and endurance are decreased compared in plain, the change rate is at around 15–25 %. It proves that people's work ability are obvious decreased in plateau environment.

60.2.3 *Explosive Forces*

The back strength and the grip force are the objective indexes to evaluate people body's explosive force. The participants are tested by electron back gauge and dynamo meter both in plateau and plain. The results are compared and analyzed. The test results are shown in Tables 60.4 and 60.5.

In plateau people's back strength and grip strength decreased slightly compared in plain. The average decline rates are 5.9 and 6.7 %.

60.3 Research of Plateau Influences on the Personnel Operation Capability

SaO2, heart rate, physical strength, explosive force response to the impact of plateau on the personnel of the single index, these indexes cannot reflect the inherent operational efficiency of personnel. According to the mission profiles and these inherent indicators, a new mission profile is designed, which can be combined with the operations and can reflect the scheme of these indicators change.

When people perform some missions, equipment preparation time is an important factor that determines the success or failure of the missions. The preparation time is also a comprehensive reflection of the people's body function and work efficiency. In this paper, some equipment operational readiness time is selected as a mission profile to test personnel work capability both in plateau and plain areas. In this profile, the same number of crew members and equipments are selected both in plateau and plain areas.

The test-specific steps and calculation scheme as follows:

1. In the equipment required storage condition, the preparation time is from receiving task to execution task;
2. The equipment preparation profile steps can reaction of mental labor and reflect the work efficiency. It commands personnel to the fastest speed.
   Mainly include:
   (a) Check the map to determine the marching solution;
   (b) Check vehicle's oil and water;
   (c) Check the weapon system;
   (d) Remove the vehicle battery replaces the new battery;
   (e) Launch the vehicle.
   The specific steps staff see Table 60.6.
3. The tests are both done two times in plateau and plain areas, the average values are calculated. Finally, the test results are analyzed according to the overall situation of processing, single preparation time is estimated. In the of single preparation time, the various elements of the work time are counted. The parallel work takes the longest time. Finally, the calculation work times are added as a single test time.
4. The preparation time point estimate value:

$$\hat{T} = \sum_{i=1}^{n} Ti/n \tag{60.1}$$

   In formula: $\hat{T}$ Single preparation time point estimate value, h;
   n Test times;
   T_i Each test of the longest working time, h;
   To calculate the test data, and then the test results are evaluated.
5. Record complete process the individual time, then calculate the cumulative time of completion, and tested in physical index of personnel.
   The test results are shown in Table 60.6.

Table 60.6 The preparation time in plateau and plain areas

Test	Driving scheme	Check weapon	Check water	Remove battery	Battery installation	Start vehicle	Total time
Plateau	3′12′	2′36″	1′36″	5′15″	7′5″	45″	13′45″
Plain	2′45′	1′56″	1′5″	4′15″	4′50″	34″	9′54″
Change%	16	25	32	19	31	24	28

From Table 60.6, the same people perform the same action in plateau and plain areas. The mission preparation time in plateau is more than 3 min and 54 s in plain, the work efficiency decline in 28 % or so. In plateau personnel's working efficiency is obviously reduced.

60.4 Construct of Personnel Capability Assessment Index System in Plateau

Through the above study, compared in plain, people's S_aO_2, explosive force, the physical endurance and work efficiency has varying degrees of decline, the average decline rates are 12.6 and 5.9 and 18.4 and 28 % in plateau (Fig 60.2).

To set up an objective and reasonable evaluation index system, not only need to consider the importance of indicators, should as far as use the quantitative indicators, to reduce the influence of subjective consciousness. In this paper, the plateau influence degree on operation personnel each index is analyzed, through investigation and expert consultation widely, and combined with the actual situation of plateau operations, the three levels evaluation index systems about plateau on staff work ability influence is established. The first class index reflects the plateau influence on staff' all aspects, including physical parameters, operation ability, work efficiency, outbreak force, plateau adaptability, etc.

Physical parameters in plateau area include heart rate, blood pressure and oxygen saturation, respiratory frequency, blood red cell, blood protein, etc. [4]. They show that the body of the maximal oxygen uptake and heart function. Work ability is mainly the labor ability, including physical strength and endurance of change, external performance and heart, lung function parameters of body organs. Mutation ability mainly includes the reaction ability, the explosive force, speed; operation ability mainly includes the efficiency, the completion of the mission profile; optimum altitude adaptation ability mainly includes the personnel to adapt to the

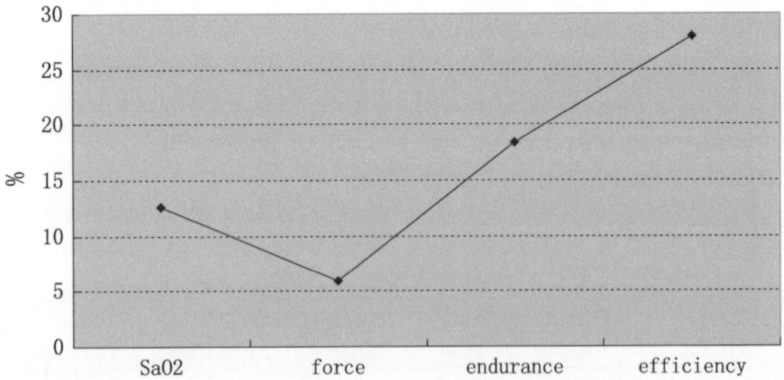

Fig. 60.2 The change of personnel effectiveness in plain and plateau areas

Table 60.7 The evaluation index system of plateau on staff influence

First index	Second index	Third index
Physical parameters	Physiological parameter	Heart rate, blood pressure, respiratory frequency
	Biochemical parameters	Oxygen, red blood cell, hemoglobin
Work ability	Physical	PWC_{150}, PWC_{170}
	Endurance	Maximal oxygen uptake
Mutation ability	Response ability	Reaction time
	Explosive force	Grip strength, back strength, speed
Combat capability	Task completion Efficiency	Combat ready time, the time to complete the task
	Completion situation	Reconnaissance, driving, shooting task complete effect
Ability to adapt to the plateau	Symptom	Mild operation without no obvious discomfort obvious discomfort, moderate no obvious discomfort, severe
	Time	Basic for 30 days, complete for 6 months
	Physical exercise	Mild exercise, moderate exercise
Improve ability	Drug	Improve cardiovascular capacity, improve lung function
	Oxygen inhalation	Low concentration of oxygen, high concentrations of oxygen
		Interval oxygen, continuous oxygen inhalation

plateau time, altitude sickness symptoms [5] in addition to the plateau symptom improvement measures. Plateau influence on operational personnel evaluation index system as described in the following table (Table 60.7).

60.5 Conclusions

Parameters evaluation of the impact of plateau on the equipment personnel involves in the complex and diverse, it needs to be analyzed from different angles, different levels. In this paper, different test plans are designed, the physiological, physical equipment, operation ability, ability to complete tasks, and change situation are compared in plateau and plain area. Based the change, the evaluation index system is established.

References

1. Pelamatti G, Pascotto M, Semenza C (2003) Verbal free recall in plateau: proper names us common names. Cortex 39(1):97–10
2. Shiyi Z, Dongfang H, Jie R (2012) the environmental impact of vehicle equipment and protective countermeasures of high temperature and high humidity. Environ Eng 16(1):265–268
3. Wenxiang G, Yuqi G (2005) Highland oxygen indication and plateau oxygen note, southwest. Def Med 5(2):228–230
4. Sodium rcgulath M (1998) Hormones at plateau: basal and post—gxercise levels. JC Lin End ocrinol Metab 83:570–574
5. MaresM BertoloC (1991) Hemorheological study in patients with coronary artery disease. Cardiology 78(2):111–116

Chapter 61
The Evaluation Standards and Methods of Human Vibration Comfort in Seated and Recumbent Position

Chen Su, Deguang Duan, Fu Niu, Xudong Ren and Xiuguo Zhao

Abstract In this paper, the evaluation standards of human vibration comfort were summarized contrastively. During the standards, the ISO 2631 (Evaluation for human exposure to whole-body vibration) was widely used to evaluate the human comfort under a vibration condition. In China, the evaluation standards were established refer to the ISO 2631 and revised by own experiment. Summing up the published standards externally and internally, the general evaluation method for human vibration comfort in seated position and recumbent position were given. The method of fatigue-decreased proficiency boundary T_{CD} was used to evaluate human vibration comfort in seated position. The method of frequency-weighted acceleration at the buttock and head a_{xbhw} was used to evaluate human vibration comfort in recumbent position. On the base of the standards, students were researching the new methods.

Keywords Human vibration comfort · Evaluate standard · Evaluate method

61.1 Introduction

People traveling by car or working in factory were always in a vibration environment. When beyond a certain intensity, the vibration environmental could affect the comfort of human body. Nowadays, the human vibration comfort had been concerned by more and more studies. The position of human in seated or recumbent had different evaluation standards and methods. So it was needed to publish a unified standard and method for the evaluation of human vibration comfort.

C. Su · D. Duan (✉) · F. Niu · X. Ren · X. Zhao
Institute of Medical Equipment, Academy of Military Medical Sciences, via Wandong Road, NO.106, Tianjin, China
e-mail: duandeguang@sina.com

C. Su
e-mail: wzssuchen@163.com

© Springer-Verlag Berlin Heidelberg 2015

S. Long and B.S. Dhillon (eds.), *Proceedings of the 15th International Conference on Man–Machine–Environment System Engineering*, Lecture Notes in Electrical Engineering 356, DOI 10.1007/978-3-662-48224-7_61

61.2 Evaluation Standards for Human Vibration Comfort

As shown in Table 61.1, the standards and methods were published to evaluate the human vibration comfort [1].

During the standards, the ISO 2631 was widely used to evaluate the human comfort under a vibration condition. For the earliest edition ISO 2631-1:1985 [2], the numerical limiting values of exposure for vibrations were given. The one-third octave comparing method and overall weighted acceleration method were

Table 61.1 The standards on human vibration comfort

Year	Number	Name	Remarks
1985	ISO 2631-1	Evaluation for human exposure to whole-body vibration: part 1: general requirements	Cancel
1992	GB/T 13441	Specification for measurement of human exposure to whole-body vibration environment	Refer to ISO2631-1:1985 cancel
1992	GB/T 13442	Reduced comfort boundary and evaluation criteria for human exposure to whole-body vibration	Refer to ISO2631-1:1985 cancel
1997	ISO 2631-1	Mechanical vibration and shock—evaluation of human exposure to whole-body vibration—part 1: general requirements	Replace ISO2631-1:1985 ISO2631-3:1985
2001	GB/T 18368	Comfort evaluation of human exposure to whole-body vibration in recumbent position	Refer to ISO2631-1:1997
2001	ISO 2631-4	Mechanical vibration and shock—evaluation of human exposure to whole-body vibration—part 4: guidelines for the evaluation of the effects of vibration and rotational motion on passenger and crew comfort in fixed-guide way transport systems	
2007	GB/T 13441.1	Mechanical vibration and shock—evaluation of human exposure to whole-body vibration—part 1: general requirements	Equal to ISO 2631-1:1997 Replace GB/T13441-1992 GB/T13442-1992
2012	GB/T 13441.4	Mechanical vibration and shock—evaluation of human exposure to whole-body vibration—part 4: guidelines for the evaluation of the effects of vibration and rotational motion on passenger and crew comfort in fixed-guide way transport systems	Equal to ISO 2631-4:2001

recommended to evaluate the human vibration comfort. The one-third octave method was that comparing the one-third octave of the vibration acceleration RMS value to the limitation in equal frequency, during human sensitive frequency range from 1 to 80 Hz. When each frequency band of the one-third octave was lower than the limitation, the human vibration comfort was acceptable. The overall weighted acceleration method considers the different influence degree on human comfort in different frequency bands. Besides the human sensitive frequency range, the weighted the acceleration was equated as the root-mean-square acceleration on sensitive frequency range.

In 1997, the standard was revised by the ISO2631-1:1997 [3]. This standard offered a new evaluation method which using weighted root-mean-square acceleration to evaluate the human comfort exposure to whole-body vibration. Contrasting the weighted curve of evaluation methods with the two standards, the new weighted values of evaluation method were greater than the older from 1 to 4 Hz and lesser from 10 to 80 Hz. Although the human sensitive frequency range on vertical was from 4 to 8 Hz, the resonance frequencies of human organs were mostly distributed on high frequency range. So the new evaluation method paid more attention to the high frequency range. Meanwhile, the relationship between values of weighted root-mean-square acceleration and feeling of human vibration comfort was provided.

In 2001, the ISO 2631-4 [4] was set up to design and evaluate the passenger comfort of fixed-guide way passenger systems. In this standard, the W_b (Frequency Weighting) was used as the appropriate weighting curve for the analysis of motions of fixed-guide way vehicles. At the same time, other factors should be taken into account for judging human comfort such as seat design, temperature, humidity, air quality, and acoustic noise.

In China, on the basis of ISO 2631-1:1985, the GB/T13441:1992 [5] and GB/T13442:1992 [6] were respectively published. Then, these standards were replaced by the GB/T 13441.1:2007 [7], which equaling to the ISO2631-1:1997. On the base of ISO 2631-4:2001, the GB/T 13441.4:2012 [8] was published. At the same time, for evaluating the human comfort in recumbent position, the GB/T 18368:2001 [9] was published independently.

61.3 Evaluation Method for Human Vibration Comfort

Summing up the published standards externally and internally, the general evaluation method for human vibration comfort in seated position and recumbent position were given (Fig. 61.1).

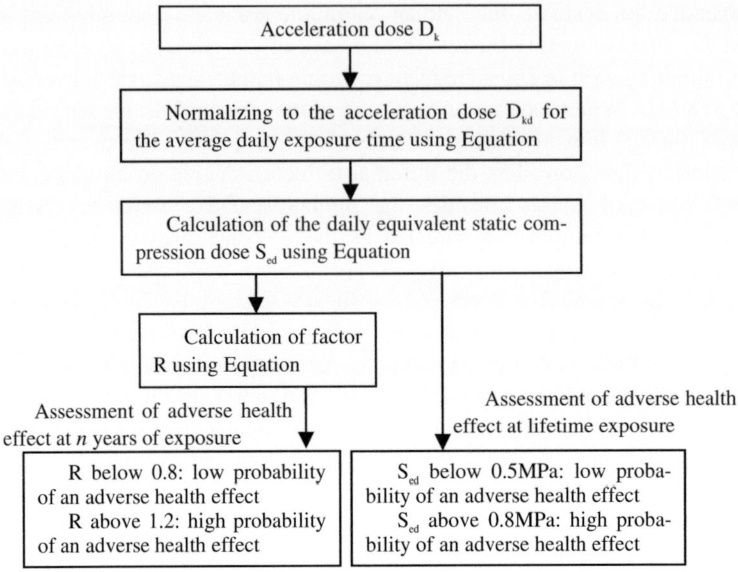

Fig. 61.1

61.3.1 Evaluation Method for Seated Position

Step 1: Calculating the RMS acceleration values of the vibration on one-third octave bands:

$$a_j = \left[\int_{f_{lj}}^{f_{uj}} G_a(f)df \right]^{\frac{1}{2}}$$

where

a_j is the RMS acceleration for the jth one-third octave band.
f_{uj} is the upper limit frequency for the *j*th one-third octave band.
f_{lj} is the lower limit frequency for the *j*th one-third octave band.
$G_a(f)$ is the self power spectral density of vibration acceleration.

Step 2: Calculating the frequency-weighted RMS acceleration on x, y, z axes:

$$a_W = \left[\sum (W_j \cdot a_j)^2 \right]^{\frac{1}{2}}$$

Table 61.2 Frequency weighting factors in one-third octaves for seated position

Frequency f_j Hz	Weighting factor x-axis y-axis	Weighting factor z-axis
1	1	0.50
1.25	1	0.56
1.6	1	0.63
2	1	0.71
2.5	0.8	0.8
3.15	0.63	0.9
4	0.5	1
5	0.4	1
6.3	0.315	1
8	0.25	1
10	0.2	0.8
12.5	0.16	0.63
16	0.125	0.5
20	0.1	0.4
25	0.08	0.315
31.5	0.063	0.25
40	0.05	0.2
50	0.04	0.16
63	0.0315	0.125
80	0.025	0.1

Notes
1 x-axis is the direction from back to chest
2 y-axis is the direction from left to right
3 z-axis is the direction from foot or buttock to head

where

a_W is the frequency-weighted RMS acceleration.
W_j is the weighting factor for the *j*th one-third octave band given in Table 61.2.

Step 3: Calculating the total frequency-weighted RMS acceleration:

$$a_{WO} = \left[(1.4a_{WX})^2 + (1.4a_{WY})^2 + a_{WZ}^2\right]^{\frac{1}{2}}$$

where

a_{WO} is the total frequency-weighted RMS acceleration.
a_{WX} is the frequency-weighted RMS acceleration on x axis.
a_{WY} is the frequency-weighted RMS acceleration on y axis.
a_{WZ} is the frequency-weighted RMS acceleration on z axis.

1.4 is the ratio of transverse vibration to longitudinal vibration during human sensitive frequency range.

Step 4: Calculating the values of the fatigue-decreased proficiency boundary:

$$T_{CD} = 0.167 \bigg/ \left(\frac{3.15}{5.6} a_{WO}\right)^2$$

61.3.2 Evaluation Method for Recumbent Position

Step 1: Calculating the RMS acceleration values of the vibration on one-third octave bands:

$$a_j = \left[\int_{f_{lj}}^{f_{uj}} G_a(f)df\right]^{\frac{1}{2}}$$

Step 2: Calculating the position-weighted acceleration at the buttock and head:

$$a_{xbhj} = W_{xbj} \cdot a_{xbj} + W_{xhj} \cdot a_{xhj}$$

where

a_{xbhj} is the position-weighted acceleration at the buttock and head
W_{xhj} is the position-weighted factors at the head given in Table 61.3.
a_{xhj} is the RMS acceleration at the head.
W_{xbj} is the position-weighted factors at the buttock given in Table 61.3.
a_{xbj} is the RMS acceleration at the buttock.

Step 3: Calculating the frequency-weighted acceleration at the buttock and head:

$$a_{xbhw} = \left[\sum_{j=1}^{23} (W_{xbhj} \cdot a_{xbhj})^2\right]^{\frac{1}{2}}$$

where

a_{xbhw} is the frequency-weighted acceleration at the buttock and head.
W_{xbhj} is the frequency-weighted factors given in Table 61.3.

Step 4: Estimating the evaluation grade of human comfort to whole-body vibration in recumbent position according to the Table 61.4.

Table 61.3 The position-weighted and frequency-weighted factors

Frequency f_j Hz	Position-weighted factors		Frequency-weighted factors W_{xbhj}
	W_{xhj}	W_{xbj}	
0.5	0.485	0.515	0.439
0.63	0.485	0.515	0.439
0.8	0.485	0.515	0.439
1	0.485	0.515	0.439
1.25	0.485	0.515	0.439
1.6	0.485	0.515	0.439
2	0.485	0.515	0.439
2.5	0.485	0.515	0.439
3.15	0.485	0.515	0.439
4	0.415	0.585	0.58
5	0.343	0.657	0.76
6.3	0.267	0.733	1
8	0.406	0.594	1
10	0.541	0.459	1
12.5	0.671	0.329	0.834
16	0.797	0.203	0.695
20	0.918	0.082	0.58
25	0.918	0.082	0.58
31.5	0.918	0.082	0.775
40	0.918	0.082	1.036
50	0.918	0.082	1.036
63	0.868	0.132	0.696
80	0.81	0.19	0.468

Table 61.4 Evaluation grade of human comfort to whole-body vibration in recumbent position

Evaluation grade	Frequency-weighted acceleration at the buttock and head a_{xbhw}
Level 1: not uncomfortable	$a_{xbhw} \leq 0.315$ m/s^2
Level 2: a little uncomfortable	0.315 m/s$^2 < a_{xbhw} \leq 0.63$ m/s^2
Level 3: fairly uncomfortable	0.5 m/s$^2 < a_{xbhw} \leq 1$ m/s^2
Level 4: uncomfortable	0.8 m/s$^2 < a_{xbhw} \leq 1.6$ m/s^2
Level 5: very uncomfortable	1.25 m/s$^2 < a_{xbhw} \leq 2.5$ m/s^2
Level 6: extremely uncomfortable	2 m/s$^2 < a_{xbhw}$

61.4 Conclusions

Nowadays countries around the world had established unified evaluation standards and methods for human vibration comfort on the basis of ISO series standards. According to these standards, the evaluation for human in seated and recumbent position was established in China. At the same time, students were researching the new methods on the base of the standards. The TANG Chuan-yin in Northeastern University proposed a fuzzily random evaluation model on the basis of annoyance rate for human body's subjective response to vibration [10]. However except the vibration acceleration, frequency, and exposure time, there were other factors influencing the human comfort in vibration environments. Such as the angle of backrest and the different subjective response between females and males. So it was the development tendency, that new influencing factors and methods were considered in the evaluation for human vibration comfort.

References

1. Niu Fu, Chen Su, Ren Xudong et al (2014) Influencing factors and evaluation standards of vibration on human comfort. Appl Mech Mater 527(2014):294–300
2. ISO 2631-1:1985 (1985) Evaluation of human exposure to whole body vibration—part 1: general requirements
3. ISO 2631-1:1997 (1997) Mechanical vibration and shock—evaluation of human exposure to whole body vibration—part 1: general requirements
4. ISO 2631-4:2001 (2001) Mechanical vibration and shock—evaluation of human exposure to whole body vibration—part 4: guidelines for the evaluation of the effects of vibration and rotational motion on passenger and crew comfort in fixed-guideway transport systems
5. GB/T 13441:1992 (1992) Specification for measurement of human exposure to whole-body vibration environment
6. GB/T 13442:1992 (1992) Reduced comfort boundary and evaluation criteria for human exposure to whole-body vibration
7. GB/T 13441.1:2007 (2007) Mechanical vibration and shock—evaluation of human exposure to whole body vibration—part 1: General requirements
8. GB/T 13441.4:2012 (2012) Mechanical vibration and shock—evaluation of human exposure to whole body vibration—part 4: guidelines for the evaluation of the effects of vibration and rotational motion on passenger and crew comfort in fixed-guideway transport systems
9. GB/T 18368:2001 (2001) Comfort evaluation of human exposure to whole-body vibration in recumbent position
10. Chuan-yin T, Tian-xia Z, Gui-qiu S (2006) Evaluation of comfortability during vibration on basis of annoyance rate. J Northeast Univ (Nature Science), 27(7):802–805

Chapter 62
The Influence Research on Operational Efficiency in Northeast Cold Regions

Honglei Li, Hualiang Xu, Zhibing Pang, He Wei, Zhengchao Li and Cheng Jin

Abstract The purpose of this paper is to summarize and analysis the natural environment characteristics, research operational efficiency between soldier and weaponry in the northeast cold regions. The methods include find documents, questionnaires and detection equipment, in-depth analysis of the characteristics of the northeast cold regions environment impact on workers and weaponry. The results summarized that northeast cold regions environment affects "man" and "weaponry" seriously. A greater impacts on workers of physical, skills and intelligence, and a greater impacts on weaponry to mechanical wear, fuel consumption, mobility, and performance. Conclusion shows that environmental characteristics in northeast cold regions severely restrict man-machine operating performance improvement. Article's findings will enhance the ability to provide for combat troops some theoretical support in cold regions, and have practical significance.

Keywords Northeast cold regions · Operational efficiency · Influence

62.1 Introduction

Cold region is the lowest average temperature is below −15 °C military operations areas [1]. The cold zone including year-round cold regions and seasonal cold region in China, an area of 4.3 million square kilometers, accounting for China's land 43.5 % [2], most of China's northeast, northwest, and north parts of the area, as well as parts of Tibet are cold regions.

H. Li (✉) · H. Xu · Z. Pang · H. Wei · C. Jin
Air Defence Forces Academy, Zhengzhou 450052, China
e-mail: fkb_mmes@163.com

H. Wei
Unit 71823 of PLA, Luohe, China

Z. Li
Unit 65585 of PLA, Dalian 116104, China

© Springer-Verlag Berlin Heidelberg 2015

S. Long and B.S. Dhillon (eds.), *Proceedings of the 15th International Conference on Man–Machine–Environment System Engineering*, Lecture Notes in Electrical Engineering 356, DOI 10.1007/978-3-662-48224-7_62

There is a large of cold area in the northeast cold regions, an area of 1,247,600 square kilometers. Northeast cold region with prominent strategic position is an important direction of our military combat future operations, special backgrounds, special battlefield, and special operations form a special case of the opponent. Harsh climate and environment affect the operational efficiency seriously, therefore, strengthen northeast cold regions environment performance of work is necessary.

62.2 Northeast Cold Regions Environment Features

Northeast China cold regions are cold temperate continental climate with cold and dry in winter, hot and rainy in summer. When its minimum temperatures up to −50 °C, with a low temperature, long cold and more cold wave features. Mainly in the following eight:

62.2.1 Low Temperatures

The feature of the cold zone climate is determined mainly by the low temperatures. The average temperature of the coldest month (January) up to −28 °C in northeast cold Regions, extreme minimum temperature: Shenyang −32 °C, Harbin −41 °C, and Mohe −52.3 °C. North of Shenyang, the average temperature up to about −20 °C. North of Harbin, the coldest in winter up to about −40 °C. Extreme minimum temperature from south to north of −20–50 °C, such as Jinzhou, Chengde, Zhangjiakou line of about −25 °C; Shenyang, Jining, Hohhot line of about −33 °C; north of Mudanjiang, Qiqihar, Manzhouli area up to −40–50 °C [2].

62.2.2 Large Temperature Difference

Year average temperature difference is about 60 °C, and day average temperature difference is about 8–15 °C in most area of the northeast regions [3]. First, large temperature difference between summer and winter. Heilongjiang border part areas reaches −40 °C in winter, the highest temperature can reach 30 °C in summer, the temperature difference is about 70 °C. Second, large temperature difference between day and night the temperature difference is about 15 °C. Third, when the invasion of the cold temperature plunged, diurnal is up to 30 °C.

62.2.3 Long Cold Time

It is up to 5 months, in the winter of northeast cold regions accounting for almost half of the time throughout the year. According to the monthly average temperature

T range, it can be divided into three phases for cold period, colder period, and arctic cold period. First, the cold period ($T \leq 0$ °C): in late September to mid May next year, the number of cold days is 240–260; Second, colder period ($T \leq -30$ °C): in mid-November to late March next year, the cold days 80–100; Third, the arctic period ($T \leq -40$ °C): late October to mid-March, arctic cold days 20–35 [4].

62.2.4 *More Cold Wave*

"Cold wave" means "within 24 h the temperature dropped more than 10 °C, the lowest temperature dropped to below 5 °C, accompanied by about 6 northerly winds." Northeast cold to cool usually in late September to early May next year in cold regions, the annual average number of 10–17 times cold, cooling margin −18.5 to −24.8 °C, each cold cool days 5–8 days [5], accompanied by strong winds and snowfall. When cold air from Siberia enter into our country, the weather sunny and cold, cold air appear 3–5 days 1, forming a "three-four warm cold" weather features.

62.3 Influences on Operational Efficiency in Northeast Cold Regions

Located in the northeast cold regions, the cold is not the avoided the natural "barrier." Northeast cold environment reduced worker mobility severely, and weaponries can easily get "cold zone and more complications." For this reason, on the basis of in summing up cold regions environmental characteristics, we must thoroughly analyze the various effects on workers and weaponry, and propose targeted countermeasures to enhance operational efficiency in cold regions.

62.3.1 *Impact on Operating Personnel*

The harsh climate environment made operating personnel adverse physiological and psychological reactions, mainly in the psychological motivation and operational capacity, It affects workers' physical, skills, and intelligence of normal performance seriously. Mainly in the following three points:

62.3.1.1 Increase Mental Pressure

Studies have shown that workers' mental exercise capacity will be affected significantly in less than −7 °C environment. Their psychological mechanism is easy to change under long-term cold regions environment, resulting in apathy, anxiety, and fear. Due to temperature changes, workers will frequent feeling tired, poor response, memory loss. In addition, the army geographically isolated and closed-end management of long term, adverse personnel psychological quantity was significantly higher than in the south, generally larger psychological pressure.

62.3.1.2 Restricted Movement Operation

Low temperature is the main factors affecting the operation of fineness. When the operator hand is at or below 12.7 °C, knuckles rigid ligaments, skin receptors reduced reactivity, decreased accuracy personnel joint flexibility, and muscle movements, resulting in personnel movement operation is limited. Studies have shown that continuous exposure to 2–3 h in the cold weather, workers' hand strength will be reduced by 20–30 %. When the temperature drops to 32 °C, the body will be numb and disoriented. Therefore, workers operate in low-temperature environment for a long time, the operation of the equipment will become unwieldy, even mistakes and errors.

62.3.1.3 Diminished Self-protection

Troops' training mission is arduous in cold regions, therefore, personnel physical exertion, systemic resistance decreased, and self-protection is weak. Frostbite is common in cold regions of our country, especially because of its longer duration of severe frostbite, high morbidity, easily to cause a large number of non-combat attrition; Long march under the cold environment, due to the heavy clothing, shoes and socks wet, heavy load and visual reduced sensitivity, prone to all kinds of trauma and soft tissue injury; At the same time, the air dry and cold easily induced upper respiratory tract infection; Restricted diet easily lead to vitamin deficiency polycythemia; March under the snow-covered areas in the sun, easy to cause snow blindness.

62.3.2 *Impact on the Weaponry*

Equipment is easy to consumables, perishable, and freeze issues outstanding in northeast harsh cold climate. Low temperatures and snowy, equipment life is relatively short; long cold period, equipment maintenance difficulties; the ice road, the troops' mobility is limited. Mainly in the following five points:

62.3.2.1 Increased Mechanical Wear

Winds of winter are more and large in northeast cold regions, dust is easily penetrated into machinery and equipment, transmission parts and the rotating' friction increase, increased mechanical wear, more easily to lead to mechanical failure. Meanwhile, 50 % of the cylinder wear occurs during startup in the engine cycle, the engine in winter wear accounted for 60–70 %. In addition, due to cold starting, lubricating viscosity, and poor mobility, the crankshaft and drive train wear and tear assembly.

62.3.2.2 Increase Operational Fuel Consumption

At low temperatures, the lubrication viscosity increased and fuel volatility variation, resulting in oil consumption increased. When the vehicle is running, the engine is in cold environment, lead to fuel incomplete combustion, the engine output power decrease, resulting in fuel consumption increased. When the coolant drops by 85–50 °C, effective power decreased by about 4 %, the fuel consumption increased by about 5 % [6]. In addition, due to the low transmission of each assembly lubricant viscosity increases, and increased resistance to various parts of the movement, so that the mechanical efficiency decline, oil deterioration accelerated, resulting in increased fuel consumption.

62.3.2.3 Restricted Equipment Mobility

Snow depth, slippery, wind, snow-covered ground, and muddy road restrict mobility forces severely in northeast cold regions. The first is climbing ability is limited. With a snow depth of 5 cm or more and more than 12° slope , heavy-duty vehicles find it more difficult [7]; The second is difficult to observe road. Highly reflective snow, it is difficult to judge the terrain, obstacles and march direction, easily out ditch, subsidence, and hit the stump and rock; The third is motorized road is muddy. When seasonal time, muddy ground Frothing, easily to cause slipping silt trap even overturned.

62.3.2.4 Reduce Equipment Operational Performance

When the temperature dropped to −20–30 °C, the equipment operational performance dropped significantly, their combat capability weaken by 40–60 % [8]. First, at low temperatures, some electronic equipment susceptible to computer startup difficulties; Second is in cold environment, equipment erection dismantling, and drawing portfolio are easily lead to the cable core, plastic insulated wire break; Third, equipment mechanical properties changed, equipment parts are easy to break, resulting in malfunctioning; four optical surfaces is easily frost affecting external observation and target shooting.

62.3.2.5 Increasing the Difficulty of Maintenance

First, snow collapsed part of the equipment may be damaged; Second, it is likely to cause a short circuit after the snow melts once infiltrated equipment inside, causing breakdown or burned electrical components. Third, the cold can make people stiff limbs, slow, and clumsy to operate, severely reducing equipment maintenance efficiency. Research shows that when people work in low-temperature environment, activity decreased more than 15, 50 % increase in physical consumption [9], when the ambient temperature is 7 °C, the efficiency is only 80 % artisanal comfortable temperature, the direct result of the maintenance reduce maintenance operations efficiency, more difficult.

62.4 Conclusions

The harsh climatic conditions affected the effectiveness of combat personnel and equipment to play seriously in northeast cold regions, high attention should be paid. Additionally, the strategic position of the northeast cold regions has become increasingly prominent, strengthening the environmental impact study is necessary for operational efficiency. To improve operating performance in cold regions, it must be comprehensive, scientific and systematic study of northeast cold regions environment for operating personnel and weaponry.

References

1. The term PLA Military Commission (2011) The PLA military terms. Military Science Publishing House, Beijing, p 366
2. Liang H, Zhao J (2013) The difficulties and countermeasures in the cold regions troops medical support operational conditions. Southwest Med 112–113
3. People's Liberation Army General Staff Department (1984) Military Training Military entry compilation. Military Science Publishing House, Beijing, pp 356–358
4. Feng T, Tang X (2011) The impact and countermeasures of electronic warfare equipment in cold alpine environment. Science and Technology, Information p36
5. Du K(2012) Soldiers' diet and dietary on plateau area. Quartermaster 58–59
6. Zhang W, Han Z(2005) Environmental Engineering. National Defense Industry Press, Beijing, p 30
7. Wei L (2001) Strengthen the management of the equipment under cold environment
8. Pang Z, Li H (2012) Operation application method of HQ-7B surface to air missile system on highland. Air Defence Forces Academy Press, Zhengzhou
9. Zhang Y(2012) Research of psychological behavior training on plateau area. Modern Arms, p 46

Chapter 63
Countermeasures to Improve Operational Efficiency in Cold Regions of Northeast China

Zhibing Pang, Haitao Zhao, He Wei, Honglei Li, Chenhui Li and Hongyan Ou

Abstract This chapter summarizes the characteristics of Northeast China's cold regions and their impact on man–machine operational efficiency. We present operational countermeasures in the cold northeast region to enhance man–weaponry operational efficiency and increase the level of combat. The methods included a document search, questionnaires, and detecting equipment to sum up measures to improve operational efficiency in the northeast's cold regions. The results summarize environmental effects on workers and weaponry in the northeast cold regions, and we make specific proposals for countermeasures. The chapter has three aspects: the human factor, the machine, and the man–machine combination. We propose ten concrete measures that can effectively improve the operational efficiency in the northeast cold regions. The chapter complies with the laws of human work, the cold zone environment and the army's combat training needs.

Keywords Northeast cold regions · Operational efficiency · Countermeasures

63.1 Introduction

Cold areas are defined as military operation areas where the average temperature of the coldest month is less than −15 °C [1]. Most of Northeast China—an area of 1,247,600 km^2—is a cold region. The northeast cold area has a very important strategic position, but its harsh climate seriously affects operational efficiency improvement. Research on operational efficiency problems in cold regions has practical significance because it can enhance combat capabilities.

Z. Pang (✉) · H. Zhao · H. Li · C. Li · H. Ou
Air Defence Forces Academy, Zhengzhou 450052, China
e-mail: fkb_mmes@163.com

H. Wei
Unit71823 of PLA, Luohe, China

© Springer-Verlag Berlin Heidelberg 2015

S. Long and B.S. Dhillon (eds.), *Proceedings of the 15th International Conference on Man–Machine–Environment System Engineering*, Lecture Notes in Electrical Engineering 356, DOI 10.1007/978-3-662-48224-7_63

63.2 Influences on Operational Efficiency in the Northeast Cold Regions

63.2.1 Characteristics in the Northeast Cold Regions

A cold zone is due to the influence of westerly circulation control and terrain, with a low temperature in winter, snow and thick ice, long and cold winters, and other characteristics. When cold wave arrives, the temperature reduces by 10 °C and the level of wind is eight. The region's annual average low temperature −52.3 °C; it is frozen up to 150–180 days, with snow thickness up to 20–38 cm and maximum permafrost of 2–3.4 m [2]. The northeast region has a summer and winter cold, with four distinct seasons, concentrated rainfall, plains wind, mountain snow depth, and other climate characteristics.

63.2.2 Effects of a Cold Region on Operational Efficiency

The environment mainly affects personnel and equipment by restricting operational efficiency in northeast cold regions. First, personnel experience adverse psychological and psychological reactions in the northeast cold regions, mainly in psychological motivation and operational capacity. The climate seriously affects workers' physical abilities, skills, and cognitive performance. Second, equipment consumables and perishables easily freeze these environments. The ability of military equipment support is restricted severely in cold regions. Equipment life is relatively short in the low temperatures and snow, equipment maintenance is difficult. Finally, the troops' mobility is limited on icy roads [3].

63.3 Countermeasures to Improve Operational Efficiency

The harsh climate in cold areas seriously limits the enhancement of operational efficiency. We should focus on the man, machine, and man–machine combination in any efforts to enhance operational efficiency in cold regions.

63.3.1 Human Aspects

63.3.1.1 Consolidating Physical Infrastructure Effectively

Winter physical training should be restructured. It mainly contains long-distance physical training. Personnel should wear a cotton cap or face mask to protect exposed areas, such as ears, nose, and face according to weather conditions.

Personnel must prepare for activities fully before running. Running distance should be 3–4 km every day in the first week, then gradually increase to 5–7 km. Exercise intensity should gradually increase, with the heart rate controlled at 120–150 beats/min, at a frequency of 5 times a week for 2 months. It can effectively enhance the physical fitness of personnel and reduce the adverse effects caused by the harsh climate.

63.3.1.2 Strengthening Psychological Training

Conducting psychological and behavioral training can effectively prevent and eliminate apathy, anxiety, and fear and other negative psychological effects that occur in cold regions. First, we can solve the problem by training, such as trust fall. The creation of auxiliary training content can also take into account the characteristics of professional operations to cultivate awareness of collaborative workers and eliminate mental apathy. Self-control training can also be implemented. Workers may use frequent positive self-affirmations to maintain peace of mind and eliminate anxiety. Third, activities such as ziplining, rope climbing, and other extreme training can help workers to overcome the psychological barriers and stimulate psychological potential to eliminate the fear of an alpine environment.

63.3.1.3 Strengthening Professional Training in Cold Regions

Experiments show that operating skill proficiency is negatively affected by low temperatures [4]. Therefore, enhancing workers' skills training is very important. The low temperature environment affects the work effect, but it does not affect learning effect. The learning efficiency at lower temperature environment (mean skin temperature dropped to 21.1 °C) is better than normal one (mean skin temperature dropped to 31.7 °C) [5]. Therefore, professional training should be improved for low-temperature environments. The advantages of combining low-temperature acclimatization and professional skills training are obvious.

63.3.1.4 Good Protection in Cold Regions

Good protection and the reduction of non-combat attrition can improve operational efficiency in cold regions. The following five points should be considered. First, low-temperature acclimatization can effectively improve temperature endurance, maintain good tactile sensitivity, and improve the accuracy of operational actions. Second, scientific training should be pursuant to regulations, with training principles being "scientifically, gradually, repeatedly, insistently" [6].

Third, even during the cold winter, staff should dress appropriately, with armbands, cuffs, trousers, shoes, and insoles, keeping shoes and socks dry. Fourth, diets should be improved. Personnel should eat hot food and drink ginger tea as much as possible to increase the intake of fat, protein, and vitamins. Finally, a reasonable rest system should be established. Depending on the type of job and ergonomic requirements, the exposure time of limbs should be minimized.

63.3.2 *Machine Aspects*

63.3.2.1 Scientifically Design the Man–Machine Interface

Man–machine interface design should address two issues: man's control of the machine and information received between machine and man. With regard to the former, the controller should be suitable for a person to operate and people should consider the configuration space and controller operation. The latter mainly refers to the configuration and how to select the display to match the controller; people should be observed during operation to determine convenient, accurate judgment. Human–computer interface design should strive to do three things: ensure system security, improve the performance of the system, and ensure the health of personnel.

63.3.2.2 Cold Region Engineering Design

Experimental results show that the flexibility of the small finger is required for many operations; however, that flexibility is decreased in the low temperatures of a cold region [7]. Compared with the hands, the area–mass ratio of fingers is large and heat dissipation is easily localized, so the finger joints become stiff, resulting in reduced operating performance. Therefore, in engineering design of equipment parts that need to operate at low temperatures, it may be appropriate to increase the volume and change the fingers needed to operate in order to maintain effective work performance.

63.3.2.3 Improve Environmental Adaptability in Cold Regions

The ability of equipment to adapt to cold regions can be approached in several ways. First, we should do more preparation in theory. Demonstrating the equipment system and the environment detailly. Second, careful and meticulous research should be carried out in cold regions. The adaptive design environment should be strengthened and environmental testing carried out in cold regions. Third, standards to ensure the quality of equipment in cold regions must be improved. Human resources management and the sense of quality should be strengthened. Reasonable choices for durable materials should be made, with attention given to quality control of the production process.

Finally, appropriate protective measures should be taken to improve equipment use in the environment according to the environmental characteristics of the cold region to reduce the impact of environmental factors on the equipment and to promote the equipment's effectiveness of play equipment. Equipment maintenance should be a focus. Training should be based on the season, climate, and intensity of use, as should the timing of weapons and equipment maintenance so that weaponry is always in good condition; feedback should be actively solicited.

63.3.3 Man–Machine Aspects

63.3.3.1 Scientific Selection of Workers

Personnel should be selected for their military post based on specific weaponry qualifications [8]. Because of individual differences in job characteristics and workers, staff may adapt differently to a post. The best selection must be made to achieve the best match between man and machine. The essence of the selection is to seek the best combination of weaponry and people. The appropriate characteristics and combinations result in efficient man–machine systems and full equipment performance, improving the performance of man–machine operations.

63.3.3.2 Optimizing and Streamlining Workflow

Workflow optimization is critical to reducing the operating time, creating efficient workloads, and reducing operator fatigue. It requires that workers take full advantage of the ability of an operating part, so that the actions of their hands and body parts are more coordinated. The improved work process eliminates complex, clumsy, unnecessary movements so that the operating method is more efficient and reasonable. Certain principles must be followed: parallel operation, action savings, operation comfort, and efficiency of the overall operation.

63.3.3.3 Strengthening Man–Equipment Combined Training

An alpine environment seriously affects the personnel who operate the equipment. It leads to increased operator error rate, decreased effectiveness of the equipment, and diminished job performance. Therefore, to properly develop professional skills in a harsh environment, the training process should focus on the discovery and solving of problems, with consideration to the necessary job precautions in cold regions. Combined training can enhance the ability of personnel to operate the equipment in cold regions.

63.4 Conclusions

Strategic position has become increasingly important in cold regions. Therefore, it is necessary to enhance operational efficiency in these cold conditions to enhance overall combat capability. To improve the operational efficiency in cold regions, we made specific recommendations for man, machine, and combined man–machine aspects. Our results will enhance the ability to support combat troops in cold regions and thus has some practical significance.

References

1. The term PLA Military Commission (2011) The PLA military terms. Military Science Publishing House, Beijing, p 366
2. Department of Beijing Military General Staff of the PLA military training entry compilation. Military Science Publishing House, Beijing, pp 356–358
3. Wang B (2013) Northeast cold regions should seize the equipment work. Armed academic, p 58
4. Li J (2010) Cold winter camping night training area nursing job readiness. J PLA Care, p 1108
5. Pang Z, Li H (2012) Operation application method of HQ-7Bsurface to air missile system on highland. Air Defence Forces Academy Press, Zhengzhou
6. Wang D, Feng J (2012) The difficulties and countermeasures of medical support to forces on plateau area. Armed Police Medical, pp 541–542
7. Tao L (2013) Research on man-machine operation optimization of HQ-16A tracking and guiding radar car. Air Defence Forces Academy Press, Zhengzhou, pp 24–25
8. Chen X (2000) Man-machine-physiological basis of environmental systems engineering. Aerospace University Press, Beijing, p 197

Chapter 64
Comfort Performances of Permeable Chemical Protection Clothing in Hot Temperature Environment

Ying Li, Xuezhi Zhang, Songtao Ding, Xiangxuan Liu and Jie Huang

Abstract Two permeable chemical protective clothing sets (clothing system 1 and clothing system 2) were tested in a hot temperature environment, in order to study the influence factors of clothing comfort. **Methods:** Using the sweating guarded hotplate and thermal manikin, two methods to measure are thermal resistance (Rct) and water-vapor resistance (Ret) values of the two clothing systems. **Results:** In sweating guarded hotplate test, clothing system 1 Rct_{11} is 14.51×10^{-3} m^2K/W and the Ret_{11} is 6.87 m^2Pa/W, clothing system 2 Rct_{21} is 22.62×10^{-3} m^2K/W and the Ret_{21} is 10.28 m^2Pa/W; in thermal manikin test, clothing system 1 Rct_{12} is 9.62×10^{-3} m^2K/W and the Ret_{12} is 5.08 m^2Pa/W, clothing system 2 Rct_{22} is 14.21×10^{-3} m^2K/W and the Ret_{22} is 8.36 m^2Pa/W. **Conclusion:** The air gap is the main factor which leads to the different results between two test methods. Optimization of chemical protective clothing structural design can effectively promote the clothing heat dissipation, which has a positive effect on the comfort and safety of chemical protective clothing.

Keywords Thermal resistance · Water-vapor resistance · Sweating guarded hotplate tester · Thermal manikin

64.1 Introduction

Chemical protective clothing is used for protective chemical damage to the human body. In addition to fulfilling the protection requirements, an important function of permeable chemical protective clothing is to regulate the heat and moisture exchange of the wearer with the environment, so that it does not lead to overheating

Y. Li (✉) · X. Liu · J. Huang
Six Department, Xi'an High Technology Research Institute, Xi'an 710025, Shaanxi, China
e-mail: li_ying_0318@126.com

Y. Li · X. Zhang · S. Ding
State Key Laboratory of NBC Protection for Civilian, Beijing 100191, China

© Springer-Verlag Berlin Heidelberg 2015

S. Long and B.S. Dhillon (eds.), *Proceedings of the 15th International Conference on Man–Machine–Environment System Engineering*, Lecture Notes in Electrical Engineering 356, DOI 10.1007/978-3-662-48224-7_64

(hyperthermia) or under cooling (hypothermia). But when the soldiers wearing the chemical protective clothing worked in the heat and humid environment, they produced large amounts of heat energy, which led to the heat balance disturbed and produced large amounts of heat stress. So far, problems of chemical protective clothing thermal physical comfort have not been a very good solution, outstandingly in the Gulf War; therefore, this issue became the most important research to the research institutions.

64.2 Physiological Comfort of Chemical Protective Clothing

In the military mission, the soldiers wearing chemical protective clothing usually produced mass heat that is particularly prominent in the tropical and subtropical environments. In these cases, there is a danger of disturbance of the thermal physiological balance (heat production in the organism = heat emission to the environment) and creation of the state of temporary extreme overheating of the body (heat stress) [1]. Therefore, chemical protective clothing must try to reduce thermal stress and increase wearing comfort without reducing the protective performance.

Comfortable chemical protective clothing needs to be able to prevent chemical stimulation from outside, along with optimal heat−wet transfer capabilities, and can move freely after wearing. There are three main factors that affect clothing comfort: thermal physiological comfort, somatosensory comfort, and ergonomic comfort. Table 64.1 shows the contents and the main characterization parameters of the three factors [2–5].

Table 64.1 Factors, contents, and main characterization parameters of physiological comfort

Factors	Content	Characterization parameter
Thermal physiological comfort	Ensure thermal balance, avoiding too cold or too hot	Thermal resistance Water-vapor resistance Moisture transfer index Air permeability Sweat transport Drying time
Somatosensory comfort	Prevent chemical stimulation from outside	Adhesion index Surface index Clothing pressure Moistening index Stiffness
Ergonomic comfort	Move freely after wearing	Adaptation rate Weight

64.3 Tests

Two chemical protective clothing sets (clothing systems 1 and 2) were tested for their usability in hot climatic zones with regard to clothing physiology. The test objectives were:

(1) Measuring thermal and water-vapor resistance of the individual textile layers of the clothing systems using the thermoregulatory model of the sweating guarded hotplate
(2) Measuring thermal resistance and water-vapor resistance of the complete clothing systems using the thermoregulatory model of the thermal manikin.

64.3.1 Test Samples

Components and mechanic capability parameters of two chemical protective clothing sets are showing in Tables 64.2 and 64.3.

64.3.2 Measuring Methods

Sweating guarded hotplate test [8]

According to GB/T11048-2008 test method, we used a sweating guarded hotplate instrument (made by SDL Atlas) to measure the thermal resistance and water-vapor resistance values of the two test clothing system fabrics.

Thermal manikin test [9, 10]

According to GB/T 18398-2001 and ISO 9920-2007, we used a thermal manikin (made by MTNW) named "Newton" to measure the thermal resistance and

Table 64.2 Test sample components

Description	Clothing system 1	Clothing system 2
Material structure	Two-layer material structure (Acrylic cotton fabric + carbon-containing fabrics)	Two-layer material structure (Acrylic cotton fabric + Spraying carbon raising fabric)
Tailoring	Two-piece	Two-piece

Table 64.3 Mechanical-technological parameters of the test samples [6, 7]

Parameter	Clothing system 1	Clothing system 2	Test according to
Surface weight (g/m^2)	467	558	GB/T 4669-2008
Air permeability (mm/s)	107	53.2	GB/T 5453-1997

Table 64.4 Sweating guarded hotplate test results

Parameter	Unit	Clothing system 1	Clothing system 2
Water-vapor resistance (R_{et})	m^2Pa/W	6.87	10.28
Thermal resistance (R_{ct}) × 10^{-3}	m^2K/W	14.51	22.62

Table 64.5 Thermal manikin test results

Parameter	Unit	Clothing system 1	Clothing system 2
Water-vapor resistance (R_{et})	m^2Pa/W	5.08	8.36
Thermal resistance (R_{ct}) × 10^{-3}	m^2K/W	9.62	14.21

water-vapor resistance values of the two test clothing system suits. Newton has mechanically movable arms and legs. In addition, Newton can simulate release heat and sweating like human. In this test, we make Newton simulate human walk, the speed of it is 60 steps per minute. The test environment parameters are same as the sweating guarded hotplate test.

64.4 Results

The results of the two tests are shown in Tables 64.4 and 64.5.

64.5 Discussion

As chemical protective clothing system is a relatively closed system, heat and water-vapor inside it cannot get out directly. In static condition, the ways of heat and water-vapor transmission of the man-clothing-environment system include heat conduction, heat convection, heat radiation, and latent heat transmission with vapor transfer. The test results indicate that the values of thermal resistance and water-vapor resistance decreased significantly in thermal manikin test due to a certain "blow-through effect". The two test clothing systems have the same structure, and the test values of them have a downward trend in thermal manikin test. That means, move creates "blow-through effect", and same structure has same "blow-through effect". Analysis of testing results show that "blow-through effect" make the thermal resistance values reduce about 33 % and water-vapor resistance values to decrease about 26 %.

Though "blow-through effect" is physiologically favorable in hot climates, that also can lead to the outside air enter the clothing system and reduce the protection performance of the chemical protective clothing.

Chemical protective clothing is a kind of special operation clothing, on the premise of meeting the protective function; it should emphasize comfort and flexibility. Nowadays, chemical protective clothing comfort researches mainly focus on clothing material thermal–wet physical performance, but it is not the only factor that affects chemical protective clothing comfort. Optimization of chemical protective clothing structural design can effectively promote the clothing heat dissipation, which has a positive effect on the comfort and safety of chemical protective clothing.

References

1. Dipl.-Ing. Hans-Joachim Töpfer Dipl.-Ing. (FH) Thomas P. Stoll Alfred Kärcher GmbH & Co. (2001) Physiological assessments of permeable NBC protection clothing for hot climate conditions, RTO HFM symposium, 2001 (10), 8–10
2. Jianhua Huang (2008) Clothing comfort. Science and technology Press, Beijing
3. Shaoguang Fan, Hao Tang, Weifeng Pan (2000) Human physiology. Beijing Medical University Press, Beijing
4. Huang JH (2007) Assessment of clothing effects in thermal comfort standards: a review. J Test Eval 35(5):455–462
5. Vogt JJ, Meyer JP, Candas V et al (1983) Pumping effects on thermal insulation of clothing worn by human subject. Ergonomics 26(10):963–974
6. GB/T 4669-2008 Textiles–Woven fabrics–Determination of mass per unit length and mass per unit area
7. GB/T 5453-1997 Textiles-Determination of the permeability of fabrics to air
8. GB/T 11048-2008 Textiles Physiological effects measurement of thermal and water-vapor resistance under steady-state conditions
9. GB/T 18398-2001 Testing method for clothing thermal resistance-Thermal manikin method
10. ISO 9920-2007 Ergonomics of the thermal environment—Estimation of thermal insulation and water-vapor resistance of a clothing ensemble

Chapter 65
Analysis of Interface Pressure During Load Carriage

Chenming Li, Yuhong Shen and Yafei Guo

Abstract *Objective* Load carriage often leads human to be uncomfortable, tired or even injured. Shoulder fatigue, as the most common local fatigue in load carriage, severely restricted the holding and plays physical fitness. This study aimed to investigate the pressure bearing and distribution characteristics of shoulder under load carriage and try to establish a fatigue model based on the biomechanics method. *Methods* overall average pressure, means of top 10 measuring points, peak pressure value and contacted area on shoulder of six participants are measured during treadmill tests under different rucksacks and loads (25, 29, 34, and 37 kg). *Results* The means of top 10 measuring points are found to differ significantly with the different width of the straps, the narrower strap is found to result in slightly higher readings. The differences in overall shoulder pressure, mean top 10 sensors and maximum pressure are found to be significant between different loads. *Conclusions* This method is sensitive to design differences between should straps and load weights and which produces results which agree with participants' ratings of perceived fatigue.

Keywords Load carriage · Interface pressure · Shoulder

65.1 Introduction

In the field of military training, field survival, outdoor sports, as well as handling and other working people generally need to walk with back load [1]. Load changes the location of gravity center of human body and influences human gait and posture. The overload, walking for too long, the unbalanced load distribution, and other

C. Li (✉) · Y. Shen · Y. Guo
The Quartermaster Research Institute of the General Logistic Department,
Beijing 100010, China
e-mail: lichenming82@163.com

© Springer-Verlag Berlin Heidelberg 2015

S. Long and B.S. Dhillon (eds.), *Proceedings of the 15th International Conference on Man–Machine–Environment System Engineering*, Lecture Notes in Electrical Engineering 356, DOI 10.1007/978-3-662-48224-7_65

factors often lead human to be uncomfortable, tired or even injured, seriously affect the maintaining and using soldier's physical [2–4]. Shoulder fatigue, as the most common local fatigue in load carriage, severely restricted the holding and plays physical fitness. A recent report of the US Army shows that the US Army infantry carry fighting load of 28.3 kg (34.9 % of body weight), marching load 43.1 kg (52.6 % of body weight), approach marching load 58.2 kg (total body weight 73.6 %) [5]. Obviously, Although the US Army constantly trying to reduce weight of soldier's load in recent years, the soldier's load still hold on a high level, because of the combat mission requirements.

At present, many foreign scholars have used the methods combined physiological with biomechanical to carry out the ergonomics-related research about walking with load, such as using the pressure sensor to analysis the forces of the main part of the trunk while walking with load; using human movement tracking system to study the joint's force conditions while walking with load; using plantar pressure sensing system to study the posture of walking with load, etc. [4, 6, 7]. Although we have done a lot of research about walking with load in domestic, mostly we used are the heart rate (HR), minute volume(VE), oxygen uptake (VO_2) [8–10], and other purely physiological indicators which are used to carry out the ergonomic evaluation of man-portable capability and individual load carrier capability. This article will use the method of biomechanics to analysis the distribution of shoulder contact forces under static conditions, different backpacks, and loads. And we will provide guidance for reducing the fatigue of load shoulder and improving the design of backpack.

65.2 Method

65.2.1 Subjects

In total, 6 healthy participants participated in five experiments. The participants had an age of (mean ± SD) 24.2 ± 3.7 years, weight of 64.5 ± 11.6 kg, height of 172.2 ± 2.4 cm, and B.M.I of 23.1 ± 1.7 kg/m^2. Prior to participation participants completed and signed a form of consent. They were asked to have a good rest and avoid smoking, alcohol, caffeine, and intense physical activity at least 24 h prior to the experiment. During the trials, the participants wore the same training shoes and cotton clothing.

65.2.2 Experimental Conditions

The temperature of laboratory environment should be controlled in 25 ± 2 °C, the wind speed should be less than 0.5 m/s, and the marching speed is 5 km/s. There are

Table 65.1 Load distribution under different load levels

Position	Load level (kg)			
	25.00	29.00	34.00	37.00
Head	1.34	1.34	1.34	1.34
Body	1.79	1.79	1.79	1.79
Chest	8.34	8.34	8.34	8.34
Back	9.86	13.86	18.86	21.86
Hand	3.67	3.67	3.67	3.67

three kinds of backpacks with four load levels which are 25, 29, 34, and 37 kg. Load distribution at different load levels is shown in Table 65.1. To ensure that the experimental data are comparable, all experiments were carried out in the morning.

65.2.3 Equipment and Measuring Parameters

We use the film-type pressure sensing chip (type: 9801, Tekscan™, USA) to measure the pressure of the subjects' shoulder. The area of each pressure sensing chip is 7.69 × 20.3 cm and the thickness is 0.1 mm. There are 96 measuring points distributed on each chip. During the test, a pressure sensing chip will be placed on each shoulder of the subject, and the pressure sensing chip is connected with the data acquisition and transfers data to the computer through a USB interface. During trials, every 3 min, it measures the numerical fatigue intensity (9–20 where "9" indicates "no fatigue" and "20" indicates "fatigue as bad as it can be") on shoulders.

65.2.4 Experimental Terminated Condition

To prevent accidental injury during the experiment, the experiment must be immediately terminated if anyone of the following situations is happened:

(1) The heart rate is over 90 % × HRmax (estimated HRmax = 220–Age);
(2) The subjective feeling is difficult to adhere to finish experiment appeared unsteady gait, pale or dizziness, chest tightness, palpitation, nausea, and other symptoms.

65.3 Results

65.3.1 Evaluation Index

Static pressure distribution of the shoulder evaluated by the overall average, the average of top 10 measuring points, the peak value, and the contacted area. The indicators' definition is shown in Table 65.2.

Using the Student's t test and repeated measures analysis of variance to analyze statistically the test results. Statistical significance level is 0.05.

65.3.2 The Influence of Shoulder Belt Width on the Stress Distribution

During the test, there are three backpack straps with three different width: standard type (type A), wider type (type B) and narrower (type C), and the weight of the load is 25 kg. Tests showed that the shoulder's overall average of B-type backpack (2.13 ±0.39 kPa) is significantly lower than the overall average of the other two groups (type A is 2.77±0.46 kPa, type C is 3.02±0.52 kPa). There are significant differences among top 10 measuring-points of three straps, and the C-type strap's pressure value (12.51±3.24 kPa) is the highest. As the maximum pressure value, there is no significant difference between B-type and A-type strap, but the test results of C-type strap (24.09±4.55 kPa) is significantly higher than the other two groups (type A is 18.61±3.89 kPa, type B is 16.46±2.83 kPa) (Table 65.3).

65.3.3 The Influence of Load Weight on the Pressure Distribution

The width of the backpack straps used in the test is standard width (type B), and the load weights are selected as the 25, 29, 34, and 37 kg. The overall average, top 10 measuring points, the peak value, the contacted area, and the subjective feeling of fatigue when the test end are shown in Table 65.4.

Table 65.2 Distribution index of shoulder pressure

Index	Definition
Overall average	The sum of sensing chip measurements divided by the number of effective measurement points
Top 10 measuring points	Average of the measured values of maximum 10 measuring points (about 5 % of the measured total points)
Peak value	Maximum for a single-measuring point
Contacted area	Area of sensing chip corresponded by the effective measurement point

Table 65.3 Effect on pressure distribution of different straps

	Standard (A)	Wider (B)	Narrower (C)
Mean overall pressure (kPa)	2.77 ± 0.46	2.13 ± 0.39*	3.02 ± 0.52
Mean top 10 sensors (kPa)	9.38 ± 1.30*	7.97 ± 0.92*	12.51 ± 3.24*
Maximum pressure (kPa)	18.61 ± 3.89	16.46 ± 2.83	24.09 ± 4.55*
Contact area (cm^2)	123.97 ± 4.43	138.99 ± 23.65	116.46 ± 12.54

*Significant difference found between the other two conditions ($P \leq 0.05$)

Table 65.4 Effect on pressure distribution of different loads

	25 kg	29 kg	34 kg	37 kg
Mean overall pressure (kPa)	2.13 ± 0.39*	2.73 ± 0.31*	4.08 ± 0.26*	4.97 ± 0.22*
Mean top 10 sensors (kPa)	7.97 ± 0.92*	10.61 ± 1.01*	14.62 ± 1.14*	18.20 ± 1.38*
Maximum pressure (kPa)	16.46 ± 2.83*	20.47 ± 2.37*	26.81 ± 2.92*	34.50 ± 2.16*
Contact area (cm^2)	138.99 ± 23.65	152.18 ± 14.31*	173.45 ± 9.65	181.13 ± 8.44
Mean fatigue rating	12.70 ± 0.73	14.40 ± 0.69*	16.41 ± 1.03	17.75 ± 0.69*

*Significant difference found between the other two conditions ($P \leq 0.05$)

As we can see from Table 65.4, with the increasing weight of load, there are significant differences in the average values of overall average, the average values of top 10 measurement points, and peak values. The contacted area and average subjective feelings of fatigue increase with the increasing weight of load.

We regard three indicators—the average values of overall pressure, the average values of top 10 measurement points, and peak values—as independent variables, respectively, and regard subjective feelings of fatigue as dependent variable. The regression equation is established using the least squares method and is expressed as:

$$Y = aX + b. \tag{65.1}$$

There, Y was mean fatigue rating, and X was mean overall pressure, mean top 10 sensors pressure, and maximum pressure. The slope, intercept, and correlation coefficient are shown in Table 65.5.

As we can see from Table 65.5, the correlation coefficients of three indicators—the average values of overall pressure, the average values of top 10 measurement points and peak values—with subjective feelings of fatigue are larger than 0.9. The strong positive correlation indicates that the shoulder pressure is an important factor causing fatigue.

Table 65.5 Slope, intercept, and correlation coefficient of Eq. (65.1)

X	a	b	R^2
Mean overall pressure	1.832	9.042	0.917
Mean top 10 sensors	0.533	8.380	0.936
Maximum pressure	0.354	0.966	0.953

65.4 Conclusions

Because shoulder fatigue is caused by a variety of reasons and the specific level of fatigue is difficult to measure objectively, this paper combines biomechanical measurement with subjective evaluation, analyzes quantitatively shoulder pressure distribution characteristics under different load weights, discusses, and analyzes the relationship between pressure and fatigue of shoulder and shoulder strap tension. The results show that: (1) as the cause of human local discomfort, even pain and loss of function, distribution of shoulder pressure can be improved to a certain extent by changing the width of the backpack straps, improving user comfort. (2) The weight of load significantly affect the average values of overall pressure, the average values of top 10 measurement points and peak values. Appropriately reducing the load weight is still focus of the study to improve the load carrying comfort.

65.5 Compliance with Ethical Standards

The study had obtained approval from the Army Key Laboratory of SEEPC Ethics Committee.

All subjects participated in experiment had signed the informed consent.

All relevant safety, health and rights had been met in relation to subject protection.

References

1. Doan JB, Bryant JT, Reid SA, Stevenson JM, Rigby WA, Ndrews D (1998) Function testing of military load carriage sub-systems. Adv Occup Ergon Saf 707–710
2. Johnson RC, Doan JB, Stevenson JM, Bryant JT (1998) An analysis of subjective responses to varying a load centre of gravity in a backpack. Adv Occup Ergon Saf 248–251
3. Knapik JJ, Reynolds KL, Harman E (2004) Soldier load carriage: historical, physiological, biomechanical, and medical aspects. Mil Med 169(1):45–56
4. Tanaka H, Monahan KD, Seals DR (2001) Age-predicted maximal heart rate revisited. J Am Coll Cardiol 37(1):153–156
5. Kenney WL, Wilmore JH, Costill D (2011) Physiology of sport and exercise, 5th edn. Human Kinetics, Champaign

6. Holewijn M, Meeuwsen T (2001) Physiological strain during load carrying: effects of mass and type of backpack. Technical Report. ADP010987
7. Whiteside RA, Doan JE, Reid SA et al (1999) Integrated patrol pack and rucksack development phase IIIB: performance testing of suspension components. DCIEM-CR-2001-087
8. Martin JL, Hooper RH (1999) Body cooling during load carriage, the evaluation of a novel military design. Contemporary Ergonomics, Taylor and Francis, London, pp 385–389
9. Patton JF, Kaszuba J, Mello PR, Reynolds KL (1989) Physiological and perceptual responses to prolonged treadmill load carriage. US Army Research Institute of Environmental Medicine, Technical Report, Natick MA, pp T11–90
10. Rigby WA, Bryant JT, Doan JB, Pelot RD, Stevenson JM (1998) Assessment of lower shoulder strap attachment points on a standard personal load carriage system. In: Proceedings of the 30th Annual Conference of the Human Factors Association of Canada, pp 221–226

Chapter 66
Analysis of Mechanical Characteristics and Fatigue on Shoulder During Load Carriage

Yuhong Shen, Yafei Guo and Chenming Li

Abstract To investigate the force bearing and distribution characteristics of shoulder under load carriage marching, and try to establish a fatigue model based on the biomechanics method, three loads (25, 29, and 34 kg) were performed on six subjects using the simulation experiment. The pressure data of shoulder, the pulling force of shoulder strap and fatigue feeling is acquired. Then a fatigue model was established by the multiple linear regression, in which the dependent variables are the values of fatigue feelings, and the independent variables are pressure, pulling force, and load.The proposed pluralistic fatigue model had a good linear fitting degree, and the simplification of model was established with the biomechanics method. It may play an important role in loaded walking mission design and the backpack ergonomic design.

Keywords Load carriage · Shoulder · Fatigue model

66.1 Introduction

There is always be a requirement for military personnel to carry a certain weight of the load in combat and training tasks, according to combat readiness, mission requirements, length of combat, terrains, and climates of combat area [1, 2]. At present, the standard field equipment of U.S. marines includes body armor, weapons, ammunition, food, water and communications equipments, of which the total weight is about 44–61 kg. In Afghanistan, the U.S. military soldier needs to carry load about 59–68 kg in 3 days tasks [3, 4]. U.S. soldier load carriage is about 75 % of their weight. Soldiers marching always easily causes the human body or local fatigue during load carriage in some situations such as too big carrying load,

Y. Shen · Y. Guo · C. Li (✉)
The Quartermaster Research Institute of the General Logistic Department,
Beijing 100010, China
e-mail: lichenming82@163.com

© Springer-Verlag Berlin Heidelberg 2015

S. Long and B.S. Dhillon (eds.), *Proceedings of the 15th International Conference on Man–Machine–Environment System Engineering*, Lecture Notes in Electrical Engineering 356, DOI 10.1007/978-3-662-48224-7_66

not scientific carrying style, not reasonable load distribution carrying, too long walk time. Shoulder fatigue seriously restricted the energy of soldiers as the most common local fatigue in the process of marching [5, 6]. At present, many foreign researchers adopt the physiological and biomechanical method to study shoulder fatigue under the condition of load carriage and established prediction method of local fatigue. Domestic research in this respect is relatively limited, and mostly by biochemical method which is analysis fatigue through the body's biochemistry index. This article is focus on subjective fatigue stress, backpack belt tension, and the shoulder change and characteristics, as well as the relations among them in static and dynamic conditions, in order to provide guidance to reduce individual fatigue and improve shoulder backpack design.

66.2 Method

66.2.1 Subjects

Six healthy male participants are participated in experiments. The participants had a (mean ± SD) age of 24.2 ± 3.7 years, weight of 64.5 ± 11.6 kg, height of 172.2 ± 2.4 cm, and B.M.I of 23.1 ± 1.7 kg/m^2. They all required to have high school education or above and strong ability to understand and cooperate with the experiment in order to accurately express themselves fatigue feelings. Prior to participation, participants completed and signed a form of consent. They were asked to have a good rest and avoid caffeine, alcohol, smoking, and intense physical activity at least 24 h prior to the experiment. During the trials, the participants wore training shoes and cotton clothing.

66.2.2 Experiment Conditions

The environment temperature of lab is controlled in 25 + 2 °C, wind speed is less than 0.5 m/s. The speed of walking is 5 km/s with load carriage of 25, 29, and 34 kg, respectively. The load distribution under different load levels is shown in Table 66.1. In order to ensure the experimental data comparable, all experiments are conducted on the morning.

66.2.3 Apparatus and Measurements

The pressures at shoulders were measured by miniature pressure sensor (Model 9801, Tekscan, USA). Two 9801 sensor pads were put on each region to detect

Table 66.1 The load distribution under different load levels

Position	Load levels (kg)		
	25.00	29.00	34.00
Head	1.34	1.34	1.34
Body	1.79	1.79	1.79
Fore breast	8.34	8.34	8.34
Back	9.86	13.86	18.86
Hand	3.67	3.67	3.67

pressure on both sides. Each 9801 sensor was plugged into a data scanner where pressure data were collected and sent to PC via USB cable where the data can be viewed, analyzed, and stored in real time with I-Scan® application software. The SMAR-P-20 micro tension sensor was used to measure pulling force of backpack belt. The Borg scale (9 points means very easily, 20 points indicates the fatigue limit) was used to record shoulder fatigue feelings every 3 min through subjective inquiry access to subjects.

66.2.4 The Experimental Termination Condition

Once found that one of the following circumstances, the experiment immediately suspends in order to prevent the accident damage in the process of experiment.

(1) Heart rate more than 90 % HRmax (estimated HRmax = 220−age);
(2) Subjective feeling is difficult to adhere to the experiment with pale face, unstable gait or appear the symptom such as dizziness, bosom frowsty, flustered, and nausea.

66.3 Results and Discussions

66.3.1 Pressure Data Pre-processing

I-Scan pressure measurement system are susceptible to interference internal or external factors to measure the pressure distribution, and the sensitivity of unit measurement point is not uniform, analog and digital conversion process, and data transmission process, and man-made factors will always produce deviation [7–9]. So it is necessary to smooth the processing for improving or eliminating the influence of the noise. Due to the smaller noise general area and small domain pressure value, big effective data area, and big domain pressure value, this paper uses the local weighted average method in image processing to smooth the image of pressure, so as to weaken the noise points of small area [10]. The method is based on the calculated pressure point as the center, setting a square area in the pressure chart, average of the pressure value of all the points in the area weighted. The average is the pressure value on the measurement point. Calculation methods are shown as formula 66.1.

$$h(i,j) = \frac{1}{M}\sum\nolimits_{(k,l)\in N} f(k,l) \times m_{k,l}, \tag{66.1}$$

where: M is sum of weighted coefficient of measurement point in the domain of N; F (k, l) is the pressure value in the domain of N (k, l); and $M_{k,\,1}$ is the weighted coefficient of point of (k, l), so $M = \sum_{(k,l)\in N} m_{k,l}$.

This smoothing algorithm can be expressed as the form of a linear operator. Because of the different choice of weighted coefficient, a variety of smooth operator could be formed. The shoulder pressure distributions that before and after the smooth processing are shown in Fig. 66.1.

66.3.2 Analysis of Should Force

Because the individual shoulder fatigue is a process of accumulation increased, the shoulder pressure value was accumulating from the concept of impulse in physics in the process of dynamic analysis by the formula $P = \int_{t_2}^{t_1} F dt$. $P = F \times (t_1 - t_2)$ if the pressure F is average value at the time (t_1-t_2). So set the pressures build and the time t = 3 min as variable to get the accumulation of pressure in time during different load carriages. Accumulations of pressure P of 6 participants are shown in Table 66.2, the average changes of should fatigue are shown in Fig. 66.2.

As can be seen from Table 66.2 and Fig. 66.2, the accumulated fatigue and pressure P have a strong correlation ($R^2 > 0.95$). Set the load weight and shoulder pressure as the independent variable, shoulder fatigue under different load and stress accumulation equation can be expressed as Eq. (66.2).

Fig. 66.1 The effect drawing of pretreatment of should pressure

Table 66.2 Accumulation of should pressure under different load carriage

Time (min)	Load 25 kg	Load 29 kg	Load 34 kg
3	9.54	14.45	20.48
6	16.9	29.71	39.51
9	25.61	45.65	59.77
12	34.47	61.28	79.55
15	43.62	77.22	99.31
18	52.96	92.9	120.02
21	61.73	109.79	141.29
24	70.73	126.74	162.43
27	79.87	143.43	183.18
30	89.54	159.9	204.84
33	98.99	176.32	225.69
36	107.86	192.24	245.8
39	116.82	207.56	265.21
42	126.18	222.49	283.95
45	136.11	237.94	301.64
48	145.84	252.5	319.03
51	155.04	267.62	336.65

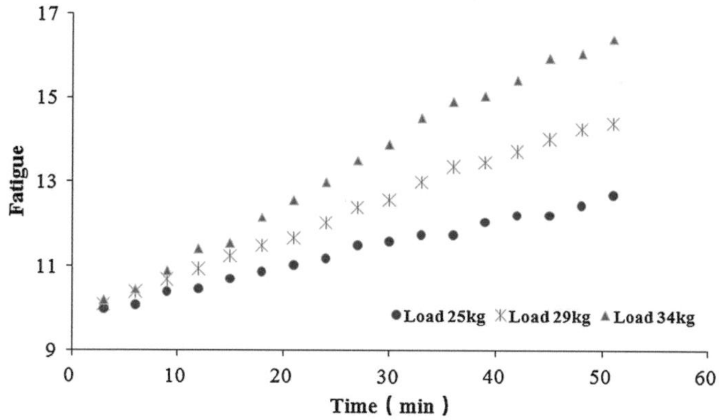

Fig. 66.2 Subjective fatigue feeling changes

$$y = 9.45 \times 10^{-5} M^2 \int_{t_2}^{t_1} F dt - 0.005M \int_{t_2}^{t_1} F dt + 0.092 \int_{t_2}^{t_1} F dt - 0.002M^2 + 0.153M + 7.876, \quad (66.2)$$

where: y is subjective fatigue feeling; F is pressure on should, kPa; M is weight of load carriage, kg; t_2 is the beginning time, minute; and t_1 is the end time, minute.

Table 66.3 Belt tension accumulation of shoulder under different load

Time (min)	Load 25 kg	Load 29 kg	Load 34 kg
3	3.88	5.59	6.73
6	7.80	11.19	13.48
9	11.76	16.77	20.37
12	15.61	22.25	27.17
15	19.42	27.63	34.06
18	23.11	33.00	40.99
21	26.86	38.33	47.89
24	30.56	43.64	54.85
27	34.26	49.05	61.78
30	37.88	54.43	68.67
33	41.56	59.78	75.31
36	45.31	65.16	81.98
39	49.02	70.45	88.59
42	52.66	75.72	95.28
45	56.20	80.98	102.01
48	59.83	86.24	108.62
51	63.49	91.50	115.33

66.3.3 *Shoulder Belt Tension and Fatigue Analysis*

The impulse theory is also used in analysis of shoulder belt tension to accumulate the shoulder pressure value, $R = \int_{t_2}^{t_1} Tdt$. Set the stress accumulation as independent variables under different load weight, and set the time t = 3 min as a division point to calculate the accumulation of pressure in time. The average shoulder belt tension accumulated R of six subjects under different load are shown in Table 66.3.

Set the load weight and shoulder belt tension as independent variables under different load weight, the shoulder fatigue and stress accumulation equation can be expressed as Eq. 66.3.

$$y = 0.001M\int_{t_2}^{t_1} Tdt + 0.002\int_{t_2}^{t_1} Tdt - 0.010M + 10.10, \qquad (66.3)$$

where: y is subjective fatigue feeling; T is pressure on shoulder, N; M is weight of load carriage, kg; t2 is the beginning time, minute; and t1 is the end time, minute.

66.4 Conclusions

The shoulder fatigue during individual walking with load carriage is caused by a variety of reasons, and it is difficult to measure the specific degree of fatigue objectively. So the dynamic analysis method were used, respectively, in this paper, combination of the biomechanical measurement and subjective evaluation to analyze quantitatively the shoulder pressure distribution under different load weight, and the relationship between shoulder fatigue with shoulder pressure and shoulder belt tension. Also, the shoulder strain mathematical model was established. The model has high coefficient of regression and shows the strong feasibility of the model. This study will provide guidance to reduce the shoulder fatigue and scientific weight-bearing walking, as well as provide the basis for the science design of load carrying equipment.

66.5 Compliance with Ethical Standards

The study had obtained approval from the Army Key Laboratory of SEEPC Ethics Committee.

All subjects participated in experiment had signed the informed consent.

All relevant safety, health and rights had been met in relation to subject protection.

References

1. Johnson RC, Doan JB, Stevenson JM, Bryant JT (1998) An analysis of subjective responses to varying a load centre of gravity in a backpack (Adv Occup Ergon Saf). In: Proceedings of the 13th annual international occupational ergonomics and safety conference, pp 248–251
2. Milivojevich A, Stanciu R (2000) Investigating psychometric and body pressure distribution responses to automotive seating comfort, SAE technical paper series, 23–25
3. Gao ZH, Xiao ZH, Li HJ (2009) Occupant position recognition with seat surface pressure distributions and support vector machines classifier. J Mech Eng 42(8):216–220
4. Zhuang YZ, Cai P, Zhou ZF (2005) Pressure mapping system and its sensor technology. J Transduction Technol 2:313–316
5. Song LH, Chen MS (2011) Research prospect of biomechanics of human loading gait. J Clin Rehabilitative Tissue Eng Res 41(4):7771–7775
6. Birrell SA, Haslam RA (2009) The effect of military load carriage on 3-D lower limb kinematics and spatiotemporal parameters. Ergonomics 52(10):298–304
7. Birrell SA, Haslam RA (2009) Subjective skeletal discomfort measured using a comfort questionnaire following a load carriage exercise. Mil Med 174(2):177–182
8. Cappellini G, Ivanenko YP, Poppele RE (2006) Motor patterns in human walking and running. J Neurophysiol 95(6):3426–3437
9. Stevenson JM, Bryant JT, Reid SA (2004) Development and assessment of the canadian personal load carriage system using objective biomechanical measures. Ergonomics 1255–1271
10. LaFiandra M, Wagenarr RC, Holt KG (2003) How do load carriage and walking speed influence trunk coordination and stride parameters. J Biomech 36(1):28–35

Chapter 67
An Experimental Study of the Effects of Low Barometric Pressure on Human Hearing Level

Junmin Du, Weiyu Sun, Huajun Xiao and Wangqiang Xu

Abstract The goal of this study is to explore the effects of low barometric pressure on human hearing level in aviation environment. A hypobaric chamber was used to simulate low barometric pressure condition in aviation. 7 participants were exposed to the hypobaric chamber at 5 simulated altitudes (1500, 2400, 3000, 4000, and 5000 m). 5 frequencies pure tones (50, 100, 400, 1000, and 8000 Hz) were played at each altitude. The hearing thresholds were recorded. It was found that the hearing thresholds ascended with the increasing of simulated altitude under all the frequencies pure tones. There were significant differences of the hearing thresholds between altitudes mostly ($P < 0.05$). It could be concluded that low barometric pressure had negative effects on hearing threshold significantly. The hearing level decreased with the altitude raising. This study is useful for the pressure design, sound design, and audition protecting in low barometric pressure condition.

Keywords Low barometric pressure · Hypobaric chamber · Hearing threshold · Hearing level · Experimental study

This research was supported by the Fundamental Research Funds for the Central Universities.

J. Du (✉) · W. Sun
School of Transportation Science and Engineering, Beihang University,
37 Xueyuan Road, Haidian District, Beijing 100191, China
e-mail: dujm@buaa.edu.cn

J. Du
Airworthiness Technologies Research Center, Beihang University, Beijing, China

H. Xiao
Air Force Institute of Aviation Medicine, Beijing, China

W. Xu
School of Aeronautics Science and Engineering, Beihang University, Beijing, China

© Springer-Verlag Berlin Heidelberg 2015

S. Long and B.S. Dhillon (eds.), *Proceedings of the 15th International Conference on Man–Machine–Environment System Engineering*, Lecture Notes in Electrical Engineering 356, DOI 10.1007/978-3-662-48224-7_67

67.1 Introduction

Low barometric pressure environment is a common condition in aviation. With the barometric changing, human ear shows the most sensitive reaction compared with other body organs. The ear tympanum pressure is influenced heavily by external low barometric pressure, which may cause hearing loss, tympanum pressure increase, heart nausea, and even aero-otitis media [1–3].

The tympanic of human ear is a cavity filled with air. The eustachian tube is the middle ear pressure adjustment device for balancing the pressure on the tow sides of the eardrum. In normal condition, the eustachian tube is close. It is only open when the pressure difference is big enough. The pressure difference between inside and outside of the tympanic will cause the eustachian tube opening or closing. When the pressure inside of the tympanic is greater than the pressure of outside, the eustachian tube is easier to open, so as to exhausts air from Eustachian tube to outside and get a new balance. However, when the pressure inside of the tympanic is smaller than the pressure of outside, it is difficult to fill air into eardrum from outside because of the eustachian tube membranous wall closing [4]. The pressure balance is essential for the normal physiological role playing of the eardrum and the ossicular. The pressure balance activity between inside and outside of the tympanic is very common in the aviation domain, which is determined by the anatomy and ventilation characteristics of middle ear, as well as the low barometric pressure environment in aviation flight.

Modern aircraft were equipped with a large number of audio devices, such as the sounds alarm device, the air-ground communications device, and the broadcasting machine. Understanding the effects of the low barometric pressure on hearing level would be helpful for dealing with the problems on getting sound information.

The goal of this study is to find out whether the low barometric pressure has effects on hearing level or not, and the effect degree if there is. The results would be useful for the design of pressure system and sound information in the pressurized aircraft cabin, so as to reduce the negative effects of low barometric pressure on hearing.

67.2 Methods

67.2.1 Participants

All participants were recruited from the undergraduate students in Beihang University. Participants were told that the general purpose was to study the effects of low barometric pressure on hearing level in a hypobaric chamber environment. An informed consent form was given and signed, which detailed the experimental procedure and risks. Seven participants participated in the experiment. They were all males. Ages were from 20 to 25. None of them had hearing or other health problem. They were living on plain and had not been plateau area in the recent past year. All participants filled out a screening questionnaire to determine eligibility for the study.

67.2.2 Experimental Procedure

A hypobaric chamber was used to simulate low barometric pressure environment in aviation, which was provided by Air Force Institute of Aviation Medicine. The participants were exposed to the hypobaric chamber with five simulated altitudes (1500, 2400, 3000, 4000, and 5000 m) and five different frequencies pure tones (50, 100, 400, 1000, and 8000 Hz). All the five frequencies pure tones were played by a computer at each altitude.

Before the participants went into the hypobaric chamber, they were guided by the experimenter on understanding the test procedure clearly. The principle of the hypobaric chamber was also introduced for them. This was helpful to make the participants build up the concept of test environment and eliminate negative psychological pressure. The methods to prevent ear injury were taught to participants, such as swallowing, yawn, pinch the nose, and blow. Informed consent was gone though.

When the experiment began, two participants were set as a group. They sat in the hypobaric chamber with pure tone player putting in the middle of him. The distance between each participant's ear and player was 1 meter. Tester sat at the most left of the participants. Oxygen masks and flight cap were prepared for emergency. During the experiment, participants' blood oxygen concentrations and heart rates were monitored. If their blood oxygen concentrations were too high or heart rates were too fast, oxygen masks would be used to ease symptoms.

At the beginning of the experiment, the simulated altitude of hypobaric chamber was set as 50 m, which is the average altitude of Beijing, where all the participants were living in the past months. Participants' hearing thresholds at this altitude with 5 frequency pure tones (50, 100, 400, 1000, and 8000 Hz) were measured. The data were recorded as ground references.

Then the simulated altitudes were adjusted from low to high, i.e., the altitudes increased to 1500, 2400, 3000, 4000, and 5000 m in turn. After reach each altitude, waited 30 s to make the pressure stable and then began testing. The staying time on each height is about 7 min. The hearing threshold of 5 frequencies pure tones were recorded at each altitude, respectively.

In order to prevent ear barotraumas, the increase rate of simulated altitude was limited less than 10 m/s. During the period of adjusting the simulated altitude, the participants were reminded to do some actions to spur the eustachian tube open and balance the pressure change between inside and outside ear. Those actions included swallowing, Valsalva pinch nose and blow, chewing and so on.

During the experiment, the background noise was about 78db, in which the primary one was caused by compression and decompression of hypobaric chamber.

67.2.3 *Statistical Analysis*

The hearing thresholds were recorded and the effects of low barometric pressure on hearing level were analyzed. For those data obtained from the experiment, the abnormal data were removed by Grubbs method. The average and standard deviation (x ± s) were calculated. The variance analysis and t-test was made.

67.3 Results

The hearing thresholds of pure tones at each simulated altitude are shown in Table 67.1, which also are represented in Fig. 67.1.

It could be seen from Table 67.1 and Fig. 67.1 that hearing thresholds ascended with the simulated altitudes increasing for the pure tones, i.e., hearing level declined with the altitudes rising. The relationship between altitudes and hearing thresholds is not linear. Hearing thresholds ascending had a slowing trend with the increasing of altitudes. Compared with the ground group data, the hearing thresholds ascended about 15–20 % at 5000 m height with pure tones frequency of 50, 400, 1000, and 8000 Hz. The hearing threshold of 100 Hz pure tone had a smaller ascend than other frequency tones, about 8.6 %.

The hearing thresholds were significantly affected by low barometric pressure at different simulated altitudes. There were significant differences ($P < 0.05$) of the hearing thresholds between the ground height and any simulated altitudes with various frequencies of pure tones, except the altitudes of 1500 and 2400 m with pure tone of 100 Hz. In most cases, the differences are very significant ($P < 0.01$) or extremely significant ($P < 0.001$). For example, the hearing thresholds at each altitude with 50, 1000, and 8000 Hz pure tones mostly had very significant differences ($P < 0.01$) comparing with the ground group. The hearing thresholds at altitudes of 4000 and 5000 m with all frequency pure tones mostly had extremely significant differences ($P < 0.001$) comparing with the ground group.

For all frequencies' pure tones, the hearing thresholds ascended significantly with the altitudes increasing when the altitudes were below 3000 m. However, the hearing thresholds had no significant differences between 3000 m and the higher altitudes above 3000 m.

There were significantly differences ($P < 0.05$) of hearing thresholds between different frequency pure tones. In a certain simulated altitude, the hearing threshold of low-frequency pure tone was significantly higher than that of high-frequency pure tone.

Table 67.1 Hearing thresholds of pure tones at each simulated altitude (Unit: db)

Altitude (m)	Frequencies of pure tones (Hz)				
	50	100	400	1000	8000
50	51.03 ± 0.16	62.70 ± 0.45	61.20 ± 3.31	63.04 ± 5.16	77.51 ± 7.21
1500	$55.57 \pm 1.71^{***}$	63.39 ± 1.14	$64.41 \pm 0.76^{*}$	$71.86 \pm 2.10^{**}$	$87.89 \pm 2.79^{**}$
2400	$57.77 \pm 2.44^{***}$	63.56 ± 2.50	$69.10 \pm 1.27^{***}$	$72.14 \pm 1.68^{***}$	$86.70 \pm 1.95^{**}$
3000	$59.19 \pm 2.62^{***}$	$66.13 \pm 3.29^{*}$	$70.16 \pm 2.34^{***}$	$74.36 \pm 3.55^{***}$	$88.61 \pm 3.96^{**}$
4000	$60.73 \pm 1.91^{***}$	$67.84 \pm 1.94^{***}$	$71.41 \pm 1.74^{***}$	$74.19 \pm 2.14^{***}$	$89.90 \pm 3.30^{**}$
5000	$61.13 \pm 2.42^{***}$	$68.09 \pm 1.76^{***}$	$70.57 \pm 3.70^{***}$	$75.69 \pm 3.19^{***}$	$91.76 \pm 2.18^{***}$

Note Compare with 50 m (ground), $P < 0.05$ mark as *; $P < 0.01$ as **; $P < 0.001$ as ***

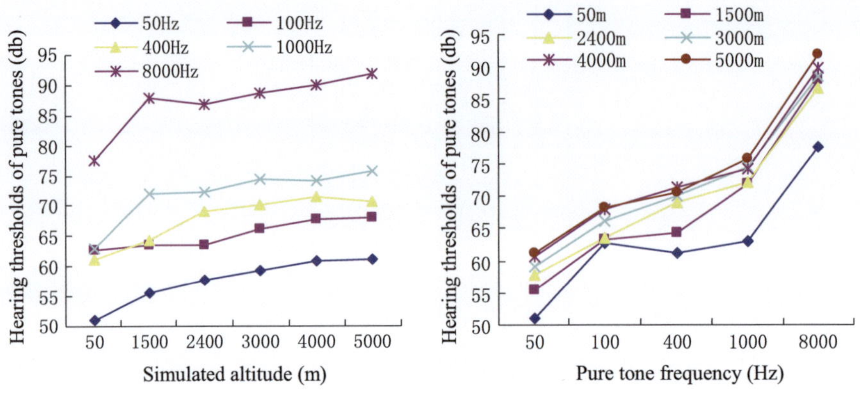

Fig. 67.1 Hearing thresholds of pure tones at each simulated altitude

67.4 Discussions

In this experiment, the simulated altitudes were adjusted to be increasing from low to high, reached to 1500, 2400, 3000, 4000, and 5000 m in turn. The altitude increasing rate was limited below 10 m/s. Because the air pressure outside of human eardrum was decreasing with altitude rising, the eustachian tube was easy to open and balance the pressure between the two sides of eardrum. There was no earache or other abnormal symptoms on participants in this experiment. The participants' hearing loss was mainly caused by ear pressure reducing.

The 5 simulated altitudes that selected in this experiment were representative. Because the participants were living in Beijing in the past long period of time, the simulated altitude for experimental control group was set as Beijing altitude (50 m). In aviation, 1500 m is the safe height of night flight; 2400 m is ergonomics ensure height; 4000 m is ergonomics permissible height and also mild hypoxia height; 5000 m is hypoxia tolerance limit height. The maximum altitude to ensure flight safety is regarded as 1500–2400 m. For example, the cabin altitude of Boeing 787 is 1800 m; airbus 350 is 1500 m. Comparing with civil aircraft, the cabin altitude of military aircraft is higher, which is 1500–3000 m [5, 6]. Therefore, in our experiment, the maximum cabin altitude was set as 5000 m.

The experimental results showed that at the simulated altitude 1500 m, which was accepted widely as a safe height, the hearing thresholds with various frequencies of pure tones had significant difference ($P < 0.05$) compared with the ground group data, except for the pure tone of 100 Hz. Therefore, the phenomenon of hearing loss would appear inevitably in the modern aircraft at cabin altitude 1500–2400 m.

When the cabin attitude was higher than 2400 m, the hypoxia symptom would appear with long-time exposure, which would give negative effect on hearing as well. In our experiment, in order to observe participants' hypoxia condition, their blood oxygen concentrations were monitored consistently. The blood oxygen concentrations of all participants were 98–100 % when the altitude was at ground level. They were 96–97 % at 1500 m; 94–95 % at 2400 m; 91–93 % at 3000 m; 88–90 % at 4000 m; less than 90 % (85–88 %) at 5000 m. During the whole experiment, for all participants, the lowest blood oxygen concentration was 85 %. Although the lowest blood oxygen concentration was not reach the alarm line yet (which is 80 %), the hypoxia symptoms would definitely impose negative effect on hearing level to some extent. The effect of hypoxia on hearing was increasing gradually with the altitude rising.

At each certain altitude, hearing threshold of high-frequency pure tones was significantly higher than that of low frequency. In other words, compared with high-frequency pure tone, low-frequency pure tone was easier to be heard no matter what the cabin altitude was. During the experiment, participants reported that the low-frequency sounds, such as 400 and 1000 Hz, were comfortable to be heard. The sound of 8000 Hz was too sharp and was not comfortable for long-time hearing.

In addition to the hearing threshold of above pure tones of 5 frequencies, the hearing thresholds of 15000 and 20000 Hz were also tested. At each altitude from ground to 5000 m, 2 out of 7 participants reported that the 15000 Hz pure tone could not be heard or could not be heard clearly. This might be due to two reasons. The first one was participants' individual differences, which meant different participant had different sensitivity to the same tone. The second reason was the noise interference by the cabin air charging and discharging. The frequency of the background noise was very similar with high-frequency pure tone. The data from other 5 participants showed that the hearing threshold of 15000 Hz was similar with 8000 Hz. For the pure tone of 20000 Hz, all participants reported that the tone could not be heard on all altitudes. This was a normal phenomenon because the upper limit of human hearing threshold range was about 20000 Hz.

67.5 Conclusions

By the experimental study, it could be concluded that low barometric pressure impaired hearing threshold significantly. The low pressure imposed negative effects on hearing level. The hearing level decreased with the atmospheric pressure decreasing or with the pure tone frequency rising. These findings would be useful for the design of pressure, sound, and audition protecting and so on in pressurization cabin or other low pressure environment. The goal of making audition environment better, improving communication quality, and reducing hearing damage would be achieved ultimately.

67.6 Compliance with Ethical Standards

The study had obtained approval from the State Key Laboratory of NBC Protection for Civilian Ethics Committee. All subjects participated in the experiment had signed the informed consent. All relevant ethical safeguards had been met in relation to subject protection.

References

1. Csortan E, Jones J, Haan M, Brown M (1994) Efficacy of pseudo ephedrine for the prevention of barotrauma during air travel. Ann Emerg Med 23(6):1324–1327
2. Jiang D, Sun J (2009) The research progress of ear baric function on military aircraft pilot. Med J Chin People Lib Army 34(2):235–236
3. Xu X (2008) Eustachian tube pathology and ear barotrauma. J Audiol Speech Pathol 16(6): 443–445
4. Fuller J (1999) Control of pressure—comfort and motion in megarise elevators. Engineering 9:148
5. Xiao H, Liu X, Xue L, Zang B (2007) The physiological basis of large aircraft cabin pressure design. In: China Aviation Society annual meeting, pp 1–5
6. Chen X, Yuan X, Jia S (1995) Man machine environment system engineering physiological basis of aircraft. Beihang University Press, Beijing

Chapter 68
Correlation Analysis Between Eye Gaze and Image Content of the Natural Scene Pictures

Yajuan Bai, Meng Yang, Yaofeng He, Yuping Luo and Guansheng Huang

Abstract Study of the eye movement characteristics of humans when watching pictures can help to improve the analysis accuracy of calculation models in the visual salient region. This study selected 100 natural images from the image library and the tested eye movement data of subjects were recorded by eye trackers, features of human eyes at gaze points were analyzed when freely viewing pictures (no task instructions), emphasizing on the study of the picture's first fixation point and second point fixation ratio which can complete characterization of picture content. Application method of eye movement data in the calculation model of visual significant area is put forward by combining with characteristics of the picture.

Keywords Eye movement · Bottom-up · Significant region · Selective attention

68.1 Introduction

Improvement of calculation models in the visual significant area using eye movement data has become one of the methods to improve effect of calculation model in the bottom-up significant regions. Eye movement databases containing a number of photos and videos have also been established abroad and are used for improving calculation models in the significant regions and validating the models' improvement results.

In computer vision model calculations [1], the human eye movement data is generally used to validate the effectiveness of salient regions from calculation models in the bottom-up significant region; application methods of human eye

Y. Bai (✉) · Y. He · Y. Luo · G. Huang
Beijing Special Vehicle Institute, Beijing 100072, China
e-mail: zhjbjs@163.com

Y. Bai · M. Yang
Beijing Normal University, Beijing 100072, China

© Springer-Verlag Berlin Heidelberg 2015

S. Long and B.S. Dhillon (eds.), *Proceedings of the 15th International Conference on Man–Machine–Environment System Engineering*, Lecture Notes in Electrical Engineering 356, DOI 10.1007/978-3-662-48224-7_68

movement data in Empirical Validation of the Saliency-based Model of Visual Attention are as follows [2]:

Record people pictures of fixation, fixation duration, focus order. Usually sets the parameters for fixation area of not less than 20 pixels, while the continuing fixation duration is not less than 100 ms.

According to participant fixation location generate a significant area of human eye diagram.

First, the fixation point is the Gaussian convolution; 3/4 width of the Gaussian filter is used to approximate the fovea of the human eye work form significantly based on human eye movement data.

Pictures of eye movement data are described in the article building process, such as subjects, the use of eye-tracking eye movement experiment guidance systems, language, and so on, but do not categorize the pictures.

In the paper "Learning to Predict Where Humans Look" a similar method of eye movement data applications was used [3], but the authors for each test section took possible effects of initial fixation point away from the center and removed the first fixation point in the scan path. However, the author took into account natural pictures in general, put the main content centers, and thus increased the advantage to identify operators of the centre.

Now it has reappeared directly using computer vision calculation model of eye movement data to improve the model's algorithm; calculation model using eye-movement data shows significant improvement; a key question is how to make rational use of eye movement data. The basic assumption is that the human eye fixation is not just a physically significantly strong regional representative, but more a picture information and semantic significance to the area. For validation above assumed, and validation people eye center preference and semantic significantly district Zhijian of relationship, we used eye moving track system. Collection was from a free watch natural scene pictures of eye moving data, eye moving data of distribution situation for statistics analysis, research has eye moving data and natural scene content characterization Zhijian of relationship, proposed has visual significantly sex calculation model in the eye moving data of application method.

68.2 Method to Set up Eye Movement Database to Freely Watch Natural Images

68.2.1 Selection of Eye Movement Data Pictures

100 nature scene pictures were selected (the picture, which comes from the public picture library, can be used for noncommercial research), principally divided into 3 categories according to images containing. It contains 36 pictures of one object and 16 pictures of 2 objects, 48 pictures of many objects, which account for 36, 16, and 48 % of the total pictures respectively. Figure 68.1 is a sample of the natural scene image.

Fig. 68.1 Sample of natural scene image

68.2.2 Analysis Method of Eye Movement Data

Tobii 120 Eye Tracker 1750 eye tracker presenting the material and eye movement data were recorded, eye tracking has a screen size of 17 in., 120 HZ refresh rate, resolution of 1024 × 768. Before the experiment started, participants were asked to adjust the position by head holding fixed head position to ensure the eye point on the middle of the screen. Eyes and screen distance is 60 cm, followed by pictures, eye movement data were recorded.

Picture looking in the range of subjects average fixation point Center as the center circle of RADIUS 56 mm, gaze fixation area picture covers an area of 20 %.

68.3 Eye-Movement Data Distribution in the Natural Picture

68.3.1 Analysis of Physical Significant Characteristics of Natural Scene Images

100 photos, classified as originals and calculation model for consistency, consistent with 75 images, poor consistency of 25 pictures, Fig. 68.2 is a nature scene pictures comparing the results consistent with the sample.

Fig. 68.2 Sample of consistency between natural images and the results of calculation

68.3.2 Fixation Distribution Discrete

On the concentration of 100 pictures, eye movement data were statistical. Statistics show that in this first fixation point area, 65 pictures first fixation point area of eye movement data do not cover all the fixation points that are relative to this area, 100 pictures of most of the pictures are in higher fixation degree of dispersion. Figure 68.3 is a natural scene sample dispersion analysis of eye movement data.

Among them, the eye movement data on a centralized, 25, 71.4 %; eye movement data on two more on the 7, 20 %; set of eye movement data on more than 3, 8.6 %. By statistical analysis we can directly see dispersion of eye movement data and picture content that are closely related to picture which contains more objects, eye movement data more discrete, single object larger, containing more information, more discrete eye-movement data; human eye movements to ensure the integrity of information extraction.

68.3.3 Relationship Between Gaze Sequence and Representation of Image Meaning

In a large number of visual cognition experiments, researchers found that when people look at computer-rendered images are always begin with the central area of the picture. This phenomenon seems to show a visual attention model in computer; increasing the simulation center operator preference can improve significantly the results and consistency of human eye movement data, and can also increase the accuracy of image information. However, experiment results show that, people eye of first watched points in pictures interest district distribution in away from pictures Center of edge location Shi, people eye of watched points still concentrated in pictures of Center regional [4], this seems to show that people eye watched points of order is not a must and image interest regional is significantly related so that eye

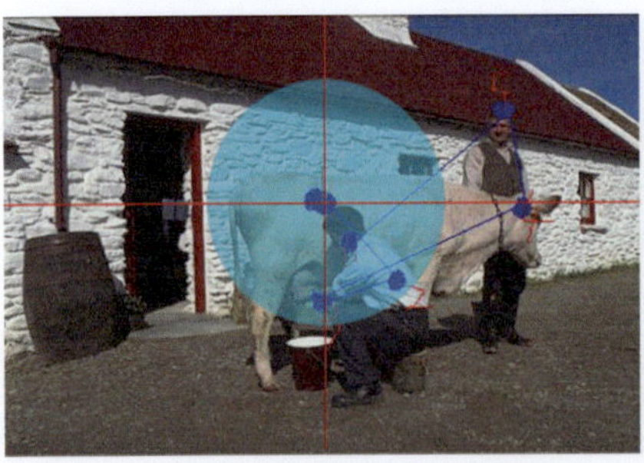

Fixation distribution discrete

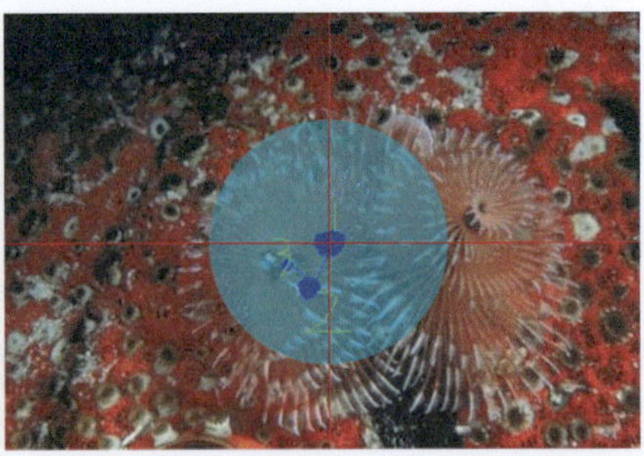

Fixation distribution concentration

Fig. 68.3 Sample of the eye movement data dispersion degree moving data of first watched points may by many factors effects is not necessarily and interest district related. On 100 site pictures eye moving data Shang superimposed the first regional watched points, on whether we can characterization pictures of main information for statistics analysis. Figure 68.4 shows natural scene pictures first watched points information characterization capacity analysis sample cases. The statistics results show that, using this species first watched points regional range, 55 site pictures of first watched points regional cannot characterize whole site pictures of main information, 45 site pictures of first watched points regional can characterization whole site pictures main information. 100 pictures of the first fixation ability to characterize the whole picture in roughly the same proportion of primary information.

100 photos by number of objects into an object, both objects and multiple objects, and statistics of various types of first fixation point and a second look at the picture if I can characterize the picture information. Figure 68.5 for classifying natural scenes pictures first fixation point and the second fixation information representation analysis sample.

Statistics show that in 36 contains an object in the picture, 18 first fixation point are meaningful, first fixation point no point 18; 33 s fixation point are meaningful, including 17 first fixation point are meaningful. In 16 pieces contains two pictures

The first fixation point cannot complete representation of image content

The first fixation point can complete representation of image content

Fig. 68.4 Analysis sample of the ability about the first fixation point representation of image content

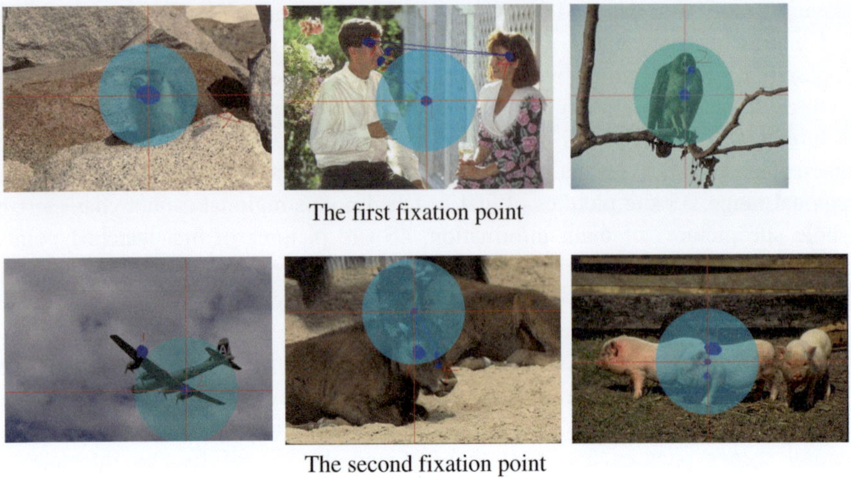

Fig. 68.5 Analysis sample of the ability about the first and second fixation point representation of image content

of objects, 16–6 in the first fixation point are meaningful, first fixation point no point 10; 13 s fixation point are meaningful, 6 of which the first fixation point are meaningful. In over 48 containing pictures of objects, 24 first fixation point are meaningful, 24 first fixation point no point; 45-s look at something meaningful, including 24 first fixation point are meaningful.

From the above statistics show that eyes secondly the probability of fixation of the characterization of the picture content is much larger than the first fixation, in the statistical process, we also found that some of first fixation point of the picture is not in the Center, but to characterize the content of the picture, and the third and the fourth look at some pictures in order to characterize the main contents of the picture.

68.4 Discussion and Conclusions

It was established that the human eyes contain 100 natural scene image database, compiled statistics separately for free watching a nature scene pictures the human eye is discrete degrees, and fixation of eye movement data relationship between sequence and characterization of picture content, results show that:

(1) According to the location of the eye fixation point, calculated according to the fovea of the human eye fixation area can effectively complete characterization of the picture content information, modified by the human eye movement data to verify computer model of attention, is to improve the consistency of results and picture content characterization of one of the feasible ways.

(2) The eye fixation point in the data and picture content in order of relevance is not exactly the same, should be combined with gaze traveled over gaze fixation duration and sequence, to comprehensive characterization of fixation of regional distinctiveness. In the application of eye movement data, all records should be used to focus the information.
(3) From the eye movement analysis of the database, we learn that in free to watch scenes of nature of first fixation does exist in characteristics of more the picture's Center, but the first fixation point may not picture has clear implications for regional, may not even have a physical significance in the picture area. Simulation of adherence to Center operator in the computer vision model, not necessarily truly simulates human visual characteristics, understanding of the picture content is not necessarily helpful.

Selected images because this trial high percentage results consistent with the computer, therefore, should be added to the computer image ratio of bad results and picture consistency, the statistical result will be more universal.

In addition, central preference occurs when human eyes freely watch nature scene pictures, its reason should also be further studied by experiments.

References

1. Itti L, Koch Ch, Niebur E (1998) A model of saliency-based visual attention for rapid scene analysis. IEEE Trans Pattern Anal Mach Intell (PAMI) 20(11):1254–1259
2. Ouerhani N, von Wartburg R, Hügli H, Müri R (2004) Empirical validation of the saliency-based model of visual attention. Electron Lett Comput Vision Image Anal 3(1):13–24
3. Judd T, Ehinger K, Durand F, Torralba A (2009) Learning to predict where humans look. In: ICCV 2009
4. Tatler BW (2007) The central fixation bias in scene viewing: selecting an optimal viewing position independently of motor biases and image feature distributions. J Vision 7(14):4, 1–17

Chapter 69
Research on Special Vehicle Crew Thermal Protection Equipment Cooling Effect Trial

Yaofeng He, Guansheng Huang, Haiyan Niu, Yuping luo, Yajuan Bai, Jianxing Bu and Rui Tian

Abstract Purpose: To supply technical support for installing and applying crew thermal protection equipment on special vehicle. Object and Methods: Analyze and evaluate the design rationality and cooling effect of crew thermal protection equipment by performance testing, crew member physiology and psychology testing and subjective sensations surveying. Results: (1) Average temperature difference (tested person of the tested group to tested person of the contrast group): chest 1.4 °C lower, left axilla nearly the same, right axillal 3 °C lower, chest temperature of the cooling clothes 5.2 °C lower to general clothes, and back temperature 5.8 °C lower; (2) No obvious difference between tested group and contrast group before/after trial; (3) 66.7 % of tested person in cooling clothes felt very comfortable, 33.3 % felt quite comfortable; (4) Tested group(three persons) fatigue degree recognition: One person normal, two persons quite ease. Compare group (three persons) fatigue degree recognition: two person quite tired, one person normal. Conclusion: Real vehicle testing manifest: crew thermal protection equipment has a good service effect. It excellently satisfied the requirement of the high temperature of special vehicle interior under the damp heat conditions of South China. It can significantly enhance the persistent working capability.

Keywords Thermal protection equipment · Special vehicle · Cooling effect · Crew work efficiency

69.1 Introduction

In South China, the climatic environment is: the temperature is high, the humidity is high and the weather is muggy. The special vehicle consists of the metallic material, which has low specific heat and fast heat conduction. Narrow interior space, closed

Y. He (✉) · G. Huang · H. Niu · Y. luo · Y. Bai · J. Bu · R. Tian
Beijing Special Vehicle Institute, Beijing 100072, China
e-mail: lei_he2010@126.com

© Springer-Verlag Berlin Heidelberg 2015

S. Long and B.S. Dhillon (eds.), *Proceedings of the 15th International Conference on Man–Machine–Environment System Engineering*, Lecture Notes in Electrical Engineering 356, DOI 10.1007/978-3-662-48224-7_69

space, poor air convection, multiple heat source and bad interior environment, which make the temperature inside special vehicle cabin reach as high as 60 °C and seriously affect the normal training of special vehicle crew [1].

Thermal protection equipment is the most effective technical mean to improve the thermal environment adaptability of crew [2, 3]. Through physiology and psychology testing and subjective sensations surveying of special vehicle crew, this research analyzes and evaluates the design rationality and cooling effect of crew thermal protection equipment, which supplies technical support for installing and applying crew thermal protection equipment on special vehicle and enhances the persistence and working capability of crew [4, 5].

69.2 Objects and Methods

69.2.1 Testing Vehicle and Person

Two testing vehicles; six crews; three sets of crew thermal protection equipment.

69.2.2 Instrument and Equipment

Instrument and equipment used for testing see Table 69.1.

69.2.3 Testing Steps and Methods

69.2.3.1 Testing Methods

There are two special vehicles for testing; one shall be installed with thermal protection equipment, and crew inside shall wear thermal protection equipment (Tested group); the other special vehicle shall be just the opposite of what was mentioned above. Two special vehicles shall be conducted a persistent driving testing for 4 h simultaneously.

Table 69.1 Trial apparatus

Trial apparatus	Model	Number	Place of origin
Dynamic temperature recorder	TMP-10	6	China
Reaction time during exercise	BD-II-505	1	China
Testing instrument			
Dynamometer	WCS-1000	1	China
Back dynamometer	BLJ-300	1	China

69.2.3.2 Testing Steps

(1) Sensor installment, testing system debugging

Sensor installment: The cooling clothes have the host inlet and outlet temperature sensor for one each and two in all; the dynamic temperature recorder sensor: T1: chest T2: back T3 left axilla T4 right axilla T5 chest of the cooling clothes T6 back of the cooling clothes T7 environment temperature

Testing system debugging: The host inlet and outlet temperature testing system of crew thermal protection equipment, and crew skin surface and colding clothes temperature testing system.

(2) Crew working efficiency index test before testing

Before the formal testing, crew working efficiency index test shall be conducted, including the psychology, the physical ability, and the physical power parameter;

(3) Driving status testing

Comparative testing: Conducting a 4 h persistent driving testing for the two special vehicles, of which one is installed with crew thermal protection equipment and the other is not.

(4) Crew working efficiency index test and questionnaire after testing

After driving testing, crew working efficiency index test, cooling effect questionnaire and self-conscious fatigue questionnaire shall be conducted.

69.3 Results

69.3.1 Temperature Testing Results

The testing has conducted 4 h, the peak and average temperature of the tested group and the contrast group, see Tables 69.2 and 69.3 respectively.

69.3.2 Crew Working Efficiency Testing Results

(1) The changing situations of crew muscle strength (grip and back strength) see Table 69.4.
(2) The testing results of time of crew reaction and movement see Table 69.5.

69.3.3 The Surveying Results of Cooling Effect

In the tested group there are three tested persons, two of them have felt comfortable and the other has felt more comfortable.

Table 69.2 Peak and average temperature of the tested group (°C)

Temperature sensor	Peak value			Average		
	Crew 1	Crew 2	Crew 3	Crew 1	Crew 2	Crew 3
T1	35.1	35.8	36.8	32.6	33.1	34.4
T2	35.6	35.1	36.0	32.9	30.1	31.7
T3	36.8	36.0	37.2	36.0	33.7	35.7
T4	36.7	35.7	36.4	35.9	34.2	34.6
T5	34.1	33.0	34.0	30.3	27.5	29.1
T6	35.1	34.3	33.3	30.2	28.0	28.5
T7	35.8	43.3	39.7	34.3	37.1	34.0

Table 69.3 Peak and average temperature of the contrast group (°C)

Temperature sensor	Peak value			Average		
	Crew 1	Crew 2	Crew 3	Crew 1	Crew 2	Crew 3
T1	36.5	36.2	37.0	34.0	34.5	35.9
T2	36.9	37.1	38.0	35.5	35.2	36.3
T3	37.2	36.9	36.7	33.8	35.9	35.8
T4	37.3	36.7	41.4	36.1	35.3	37.3
T5	34.2	36.0	39.1	32.2	33.6	36.7
T6	36.1	37.8	38.3	32.5	35.5	36.0
T7	41.6	39.5	38.6	33.0	36.0	36.2

Table 69.4 Grip and back strength of crew

Project	Number	Before trial		After trial	
		Tested group	Contrast group	Tested group	Contrast group
Grip	3	43.3	39.2	43.1	39.9
Back strength	3	96.8	96.2	106.3	97.0

Table 69.5 Time of reaction and movement of crew

Project	Number	Before trial		After trial	
		Tested group	Contrast group	Tested group	Contrast group
Time of reaction	3	0.409	0.723	0.380	0.933
Time of movement	3	0.400	0.560	0.421	0.755

69.3.4 The Surveying Results of Self-conscious Fatigue

In the tested group there are three tested persons, one of them has felt generally tired, and the other two have felt more relaxed. In the contrast group there are three tested persons, two of them have felt more tired, and the other one has felt generally tired.

69.4 Discussion

69.4.1 Temperature

In the course of testing, the meteorological temperature is 35.4 °C, and the humidity is 52.6 %. Comparing the testing datum of the tested group and the contrast group, we can see that: Average temperature difference (tested person of the tested group to tested person of the contrast group): chest 1.4 °C lower, the back 4.1 °C lower, left axilla nearly the same, right axilla 1.3 °C lower, chest temperature of the cooling clothes 5.2 °C lower to general clothes, and back temperature 5.8 °C lower [6].

69.4.2 Crew Working Efficiency Testing Results

There is no distinct difference between the tested group and the contrast group before and after testing.

69.4.3 The Surveying Results of Cooling Effect

In the tested group there are three tested persons, two of them have felt comfortable and the other one has felt more comfortable. 66.7 % tested persons in cooling clothes have felt comfortable and 33.3 % tested persons in cooling clothes have felt more comfortable [7].

69.4.4 The Surveying Results of Self-conscious Fatigue

In the tested group there are three tested persons, one of them has felt generally tired, and the other two have felt more relaxed. In the contrast group there are three tested persons, two of them have felt more tired, and the other one has felt generally tired.

69.5 Conclusion

Real vehicle testing manifest: crew thermal protection equipment has a good service effect. It has excellently satisfied the requirement of the high temperature of special vehicle interior under the damp heat condition of South China. It can significantly enhance the persistent working capability and efficiency.

69.6 Compliance with Ethical Standards

The study had obtained approval from the State Key Laboratory of NBC Protection for Civilian Ethics Committee.

All subjects participated in the test had signed the informed consent.

All relevant ethical safeguards had been met in relation to subject protection.

References

1. Vincent I, Joseph C, Michael K et al (1982) Development of microclimate cooling system for combat vehicles. Technical Report No.T10/82. USARIEM, Natick, MA 01760
2. Goldman RF (1963) Tolerance time for work in the heat when wearing CBR protective clothing. Mil Med 128:776–786
3. Toner MM, White RE, Goldman RF (1981) Thermal stress inside the XM-1 tank during operations in an NBC environment and its potential alleviation by auxiliary cooling. Technical Report No.T4/81. USARIEM. Natick, MA 01760
4. Robbert SH (1997) Cycles of war. Armed Forces J (7):38–42
5. Shvartz E (1972) Efficiency and effectiveness of different water cooled suits. A review. Aerospace Med 43:488–491
6. Walter T (2006) Microclimate cooling chemical biological individual protection conference. U.S. Army Natick Soldier Center Individual Protection Directorate
7. Schutte PC, Klerk C, Matesa J et al (2002) Body-cooling systems. Final Report To the Salty in Mines Research Advisory Committee. CSIR Mining Technology, 2002

Part VI
Research on the Machine–Environment Relationship

Chapter 70
Electronic Equipment Service Life Prediction Under the Southeast Coastal Environment

Xiaoyang Li, Zuobin Yang, Zhibing Pang, Yangke Liu, He Wei and Jiang Luo

Abstract *Objective*: To improve the study of life under hot and humid southeast coastal areas of high salinity conditions of electronic equipment. *Methods*: This paper analyzes the characteristics of electronic equipment, as well as the impact of high temperature, humid, and high salinity environment for electronic equipment, the use of neural network method to establish hot, humid, and high salinity environment of electronic equipment life prediction models. *Results*: The predicted life of the equipment at a specific temperature and humidity conditions of electronic equipment, and provide a basis for preventive maintenance. *Conclusion*: The method can be workable, effective solution to the problem of equipment life prediction of high temperature, and high salinity humidity conditions for electronic equipment.

Keywords Southeast coastal environment · Electronic equipment · Life

70.1 Introduction

With the rapid development of information technology, electronic products gradually applied in various military fields, in modern warfare is growing. However, the complex and harsh battlefield environment of electronic products will have a tremendous threat to work safely and reliably. The direction of the southeast coast is one of the main operational direction of China, which has a hot, humid, high salinity environment, etc. In the hot, humid and high salinity environment, the service life of the electronic equipment, a downward trend, study the variation of life is of great importance.

X. Li (✉) · Z. Yang · Z. Pang · Y. Liu · H. Wei · J. Luo
Air Defence Forces Academy, Zhengzhou 450052, China
e-mail: 18530092687@163.com

© Springer-Verlag Berlin Heidelberg 2015

S. Long and B.S. Dhillon (eds.), *Proceedings of the 15th International Conference on Man–Machine–Environment System Engineering*, Lecture Notes in Electrical Engineering 356, DOI 10.1007/978-3-662-48224-7_70

70.2 Southeast Coastal Climate

70.2.1 *Humidity*

Southeast coastal maritime climate, cold, and warm air encounters relatively long duration; there is a long rainy season. 2–4 month of the rainy season, rainy; 5–9 months for the typhoon season rain, humidity, and average monthly relative humidity of 85 %.

70.2.2 *High Temperature*

The average temperature in the southeast coastal areas 20–30 °C, the higher the south, the north is low, the extreme maximum temperature up to 39.7 °C.

70.2.3 *High Salt Content [1]*

Because seawater salt molecules with waves and airflow, it being brought into the air, into the land by the monsoon or typhoon. The salt content in the air and off the coast of distance, 200 km from the sea in the air, decreased rapidly.

70.3 Southeast Coastal Environment Impact Analysis of Electronic Equipment

70.3.1 *Structural Features of Electronic Equipment*

Electronic components and equipment used in many types of metallic and non-metallic materials. According to statistics, there are copper and copper alloys, nickel and nickel alloys, gold and gold alloys, silver and silver alloys, aluminum, galvanized steel, etc., were used as the conductive material of the printed circuit board, low contact resistance of contacts or contact materials, and welding excellent-coating materials. Therefore, these metals, alloys, metal coatings, and polymers corrosion resistance and stability in various environments will directly affect the reliability and service life of the electronic equipment, and determine the environmental adaptability of electronic equipment. Electronic equipment components highly integrated, complex structure, space is small, highly integrated components and printed circuit board constitute a variety of metals and alloys system. This system will not only happen galvanic corrosion corrosion damage, and because of flaking and whisker generation conductive corrosion products, but also tend to make the device and short circuit occurs in a small space. The use of electronic equipment in the southeast coastal areas affected by hot and humid environment, making equipment changes occur in the course of life [2].

70.3.2 *High Temperature and Humidity Effects on Electronic Equipment*

70.3.2.1 The Impact of High Temperature Electronic Devices

In the environment of high temperature and electronic equipment will generally overheating. Overheating is one of the major electronic equipment failure. Failure to make the device performance degradation caused by chemical or physical change. Because as the temperature increases, the movement speed of electrons, atoms, and molecules accelerates, so that the performance of the electronic device changes [3]. With the aging of the device, these changes occur gradually; when it reaches a certain stage it caused serious failures.

High temperature stimulated major failure modes are: first, the expansion coefficients of different materials bonded together inconsistent with the parts; the second is to reduce the viscosity of the lubricant. Lubricants outflows connection loss lubricating ability; the third is packed village pads, seals, bearings, and shaft deformation occurred, bonding and failure, caused by mechanical failure or damage to the integrity; fourth, different materials and different temperature gradient expansion inconsistency makes the stability of electronic circuits changed; fifth, organic material fade, crack or cracks; six is overheated transformers and electromechanical components; seven relay and magnetic or thermal means moving on/off range.

70.3.2.2 Moist Environment for Electronic Devices

Effect humid environment refers appearance or physical, chemical, and electrical properties of the product or material degradation occurs under wet conditions and lead to a functional failure of the combined effects of the device [3]. There are a variety of air pollutants may strengthened humid climates, such as corrosive gases and moisture interaction will increase the rate of corrosion of gold in the work environment, and some dust will contribute to readily absorb moisture condensation test sample surface or moisture absorption, thereby increasing the surface of the insulation performance degradation.

Humidity-inspired failure modes are: first, the surface absorption. When the risk of surface contamination related to the mechanism of failure is caused by the absorption of particular importance. Second, the capillary condensation. When the risk of fracture, cracks, and pores related to capillary condensation is the main mechanism of failure. If the water in such a way to penetrate the surface of the protective layer and acts on the material to be protected, then it may be faulty. Third, through the proliferation of loose material is the third failure mechanism. It is usually not directly related to or less concerned with hazards. This mechanism is very slow, to a few days to several months.

70.3.3 Effect of High Salt Environment of Electronic Equipment

The main effect is to produce salt spray corrosion, accelerate corrosion of metals in weaponry, and decrease reliability; because electrical degradation of electronic components, the insulating material is etched to produce mildew environment.

70.4 BP Neural Network-Based Electronic Equipment Life Prediction Modeling

Due to the impact of high salinity environment of high temperature and humidity, the electronic equipment is nonlinear trends lifetime characteristics, and it is difficult to describe with mathematical formulas, and the neural network has a strong self-learning ability of parallel computing power and better able to predict the electronic equipment life in hot and humid conditions.

70.4.1 BP Neural Fundamentals

BP neural network [4] is widely used neural network model because of its ability to become a good nonlinear. It has a wide range of applications in classification, prediction, fault diagnosis, and parameter detection. BP neural network forecasting capabilities are learning algorithm by errors of reverse spread to achieve. The main idea [5]: for q learning samples: $p_1, p_2, \ldots, p_q$, the corresponding output samples: $T_1, T_2, \ldots, T_q$. The purpose of learning is to use actual output $A_1, A_2, \ldots, A_q$ and output samples $T_1, T_2, \ldots, T_q$ between to modify its weight, so that $A_1, A_2, \ldots, A_q$ and the desired $T_1, T_2, \ldots, T_q$. As close as possible, even if the network output layer and the minimum-squared error. It continuously calculates the error function with respect to the direction of change in the slope of the decline in the value of network weights and bias by so close to the goal. Every change weights and deviations are proportional with the network errors, and by way of counter-propagating transmitted to each layer. Input set to p, the input neurons having r, s_1 inner hidden layer neurons, activation function for T; s_2 neurons have the output layer, corresponding to the activation function, output A, output samples T. The steps [5] as follows:(1) ith hidden layer neuron output:

$$a_{1i} = f_1\left(\sum_{j=1}^{r} w_{1if}p_j + b_{1i}\right), i = 1, 2, \ldots, s_1 \tag{70.1}$$

(2) the output of the output layer neurons k as:

$$a_{2k} = f_2\left(\sum_{i=1}^{s_1} w_{2ki}a_{1i} + b_{2k}\right), k = 1, 2, \ldots, s_2 \tag{70.2}$$

(3) the error function is defined as:

$$E(w, B) = \frac{1}{2}\sum_{k=1}^{s_2}(t_k - a_{2k})^2. \tag{70.3}$$

(4) Find the value of the change in the output layer with the gradient method: from the ith input to the variations in weight of the kth output is:

$$\Delta w_{2ki} = -\eta\frac{2E}{2w_{2ki}} = -\eta\frac{\partial E}{\partial a_{2k}}\frac{\partial a_{2k}}{\partial w_{2ki}} = \eta(t_k - a_{2k})f_2 a_{1i} = \eta\delta_{ki}a_{1i}, \tag{70.4}$$

$$\delta_{ki} = (t_k - a_{2k})f_2^t. \tag{70.5}$$

Similarly available:

$$\Delta b_{2ki} = -\eta\frac{\partial E}{\partial b_{2ki}} = -\eta\frac{\partial E}{\partial a_{2ki}}\frac{\partial a_{2kt}}{\partial b_{2ki}}, \tag{70.6}$$

(5) changes in the hidden layer weights. Input from the jth to the ith output power values are:

$$\Delta w_{1ij} = -\eta\frac{2E}{2w_{1ij}} - \eta\frac{\partial E}{\partial a_{2k}}\frac{\partial a_{2k}}{\partial w_{1i}}\frac{\partial a_{1i}}{\partial b_{1ij}} = \eta\sum_{k=1}^{s_2}(t_k - a_{2k})f_2 w_{2ki}f_1 p_j, \tag{70.7}$$

$$\delta_{ij} = \sum_{k=1}^{s_2}\delta_{ki}w_{2ki}f_1. \tag{70.8}$$

Similarly available:

$$\Delta b_{1i} = \eta\delta_{ij}. \tag{70.9}$$

70.4.2 Electronic Equipment Life Prediction BP Neural Network Design

(1) Enter the design of the output layer. BP neural network based on design features, without regard to the mutual relationship between the factors affecting that there is no connection between neurons at all levels within the lifetime of electronic equipment that can be selected to predict the temperature, humidity, and salinity as input layer [6], the input nodes is 7. Output layer is the actual life of electronic equipment, the output layer node 1. For the hidden layer node selection should follow nodes as small as possible, otherwise it would be too long-training time, errors may not achieve the desired requirements. Therefore, the number of nodes in the hidden layer to select according to experience, the calculation formula is:

$$n_1 = \sqrt{n + m + a} \tag{70.10}$$

where n is the number of nodes in the hidden layer; n is the input layer nodes; m is the output layer nodes; a 10 to choose between a constant 1.Optimal tradeoff may be determined according to Eq. (70.10) the number of nodes of the hidden layer. As electronic equipment design life prediction neural network model, the actual lifetime with temperature and humidity and salinity changes are nonlinear. Thus, the initial weights [6] to select whether to achieve the predicted life-cycle cost or whether convergence length of training time and so have a great relationship. So for electronic equipment life prediction layer 2 network, in order to prevent emergence of a local minimum, no convergence or training time is too long, etc., can be used to give Luo Wei initial weights of selected strategies: the right to choose the value of the order of $r\sqrt{s_1}$, where s_1 is the number of neurons in the first layer.

(2) the selection of the target. Beginning in the design of electronic equipment, which will deal with the high temperature, high humidity, and high salinity environments demonstration. According to the case demonstrated, using multiple regression method, the cost parameters, analogy, and other methods to estimate the expected target electronic equipment time. As the target output.

(3) the establishment of the model. In the course of this BP neural network training, the first to take a certain number of samples, to learn their initial weight is selected, then the results of its output layer is the life-cycle cost.

(4) computing model. Model calculated using Matlab software, the calculation process is similar to the first electronic equipment according to previous data as sample data for training, adjust the weights, the final estimate of the actual life of the corresponding data based on estimates nice.

70.5 Conclusion

Hot and humid in the southeast coastal areas of high salinity conditions, the life of weaponry downward trend, and the high temperature and humid environment with high salt content of environmental impact, changes in life is to present a nonlinear characteristic; there is no effective method its service life prediction. In this paper, electronic equipment mechanism induced by analyzing heat and humidity was failures and stimulate major failure mode, clear hot, and humid environment of electronic equipment life decline. Using Matlab software tool to establish a neural network method, that the temperature and humidity is used for the input and the output of service is life of the equipment. This method can be workable, effective solution to the problem of equipment life prediction hot, and humid conditions of electronic equipment.

References

1. Zhang C (2004) The southeast coast of the natural environment on the ground equipment and preventive measures. Equipment environmental engineering, general armament department of electronic information technology infrastructure bureau, Beijing
2. GJB 150-1986 (2010) Military equipment, environmental test methods
3. Zhu X (2012) Electronic environmental test equipment failures influencing factors. Electronic Technology
4. Wen X (2000) The design to the application of MATLAB nerving network.Science Press, Beijing (in Chinese)
5. Yuan Z (1999) Manual nerving network centre and its application.Tsinghua University Press, Beijing (in Chinese)
6. Zhu H (2006) The estimated method and application based on the dynamic bp nerving network. Comput Netw 9(5): 11—16 (in Chinese)

Chapter 71
Analyses on Influences of Humiture upon Ammunition Storage and Countermeasures

Rongzhi Yang, Zhi Li, Meng Kang, Xuming Sun and Haoran Zhu

Abstract **[Purpose]** Based on the status quo of ammunition storage in the army, this paper starts with studies on influence of humiture, to explore countermeasures of high-quality ammunition storage. **[Method]** It finds out various factors influencing ammunition storage environment, followed by an analysis of how those factors function, and points out that humiture is the primary one that leads to ammunition quality decrease. **[Result]** Through studies, it obtains a favorable humiture range for ammunition storage. **[Conclusion]** In accordance with the guiding principle of *solving key problem by grasping principal contradiction*, this paper combines the functional mechanism of the primary factor, puts forward countermeasures of humiture influences, and affords army units references to improve ammunition storage quality.

Keywords Army units · Ammunition storage · Humiture · Countermeasures

Central Military Commission Chairperson XI Jinping pointed out that significance of weapons and equipment should never be underestimated although human remains the decisive factor in modern warfare [1]. Ammunition is the fundamental element of weapon system, whose performance shall have immediate impact on combat effectiveness and the result of a battle. It is a basic requirement to strengthen the scientific management of ammunition so as to maintain ammunition in good condition. In this paper, basic cognition and consideration are discussed with regard to influences of humiture upon ammunition storage.

R. Yang · Z. Li (✉) · M. Kang · X. Sun · H. Zhu
Air Defence Forces Academy, Zhengzhou 450052, China
e-mail: lizhi201424002@163.com

© Springer-Verlag Berlin Heidelberg 2015

S. Long and B.S. Dhillon (eds.), *Proceedings of the 15th International Conference on Man–Machine–Environment System Engineering*, Lecture Notes in Electrical Engineering 356, DOI 10.1007/978-3-662-48224-7_71

71.1 Basic Cognition of Ammunition Storage

71.1.1 Function of Ammunition

Ammunition is understood as a general term for such explosives as rifle cartridge, cannonball, grenade, bomb, and landmine [2]. As an important component of weapon system, ammunition is frequently used in training and largely consumed in operation. Ammunitions may hit the target or target zone by means of weapon or other equipment to ensure army units to fulfill a task and win a battle. Precision strike in future informative wars demand high-quality ammunition support to achieve information combat intentions.

71.1.2 Storage of Ammunition

After being produced, ammunition is mainly stored in depots of military units for regular training and combat. Such depots are often seen as cave depots, ground depots, and temporary depots [3]. A cave depot refers to an army depot built in the side of a cliff or hill, or one reconditioned out of a natural depot; a ground depot means an army depot built on the ground; a temporary depot is a temporarily established depot made of tent material or waterproof cloth.

The ammunition is usually stored in cave depots and ground depots.

71.1.3 Demand upon Ammunition Storage Environment

The ammunition is inflammable and combustible. Accidents are very likely to happen in conditions such as heat, friction, shock, collision, and direct sunshine. The ammunition may also spoil or malfunction if affected with damp or erosion [4].

71.1.3.1 Natural Environment

The natural environment of ammunition storage is primarily understood as the geographical environment of the ammunition depot. An ammunition depot should be located in mountainous area of good natural disguise of round mountains, ravine or deep mound, with favorable geological conditions of lithic or rock mass, and away from large and medium-sized cities, important transportation hub and large and medium-sized reservoirs in the mean time.

71.1.3.2 Internal Environment

The ammunition depot should satisfy relevant technical demands of construction with complete auxiliary facilities, as well as protective measures against stealth, fire, thunder, force attack, and static electricity. Environmental monitoring devices and humiture adjusting devices should be quipped to regulate the internal storage environment.

71.1.4 Social Environment

Ammunition depots should meet specified safety requirements, lest civilians be injured if unexpected accidents happen. Army units should therefore keep close relationship with local government for military-civilian cooperation and coordination to protect the safety of natural defense facilities.

Through study, we believe there are many external factors that may lead to ammunition deterioration but humiture is the main factor.

71.2 Analyses on Influences of Humiture upon Ammunition Storage

At present, army units pay much attention to environment monitoring on temperature and humidity, survey relevant data and make records, and take effective measures to improve substandard work of humiture control. However, problems still exist and cannot be ignored.

71.2.1 Existing Problems

(1) Administrative vulnerabilities. Humiture administration is found substandard in some ammunition depots, for instance: officers and men attach insufficient importance to humiture control; officer and men are not adept at business; protection capacities are poor, etc.
(2) Old depots. Some depots were built in 1950s and are of outdated structure. Owing to geological movement and other external force, cracks and leakage appeared and storage condition went bad; some others are lack of moisture-proof measures, the ground and walls of which are not specially processed against dampness.

(3) Backward environmental monitoring devices. The humiture monitoring system of some ammunition depots are large and complex in structure, and of low accuracy and short service life.

71.2.2 *Analysis of Reasons*

(1) Poor awareness of combat preparation. A long time of over 3 decades free of war has slackened some people's vigilance of war and inactivated their cognition of combat readiness to manage ammunition depot carefully.
(2) Lacking of responsibility. The main reason of problematic unit and individual is lacking of sense of responsibility, duty, and honor. The problem in humiture management of ammunition storage also results from the same fact.
(3) Poor subjective initiative. Failure to bring subjective initiative into full play, to solve such problems as serious influences of external storage environment, outdated devices, backward techniques, prolonged service of equipment, has posed increasing adverse influence upon the quality of ammunition [5].

71.2.3 *Suitable Humiture Range*

By consulting a large number of documents, we find military experts have different ideas about the suitable humiture range for ammunition storage. Through comprehensive studies, we believe the suitable temperature range is 15–25 °C, never higher than 30 °C and lower than −12 °C; the suitable relative humidity range is 55–65 %, never higher than 70 % and lower than 40 %.

71.3 Management Countermeasures to Influences of Humiture upon Ammunition Storage

The key link of studying the mechanism of humiture influence upon ammunition storage is to reduce possible adverse influences and to promote the initiative, pertinence and effectiveness of ammunition storage management in army units.

71.3.1 *To Strengthen Leadership*

The safety management work of ammunition depots is relevant to the general situation of combat readiness building and informative operation, and thus the important influences of humiture upon ammunition storage should be fully emphasized. Leading officers on various levels should strengthen the safety management and humiture monitoring of ammunition depots, persist in implementing ammunition depot safety management work as a serious political project and *chief officer's project* around the influences of humiture, take new situations and new problems that occur along with the development of weapons and equipment into consideration, follow the principle of *paying close attention to key points, consolidating strengths and making up for weaknesses*, and start from the overall situation of military preparedness to make progresses in ammunition depot building and management. Officers and men should correspondingly be educated to act in the same way in peace time as in war time.

71.3.2 To Reconstruct Depots

Chairperson XI pointed out that all such work as military training, politics, logistics, and armament should be conducive to promoting the army's fighting capabilities or otherwise it will be as good as nothing [1]. Ammunition depot construction is an important part of armament building, therefore the party committee should spare no effort to invest in depot reconstruction and persist in the principle of technology advanced, information-led, overall planning, and stressing implementation to vigorously refurbish part of old depots or build new ones. Make sure that refurbished and newly built depots are of new design, new material, and new technology, with more reasonable structure, improved function and better humiture environment. In the meantime, develop or purchase advanced devices to intelligentize ammunition depot humiture monitoring and controlling, to ensure timely monitoring data upload, real-time data display, data graph display and historical data storage, and realize warning function if relevant parameters beyond limit so as to master the all-time-space and all-process changes of temperature and humidity to better depot humiture controlling.

71.3.3 To Practice Rules and Regulation Strictly

There is a complete set of rules and regulations for ammunition depots. Depot officers and men should establish the mind of observing disciplines and create the atmosphere of all knowing the rules, all obeying the rules, all being expert at business and all stressing safety. Ensure that all depot people are guardians and safety supervisors, all work is in the charge of someone, and avoid dead ends and dead zones. Prevent humiture from exceeding standard and avoid hidden dangers to guarantee the quality and effectiveness of humiture controlling. Further perfect responsibility system and lay solid foundation for safety management, by multi-layer joint controlling create the atmosphere of ensuring safety collaboratively and emphasize implementation at terminals.

71.3.4 To Promote the Professional Quality of Officers and Men

Persist in promoting the professional quality of depot officers and men as foundation engineering and keep it on by means of encouraging self-study, off-job training and in-post intensified training. Organize officers and men to study technologies and subjects of depot management, particularly with regard to the change rules of humiture and countermeasures to situations of humiture beyond limits. In addition, focusing on the changes of ammunition depot environment, research, study and demonstrate the organic combination among man, ammunition and environment to create conditions to improve depot environment in good time.

71.3.5 To Promote Ability of Safety Protection

The ammunition depot safety protection is a complex system, which requires both prevention from excessive internal humiture and adverse influences of severe natural conditions and man-made destruction. Effective precautionary measures should therefore be taken, for example, to join up computerized telephone, pulse power network, IR warning system, dehumidifying device as well as monitoring system into an automatic, visualized and networked modern protection system of grade separation junction and complementary warning, so as to promote the ability of safety protection of the ammunition depot.

References

1. Excerpts of XI Jinping's (2014) Important discourses on building a strong army in new situation. PLA Press, Beijing
2. Modern chinese dictionary (5th edition) (2005). The Commercial Press, Beijing
3. Jiumin QI (2010) Cannonballs and Ammunition. Aviation Industry Press, Beijing, p 7
4. Shoujiang T (2013) A study on the influences of humiture upon ammunition quality change. J Hubei Sci Technol Inst
5. Yulong G (2010) Influences of storage environment upon ammunition quality

Part VII
Research on the Overall Performance of Man–Machine–Environment System

Chapter 72
Evaluation on PPy-Coated Conductive Fabric for Human Motion Monitoring

Xiaofeng Zhang, Guohao Li, Jiyong Hu and Xudong Yang

Abstract The wearable human motion monitoring system benefits the prevention and treatment of some diseases. Because conductive fabric can be well integrated with the garment, it is ideal sensing elements of wearable human motion monitoring system. However, the electrical resistance of conductive fabric is lack of systematic evaluation. This paper prepared PPy (polypyrrole) conductive fabric by in situ polymerization, and the anisotropic property of the fabric resistance as well as the resistance-strain relationship were measured. And also, the relationship between the resistance and the human knee and elbow movements were discussed preliminarily.

Keywords Body motion · Conductive fabric · Ppy · Anisotropic · Resistance-strain

72.1 Introduction

Measuring human movements is beneficial to rehabilitation, training, or exercises. Motion capture system, which consists of cameras, accelerometers, and flexible electrogoniometers, has been used to monitor human motion [1]. Although motion capture system is capable of accurate measurements, their sensors fixed on the cloth by strap or other methods make uncomfortable to use. During the flexion of a limb,

J. Hu (✉) · X. Yang
Key Lab of Textile Science and Technology, Ministry of Education Donghua University, Shanghai 201620, PR China
e-mail: hjydhu@126.com

X. Yang
e-mail: xdyang@dhu.edu.cn

X. Zhang · G. Li
College of Textiles, Donghua University, Shanghai 200237, People's Republic of China

© Springer-Verlag Berlin Heidelberg 2015

S. Long and B.S. Dhillon (eds.), *Proceedings of the 15th International Conference on Man–Machine–Environment System Engineering*, Lecture Notes in Electrical Engineering 356, DOI 10.1007/978-3-662-48224-7_72

the skin around the joint stretches and the skin around the knee stretches lengthwise about 40 % of its normal length require 25–30 % stretchability from the fabric to ensure general comfort [2]. In terms of the well integration with clothing, textile solutions are well suited for constructing a comfortable sensing system for the wearer.

Textile sensors based on electrical resistance for wearable devices can be produced by conductive elastomer coating or attached on conventional fabrics, or by knitting and weaving conductive yarns. In recent years, the technology for conductive fabric sensing elements from conductive polymer has been extensively studied. These materials offer several advantages with respect to other sensors: lightness, large elasticity and resilience, resistance to corrosion, flexibility, etc. [3]. Among several typical kinds of conductive polymers (polypyrrole (PPy), polyaniline (PANI), and polythiophene (PTh)), PPy has some advantages such as high conductivity, good environmental stability, ease of synthesis, adhesion, and non-toxic. The surface resistance of PPy conductive fabric shows a high sensitivity with its single-axis elongation [4], and the single-axis strain sensitivity coefficient of PPy coating Lycra fabric is up to 25 [5].

The body joint motion was detected by monitoring the resistance change of the conductive fabric. For human joint movement, its movement is usually multi-direction, such as the torsional flexion of elbow joint. In this case, the fabric sensor around the joint is not only stretched along the axial direction of extremities, the sensor can deform in the other direction, as is usually neglected in previous studies. To characterize the joint movement, the resistance in all directions should be monitored. Therefore, it is necessary to understand the directional resistance distribution of the conductive fabric. Due to more or less anisotropic structure of the fabric, the PPy-coated fabric will show to some degree anisotropic electrical resistance [6]. In this sense, it is significant to investigate the dependence of the anisotropic resistance on fabric structure.

In this paper, the electrical conductivity of PPy woven fabric was studied with respect to the surface resistance and its directional distribution, as well as the elongation-resistance relationship. And then, the PPy-coated conductive fabric is used to monitor the extension angle of the knee and elbow, and the angle-resistance relationship is discussed.

72.2 Preparation for PPy-Coated Fabric

72.2.1 Materials

Three kinds of fabrics were chosen, and their parameters are summarized in Table 72.1, and the preparation techniques of PPy-coated fabrics can referred to [7].

Table 72.1 Constructional properties of three kinds of fabrics

No.	Weave	Component	Warp (tex)	Weft (tex)	Warp density (threads/10 cm)	Weft density (threads/10 cm)
1	Plain	93 % C7 % SP	17	17	900	880
2	Plain	93 % T7 % SP	8	8	360	240
3	Twill	80 % C20 % SP	25	25	460	350

72.3 Characterizations

72.3.1 *Surface Resistivity*

The fabric surface resistivity was measured according to AATCC 76-2002. The assessment of the anisotropy of conductive fabric resistance is based on the anisotropy function [6], which defines the dependence of the resistance *R* on the angle α related to the sample resistance value [8]. The anisotropic degree of resistance was measured by Rmin/Rmax. When the ratio is in the interval [0.95, 1], the sample is called as isotropic resistance, otherwise anisotropy. The test procedures are depicted in Fig. 72.1. A customized fixture is used to fix the round specimen by 16 electrode holes evenly distributing along the edge of the sample. Each sample is cut into a circle with a radius of 10 cm, then 4 electrodes was arranged in the vertices of the square inscribed in the circle. The tests for the anisotropy of resistance of samples were conducted in the range of $\alpha \in [0, 2\pi]$ with an angle step of $\pi/8$. Each measurement was repeated five times.

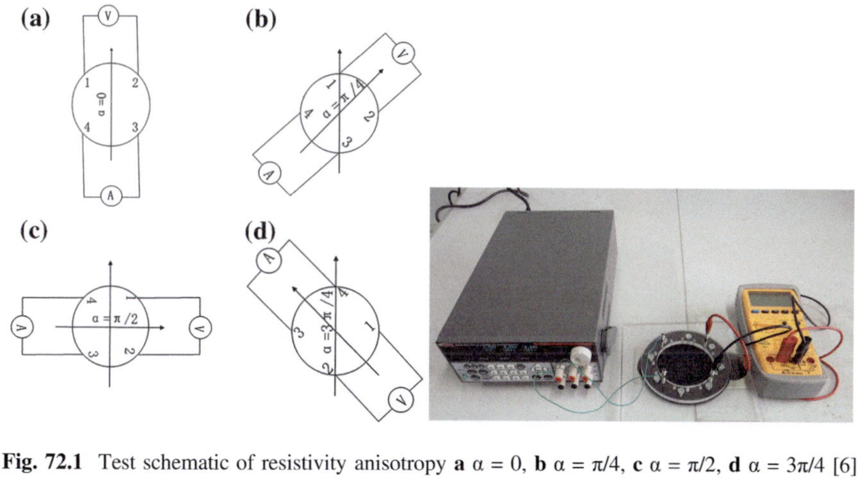

Fig. 72.1 Test schematic of resistivity anisotropy **a** $\alpha = 0$, **b** $\alpha = \pi/4$, **c** $\alpha = \pi/2$, **d** $\alpha = 3\pi/4$ [6]

72.3.2 Strain-Resistance Relationship

A universal testing machine is used to stretch the conductive fabric, and the change of electrical resistance is recorded. Five samples were tested, and the average value of the resistance was calculated. The fabric size is 5 × 15 cm. The testing machine's running speed is 10 mm/min, and fabric resistances under elongation of 5, 10, 15, 20, 25 % are tested, respectively.

72.3.3 Extension Angle-Resistance Relationship

The change of the fabric electrical resistance under different bending angle of human knee and elbow was tested. The bending movement of human knee and elbow joint is shown in Fig. 72.2, and the bending angle is successively increased from θ1 to θ4. Such a process will be accompanied by the elongation of conductive fabric around the joint. The conductive fabric was fixed on the knee and elbow pads during the test process, and two copper electrodes are used to connect a multimeter for measuring the resistance of conductive fabric. The size of PPy conductive fabric is 2 × 18 cm, and the distance between the two copper electrodes is 15 cm. After that, the knee and elbow pad is worn in the corresponding parts of the subject, and the real-time resistance value is recorded by the digital multimeter. The detailed test procedures are shown in Fig. 72.3, and the movements of the elbow joint and the knee joint are in Fig. 72.3a, b, respectively. The subgraphs indexed by a, b, c, and d are corresponded to Fig. 72.2 sequentially θ1, θ2, θ3, and θ4. Each bending test is repeated with five times.

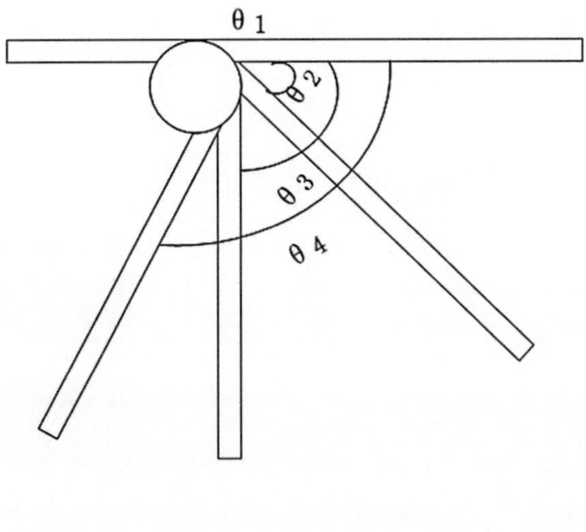

Fig. 72.2 Illustration for different bending angle of *knee* and *elbow* joint

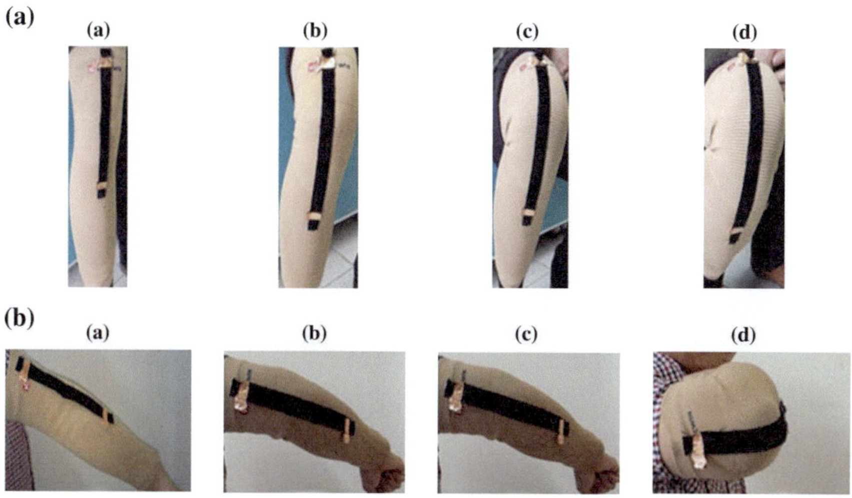

Fig. 72.3 Bending tests of human *knee* and *elbow* joint

72.4 Results and Discussion

72.4.1 Anisotropy of Electrical Resistance

For fabric 1, the measured conductive resistances were in the range of [2.39, 3.83] kΩ, fabric 2 in [1.23, 3.21] kΩ, and fabric 3 in [0.67, 1.84] kΩ. The obtained values by Eq. (2) were equal to 0.63 for fabric 1, 0.38 for fabric 2, and 0.36 for fabric 3, respectively. The equation is $R_{\max}/R_{\min}$, the R is the resistance of the fabric. Their magnitudes indicate the anisotropy of electrical resistances. The lowest anisotropy degree is fabric 1. Fabric 1 is plain fabric and has similar warp to weft yarn density, and intrinsically is more uniform than fabric 2 and fabric 3. Therefore, the anisotropy of PPy conductive fabric resistance is highly related to fabric structure.

The directional distribution of electrical resistance of three fabric samples is schemed in Fig. 72.4. Each of three fabrics shows a periodic anisotropy and a symmetry at about $\alpha = \pi$. In the case of fabric 2, the maximum resistance, 3.21 kΩ, was achieved for an angle of 0 rad, and the lowest resistance value, 1.25 kΩ, was obtained for an angle of $\pi/2$ rad. This characteristic distribution of electrical resistance is attributed to the regularity of fabric structure, i.e., the warp and weft yarn in woven fabric are orthogonal. When the electrodes are parallel to the warp or weft direction, as shown in Fig. 72.1a–c, the extreme values occur. Thus, the interval between the maximum and minimum is $\pi/2$ rad. After rotating π rad, the test is only to exchange the voltage electrodes and current electrodes, and the values along every direction have no significant change. Of course, a little difference was observed, and it is due to the not absolutely

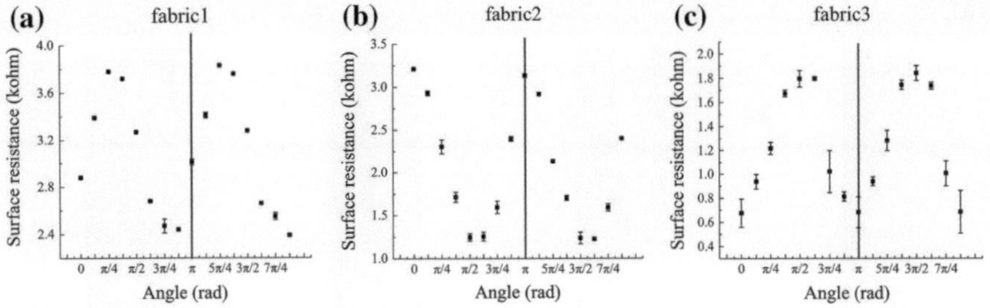

Fig. 72.4 Values of the anisotropy function for the fabrics

uniform fabric structure, such as yarn spacing, warp, and weft contact area. When the current is input from different terminal of one transmission line, the non-uniform structure features will affect the current transmission, and the resistance will change with the exchanged positive and negative electrodes. To note, test results show that the distribution trend of electrical resistance in each direction is irrelevant to the fiber contents of the fabric and the base weaving texture (plain or twill).

72.4.2 *Strain-Resistance Relationship*

As an example, test results of fabric 3 are shown in Fig. 72.5. It indicates that the resistance slowly increases with the elongation up to 15 %. After the elongation reaches 25 %, the resistance almost keeps constant. This behavior is

Fig. 72.5 Resistance-strain relation of fabric

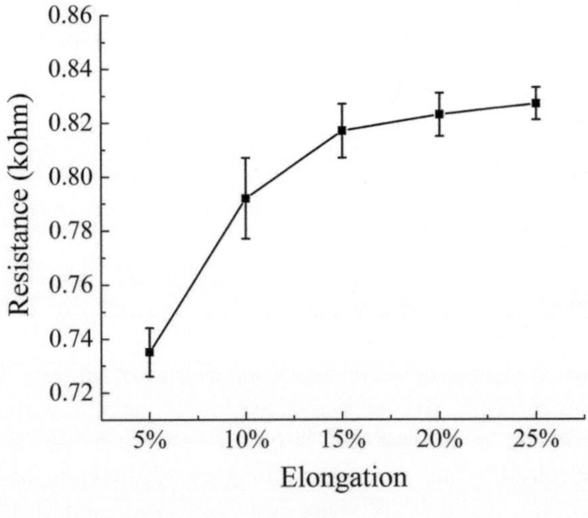

attributed to the electrical resistance nature in woven fabric. The electrical resistance consists of two parts [9]: contact resistance, Rc, which holds the resistance between the crossing points, and yarn resistance R, which comes from the yarn segment.

During stretching, the elongation of the conductive fabric will lead to variation in length and cross-section of the yarns in the fabric. From formula $R = \rho L/A$, where ρ is electrical resistivity of a material, L the length of the conductor, and A the cross-section area of the conductor, with the increase of the elongation the length of the yarns increase and their cross-section decrease will increase electrical resistance. On the other hand, the elongation of the woven fabric decreases the contact area between the warp and weft, thus the contact resistance Rc will increase. Therefore, in the beginning of stretching conductive woven fabric, the fabric resistance will increase dramatically. When the elongation reaches a certain size, yarn elongation has reached its limitation, thus the fabric resistance will not change significantly.

72.4.3 Human Motion Test

Fabric 3 is chosen to take preliminary tests of the capability in monitoring human movement state. Test result is shown in Fig. 72.6. Figure 72.6a indicates that the resistance increases with the increase of the degree of elbow bending (see Fig. 72.3). Similarly, the knee bending angle can be monitored by the surface resistance in Fig. 72.6b. In this study, the conductive fabric was attached to the kneepad (elbow pad) and skin tight fit. When the body joint bends, the fabric around the joint has a certain degree of elongation. The resistance-strain relations of the conductive fabric show that the fabric resistance increases with the strain.

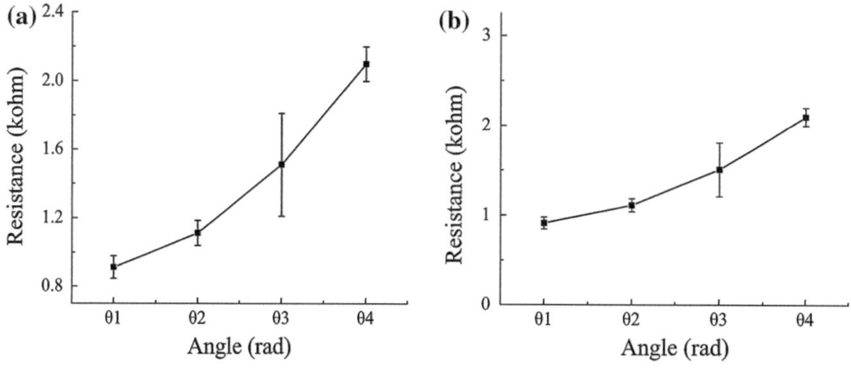

Fig. 72.6 Dependence of electrical resistance on joint bending angle: **a** *elbow*; **b** *knee*

72.5 Conclusion

The conductive resistance of PPy-coated woven fabric is anisotropic, and the anisotropy feature is relevant to structure of the fabric itself. And the trend of resistance distribution is only related to the orthogonality of the fabric sample structure, and the maximum or minimum value occurs at the direction which parallels to the axial direction of the warp or weft yarn. The size of electrical resistance change is limited and also depends on the fabric structure. The preliminary angle-resistance test showed the capability of PPy-coated fabric in monitoring the movement of human joints.

Acknowledgments This work is supported by the National Natural Science Foundation of China through project (No. 51405079, No.51175076), the Natural Science Foundation of Shanghai through project (No. 12ZR1400500) and the Fundamental Research Funds for the Central Universities.

References

1. Banabic D (2009) Sheet metal forming processes. Springer, Heidelberg
2. Corbman BP (1983) Textiles: fiber to fabric. Gregg/McGraw-Hill Marketing Series; McGraw-Hill. Gregg Division, New York
3. Scilingo EP, Lorussi F, Mazzoldi A, De Rossi D (2003) Strain-sensing fabrics for wearable kinaesthetic-like systems. IEEE Sens J 3(4):460–467
4. Li Y, Cheng XY, Leung MY, Tsang J, Tao XM, Yuen CWM (2005) A flexible strain sensor from polypyrrole-coated fabrics. Synth Met 155(1):89–94
5. Wu J, Zhou D, Too CO, Wallace GG (2005) Conducting polymer coated lycra. Synth Met 155 (3):698–701
6. Tokarska M, Gniotek K (2015) Anisotropy of the electrical properties of flat textiles. J Text Inst 106(1):9–18
7. Lee JY, Park DW, Lim JO (2003) Polypyrrole-coated woven fabric as a flexible surface-heating element. Macromol Res 11(6):481–487
8. Tokarska M (2013) Measuring resistance of textile materials based on Van der Pauw method. Indian J Fibre Text Res 38(2):198–201
9. Banaszczyk J, De Mey G, Schwarz A, Van Langenhove L (2009) Current Distribution modelling in electroconductive fabrics. Fibre Text East Eur 17(2):28–33

Chapter 73
Analysis of Human Factors on Troops Training Accidents and Research on Precaution and Control Countermeasures Based on CREAM

Peng Gong, Zhenguo Mei and Chang Mei

Abstract With a strong military practice deeply advancing and the reality-based military training continuously developing safety risk of troops training is increasingly widening. How to perform safety precaution and control well in military training has been a real urgent subject. At present, research on troops training accidents mainly depends on qualitative analysis and is short of scientific guide, so it is difficult to find the root cause of accidents occurrence correctly. Applying CREAM, this essay constructs retrospective analysis mode of the root cause of human factors error in troops training accidents, performs retrospective analysis of the accidents root cause, and puts forward the strategy and idea of precaution and control in troops training accidents. It provides scientific basis and effective means based on the analysis of the 158 troops training accidents.

Keywords MMESE · Troops training accidents · Human factors analysis · CREAM · Safety precaution and control

At present, research on troops training accidents mainly depends on qualitative analysis and is short of scientific guide, so it is difficult to find the root cause of accidents occurrence correctly. Applying CREAM, this essay constructs retrospective analysis mode of the root cause of human factors error in troops training accidents, performs retrospective analysis of the accidents root cause, and puts forward the strategy and idea of precaution and control in troops training accidents. It provides scientific basis and effective means based on the analysis of the 158 troops training accidents in recent years.

P. Gong (✉) · Z. Mei
Air Defense Forces Academy, Zhengzhou, China
e-mail: xwy@zzu.edu.cn

C. Mei
Beijing Foreign Studies University, Beijing, China
e-mail: pamelabfsu@163.com

© Springer-Verlag Berlin Heidelberg 2015

S. Long and B.S. Dhillon (eds.), *Proceedings of the 15th International Conference on Man–Machine–Environment System Engineering*, Lecture Notes in Electrical Engineering 356, DOI 10.1007/978-3-662-48224-7_73

73.1 Brief Introduction on CREAM

CREAM is a short form for Cognitive Reliability and Error Analysis Method. It is a newly human-factor reliability analysis method put forward by Eric Hollnagel. There are two characteristics for it: the first is to emphasize important influence of situational environment on human behavior. The second is to bring up special model frame of retrospective analysis of the root cause of human factors error, see Fig. 73.1 [1]. CREAM regards error mode as the beginning point of analysis, confirms cause, and begins retrospective analysis according to error mode cause table. It regards error mode cause as effect to find the related general cause and concrete cause in effect-cause chain table. As effect, the general cause continues to be analyzed in effect-cause chain table. Finally the root cause of the accidents is found.

73.2 Construction of Parsing Table of Retrospective Analysis of the Root Cause in Troops Training Accidents

73.2.1 Mode Classification of Human Error in Troops Training Accidents

Human is a complicated complex and human behavior is influenced by all kinds of factors. In troops training, the behavior of officers and soldiers is not only limited by their thought, psychology, physiology, knowledge, technique, and other inner factors, but also influenced by military training features, training organization, armament technology, training environment, and other external factors. In this way, their behaviors take on some uncertainty, will easily lead to error behavior, and cause training accidents.

American scholar Reason divided human unsafe behavior into two categories: the first category is to ineffectiveness in the process of implementing the intention;

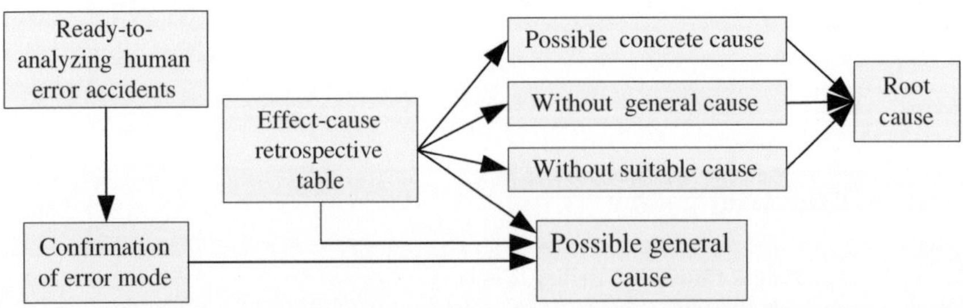

Fig. 73.1 Retrospective analysis frame of the root cause in accidents CREAM

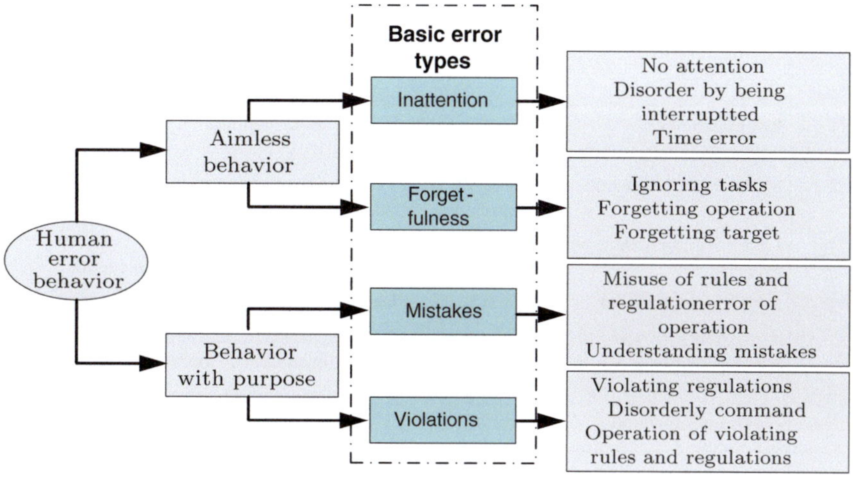

Fig. 73.2 Classification diagram of human factors error mode in troops training accidents

it is called inattention and forgetfulness. The second category is ineffectiveness in setting up the intention; it is called mistakes and violations [2]. In combination with troops training, the error behavior of officers and soldiers is classified into inattention, forgetfulness, mistakes, and violations, see Fig. 73.2.

73.2.2 Construction of Classification Table of Human Error Cause in Troops Training Accidents

Through analysis of the 158 troops training accidents in recent years, human error that leads to accidents occurrence is divided into five aspects: person, unit, troops, equipment, and finally environment. Correspondently the cause of human error in training accidents is also divided into five categories: the cause related to persons, the cause related to units, the cause related to troops, the cause related to equipment and technology, and the cause related to environment. They are separately expressed in letter "P, U, T, ET, E." Every group is also divided into several causes and distinguished in number. (see Table 73.1)

73.2.3 Construction of Possible Basic Cause Table of Human Error in Troops Training Accidents

In combination with troops training practice, possible general cause and concrete cause of the 4 human error modes in troops training accidents are summarized and

Table 73.1 Classification table on causes of human error in troops training accidents

Cause classification		Definition/contents
Code	Name	
P1	Insufficiency of cognition	Lack of knowledge and understanding
P2	Mistakes of judgments	Incorrectly identify accidents symptom, mistakenly or incompletely judge hazard
P3	Lack of implementation capacity	Lack of working enthusiasm or insufficient capacity, behavior error or untimely correcting unsafe behavior
P4	Poor technique level	Operation procedure cannot be implemented correctly, failure in identifying hazard potential
P5	Memory mistake	Information is forgot, information is looked back partly or mistakenly
P6	Fluke luck in mind	Operating without wearing safety protection apparel, refuse to abiding by procedure, command beyond one's power, act recklessly
P7	Psychology/physiology stress	The action is stopped or the reaction is extremely slow because of stress from fatigue and fear
U1	Insufficient mutual support	Within operating units cooperation between officers and soldiers is uncoordinated and lack of cohesive force
U2	Confusion of responsibilities	Confusion of responsibilities, unclear task allocation, or vague order
U3	Overloaded task	Urgent time, heavy training task, insufficient, available resource etc.
U4	Lack of communication and exchange	Lack of understanding and exchanging, mutual supervision relatively or untimely between officers and soldiers
T1	Ineffective safety education	Lack of or poor safety education and lack of safety consciousness for officers and soldiers
T2	Inefficient technique training	Officers and soldiers are short of knowledge; skills and practical experience that they should know and understand, loss of understanding toward the current situation
T3	Flaws in management and control	Unreasonable organization and structure, incomplete safety supervision, weak implementation capacity etc.
T4	Command error	Commanders make policies with prejudice or error because of defect of knowledge or being interrupted
T5	Unreasonable planning	The plans are not complete or emergent schemes are not laid down
T6	Slow reaction	Early-warning systems are not good and its feedback is blocked so that officers and soldiers are full of hesitation and cannot manage it timely when risk occurs
T7	Imperfect safety cultural education	Good atmosphere of safety culture is not formed and safety cultural level is low
ET1	Ineffective armament and instruments	Working state is abnormal because of aging equipment and its breakdown and maintenance
ET2	Imperfect rules and regulation	Unclear, incomplete and incorrect rules and regulations do not match with current situation.

(continued)

Table 73.1 (continued)

Cause classification		Definition/contents
Code	Name	
ET3	Unreasonable design	Unclear marking or no marking for equipment interface, button, monitor etc., there exists design flaw
ET4	Failure of communication liaison	Information in training activities or the transmitted information are not received, or are not understood mistakenly
ET5	Loss of information	Information is not provided, incomplete or incorrect information
E1	Poor operating circumstance	Narrow space when training and operating, poorly ventilated, dark and cold
E2	Terrible environment	Stormy, extremely cold and hot weather, bad visibility etc.

Table 73.2 Possible basic cause table of human error modes in troops training accidents

Error mode	Basic cause
Inattention	P1, P3, P5, P7, U2, U4, T1, T4, T7, T5, E1, E2
Forgetfulness	P1, P2, P3, P5, P6, P7, U4, T2, T3, T7, T6, ET1, ET4, ET5, E2
Mistakes	P1, P2, P3, P4, P7, U1, U4, T2, T3, T4, T7, ET1, ET3, ET5, E1, E2
Violations	P3, P5, P6, U3, T1, T3, T4, T7, ET1, ET2, ET3

put forward based on the 158 troops training accidents analysis in recent years. (see Table 73.2)

73.2.4 Construction of Effect–Cause Retrospective Table of Human Error in Troops Training Accidents

In Table 73.2, there are five error reasons contained in: person, unit, troops, technology, and environment. There exists some cause-effect connection among these causes. As effect, every cause may analyze general cause and concrete cause that lead to effect. Concrete cause may be directly regarded as root cause, while as effect general cause continues to retrospect. The model circulates and brings a series of effect-cause chain groups. These chain groups are synthesized to constitute an effect-cause retrospective table of human error in troops training accidents [3]. (see Table 73.3)

Table 73.3 Effect-cause retrospective table of human error in troops training accidents

Code	Effect	General cause	Concrete cause
P1	Insufficient cognition	E1, P4, P7, T2, ET5	Complicated situation, sense error, signal vagueness
P2	Mistakes of judgments	P1, U4, T2, ET4, ET5	Subjective imagination, lack of experience
P3	Insufficient implementation capacity	P1, P4, P7, T1, T4, ET4	Lack of knowledge and experience
P4	Poor technique level	P1, P7, T2	Poor training quality of officer and soldiers
P5	Memory mistake	P7, U3, T1	Poor memory and study performance
P6	Fluke luck in mind	T1, T3, T7, ET2	Gaining advantage by trickery, weak safety consciousness
P7	Psychology/physiology stress	P5, E1, E2, T2	Poor health, panic, extreme tension, and stress
U1	In sufficient mutual support	T3	Defect of character, incomplete knowledge structure
U2	Confusion of responsibilities	T3, ET2	Vague responsibilities allocation
U3	Overloaded task	T3, T5	Irrational training scheme and radically unexpected task
U4	Lack of communication	U1, T5, ET4	Weak safety circumstance
T1	Ineffective security education	T3, T7	Out-of-date education contents and insufficient time
T2	Insufficient skill training	T3, T7, T5	Out-of-date training contents and methods, insufficient time
T3	Flaws in management and control	P1, U2, T4, T5	Flaws in rules and regulations, incomplete supervision
T4	Command error	P1, P2, P4, U4, T3, ET5	Lack of technique and knowledge. Mode mistake, being interrupted
T5	Unreasonable plan	P1, T1	Escaping reality-based task training
T6	Slow reaction	P4, U1, U4	Imperfect emergency mechanism
ET1	Ineffective armaments	ET3	The size of armaments and instruments is not suitable, damaged and maintained, out-of-date
ET2	Imperfect rules and regulations	T3, T7, ET3	The situation on the spot is beyond the regulation and procedure
ET3	Unreasonable design	E1, T3, T7	Incorrect design target of armament and instruments
ET4	Failure of communication and liaison	E2, ET1, P4, T6	Equipment is damaged and interrupted by surroundings
ET5	Lack of information	T3, ET1, ET3	Problems of equipment monitoring, defect of design, no information record

73.3 Retrospective Analysis of the Root Cause of Human Error in Troops Training Accidents

73.3.1 Basic Course of Retrospective Analysis of the Root Cause in Troops Training Accidents

The basic course of retrospective analysis of the root cause of human error in troops training accidents includes 5 steps: (1) to confirm error mode types according to outer performance of human error accidents; (2) to choose possible cause as the beginning point of retrospective analysis in the related error mode cause table; (3) to regard the chosen cause as effect, and analyze and find possible general cause and concrete cause in effect-cause chain. For the branch with concrete cause, get root cause directly and not necessarily to analyze; for the branch with general cause, return back effect-cause chain and continue to analyze; (4) when choosing a lot of causes, add new branch; if there is no suitable cause the analysis of the branch will be stopped; and (5) every time having analyzed a branch, we will return back step 3 to have analysis of next branch, until the analysis of all branches is finished. The whole course is as follows [4]. (see Fig. 73.3)

73.3.2 The Result of Retrospective Analysis of the Root Cause of Human Error in the 158 Training Accidents

Applying methods and mode of retrospective analysis of root cause of human factors error in troops training accidents, we get the result of root cause based on retrospective analysis of root cause in 158 troops training accidents in recent years. (see Table 73.4)

Table 73.4 illustrates that the root causes in troops training accidents mainly concentrate on seven aspects: lack of cognition, fluke luck in mind, psychology/physiology stress, shortage of safety education, insufficient technique training, ignorance of management and control, and command error. In them, the root causes related to individuals include 4 aspects: lack of cognition, fluke luck in mind, psychology/physiology stress, and command error; the root causes related to troops and organization includes 3 aspects: shortage of safety education, insufficient technique training, and flaws in management and control.

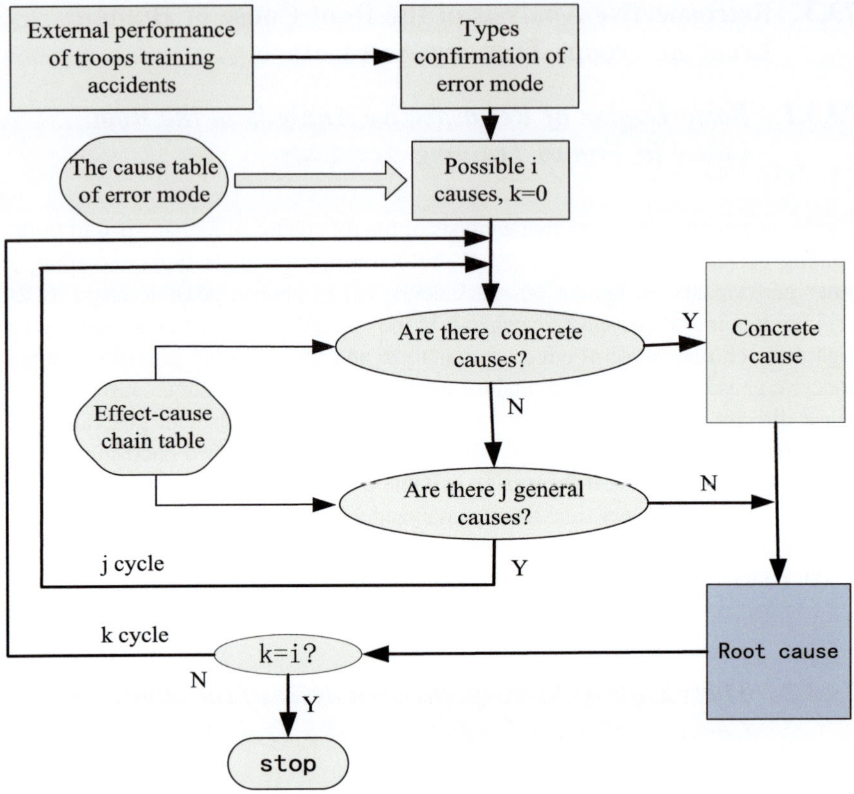

Fig. 73.3 Flow chart of retrospective analysis diagram of root cause of human error in troops training accidents

Table 73.4 The result of retrospective analysis of the root causes in the 158 troops training accidents

Root cause	Accidents quantity	Percentage (%)
Lack of cognition	52	32.9
Fluke luck in mind	112	70.9
Psychology/physiology stress	105	66.5
Loss of safety education	150	94.9
Insufficient technique training	82	51.9
Flaws in management and control	133	84.2
Command error	101	63.9

73.4 Precaution and Control Countermeasure in Troops Training Accidents

73.4.1 Basic Strategy of Precaution and Control in Troops Training Accidents

There are five aspects on the basic strategy of accidents precaution and control in troops training: the first is to reinforce safety idea in officers' and soldiers' minds. Through formal education guide, a series of activities and familiar environment to lead officers and soldiers to firmly establish the idea that focusing on security is to attach importance to firm politics, development, key matters, and the fighting capacity. The second is to improve safety capability of officers and soldiers. Through education, supervision and training to promote the capability of identifying solving and avoiding risk capability. The third is to develop safety habits of officers and soldiers. Through the means of behavior norms, a series of connection and activities and praising and punishing mechanism to develop good safety habits of officers and soldiers. The fourth is to perfect working safety mechanism. Through perfecting mechanism of responsibilities in different levels, mechanism of management and precaution from participation of all the working staff and mechanism of technical precaution to perform safety responsibility of officers and soldiers well. The fifth is to remedy error behavior of officers and soldiers. Through timely check and guide, the whole analysis and judgments, measure correctness and suitable evaluation to find and correct error behavior of officers and soldiers [5].

73.4.2 Concrete Measures of Precaution and Control in Troops Training Accidents

Directing at the seven root causes of retrospective analysis in the 158 troops trainings accidents, we should attach importance to four points in order to perform precaution and control in troops training accidents. The first is to reinforce the basis of thought on officers and soldiers training safety. Through a series of educations such as safety situation, common security knowledge and typical cases to create a good circumstance in which every people pay high attention to safety. The second is to cultivate personal quality of right behavior for officers and soldiers. Through reinforcing basic military skills of officers and soldiers, fortifying the fully psychological quality, progressing safety quality to develop good ability in troops training under the complicated circumstance; the third is to enhance organizing, training, and commanding capacity for officers and other backbone. Through learning and researching theory, summing up and exchanging experience, exploring methods and natural law to advance organizing and training capacity of officers and others scientifically, leading and commanding capacity, and avoiding command error and strategic failure of leaders. The fourth is to raise scientific level of safety management and control in troops training. Through analyzing information correctly, evaluating risk, strengthening risk management and control, and correcting error timely to fortifying management and control in troops training scientifically, purposely, and effectively.

References

1. Hollnagel E (1998) Cognitive reliability and errors analysis method. Elsevier Science Ltd, Oxford (UK)
2. Reason J (1990) Human Error. Cambridge University Press, Cambridge, UK
3. Shen Z, Wang Y, Gao J (2005) Effect and cause retrospective table of human error. Sci Technol 45(6):799–802
4. Wang Y, Shen Z (2005) Analytical methods of the second generation human factors reliability. Indus Eng Manag 3:17–21
5. Miltenberger RG (2001) (Trans: Hu P). Principles and methods of behavior modification. Chinese Light Industry Press, Beijing

Chapter 74
Mechanisms and Human Factors of Military Accidents and a Prevention Strategy

Zhenguo Mei, Jun Qi, Zaochen Liu and Chang Mei

Abstract Military activities are full of risks. The whole process is very comprehensive and complicated; in the process, all kinds of unsafe factors struggle with each other. In today's world, the risks are continuously increasing, with deep contradictions and questions of safety management gradually emerging. All of these factors necessitate new and greater requirements for the management of troops' safety management. Troop commanders must understand the mechanisms of accident occurrence, as well as the human factors of the accidents. In addition, commanders must determine and firmly enforce an appropriate prevention. Through a series of efforts, the commanders provide safety support for building a strong military and advancing troops' safety. This chapter focuses on the mechanisms, human factors, and prevention of military accidents.

Keywords MMESE · Mechanism of troops' accidents occurrence · Analysis of human factors · Safety precaution · Prevention strategy

At present in China, the informed construction of the armed forces is accelerating, a strong military practice is being promoted, and reality-based military training is being used extensively. Thus, profound changes have taken place in internal and external environmental conditions. The security risk in the fighting capacity construction has continuously increased and some deep contradictions and problems in safety management have gradually appeared. All of these factors have increased the requirements for safety management [1]. Troop commanders must understand the mechanisms of accident occurrence, highlight human factors, determine a proper prevention strategy, firmly enforce safety precautions, and provide support for safety development.

Z. Mei (✉) · J. Qi · Z. Liu · C. Mei
Air Defense Forces Academy, Zhengzhou 450052, China
e-mail: xwy@zzu.edu.cn

© Springer-Verlag Berlin Heidelberg 2015

S. Long and B.S. Dhillon (eds.), *Proceedings of the 15th International Conference on Man–Machine–Environment System Engineering*, Lecture Notes in Electrical Engineering 356, DOI 10.1007/978-3-662-48224-7_74

74.1 Mechanisms of Military Accident Occurrence

Judging from the whole course of accident occurrence, accidents mainly result from many unsafe factors. These unsafe factors can be classified into three aspects—unsafe human behavior, unsafe state of materials, and unsafe environmental conditions, as follows [1]:

(1) *Unsafe human behavior* refers to a situation that officers and soldiers consciously or unconsciously display themselves during the course of work. They may not abide by regulations or surpass the limit of operational principles of armaments, causing them not to finish the expected task precisely and appropriately. It is characteristic of operating errors and commanding mistakes, including breaking rules and regulations for organization, management, and other unsafe habits.
(2) *Unsafe state of materials* refers to the hazard of equipment, materials, and facilities, as well as their unsafe state caused by natural wear or quality defects. It includes some things with inherent hazards, such as munitions, poisonous substances, flammable and explosive materials, causative organisms, vehicle of infection, and electrified bodies. It also includes hazards that are caused by some equipment, facilities, and vehicles because of improper maintenance and management.
(3) *Unsafe environmental conditions* includes poor work sites and adverse weather conditions. Poor work sites mainly refer to rivers, lakes, seas, ponds, complicated road conditions, and poor topographical circumstance. Adverse weather conditions mainly refer to high temperatures, extreme cold, strong winds, heavy rain, snowstorm, heavy fog, and lightning.

These three aspects are *hazard installations*—roots and states that may lead to bodily harm, property loss, work environment damage, and the combination of bad results [2]. In a broader sense, a hazard installation is composed of three aspects: unsafe human behavior, unsafe state of material, and unsafe environmental conditions. In a narrow sense, a hazard installation is composed of two aspects: unsafe state of material and unsafe environmental condition. A hazard installation is also called an inherent hazard installation because it is characteristic of the inherent objective existence of hazard.

Through an analysis of military accidents, we can conclude three points. First, inherent hazard installation is not a certain factor but a basic factor of accident occurrence. In developing the process of objects, inherent hazard installation exists universally, but not all inherent hazard installation may lead to accidents. That is to say, a certain and direct causal relationship does **not** exist between inherent hazard installation and accidents. Second, the unsafe behavior of officers and soldiers is a direct cause of accident occurrence. From the nature of military accident occurrence, it is closely related to troops' activities. For any military activity, the three indispensable basic conditions are officers and soldiers, weapons and armament, and the related external environment. Evidence has shown that the unsafe behavior of officers and soldiers is a direct cause of accidents occurrence. Third, a combination of potential accidents types is a necessary condition. Viewing from overall situation an accident, the probability of unsafe behavior of officers and soldiers is far beyond the probability of accident occurrence—they are not equal. This shows that accident occurrence is conditional: The relationship between potential hazards and accident occurrence is not a simple one-to-one causal relationship; rather, the condition is the combination of potential accident types. That is to say, the accidents cannot occur immediately only because of an existing inherent hazard installation. Only when it meets another potential accident condition (unsafe behavior of officers and soldiers) can the combination of potential accidents types really lead to accident occurrence.

To sum up, the mechanism of accidents occurrence can be explained as the inherent hazard installation and unsafe behavior of officers and soldiers leading to the occurrence of potential accidents. In the proper sequence, the combination of potential accident types causes accident occurrence (Fig. 74.1).

The mechanism of troops' accidents occurrence tells us that a potential accident is the basic condition of accident occurrence. Effective precaution and management is the key point of prevention. However, it is also pointed out that unsafe human behavior, unsafe state of materials, and unsafe conditions of the environment are not fully identical in their basic features and effectiveness of interaction. According to an analysis of 75,000 accidents done by American scholar H.W. Heinrich, natural factors account for 2 %, material factors 8.2 %, and unsafe human behaviors 89.8 % in the interaction of all factors [3]. Therefore, accident prevention should be focused on human factors as the key to prevent unsafe behaviors of officers and soldiers.

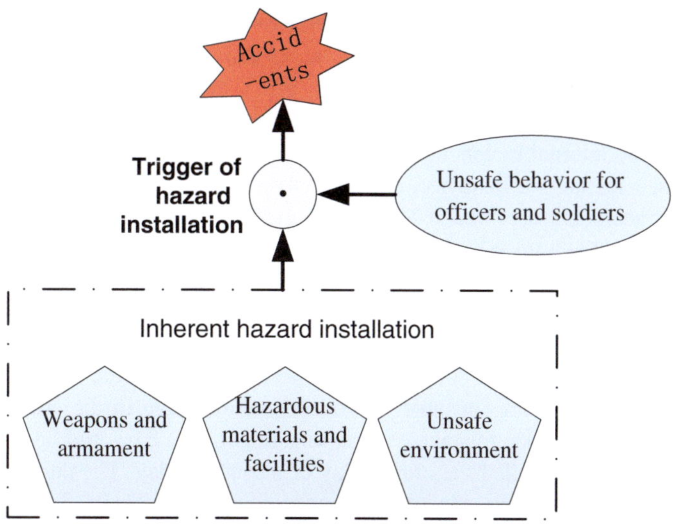

Fig. 74.1 Diagram of the mechanism of troops' accidents occurrence

74.2 Analysis of Human Factors

Unsafe human behaviors are the key factors that lead to accident occurrence, so they are the starting point to preventing accidents; thus, we analyze and explore the reasons for officers and soldiers' behavior errors. However, their behaviors are comprehensively influenced and limited because of internal factors and external conditions.

74.2.1 Internal Decisive Factors of Officers' and Soldiers' Behavior Errors

Internal factors behind officers' and soldiers' behavior errors include thought, psychology, physiology, and knowledge technology.

(1) *Thought* is the pioneer and engine of behavior. The level of thought comprehension plays an important restrictive function on behavior. An individual who lacks firm belief, strong legal ideas, good thoughts, and proper safety consciousness will easily participate in deviant and unsafe behavior.
(2) *Psychology* is the intrinsic motivation of behavior. Anyone with a certain psychological state will take on corresponding behavior features. Character is the core of personal psychology. It determines the attitude and behavior of officers and soldiers towards certain objects. Characters such as rudeness, ignorance, and laziness usually cause unsafe behavior. Irrational behaviors account for a higher proportion of unsafe behaviors that trigger accidents. Violating rules and discipline demonstrates irrational behavior, especially in taking risks irrationally. Irrational behavior is mainly attributed to psychological features such as trusting to luck, showing off, rebelliousness, and blindly following a crowd.
(3) *Physiology* refers to the functions of human organs and their natural state. It is one of the internal factors that determine human behavior. It can make humans produce instinctive behavior, and it is also known as unconscious behavior. Many accidents are related to physiological factors. Scientific research has shown that the human body has a biological rhythm. Human physical strength, emotion, and intelligence are periodically changing on a sinus curve. The cycle of human physical strength is 23 days, the cycle of emotion is 28 days, and the cycle of intelligence is 33 days. Biological rhythm is divided into a climax period, low period, and critical period [4]. Emotion, intelligence, and physical strength are unstable and quickly changeable in the critical period. People easily make mistakes when finishing tasks. The middle day of the critical period is called the critical day or hazardous day. The days before and after the critical day are called near-critical days. These 3 days form a

hazardous period. Research has also found that 60–70 % of accidents occur in the critical period. Many countries, such as the U.S., Sweden, and Japan attach great importance to applying biological rhythms to guide safety at work and clearly regulate national safety production at legal levels. In China, the Guangzhou Railroad Company and some troops apply biological rhythms to guide management with remarkable effects: the rate of accidents declined by 40–50 %.

(4) *Knowledge and technology* and the cognitive level of officers and soldiers both play important restrictive functions on their behaviors. The behavior of a well-educated person with good technological knowledge is much better than a poorly educated person with limited technological knowledge. If an individual lacks safety knowledge, he or she will work with ignorance and blindness. When hazards occur, an unknowledgeable person is inevitably confused and cannot properly manage them.

74.2.2 External Influencing Factors on Behavior Error

External factors influencing behavior error include management, material, and environment:

(1) *Management.* Questions are rendered by officers and soldiers, and the roots lies in commanders and the management system. If commanders have weak dedication and responsibility, loose management requirements, poor performance of regulations, and informal troop management order, this will easily lead to problems and defects, and finally cause unsafe behavior such as violating regulations and disciplines.

(2) *Material.* The technological performance of armament and instruments, engineering characteristics of the human body, and the degree of operational complexity can have certain effects on officers and soldiers. Sophisticated armament technology and poor engineering characteristics can add to the difficulty of armaments operation, which will easily lead to mistakes in judgments and operational errors.

(3) *Environment.* The behavior of officers and soldiers is deeply influenced by the internal and external environment. Certain environments with bad customs and habits, as well as a lack of friendships and unity, will stimulate the thought of officers and soldiers. These environments have a corrosive effect on the spirit of officers and soldiers, as well as an accelerating role in the violation of regulations and disciplines. Complicated topographical, geographical environments, and bad climate conditions will have poor influence on their physiology, psychology, and behaviors and trigger behavior error.

74.3 A Basic Strategy for Accident Prevention

Based on an analysis of the two aspects mentioned above, the following three basic strategies should be implemented for accident prevention.

74.3.1 Strict Management and Control of Inherent Hazard Installation: Establishing the Basis of Accident Prevention

Troops are armed groups who carry out military missions. Inside troops, there are a lot of hazardous materials, such as guns, bullets, vehicles, and chemical substances. In the meantime, military activities are done under quite complicated circumstances, so there are complicated unsafe conditions; thus, military activities are often performed along with a lot of inherent hazard installations. In this way, strict management and control of hazard installation has been an important basis in perform accident prevention for troops.

To manage and control hazard installation, three pieces of work should be valued: the first is hazard identification. Recognizing hazard installation is the first step to manage and control it. The contents usually include physically hazardous factors, chemically hazardous factors, and objective environmentally dangerous factors. Troops can identify the existing hazard installation in some military activity by using observation on the spot, task analysis, safety surveys, and consulting the related record and information.

The second aspect is hazard assessment. The identified hazard installation should be graded into different hazard levels through analysis of the combination of qualitative and quantitative means, and confirming the grade of hazard installation. The grade of hazard installation is generally on a seven-level system, where level 1 is the top hazard installation and level 7 is the lowest one. The lower the grade is, smaller the hazard level is, and vice versa.

The third aspect is hazard control. The target of hazard control is to prevent accidents and avoid personal injuries and property loss. Based on the hazard level, technology and management means are used to control and eliminate the hazard sources. Hazard control technology includes the safety technology of preventing accident occurrence and decreasing (or avoiding) accident loss. Management controls hazard installation effectively by a series of planned and organized safety management activities [5].

74.3.2 *Wiping Out Unsafe Behavior: Grasping the Key to Safety*

Heinrich's research showed that unsafe human behavior has played a critical role in accident occurrence. Therefore, the troops need attach great importance to human factors and wipe out the unsafe behavior of officers and soldiers in time to improve the troops' safety.

To wipe out unsafe behavior, three key points should be noted. The first is to enhance the position adaptability of working staff. The troops need encourage officers and soldiers to strengthen the consciousness of adapting to their position. The troops should also arrange suitable working positions according to their psychology, physiology, and quality to prevent hidden accidents. This point is very important, especially in some positions with special tasks, complicated technology, and important statuses.

The second point is to strengthen the standardization of educational training. Safety education training should be brought into political education and military training programs. The foundation of thought and consciousness and the basis of capability and quality should be reinforced through scientific and formal education and training.

The third point is to improve the effectiveness of supervision and management. Commanders should reinforce the sense of responsibility and meanwhile check and supervise officers' and soldiers' behaviors in time. Once unsafe behaviors are found, commanders should correct them immediately. At the same time, according to the situation, mission, and environment, laws and regulations should be revised, the management environment should be optimized, the methods and means of management should be adjusted, and the achievement of educational training should be continuously strengthened, thus guaranteeing that the troops run well in accordance with the law.

74.3.3 *Preventing the Combination of Potential Accident Types*

To advance the effectiveness of safety precautions, preventing the combination of potential accident types must also be carefully considered. This is because troops' hazard installations exist objectively and most of them can only be controlled, not eliminated. In addition, some unsafe behaviors are inevitable due to ignorance of the management system and the quality of officers and soldiers. Therefore, preventing the combination of potential accident types is the utmost means of accident prevention.

To prevent the combination of potential accident types, the four aspects should be addressed. The first is to improve the aim and accuracy of safety analysis, in accordance with the information. Effective segregation and shielding measures should be implemented to prevent potential accidents at the same time and place. The second aspect is to promote the special ability of officers and soldiers in emergent situations, as well as to prevent unsafe behavior; these behaviors lead to accident occurrence under the condition of the existing hazard installation. The third aspect is to emphasize connecting the features with the requirements of the military mission, as well as to strengthen the forecast and prediction of environmental conditions, to purposely conduct military exercises, continuously improve the risk avoidance of officers and soldiers, increase the adaptability of environmental risk, and ensure the effectiveness of accident prevention. The fourth aspect is to fully consider safety requirements during the course of developing and upgrading weapons and armaments, as well as to focus on the design of errors and continuously advance the stability and reliability of weapons and armaments. In this way, commanders can minimize the probability of unsafe states of armament and unsafe human behavior as soon as possible.

References

1. Wang C, Chen A, Chen G (2011) Research on troops management under the new period. Liberation Army Press
2. Luo Y (2013) Research on important hazard origin of air defence forces troops and its countermeasures of management and control. Zhengzhou Air Defence Forces Academy
3. Heinrich HW, Petersen D, Roose N (1980) Industrial accident prevention, 5th edn. McGraw Hill, New York
4. Ye L, Li S (2005) Safety behavior science. Tsinghua University Press, Beijing
5. He Z (2010) Identification of hazard installations, risk assessment and its control in drilling engineering techniques. Chin Sci Technol Saf Prod 6:177–181

Chapter 75
Design and Implementation of Cockpit Virtual Prototype Based on Virtools

Xue Shi, Zhenqi Xu, Fang Xie, Qun Wang, Ruyuan Liu, Sijuan Zheng and Zhongliang Wei

Abstract The virtual prototype technology is a comprehensive technology and Virtools is a tool used for virtual reality, which has a user-friendly graphical interface and a powerful 3D engine. This paper introduced the characteristic advantages of virtual prototype technology and tried to build a virtual prototype of the cockpit. With the help of Virtools and multimedia technology, all modules' functions in cockpit were demonstrated and exhibited. All the work was applied in a multi-channel virtual reality system. To build a virtual prototype, cockpit based on Virtools is helpful to improve the efficiency of scientific research and to reduce the cost and the difficulty of assess.

Keywords Virtual prototype · Virtual reality · Human–computer interaction · Virtools

75.1 Introduction

The virtual prototype technology is a comprehensive technology which takes the kinematics, dynamics, finite element analysis control theory as the core and uses mature computer graphics technology. It integrates the design and analysis of product component together, and builds the digital model of mechanical system before the first prototype. At the same time, this technology provides a virtual reality-based research platform for design, study, and improvement. Therefore, virtual prototype is called digital function prototype [1].

Virtual reality, which is an advanced human–computer interface technology, generates a 3D, information-based virtual environment. It realizes the interaction between user and virtual environment. This technology has been applied broadly to

X. Shi (✉) · Z. Xu · F. Xie · Q. Wang · R. Liu · S. Zheng · Z. Wei
China North Vehicle Research Institute, Beijing 100072, China
e-mail: xuezi911@qq.com

© Springer-Verlag Berlin Heidelberg 2015

S. Long and B.S. Dhillon (eds.), *Proceedings of the 15th International Conference on Man–Machine–Environment System Engineering*, Lecture Notes in Electrical Engineering 356, DOI 10.1007/978-3-662-48224-7_75

military simulation, scene simulation, virtual tour, aircraft, and automobile manufacture and scientific visualization.

Virtual prototype is not only a successful computer technology application in engineer, but also a totally new mechanical product design method. On one hand, traditional simulation technology focuses on single sub-system, but virtual prototype technology pays more attention to optimization of whole system. By the cooperation of virtual prototype and virtual environment, it tests and accesses multiple solutions and improves design until availability of the most optimal performance. On the other hand, with the combination of virtual prototype and virtual reality, human factor can be taken into consideration when we talk about virtual prototype, making human an important factor and all these things play a vital role in accessing a product [2].

With the development of technology, many software and tech can be used for the creation of virtual scene. This paper uses Virtools as a developing platform. An virtual prototype simulation platform makes 3D model browsing and real-time interaction with control of keyboard and mouse available by Virtools. This method has a lot of advantages, such as realty, interaction, simplicity, and easy completion. And it has been applied widely in the field of automobile design.

75.2 Virtools

Virtools, a tool used for virtual reality, is more and more chosen by developers with its user-friendly graphical interface and powerful 3D engine. It is an ideal tool for the research and development of virtual prototype. Developers can manage 3D model easily with Graphical interface.

With 700 interactive Behavior Building Blocks, an interactive product display system can be created promptly. This method gives CAD digital prototype more possibility, making initial virtual prototype close to real product and providing designers a ideal visualization research platform [3].

75.3 Overall Design

Cockpit virtual prototype is a simulation system created at the very beginning of design. In the system, designers can construct original interface and plan project based on driver's task. They can simulate human–machine interaction and verify function with virtual interface at the same time. The cockpit virtual prototype system includes mainly mechanic structure, human–machine interface (such as display device and operation panel), seats and environment control system. Designers can have a good understanding of overall layout, operation interface, and sub-system's function. A single-channel and a multiple-channel cockpit virtual prototype were made.

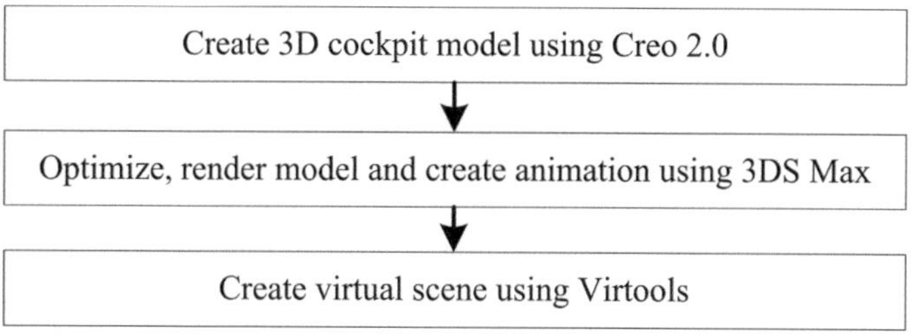

Fig. 75.1 The process of model creation

75.3.1 Model Resource Construction

Although Virtools has a good interaction, user-friendly interface and a relatively high expansibility, it has not functions such as 3D modeling and scene modeling. So other softwares are needed to create high-fidelity model.

To create cockpit model, Creo, a 3D modeling software, was used. After the model was created, a third-party tool was needed to transform the model to other type which could be used in 3DS Max. With 3DS Max, geometry of the model was modified and optimized. To get a more realistic and more immersive scene, the texture coordinate was re-assigned in 3DS Max and the effect of light was adjusted with the help of VRay. At last, the model was rendered and baked, and the high fidelity was obtained. Besides, because of the display of cockpit function, 3D animations were created in 3DS Max and all models and animations were exported as nmo, a file format supported by Virtools. The process of model and scene creation is illustrated as below (Fig. 75.1).

75.3.2 Function Design

Based on different application environments, virtual prototype must work in two situations: designer-oriented research and discuss mode, user-oriented report mode. Two modules, interactive module and display module, were defined in top level layout.

Three sub-levels were included in top level layout. The first sub-level was mode selection interface, the second was chapter selection interface, and the third was device detail display interface. Selection button and return button were used for different sub-level switches. All these things are illustrated as below (Fig. 75.2):

75.3.2.1 Interface Setting

Interaction module was the second level, where users can click certain buttons to enter corresponding chapters. Seven second levels were defined in the system and were illustrated as follows (Fig. 75.3).

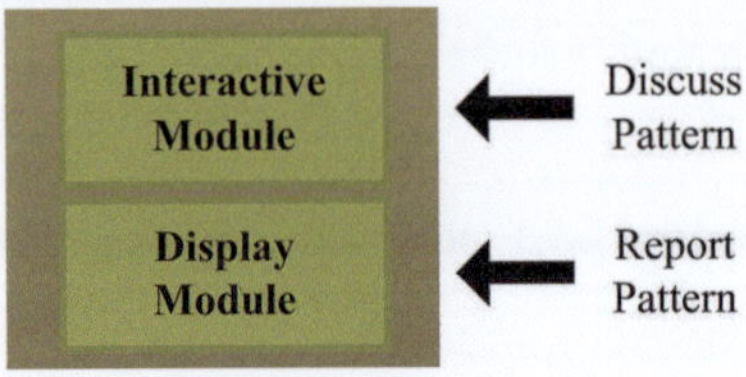

Fig. 75.2 The top level menu

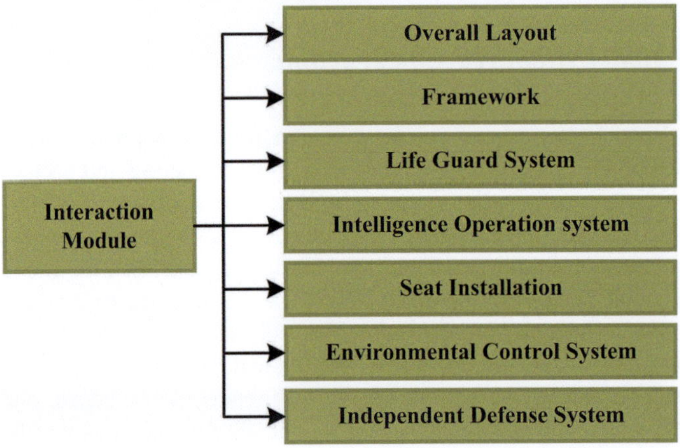

Fig. 75.3 The second level menu

Independent buttons, realizing enter and exit functions, were designed. Some sub-chapter buttons can be shown or hidden based on users' operation, which makes the interface simple. Some buttons were illustrated as follows (Fig. 75.4).

75.4 Script Implementation

Based on the developing task mentioned above and the various building blocks in Virtools, this chapter summarized common script implementation.

75.4.1 Select Button

The main purpose of button selection is to enter submenu and to display corresponding content after users click certain button. The detailed behaviors building block is shown in Fig. 75.5.

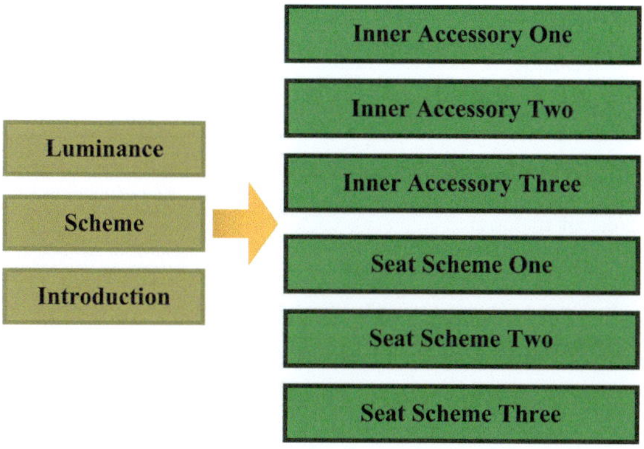

Fig. 75.4 The third level menu

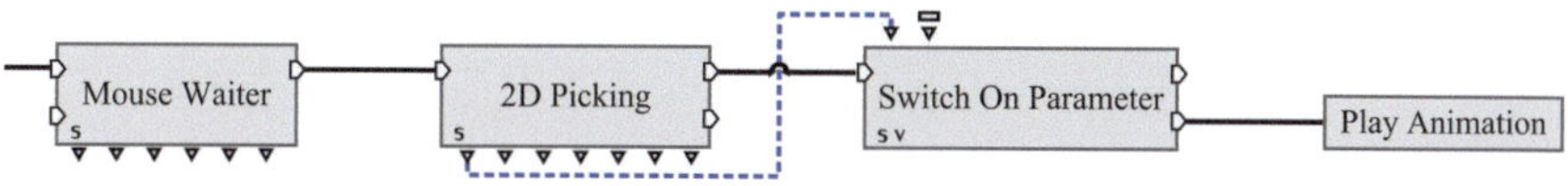

Fig. 75.5 *Script* mouse click

Mouse Waiter building block can identify mouse left click event, and 2D Picking building block can get the name of clicked object; then, Switch On Parameter building block compares the button name with all current button names which are qualified. When the condition is met, the logic outputs will connect the corresponding functions.

75.4.2 *Change Material*

The main function of changing material is to highlight button material when mouse moves on the button. Users can get the hint and then click certain button by this way, and the detailed implementation is shown as below.

Firstly, get the button name, modify the button's material and texture, change to the selected status, and change other buttons' material and textures to default. If the 2D Picking building block gets nothing, its second logic output is activated and change all buttons' material and textures to default (Fig. 75.6).

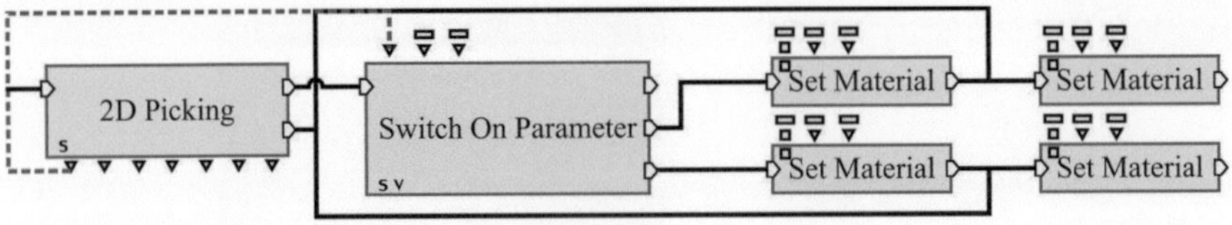

Fig. 75.6 The change of material

75.4.3 Obtain and Restore Camera Matrix

Camera matrix obtainization and restoration are to display objects correctly in the scene. When switching between different chapters, different models should be loaded. The location and scale of different models are not the same, so camera matrix should change according to specific model.

The Op building block can get new camera matrix, Set World Matrix building block can set main camera matrix parameter. Before setting camera matrix, main camera matrix should be saved at first (Fig. 75.7).

75.4.4 Load, Play, and Unload Animation

Because external animation should be imported in some chapters, main program should appear after animation play. So animation load, play, and unload should be taken into account.

With Object Load building block, animation will be loaded in main program. After this, Camera matrix parameter will be set and then Play Global Animation building block will play the animation. At this point, several situations should be considered (Fig. 75.8):

(1) Because the NMO file is loaded after main program starts, it cannot be used before. A global variable should be created to call corresponding animation. The global variable's type is String and value is animation's file name [4].
(2) After a display unit is finished, main interface will appear, and the corresponding NMO file should be unloaded. All these can be achieved by Wait Message building block. After receiving specified message, the second parameter of Object Load building block will be activated and the corresponding NMO will be unloaded.
(3) Users can double click "Play Global Animation" to change loop parameter to control the loop of animation.

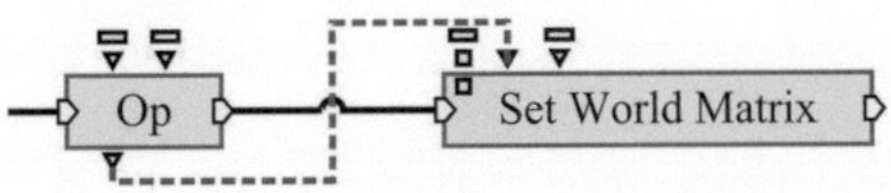

Fig. 75.7 *Script* matrix

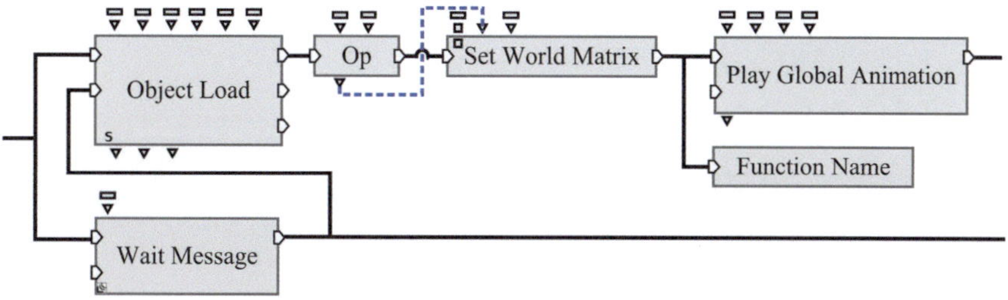

Fig. 75.8 *Script* the parameter setting of animation

75.4.5 Judge Cycling

During display loops, a condition control was needed in case that the loop is still running after exiting under which condition error information will be displayed or computing resources will be wasted [5]. Binary switch can be used to deal with such problems. This building block has a boolean-typed output parameter. If the output parameter is true, the first logic output will be activated, or the second logic output will be activated.

75.4.6 Rotate Camera

The main purpose of camera rotation is to display all objects in the scene comprehensively and to give users a overall concept. Concrete implementation is as follows.

After mouse left button is clicked, the Getting Mouse Displacement building block will obtain mouse's distance in x-direction. Rotation angle will be got multiplied by the distance with certain value. Then camera will rotate around specified object assigned by Look At building block and the rotation function will be accomplished [6] (Fig. 75.9).

75.4.7 Display Resource

During the development, different scripts were implemented and grouped based on different requirements. And at the same time, resources such as videos and textures were created.

In multi-channel edition, all animations, which were not loaded at first, were loaded dynamically to obtain a comprehensive stereo display.

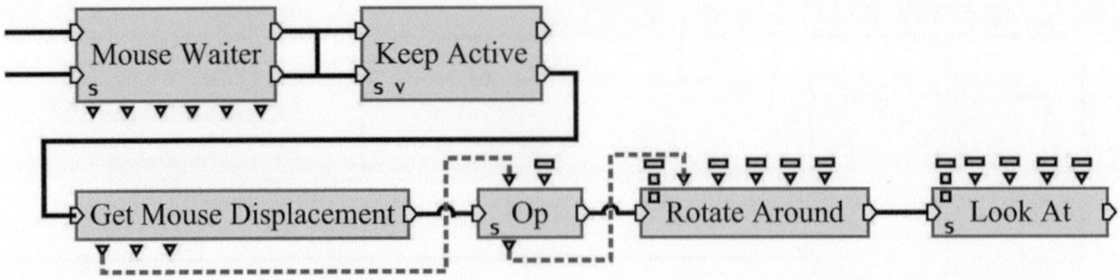

Fig. 75.9 *Script* the BB of camera

75.5 Conclusion

This paper constructed a vehicle cockpit virtual prototype system, which realized all modules' display and achieved the meaning of the system with Virtools and multimedia technology. This system has been demonstrated successfully in single-channel virtual reality system at the same time. Furthermore, the system will be transplanted to five-channel CAVE system and be debugged, and real driver seat and actuating mechanism will be combined to achieve a super fidelity immersive experience platform.

References

1. Zhu T (2006) Virtual maintenance simulation. Anhui University of Science and Technology, Anhui
2. Dingfang C, Yabo L etc. (2002) Virtual design. China Machine Press, Beijing
3. The Help Documentation of Virtools Dev3.0
4. Wendy S (2004) Beginning math and physics for game programmers. New Riders Games
5. Wang Z (2008) The design of 3D VR. Wen Kui Zi Xun, Taiwan, p 7
6. Mingkun L (2005) The design of 3D games. Sichuan publishing group, Sichuan

Chapter 76
Extravehicle Activity (EVA) Ergonomics Research in the Space Station

Yulin Xu, Chunhui Wang, Li Wang, Shihua Zhou and Changhua Jiang

Abstract Extravehicle activity (EVA) plays an important role in the construction and operation of the space station. Ergonomics research is crucial to the accomplishment of EVA tasks. This chapter discusses the common ground-based zero-gravity simulation techniques and platforms and summarizes achievements in the field. We discuss relevant research achievements worldwide, China's existing research, and the technological status of the space station. We also analyze the baseline of the work capacity, ergonomics requirements, and evaluation methods, as well as the direction and key points of the evaluation technology research of human–extramobility unit (EMU) systems, in which way the research and development of the space station may be effectively guided.

Keywords EVA · Space station · Ergonomics

76.1 Introduction

The history of manned spaceflights has demonstrated that extravehicle activity (EVA) plays an important role in the construction and operation of the space station and outboard scientific experiments. EVA has extremely high risks. To guarantee that an astronaut can complete EVA smoothly, the early stages of engineering research and development should include ergonomics requirements for the human–machine interface of an EVA system. The research on ergonomics evaluation technology should investigate rational and feasible evaluation methods. Scientific and effective ergonomics evaluation should be implemented to assure the effectiveness of the human–machine design. This approach helps to ensure that clothes,

Y. Xu · C. Wang (✉) · L. Wang · S. Zhou · C. Jiang
National Key Laboratory of Human Factors Engineering, Astronaut Centre of China, Beijing, China
e-mail: xuyulin014@126.com

© Springer-Verlag Berlin Heidelberg 2015

S. Long and B.S. Dhillon (eds.), *Proceedings of the 15th International Conference on Man–Machine–Environment System Engineering*, Lecture Notes in Electrical Engineering 356, DOI 10.1007/978-3-662-48224-7_76

hardware equipment, payload, and work procedures in EVA are designed rationally and the outboard workflow and workload are determined scientifically.

Due to the speciality of EVA equipment and outboard working environments, ground-based zero-gravity simulation test technology of EVA research should be considered first in EVA ergonomics research. Secondly, because EVA research is costly and demands extra resources, such as personnel and time, it is important to fully investigate the relevant research methods and achievements, consider the actual needs of the proposed task closely, and rationally determine the key research points before carrying out the research.

This chapter introduces several common zero-gravity simulation test technologies; presents the EVA ergonomics research achievements from around the world; and analyzes the system, content, and key direction of the EVA ergonomics research in the tasks of China's space station.

76.2 Common Zero-Gravity Simulation Test Technology/Platform

Ground-based zero-gravity simulation test technology consists of physical simulation (e.g., high tower, drop tube, airplane, balloon) and effect simulation (e.g., neutral buoyancy water channel, air-bearing floor, inclined guide rail). According to the application of various test technologies in EVA research, the neutral buoyancy water channels, zero-gravity airplane, air-bearing floors, and suspension are the most commonly used in research.

76.2.1 Neutral Buoyancy Water Channels

The neutral buoyancy water channel is one of the most important and common technological means for carrying out EVA ergonomics research. It obtains a simulated zero-gravity effect by increasing and decreasing the balance weight and levitron to realize equal gravity and buoyancy of a person's body and articles immersed.

When carrying out tests with a neutral buoyancy water channel, the vehicle and test equipment should be waterproof; in addition, the maintenance cost is high. However, the neutral buoyancy water channel allows for long simulation times and has a large test space. Therefore, a full-size space vehicle can be immersed. The test may be carried out at 6 degrees of freedom, and the kinetic and dynamic feeling in the simulated zero-gravity environment are very real. A large amount of EVA ergonomics research has been carried out in neutral buoyancy water channels worldwide, and the achievements have been verified and applied during flights.

76.2.2 Zero-Gravity Airplanes

Zero-gravity airplanes generate zero-gravity environments through parabolic flight. During each parabolic flight, the zero-gravity airplane may generate 15–40 s of microgravity time for operation performance tests and evaluation of various equipment and new extramobility unit (EMU), as well as dynamic and kinetic research of astronauts.

A zero-gravity airplane has advantages such as high simulation accuracy of a microgravity space environment, repeated use, and the capability for three-dimensional zero-gravity simulation tests. Its disadvantages include high costs, limitation of the structure size of the tested equipment, and short duration times of single zero-gravity simulations, which limits the selection of testable items.

76.2.3 Air-Bearing Floors

An air-bearing floor is a horizontal smooth plane supported by high-pressure gas. It achieves microgravity simulation by a lifting-gravity offset and may provide friction-free movement of two translational degrees of freedom and one revolving degree of freedom to the tested object. The research on kinetic characteristics under zero-gravity environments can be carried out via an air-floating platform. Its advantages include the short construction period, low cost, high accuracy, simple construction and maintenance, no test time limitations, and less structural limitations on space vehicles/test equipment. However, it can only achieve horizontal-plane zero-gravity simulation test, not kinetic simulation within a three-dimensional space [1].

76.2.4 Suspension Systems

Suspension offsets some gravity effects by applying force to some parts of the human body in the opposite direction of gravity to achieve zero-gravity simulation. Suspension systems are commonly used in kinetic and dynamic characteristic research under zero gravity and partial gravity. However, its simulation effect has a strong relationship with the position and strength of the force enforced on the human body.

At present, springing devices, such as hammocks and slings, jointly with a slideway and/or air-floating guide rail, are generally adopted to achieve suspension. The hammock and sling are used to exert force in the opposite direction of gravity, while the slideway and/or air-floating guide rail may guarantee that humans may receive minimum frictional resistance and move freely in the horizontal plane.

76.3 Relevant Research Worldwide

The United States and Russia have many years of history on manned space flights. They have carried out hundreds of EVAs to complete a huge amount of tasks, such as the construction and maintenance of space vehicles and scientific tests. The successful implementation of these EVAs could not be achieved without years of continuous EVA ergonomics research.

76.3.1 Kinetics and Dynamics of Astronauts in EMU

The National Aeronautics and Space Administration (NASA) has tested the joint flexibility, view, and reachable field of astronauts in EMU. The test results are publicized in human-EMU system integration standards as important guidelines for outboard product and task design.

In addition to basic working capacity characteristics, NASA also measured kinetic and dynamic characteristics of the astronauts in SAFER (Simplified Astronauts Fast Emergency Rescue). After the space shuttle Columbia accident, NASA proposed many inspection and maintenance methods for thermal protection systems of space shuttle, including that astronauts in SAFER get close to the damaged thermal protection tile to fix it. Therefore, it was necessary to research the kinetic and dynamic characteristics of astronauts in SAFER.

Personnel completed the research using an air-bearing floor (left on Fig. 76.1) and zero-gravity airplane (right on Fig. 76.1). A pressure test unit was used to test the dynamic characteristics of astronauts in SAFER. A Vicon movement capture system was used to acquire the kinetic characteristic data so that the feasibility of the maintenance project could be verified for all aspects [2].

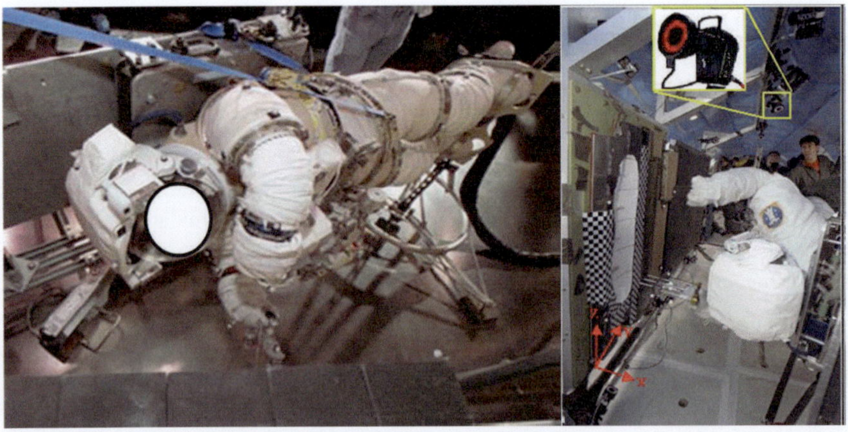

Fig. 76.1 Kinematic and dynamic characteristic tests of astronauts in SAFER. Reprinted from Ref. [2], Copyright 2005, with permission from AIAA

76.3.2 Test, Comparison, and Verification of EMU

EMU is the astronaut's essential equipment in EVA. Its ergonomics indicators such as adaptability, activity, and operating capacity play a decisive role in the completion of an outboard task. Therefore, during the research and development of EMU, zero-gravity simulation test technology must be adopted to verify its ergonomic performance. For instance, the Space System Laboratory of the University of Maryland carried out underwater integrated testing and experiments for MX-1 and MX-2 EMU in neutral buoyancy water channels, including a test of the capacities of precise operations, opening the simulation door of the Hubble Space Telescope, coordinated repair with robotic and operating controllers, and EVA tools in EMU, as well as a comparative test of the operating ergonomics of three gloves (Mark1, Mark2, and Mark3) [3, 4].

76.3.3 Design, Research, and Verification of Space Vehicles, Large Structures, Payload/Equipment, Tools, and Auxiliary Devices

In the early stages of research and development for space vehicles, large-scale space frameworks, important outboard equipment/payload, tools, and auxiliary devices, the verification of ergonomics design should be carried out using zero-gravity simulation technology. Usually, full-size space vehicles, outboard equipment/payload, and other large parts (including whole cabins and robotics) are put in a neutral buoyancy water channel for ergonomics assessment.

The United States found that there was a design problem in the solar panel of the No. 1 and No. 2 Space Labs through verification and testing in a neutral buoyancy water channel, after which they completed technological improvements. The size and installation method of the reflector panel of the James Webb Space Telescope were also verified both in the laboratory and neutral buoyancy water channel (Fig. 76.2). It was proven that astronauts may complete the assembly of 37 reflector panels within 5 h, which verified the feasibility of the design project.

NASA also carried out an ergonomics assessment test of the ability of the astronauts to quickly enter into the re-entry capsule, as well as the operability of the equipment in the re-entry capsule under zero-gravity status in a zero-gravity plane. The kinetic characteristics of full-size and large-weight equipment in the space shuttle and International Space Station were obtained through an air-bearing floor. By combining an air-bearing floor and suspension system, ergonomics research was implemented to assess the operating characteristics of equipment [6].

The ergonomic design of the airlock module, EVA handhold, restraint, and outboard tools, among others, has important influence on whether EVA tasks succeed. It is also necessary to carry out relevant ergonomics design requirement research, ergonomics assessment, and verification using zero-gravity simulation technology.

Fig. 76.2 Assembly test of reflection mirrors in the web space telescope. Reprinted from Ref. [5], Copyright 2000, with permission from NASA

76.3.4 Robotics

Ergonomics design requirements and assessment methods for robotics are also significant aspects of EVA ergonomics research. NASA has carried out a huge amount of research on astronauts' ability to perform payload transit and outboard maintenance using operating robotics, the display of robotics controls, and the layout of outboard cameras both in ground and neutral buoyancy; they subsequently propose rational and effective ergonomics design requirements. Meanwhile, an enormous amount of research effort went into ergonomics assessment of the robotics so that the human–machine interfaces of a display-control circuit could reach a rational level. In 2014, the European Space Agency (ESA) cooperated with Russia to establish underwater robotics. They carried out research on the robotics workflow, tools and auxiliary devices, and the effectiveness of outboard operation, and then proposed relevant ergonomics requirements and advice.

76.3.5 Verification and Optimization of EVA and Outboard Work Procedures

EVA and outboard work procedures include the operation procedure under normal EVA conditions, outboard work procedures, procedures for return to the airlock module under abnormal conditions, repair and maintenance procedures for space vehicles under fault conditions, and joint butting and assembly procedures for large-scale space structures. Before astronauts' training on EVA and outboard work procedures, the procedures should be verified and assessed to ensure their design rationality and astronauts' outboard operating safety.

The repair of the Hubble Telescope is a very complex outboard task and the design of its repair procedure is extremely vital. The University of Maryland verified and optimized the repair and maintenance procedure for the Hubble Telescope in a neutral buoyancy laboratory. The procedure was deconstructed into target, subtarget, and basic operations. The operating performance data for the procedure were collected in a neutral buoyancy water channel. The research results proved that robotics can help astronauts to increase operating performance by 60 % through rational human–robotics function distribution. Therefore, the procedure can be optimized to the enhance performance of a human–robotics system [7].

76.3.6 EVA Work Load

EVA always brings a large workload to astronauts. It was once happened that astronauts suffered too large of workload to complete a whole EVA. Therefore, the assessment of outboard workload is very important to the design of EVA and its work procedures. ESA has made use of a neutral buoyancy laboratory to carry out design research on EVA workload and the human–machine interface for Columbus Cabin.

The workload for astronauts can be assessed and tested through visibility, accessibility, and operability. Underwater workload can be determined according to basic physical signs (vital capacity and heart rate) during underwater operation. The test results proved that the heart rate of subjects increased by almost 50 % during underwater work over resting time and the vital capacity increased by 200 %. Furthermore, the change of the physical signs of subjects has direct relevance to the workload caused by test tasks [8].

76.4 Research on EVA Ergonomics in China's Space Station

It is extremely necessary to carry out EVA ergonomics research in all design phases for China's space station. According to other country's experiences, the test technology should emphasize the construction of various ground-based zero-gravity simulation test technologies/platforms, exploration of the joint use of various simulation test technologies/platforms, and enrichment of the research technology and methods. With regard to research content, the research should be carried out step by step according to existing research data and the actual technological status of the space station. The research content of EVA ergonomics mainly includes the baseline research on work capacity of the human-EMU system, research on ergonomics requirements, and research on assessment methods and technology.

76.4.1 Research on the Work Capacity Baseline of Human-EMU Systems

The main body of EVA is the astronaut in EMU. The capacities of an astronaut wearing EMU (e.g., movement scope, accessibility, operability, view) greatly differ from that of an astronaut without EMU. These characteristics are the baseline of EVA. Therefore, it is necessary to test the baseline of the working capacities of astronauts in EMU; then, follow-up ergonomics requirements and assessment research should be carried out on this basis.

Based on foreign research achievements, China has carried out research on the working capacity of a human (astronaut)-EMU system. Then, researchers defined that working capacities should include view scope, joint motion scope, reach area of upper limbs, operating strength of various types, precise operating ability of hands, moving speed, and payload translating capacity of astronauts in EMU. Research plans should be carefully drawn up after deep analysis and exploration on existing researching data.

76.4.2 Research on EVA Ergonomics Requirements

According to the key points and direction of existing EVA research and by considering the present EVA task limitations of the Chinese space station, ergonomics requirements research should focus on the following items:

- *Human–robotics system*: The importance of robotics in EVA has been proven by the practice of EVA by other countries. Robotics will be also equipped in our space station to assist astronauts in completing various outboard tasks. Therefore, it is necessary to research human–machine interfaces, such as the internal operating and control interface, display and control circuits, and outboard end port interface, then propose rational ergonomics design requirements.
- *Layout of outboard work stations and auxiliary devices*: The layout of outboard work stations and auxiliary devices should be researched to ensure their visibility, accessibility, and operability so that astronauts can arrive at the appointed work stations through auxiliary devices and complete replacement, repair, and maintenance tasks or scientific experiments successfully.
- *Outboard equipment, tools, and payload*: The design of outboard equipment, tools, and payload directly decides whether an astronaut can successfully complete an operation and his or her operating performance. To propose rational design requirements, relative research must be reinforced.
- *Outboard precise operating capacity*: The operating capacity of astronauts will be greatly reduced when wearing EMU. However, EVAs involve many complex and precise operations. Therefore, it is necessary to carry out comprehensive research on the precise outboard operating capacity of astronauts to support the ergonomic design requirements of outboard equipment, tools, loads, and tasks.

- *Delivery and installation capacity of payloads*: The delivery and installation of payload is a basic task and operation in EVA. It also has direct relevance to the design limit value of the size, quality of outboard equipment/load, and human–robotics functional distribution of outboard tasks. The delivery and installation capacity depends on the characteristics of EMU. Therefore, foreign EMU data cannot be used directly. Communication and coordination with the EMU manufacturer should be carried out as early as possible to develop relevant research, aimed at guiding the design of EVA payload/equipment and tasks.
- *Outboard work procedure*: EVA has high risk and very valuable time resources. Therefore, the design of outboard work procedures should be rational and high efficiency. To propose corresponding ergonomic requirements, research in time resources, movement routes, mental and physical loads, task distribution of astronauts, and safety should be carried out.

76.4.3 EVA Ergonomics Assessment Method and Technology Research

Based on previous ergonomics assessment technology and by considering EVA types for the Chinese space station, research on EVA ergonomics verification and assessment should be carried out. In addition, assessment of EVA ergonomics of indicator systems, technology, and comprehensive methods should be proposed to provide guidance and support for the design verification of relevant manufacturers. The key research points include the following:

- *Human–robotics systems*: The assessment of human–robotics systems is a new assessment task. Its display-control circuit involves inboard and outboard at the same time and its interface is complex. The operating tasks of astronauts are also complex and there are no relevant experiences. For subsequent ergonomics verification and assessment of human–robotics systems, research on ergonomics requirements, indicators, and methods should be carried out at the same time.
- *Outboard work procedures*: Outboard work procedures involve outboard tour inspection, maintenance and replacement, and science experiment procedures, and they include many factors. It will be worthwhile to carry out extensive research to comprehensively assess the rationality and executive efficiency of outboard work procedures considering all limitation factors.
- *Outboard workload*: Outboard workload has a significant influence on the accomplishment of EVA. In order to ensure the safety and working efficiency of astronauts, research on outboard workload assessment methods should be carried out to propose rational ground assessment methods. Meanwhile, feasible on-orbit workload supervision methods/equipment should be studied to verify the accuracy of ground workload assessment.

76.5 Conclusions

This chapter has presented several common zero-gravity simulation test technologies and platforms and summarized the achievements in EVA ergonomics research, also in regard to the development of the Chinese space station. Key points for future research on the work capacity baseline of human-EMU systems, EVA ergonomics requirements, and assessment of methods and technology were analyzed, which could help to guide the research and construction of the Chinese space station.

References

1. Qi N, Zhang W, Gao J et al (2011) The primary discussion for the ground simulation system of spatial microgravity. Aerosp Control 29:95–100
2. Scoville ZC, Rajula S (2005) SAFER inspection of space shuttle thermal protection system. In: AIAA 2005-6722
3. Jacobs SE, Akin DL, Braden JR(2006) System overview and operations of the MX-2 neutral buoyancy space suit analogue. In: SAE 2006-01-2287
4. Braden JR, Akin DL (2004) Development and testing update on the MX-2 neutral buoyancy space suit analogue. In: SAE 04ICES-235
5. Sanchez MJ (2000) A human factors evaluation of a methodology for pressurized crew module acceptability for zero-gravity ingress of spacecraft. In: NASA/TM-2000-209764
6. Ney ZA, Looper CA (2006) Developing the infrastructure for exploration EVA training. In: AIAA 2006-7451
7. Akin DL, Roberts B, Pilotte K, Baker M (2003) Robotic augmentation of EVA for hubble space telescope servicing. In: AIAA 2003-6274
8. Toscano E, Fubini E, Gaia E (2004) Microgravity simulation: physical and psychological workload evaluation tests in an underwater environment. Appl Ergon 35:383–391

Chapter 77
Design of Non-commissioned Officer Cadet Management Information System

Huili Yu

Abstract Since non-commissioned officers of all levels shall receive training before promotion, the quantity of members of non-commissioned officer college increases sharply, and various non-commissioned officer cadet information is doubled and redoubled. Traditional management is low-efficiency and poor reliable. The development of non-commissioned officer cadet management information system based on the campus network of the military academy may abstract the labor management method to automatic mode of processing on computer. The system design shall fully consider the characteristics of the military academy, carry out demand analysis and system analysis, work out the purpose of the system development and design, determine suitable system structure, and complete function division and module design of the system based on careful investigation and research.

Keywords Non-commissioned officer cadet · Management information system · C/S mode

77.1 Introduction

Management information system is a human-computer system that consists of human, computer, and database and may carry out information collection, transfer, storage, processing, maintenance, and use. Through the system, people may conveniently check various information, forecast the future according to the stored huge amount of data, make auxiliary decision in overall situation, and further help the employer to realize the planned purpose [1].

There are four common development ways of management information system: joint development, entrusted development, software package purchase, and

H. Yu (✉)
PLA Special Operations University, Guangzhou 510500, China
e-mail: tomato8350@sina.com

© Springer-Verlag Berlin Heidelberg 2015

S. Long and B.S. Dhillon (eds.), *Proceedings of the 15th International Conference on Man–Machine–Environment System Engineering*, Lecture Notes in Electrical Engineering 356, DOI 10.1007/978-3-662-48224-7_77

independent research and development. The non-commissioned officer cadet management information system is mainly used to manage the teaching information and daily administrative information of the cadets. Since certain cadet information is relatively sensitive, some daily administrative information such as internal affair sanitation inspection and various trainings is as well different from common universities. Therefore, the system developed via the first three systems cannot satisfy the actual demands of the military academy. Independent research and development shall be the unavoidable way of the system design.

77.2 Purpose of the System Development and Design

After the fifteenth conference of the military academy, the ratio of the non-commissioned officers to the soldiers has increased greatly over before. The non-commissioned officers of each level shall receive training first before selection and promotion, which caused that the quantity of the cadets of the academy that bears the non-commissioned officer training task increases sharply too, and the volume of the cadet information is doubled and redoubled. Traditional handwork management or half-handwork management does not only consume huge amount of time and energy of the management personnel, but also accompany with advantages such as bad real time and liability. As the base population of the training cadets increases unceasingly, the weakness of such management mode becomes increasing extruded with each passing day.

According to the careful investigation and research, the precise demand analysis and system analysis on the system aims at developing a set of non-commissioned officer cadet management information system that may run in campus internet of the military academy and realize the data processing on computer such as collection, processing, storage, and analysis. The design of system framework and data structure may ensure the safety and consistence of the non-commissioned officer cadet information. The system is mainly used in relevant teaching departments and management departments and can carry out comprehensive management to the education information and daily administrative information of the non-commissioned officer cadets.

77.3 System Framework

The system framework of the management information system refers to the information running mode of the information system in computer internet and is one of the important contents of the design of the information system platform. The selection of the system framework may generate important influence on the development, actual application, and follow-up maintenance of the system. The system framework is generally determined according to the result of the system investigation, demand analysis, and system analysis, meaning to first consider the business demand and second consider the actual geological position where the business runs and hardware fitting and management maintenance demand of the system. Of course, during actual implementation, there are other factors. For instance, whether the system runs in LAN or WAN; what is the safety performance requirements; whether the maintenance is convenient, which shall be taken into account all. At present, two main-stream frameworks in common use are Client/Sever (C/S) and Browser/Server (B/S) [2].

C/S mode is two-level framework and only completes data exchange between client and server. The data transmission quantity is small and the server load is less. Point-to-point structure is adopted. Generally speaking, the client group is relatively stable and the safety can be guaranteed well. In addition, the client and server work independently under C/S mode. The server stores data according to the rules and the client is unnecessary to watch the work course of the server. The server on contrary has strict safe authority review to the visit and checking to the client. Therefore, the system safety may be further protected. The system developed under C/S mode has as it been inherent superior information safety control ability than that under B/S mode. The data memory access is much safer.

Then we come to analyzing the shortages of the C/S mode. First, since the application procedures of the client need to install separately, the cost on software installation and maintenance is higher if the client quantity of the C/S mode is huge or the client distribution is scattered; second, C/S mode stores the data of the application procedure and the supporting application procedure separately on client and server, and the calculation correctness can be guaranteed only when they are synchronic. If the running environment of the client's machine is different, for instance, adopt different operating systems and support software of different editions, it may cause unblocked communication between the client and the server and even cause calculation error due to the inconsistency of the procedure editions [3]. Then, what influence do these shortages may cause on the system that we develop?

The use units of the non-commissioned officer cadet management information system mainly include teaching affair department, teaching and research office, cadet team and departments, etc. The are not many clients to be installed, and most clients are distributed in the campus and are not far away. Thus, no matter whether from the quantity of the installed systems or the installation path, the cost is relatively lower. The problem of C/S mode caused by client condition has been greatly weakened. On the other hand, the use scope of the system is merely limited to the campus network. The campus network is internal network. According to the requirements, the operating system shall be installed by special person appointed by relevant business departments. The client has the same operating system of the same edition and the running environment of the client is highly unified, which has again overcome the inherent flaws of the C/S mode in aspect of allocation.

To sum up, the C/S mode may adequately guarantee the safety of the information memory, transmission, and use. Although the B/S mode is applied broader in many fields, we adopt C/S mode in developing non-commissioned officer cadet management information system.

77.4 Function Division and Module Design of the System

The members of the research team have spent a long time on carrying out initial investigation and knowing the organization and structure condition, environmental condition and business flow condition of the system in the ways of: questionnaire investigation, discussion, looking up archives and materials, personal practice to get deep understanding about various rounds and running conditions of the investigated object.

77.4.1 Function Division of the System

First, the system needs to set a user's password table. On the log-in interface, the system may automatically judge whether the user is "system manager," "super user," or "common user" according to the user's password table. Three verifications are allowed. If you fail the verification for three times, you will be forced to quit the system directly. The "system manager" has the highest level and can carry out any operations to all information of all units. For instance, addition, revision, deletion, checking and statistics, "super user" is the second and can carry out any operations to the management information of the unit; "common user" can be only allowed to check the management information of the unit. The function division of main use units is as below [4]:

77.4.1.1 Departments of Institution

Lead-in of cadet's basic information: lead the data relevant to cadet's basic information in DBF database generated by recruitment system into SQL Server database of the system.

Type-in of cadet's basic information: the information of the cadets who get access to the academy with special conditions shall be typed in one by one.

Information maintenance of subordinate units and personnel: maintain the information of cadet team, teaching and research office, cadres of the team and faculty.

Checking and statistics: can be carried out according to key words such as year, cadet number, and class.

77.4.1.2 Cadet Team

Manage basic information of the cadet team: only allowed to add, revise, and delete the information of the cadets of the team.

Manage curriculum information of the cadets: including curriculum name, faculty information, class hour, credits, etc.

Manage administrative items: such as business trip, attendance, and leaves.

Manage awards and punishment items: such as study quality awards and punishment.

Checking system: carry out the checking and statistics of the cadet's study and administrative management condition and guide follow-up work.

77.4.1.3 Teaching Affair Department

Master the conditions of the teaching faculty: for the convenience of coordinating teaching affairs.

Master the conditions of the cadet team: including the information of the cadres and cadets of the team.

Checking and statistics: carry out statistics and filing of the cadets' credits typed by the faculty.

77.4.1.4 Other Departments

Take the department in charge of cadet's internal affair and team condition as example. They type in the results or conditions of various inspections and examinations, and the cadets can check at real time and timely carry out analysis and statistics in order to rectify fast.

77.4.2 Function Module Design of the System [5]

According to the division of the function requirements, the function module design of the non-commissioned officer cadet management information system mainly includes three parts: basic information management module, cadet team management module, and checking and statistics module.

Basic information management module is mainly used to manage the basic information of cadet team, teaching and research office and faculty, including sub-modules such as team information maintenance, team leader's information maintenance, information maintenance of teaching and research office, faculty information maintenance, annual cadet number maintenance, semester schedule maintenance, and user management.

Cadet team management module is mainly used to manage cadet information, curriculum of cadet team, daily administrative awards and punishment, including sub-modules such as cadet management, curriculum plan management, administrative items, administrative information management, administrative awards and punishment items, awards and punishment information management and credit information management.

Checking and statistics module is mainly used to satisfy daily checking and statistics of various users, including sub-modules such as statistics of cadet information, administrative information, awards and punishment, cadet's credits and cadet's post condition.

77.5 Conclusion

The system running overcomes various shortages caused by inherent leakiness of the labor management and has greatly increased the data accuracy degree. The community network realizes interconnection and sharing of various data, which not only saves management personnel from fussy data processing but also optimizes management system and enhances the transparency and interaction of the management work. Flexible checking function and powerful statistics function help the users to conveniently and swiftly complete statistics and analysis and further affirm the inner link of the cadet information and reach the purpose of auxiliary decision. The efficiency of relevant works of the unit has been obviously enhanced.

References

1. Xing XR (2010) Management information system. Beijing Electronic Industry Press, Beijing
2. Yu HL (2011) Research and development of the non-commissioned officer education management information system based on military campus network. Guangdong University of Technology, Guangzhou
3. Deng T (2010) Design and implementation of information management system of College students based on C/S mode. Software Publication
4. Yang XJ, Song XM (2011) Structure design standardization management information systems linked by mode of C/S and B/S. Equipment for Electronic Products Manufacturing
5. Niu HQ (2009) On the building strategy of information system of university based on Nolan stage-of-growth model. J N Univ China

Chapter 78
Design of a Virtual Training System and Application of an Evaluation Scheme for Orientation in a Spacecraft Cabin

Xiuqing Zhu, Yuqing Liu, Bohe Zhou, Ming An, Xuewen Chen, Fuchao Hu and Guohua Jiang

Abstract In manned spaceflights, weightlessness has great influence on the astronaut`s spatial orientation ability, especially in large spacecraft with complicated structures, such as space stations. Lack of gravity cues and a variety of body postures often result in astronauts' spatial disorientation. In a mockup on the ground, astronauts are only able to conduct spatial orientation training in erect postures and cannot simulate the variety of body postures. A virtual reality training system for coping with spatial orientation training was proposed, and the training methods and courses were devised. Research on applying evaluation approaches for the virtual training system were conducted. A preliminary evaluation scheme including the relationship of human physiology, cognitive mentality, space orientation ability, and the usability of the virtual training system is presented in this chapter.

Keywords Spatial orientation · Virtual reality · Training · Evaluation

78.1 Introduction

A space station is generally connected together by several modules, with different directions and a large interior space. When working in a space station, astronauts often lose their orientation in space. In conditions with low visibility, such as fire and loss of pressure in the module, orientation in the modules is even more difficult, which can directly affect the safety of the space station and astronauts. Hence, orientation in modules and emergency operations are important life-saving skills that astronauts must master and thus are important components of astronaut training [1, 2].

X. Zhu (✉) · Y. Liu · B. Zhou · M. An · X. Chen · F. Hu · G. Jiang
National Key Laboratory of Human Factors Engineering,
China Astronaut Research and Training Center, Beijing, China
e-mail: zz_xx_qq@126.com

© Springer-Verlag Berlin Heidelberg 2015

S. Long and B.S. Dhillon (eds.), *Proceedings of the 15th International Conference on Man–Machine–Environment System Engineering*, Lecture Notes in Electrical Engineering 356, DOI 10.1007/978-3-662-48224-7_78

Astronauts perform spatial orientation mainly based on visual information under microgravity conditions. Generally, astronauts become familiar with layout of the module and the environment before performing missions via the ground simulator. The ground simulation training is done under 1 g gravity while keeping the body upright. Astronauts' awareness of the module layout takes a positive dimensional axis as a reference; generally, the body axis is referred to as the partial visual vertical positive axis. However, in orbit, astronauts float in the module, and their body postures may be in any direction relative to the module environment. Therefore, a complex cognitive transition is needed for identifying their body axes relative to the orientation of the surrounding environment.

If orientation errors occur, astronauts may mistakenly regard the floor above their body as the ceiling; this phenomenon is called "visual reorientation illusion." Moreover, each module's visual vertical positive axis sometimes has an inconsistent definition due to design problems. Therefore, astronauts need to become familiar with all modules, which are not actually connected with each other; thus, it may be difficult to identify or remember spatial correspondence relations between the neighboring modules. When astronauts float in these modules, changes in the local visual vertical positive axis of each module will cause astronauts' disorientation.

Common solutions are setting directive signs in each module of the space station. When arriving in the node module that connects multiple modules, astronauts can choose the right path according to the path policy by examining each module's hatch logo and choose the entrances. However, astronauts who only depend on identifications and path policies in the module may become dependent on the identifications and paths. When visibility is decreased in the module, identifications in the module may not be seen clearly, and astronauts are prone to get lost. Hence, it is very important to maintain the overall directional map of the space station and instinctively know how to reach the goal modules.

Astronauts need to view the space station as a whole to set up a global reference coordinate system that connects every module, so as to achieve orientation awareness of one related to the entire spacecraft. For modules with inconsistent definitions of visual vertical positive axes, the relationship between the module and the global coordinate system should first be established, then a cognitive transition can be performed to re-establish the cognitive map of inconsistent modules and set the movable direction in the module. In fact, it is very difficult to establish overall directional sense in the space station; astronauts cannot see the spatial corresponding relation between different modules due to the bulkhead blocking. Moreover, the local visual vertical positive axis will have a strong influence on astronauts' spatial orientation reference. If it is inconsistent with the global coordinates, it is not advised for the astronauts to form a cognitive map of the space station based on the global reference coordinate system.

The main reasons for astronauts' spatial disorientation include the absence of gravity, astronauts' body posture changes, inconsistent module visual vertical positive axis, and difficulties in obtaining and maintaining a comprehensive cognitive map of the entire space station. Virtual reality may provide astronauts with spatial orientation training in a virtual environment before flight. In the virtual

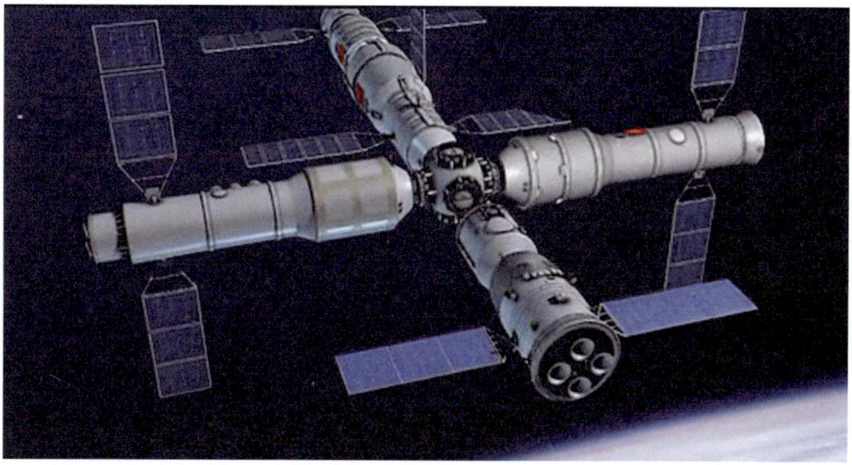

Fig. 78.1 The imaginary configuration of the space station

environment, astronauts can experience their surroundings in various body postures, so as to reduce spatial orientation capacity constraints that are caused by single body posture training. Moreover, the astronauts can experience moving through modules with inconsistent definitions in the visual vertical positive axis, which can help them to build overall cognitive maps of the spatial relationships of a space station. Otherwise, it can simulate the module environment with decreased visibility in virtual reality, as well as offer astronauts emergency evacuation training under failure emergency conditions [3, 4] (Figs. 78.1 and 78.2).

Fig. 78.2 Simulation of passing through various modules of the space station by different body postures and viewpoints

78.2 Design of Virtual Training System

78.2.1 Design of Virtual Environment Within the Modules of the Space Station

This study constructed the module segment models of a virtual space station, and the participants simulated walking through the space station in different body postures and viewpoints. Astronauts were trained to become familiar with the module interior layout and orientation reference marks. In addition, interactive roaming was done via data gloves or other interactive devices to learn the working environment and the moving paths in the space station, as well as to reduce cognitive gaps of space orientation caused by body training in a single posture. Space sickness confrontational training was also conducted. The virtual reality system can easily simulate environments in the module with low visibility caused by smoke or fog; training in contingency failure situations of in-vehicular emergency evacuation can also be simulated.

78.2.2 System Interactive Design

To achieve space station in-vehicular orientation and mobility training, the training system should support the motion tracking of a user's hand and upper limbs, helmet stereo display, and force feedback. Participants wore interactive devices (head-mounted display, force feedback devices, location tracking meter, and data gloves) to interact with the system; the movement of head, hand, and upper limbs were tracked; client computers acquired action data and passed the data to the server via a network; the server computed virtual persons' movement, collisions, and gripping gestures; and then the force feedback information and virtual scene status were transmitted to the simulation host, which controlled output of the force feedback devices plus the image signals, which were output to the head helmet display (Fig. 78.3).

78.2.3 Training Subjects and Methods

Virtual reality environments were used to realize in-vehicular orientation training subjects, including regular training and emergency evacuation training. In the regular training, trainees simulated different body postures and perspectives to pass through the space station, became familiar with the module layout and local orientation reference marks, and understood the working environment of the space station and the moving paths. With the help of a virtual guide, trainees controlled virtual persons' movements through data gloves, and a virtual human body posture

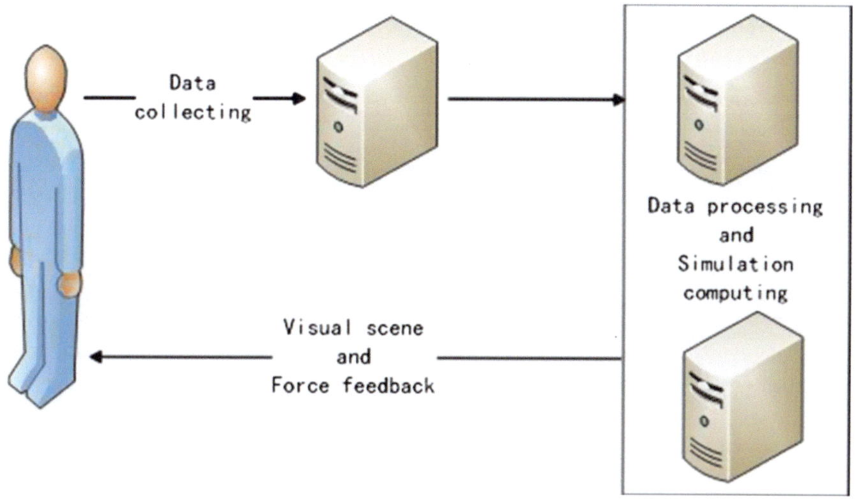

Fig. 78.3 The interactive loop of human and machine

was kept consistent with the virtual guide. Specific training methods divided the participants into three groups. Each group chose one of three training modes below with different body postures:

(1) Module segment local posture: the trainees' postures remained consistent with the local module's visual positive axis
(2) Global postures in the space station: the trainees' posture remained consistent with the station's allocentric reference frame
(3) Mixed postures: the trainees' posture remained consistent with a local reference frame in the first half training and the all-around reference frame at the remaining half of training.

In the training, every trainee had a performance index to complete tasks to be tested, including time to complete tasks, turn times in each module-passing experiments, error rate of indication, error rate of description of space layout, etc. By statistical analysis of the measurement data, the training effects were evaluated, and the results can be used as a basis for optimal training strategies.

In the training for emergency evacuation, the virtual guide was removed. Trainees depended on their own spatial cognitive abilities to complete passing through the module from the starting point to the destination segment through the process. In the first half of emergency evacuation training, module visibility settings were consistent with regular training; in the latter half, training was completed in the setting under low visibility conditions.

78.3 Application Evaluation Design of a Virtual Training System

Spatial orientation abilities have a close relationship to visual, vestibular, and proprioceptive sense information [5]. Cognitive factors also play an important role in that. By designing a few evaluation tests of a virtual training system, including physiological, cognitive skills, training system suitability, and performance evaluation, and the applied value of the virtual training system can be explored.

78.3.1 Evaluation of Physiological Influence Factors

Experimental studies on the correlation between spatial orientation abilities and vestibular function have been carried out to design changes of spatial orientation ability under different vestibular stimuli-response experiments. Based on the virtual training system of orientation in a space station module, factors that influence human spatial orientation ability of vestibular function have been investigated by the distinction between dimensions of vestibular function, exploring different groups of spatial orientation in a virtual reality environment changes. Experimental design includes the following:

(1) Study of the role of sensorimotor control in the spatial orientation of a virtual reality, by documenting and comparing data of human motion change of postures, gait, and Electromyography (EMG) activity before, during, and after spatial orientation training.
(2) Effects of research on vestibular function sensitivity differences and space directed approaches on space orientation capacity, grouping subjects by different vestibular function, in space-directed capacity training; dynamically recording subjects' heart rhythms, fingertip pulse, and eye movements; comparing different vestibular function for subjects in a virtual reality directed task in the appearance of independent neural symptoms; visual illusion and lost direction of the situation; and the role of vestibular function factors in virtual reality space directed task training.
(3) Training on vestibular function and vestibular stimulation for the dynamic changes of spatial orientation abilities; quantitative analysis of vestibular stimulation, dynamic changes in spatial orientation abilities.

78.3.2 Evaluation of Cognitive Influencing Factors

Evaluation of spatial information processing abilities should include human factors (especially the ability of mental rotation), explore different groups of spatial orientation in a virtual reality environment, and corresponding virtual environment training effects and fitness evaluations. The experience will accumulate initial data for spatial orientation tasks, astronaut selection standards, and virtual reality training assessment. Design of experimental contents includes the following:

(1) Test of spatial information processing abilities

Based on theoretical analysis, this may have a significant impact on astronauts' spatial orientation task performance of spatial information processing abilities, including mental rotation and spatial perspective transformation abilities. According to the paradigm of psychology experiments, computerized testing software could be developed on the subject of mental rotation to test the capacity and spatial perspective transforms, as well as major indexes including the subject's reaction time and accuracy. Through a clustering analysis method, human subjects could be classified according to their spatial information processing abilities, using virtual reality technology research in human subjects.

Mental rotation ability is the capability in the absence of external aid through mental imagery on the rotation of two-dimensional or three-dimensional graphics. It is often used in the study of psychology paradigms (Shepard and Metzler, Cooper, among others) and the development of two-dimensional and three-dimensional mental rotation tests. Spatial perspective transformation refers to a person's ability to identify the same object in a different perspective or according to current observation maps to determine the object's perspective. Based on the above test paradigm, test material can be developed and corresponding software designed, ensuring the reliability and validity of the test.

(2) Influence of the research on spatial information processing capability for spatial orientation

Here, subjects perform the same tasks but have different information processing abilities for spatial orientation. Spatial task performance and the appropriate physical indicators Electroencephalography (EEG), Electrooculography (EOG), and Electrocardiogram (ECG) are tested through the unified analysis of experimental data to explore the spatial information processing capacity, correlation between spatial orientation abilities, and the application of virtual reality technology to adapt to the situation. Response time, accuracy, error rate, and other indicators are used to assess the spatial orientation in task performance by subjective evaluation; physiological measurement methods assess the operational load. The spatial information processing abilities of the subjects, as well as spatial orientation task performance and workload, are measured, respectively, by correlation analysis and clustering methods. The spatial information processing abilities' influence on the spatial orientation of task performance and workload are assessed.

(3) Influence of research on virtual reality training for spatial information processing functions

After subjects received some training in the virtual reality system, subjects' mental rotation and spatial perspective transforms, such as spatial information processing proficiency, were tested again. Test results from before and after training were compared. Virtual reality training associated effects on spatial information processing abilities were studied.

78.3.3 Evaluation of System Applicability

Orientation in the space station module and typical operational tasks are taken as test examples. User experience testing, questionnaires, evaluation scale tables, physical parameter measurement, and statistical analysis of test data and other forms are used to evaluate the application effect of the virtual training system [6, 7]. Research has been done to apply virtual reality in astronaut training, including for applicability and applied effects, advantages and limitations, and limitation factors. Moreover, training effectiveness and applicability of the virtual environment have been valuated.

(1) Experience evaluation of human–computer interfaces
Visual experience evaluates the verisimilitude and adaptability of sight. Experience of force feedback devices evaluated the verisimilitude and adaptability of force feedback effects. The applicability of interaction rules evaluated the development and use experience through collection and analysis by evaluation questionnaires.

(2) Evaluation of user acceptance
Users' comfort evaluation in usage included comfort wearing the device, realism of virtual information awareness, sensory stimulation, the degree of adverse effects on humans, etc. Evaluation includes the helmet, gloves, comfort, and force feedback devices, such as interactive devices to determine the visual experience and human feelings. User feelings included a fatigue evaluation, collection of user training, and evaluation data of when the body felt tired and uncomfortable. The data from the user experience questionnaire were analyzed.

(3) Design of user evaluation questionnaire
Subjects complete the user questionnaire after completing the training missions. The questionnaire uses a Likert scale form, which is composed of question statements, and with answers recorded as approval or disapproval. Answers are divided into different levels; they can measure users' satisfaction and personal psychological feelings.

(4) Objective evaluation method
Heart rate variability (HRV) and other indicators are taken during the test/training period according to the training content; evaluation methods for objective indicators

have been explored. A difference in rhythm changes for normal and emergency training may be compared, and changes in users' physiological loads can be verified from an objective point of view.

78.4 Conclusion

This study examined space station missions as the application background, using virtual reality technology for the astronauts' spatial orientation training. An orientation virtual training system was built in a module, which provides a new training means for space station module orientation training. A systematic design scheme was proposed. The design of the application evaluation research program on the system has been carried out from the physical, cognitive, and applicability viewpoints. The main research and program ideas have been proposed; they will be further refined and the test scheme will be revised accordingly.

Acknowledgments This study is funded by Foundation Scientific Research Plan of National Defense (B1720132001).

References

1. Charles MO (2006) Spatial orientation and navigation in microgravity. In: Mast F, Jancke L (eds) Spatial processing in navigation, imagery and perception. Springer, New York, pp 209–247
2. Ronald LS, Charles MO, Thomas DJ (2012) Space shuttle flight crew spatial orientation survey results. Aviat Space Environ Med 83(4):383–387
3. Stroud KJ, Harm DL, Klaus DM (2005) Preflight virtual reality training as a countermeasure for space motion sickness and disorientation. Aviat Space Environ Med 76(4):352–356
4. Hirofumi A, Charles MO, Alan N (2007) Virtual reality-based 3D navigation training for emergency egress from spacecraft. Aviat Space Environ Med 78(8):774–783
5. William HP, Charles MO, Jacob JB et al (2008) Risk of sensory-motor performance failures affecting vehicle control during space missions: a review of the evidence. J Gravit Physiol 15 (2):1–29
6. Shannon RB, Sittichai K, Mohammad TK (2003) Evaluating the use of virtual reality for training in aircraft inspection. In: Proceedings of the 8th annual international conference on industrial engineering—Theory, applications and practice Las Vegas, Nevada, USA, 10–12 Nov 2003
7. Ronkko J, Jussi M, Raimo L et al (2006) Multimodal astronaut virtual training prototype. Int J Hum Comput Stud 64(3):182–191

Part VIII
Theory and Application Research

Chapter 79
Application of Man–Machine–Environment Systems Engineering in Financial Innovation and the Transition of Central Banks

Yinying Huang and Zhihao Hu

Abstract To analyze the financial innovation and the transition of central bank, the research framework of man-machine-environment system engineering (MMESE) on this field is proposed in this paper. The main conclusions are: central banks ("man") needs to comply with the long-term trend of marketization of financial market ("environment"), to change its temporary and passive crisis aid system afterwards into active continuous innovation of monetary policy tools ("machine"), and policy transmission mechanism ("machine-environment relationship"). With breakthrough in its traditional function, the central bank can create a new "man-machine-environment system" to control financial risk and ultimately achieve its "safety, efficiency and economy" objectives.

Keywords Man-machine-environment systems engineering · Financial innovation · Central banks · Monetary policy · Conduction mechanism

79.1 Introduction

Man-machine-environment system engineering is a comprehensive boundary technological science [1] born in 1981 under the personal guidance of Qian Xuesen —a great scientist. Mr. Qian spoke highly of this emerging science and pointed out

Y. Huang (✉)
Visiting Scholar of Columbia Business School, New York, USA
e-mail: yh2610@culumbia.edu

Z. Hu
Professor of Economics, Finance and Banking, Institute of Chinese Academy of Social Sciences, Beijing 100836, China

Y. Huang
Postdoctoral of Economics, Institute of Chinese Academy of Social Sciences, Beijing, China

© Springer-Verlag Berlin Heidelberg 2015
S. Long and B.S. Dhillon (eds.), *Proceedings of the 15th International Conference on Man–Machine–Environment System Engineering*, Lecture Notes in Electrical Engineering 356, DOI 10.1007/978-3-662-48224-7_79

in the letter to Mr. Long Shengzhao on October 22, 1993: **"You are developing this important modern science and technology in socialism China!"**

Man-machine-environment system engineering is a science employing systematic scientific theory and engineering approaches, dealing with the relationship between three factors, namely man, machine, and environment and doing further research on the optimal combination of man-machine-environment system. In general, man-machine-environment can be divided into three categories: simple man-machine-environment system, complicated man-machine-environment system, and generalized man-machine-environment system.

Man-machine-environment system engineering can be widely used not only in the field of natural science, but also has a strong theoretical and practical guiding significance in the field of social science. This article defines financial system as a generalized man-machine-environment system and attempts to state modern financial innovation and transition of monetary policy system of central banks by the research method of man-machine-environment system engineering and discusses the application of MME in research of monetary policy of Central Bank.

79.2 Man-Machine-Environment System in Research of Monetary Policy

79.2.1 "Man" in the System—Central Banks

"Man" in the system refers to the "man" as the subject of work, such as operators and decision makers [2]. In the research of monetary policy, we can regard the decision makers of monetary policy—central banks as "man," who also operate macro-control. Central banks as monetary, financial management institute of highest level in one country plays the leading role in the development of financial system in every nation. Central banks are granted with the rights of making and implementing monetary policy, performing macro-control in national economy and supervising the financial market.

79.2.2 "Machine" in the System—Monetary Policy Tools

"Machine" in the system refers to the general term of all objects that can be adjusted and controlled by "man." It is defined as monetary policy tools, such as interest rate, reserve fund, supply of money, discount rate, and otherwise. Monetary policy tools ("machine") are tools and means used in implementing monetary policy by central banks ("man") to achieve policy objectives. Tools of monetary policy used by central banks are not invariable. For example, the Federal Reserve used "discount rate" as tools in 1913 and then employed "lawful reserve" in 1935 and brought "open market operation" in its basic operation tools in 1951.

79.2.3 "Environment" in the System—Financial Market

Compared with relevant science (such as man-machine system), man-machine-environment system engineering brings the factor "environment" as a proactive factor of great importance into the system, other than a passive interference factor [3]. "Environment" in the system refers to specific working conditions where man and machine exist together. Due to "Financial Market" as the background of central banks implementing monetary policy and the base of macro-control, it has great influence on central banks' making monetary policies and the implement of these policies. Thus, we take "Financial Market" both home and abroad as the main "Environment" factor of monetary policies into consideration, as it is shown in Fig. 79.1.

79.2.4 "Man-Machine-Environment System" in Research of Monetary Policy

"**Man-machine relationship**" refers to the relationship between "man" (Central Bank) and "machine" (monetary policy tools), which means how the central banks choose policy tools and how policy tools influence the self-location and functions of central banks; "**Man-environment relationship**" refers to the relationship between "man" (Central Bank) and "environment" (Financial Market), which means how central banks regulate and control financial market and financial market influence the effect of that adjustment and control; "**Machine-environment relationship**" refers to the relationship between "machine" (Monetary Policy Tools) and "environment" (Financial Market), which mainly means transmission mechanism of central banks' monetary policy, and that is by which mechanism monetary policy tools influence financial market and financial market influence the actual effect of policy tools' implementation. "Man-machine-environment System" in research of monetary policy is the organic system formed under the mutual

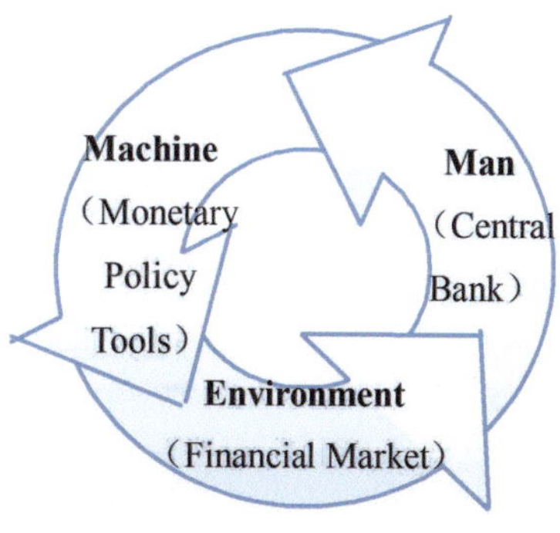

Fig. 79.1 The simple diagram of man-machine-environment system in monetary policy study

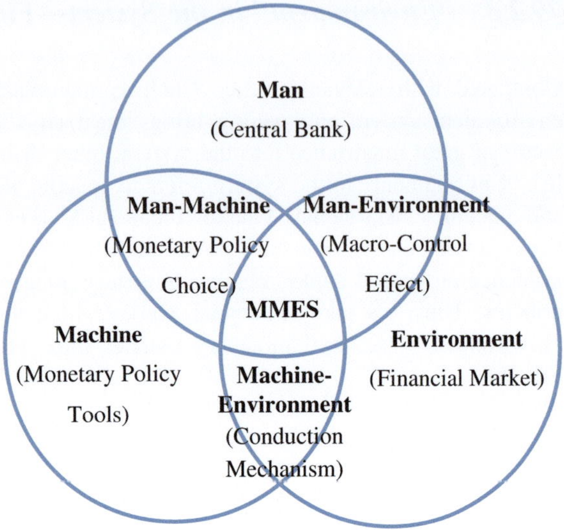

Fig. 79.2 The complex diagram of man-machine-environment system in monetary policy study

influence and interaction between three factors, namely central banks, monetary policy tools, and financial market, as it is shown in Fig. 79.2.

79.3 Traditional Monetary Policy System ("Man-Machine-Environment System") Will Be Broken

The article below is exampled by transition of the USA Central Bank (the Federal Reserve) since global financial crisis. It employs the frame of man-machine-environment system to state the mutual evolution and development relationship of "man" (Central Bank), "machine" (Monetary Policy Tools)and "environment" (Financial Market). It highlights the importance of environmental factor (Financial Market) in the evolution of "man-machine-environment system" (Monetary Policy System).

79.3.1 Evolution of "Environment" (Financial Market)

Since 1970s, at the background of financial innovation and rapid development of shadow banks, the structure and scale of financial market, as the main "environment" factor in the monetary policy system, have experienced significant change. Traditional financial market takes commercial banks as principle part, and since the new century, shadow banks have developed far beyond commercial banks whether in scale or in significance. Compared with traditional bank industry, shadow banking system has accumulated competitive advantages and becomes gradually more powerful main part of market than traditional banks. Shadow banks are the inevitable products of financial innovation, marketization of financial market and globalization, the rapid development of which marks the evolution of market-oriented financial system. Under the background of globalization, all main financial markets extend and scale of shadow banks in all nations grows continuously. As a result, it forms new American financial market and international financial market of extreme prosperity (new "environment") [4] which takes shadow banks as main market players.

79.3.2 Traditional "Man, Machine, Environment" Relationship Is Broken

Taking the Federal Reserve as an example, in the traditional financial market "environment" centered on commercial banks, traditional transition of monetary policy (traditional "machine-environment" relationship) is adjusting Federal Funds Rate (traditional "machine") by adjusting and controlling the supply and demand of federal funds, and regulating and controlling financial market (traditional "environment") indirectly by adjusting and controlling interest rate of commercial banks. However, federal funds are used for commercial banks registered in the Federal Funds to trade reserve fund, while shadow banks, as main part of emerging market, are not in the commercial banks' range of trade, adjustment and control, and free from restriction of monetary policy. It means that in the new "environment" where shadow banks develop rapidly, transmission mechanism of traditional monetary policy (traditional "machine-environment" relationship) from monetary policy tools ("machine") to financial market ("environment") will not work out any more. Traditional stable mechanism of "man-machine-environment" system aimed at "safety, efficiency and economy" will be broken and systematic risks expand rapidly. Finally, subprime crisis burst in 2008 and provoked global financial crisis.

79.4 Transition of Monetary Policy System Based on New "Man-Machine-Environment System"

79.4.1 Transition of New "Machine"

Since global financial crisis in 2008, the Federal Reserve ("man") has created a series of new monetary policy tools (new "machine") to supplement the deficiency of traditional policy tools (traditional "machine") and transmission mechanism of traditional monetary policy (tradition "machine-environment" relationship). From

November in 2008 to the end of October in 2014, the Federal Reserve has implemented four rounds of quantitative easing policy. Their large-scale property purchasing plan has gone far beyond bond scale and types [5] permitted to be purchased by central traditional banks according to original relevant laws.

Table 79.1 Fed's new monetary policy tools

Almsman	New tools	Introduce time	Exit time	Collateral	Maturity
Depository institution	Term discount window	2007.8.17	Going on	Discount window collateral	<90 days
	Term auction facility	2007.12.12	2010.3.8	Discount window collateral	28–84 days
	Currency swap agreements[a]	2007.12.12	Going on	National currencies	Overnight-3 months
Dealer	Primary dealer credit facility	2008.3.16	2010.2.1	Tri-party repo collateral	Overnight-14 days
	Term securities lending facility	2008.3.11	2010.2.5	Government security, investment bonds	28 days
	Term securities lending facility options program	2008.7.30	2009.6.25	Investment bonds	<14 days
Money market	The ABCP money market fund liquidity facility	2008.9.19	2010.2.1	High-quality ABCP	<270 days
	Money market investor funding facility	2008.10.21	2009.10.30	Dollar deposit commercial paper	–
Certain enterprise	Commercial paper funding facility	2008.10.7	2010.2.1	3 month ABCP	3 months
	Term asset-backed securities loan facility	2008.11.25	2010.2.1	AAA-ABS	Over 1 year

[a]In a sense, bilateral swap agreement can be regarded as a kind of relieving facility from overseas central banks to their depository institutions

Moreover, in view of limitation of traditional transmission mechanism of monetary policy (traditional "machine-environment" relationship), the Federal Reserve has begun to step over traditional central banks gradually by creating a series of system liquidity facilities (new "machine"), employing monetary policy tools [6] capable of directly adjusting and controlling various kinds of financial markets (new "environment") besides of commercial banks, as it is shown in Table 79.1.

79.4.2 New "Man-Machine-Environment System"

Since global financial crisis in 2008, various kinds of operation tools created by the Federal Reserve has gone far beyond the traditional functions that central banks can only provided financial support to commercial banks. The Federal Reserve has begun to compensate for the limitation of effect of traditional policy tools including federal funds rate (traditional "machine") by creating new tools such as term deposit facility to fill in the liquidity directly to financial market. Its monetary policy reports have also restated for several times, "Since international financial crisis, macro-control means by adjusting scale of reserve funds to control interest rate of federal rate through traditional open market operation has failed gradually and degrees of sensitivity from federal funds rate on adjustment of reserve funds has fallen to the lowest level in history." Thus, change of monetary policy tools ("machine") and that of traditional mechanism ("machine-environment" relationship) provoked by change of base of central banks' macro-control reflects that the current regulatory mechanism of the Federal Reserve has been transmitted to the new paradigm of regulating market liquidity directly (new "machine-environment" relationship).

79.5 Conclusion

In the marketization process of financial system, under the background of continuous change of macro-control "environment" (as financial market), central banks are required to comply with the long-term development trend of marketization, to change its temporary and passive crisis aid system afterwards into active continuous innovation of monetary policy tools ("machine") and policy transmission mechanism ("machine-environment relationship"). With proactive breakthrough in its traditional functions, expected efficient control of financial risks can be reached. Central banks are also required to control financial risks efficiently and keep the financial crisis away [7], which means to achieve the final objectives of "Safety, efficiency and economy" in man-machine-environment system.

Acknowledgments This paper was supported by "The structure and mechanism innovation for improving rural governance as base of National comprehensive security", a major project funded by the National Social Science Foundation of China (No.14ZDA064).

References

1. Long S et al (2004) Foundation of theory and application of man-machine-environment system engineering. Science Press, Beijing
2. Long S (1999) Analysis of the status and prospect of application of man-machine-environment system engineering in the development of science and technology. In: The research progress of man-machine-environment system engineering, vol 4. Ocean Press, Beijing
3. Long S (2003) The relationship between man-machine-environment system engineering and related disciplines. In: The research progress of man-machine-environment system engineering, vol 6. Ocean Press, Beijing
4. Li Y (2011) The development of shadow banking system and financial innovation. China Finance Press, Beijing
5. Gagnon J, Raskin M, Remache J et al (2010) Large-scale asset purchases by the federal reserve: did they work? Staff Report, Federal Reserve Bank of New York
6. Huang Y (2012) Analysis of the practical effect of the fed's QE monetary policy to American economic recovery from the perspective of the transmission mechanism of monetary policy. Dyn Econ 621(11):10–16
7. Huang Y (2014) The fed in transition: analysis of the fed's monetary policy since the international financial crisis. Doctoral thesis of the Graduate School of Chinese Academy of Social Sciences, pp 113

Chapter 80
Application Research on Man–Machine–Environment System Engineering Theory in the Development of Simulation Training System

Junfeng Liu

Abstract Man-machine-environment system theory provides important theoretical guidance and practical method to develop simulation training system; it is of important realistic significance for the development of simulation training system to apply these research results. This thesis is about the application of man-machine-environment system engineering theory in the development of simulation training system, introduces the functions and roles of simulation training system, explains the guiding significance of man-machine-environment system engineering theory to the development of simulation training system, and puts forward the basic thoughts to use man-machine-environment system engineering theory research results to guide the development of simulation training system.

Keywords Man-machine-environment system · Simulation training system · Development and application

80.1 Introduction

The simulation training system was first applied in the military field, and with the development and popularization of simulation technology, the simulation training has been widely applied in the aviation flight training, ship sailing training, and large-scale equipment driving training. At present, the development and application of simulation training system have problems like not very advanced technology, not very wide application range, and not very high working efficiency; especially, the research and application of correlation of man, machine and environment are always like "attend to one and lose another," leading to the defects in the developed simulation training system and the training effect inconsistent with the actual equipment,

J. Liu (✉)
Troop 65066, Dalian, China
e-mail: dl.ocean@gmail.com

© Springer-Verlag Berlin Heidelberg 2015
S. Long and B.S. Dhillon (eds.), *Proceedings of the 15th International Conference on Man–Machine–Environment System Engineering*, Lecture Notes in Electrical Engineering 356, DOI 10.1007/978-3-662-48224-7_80

actual environment, actual operation, and training command. This thesis puts forward opinions on how to use man-machine-environment system engineering theory research results to guide the development of simulation training system [1].

80.2 Functions and Roles of Simulation Training System

There are many types of simulation training systems, including aviation flight simulation training system, ship sailing simulation training system, engineering equipment operating simulation training system, arm weapon equipment technology simulation training system, and troop commanding simulation training system, and their functions and roles are as shown in the followings (Fig. 80.1).

80.2.1 To Innovate the Means and Modes of Military Training and Accelerate the Pace of New Military Reform

The simulation training methods and systems developed based on modern information technology have become the necessary foundation for the combat training of troops in many countries, which significantly extends the training scope and improves the training efficiency and level. And the modern computer combat simulation training has created the best environment for the future commanders and combatants. At the moment, the troops of many countries have widely adopted the computer simulation training system to carry out highly emulational battle tactical combat training, weaponry skill training, team tactical action training, etc. With the

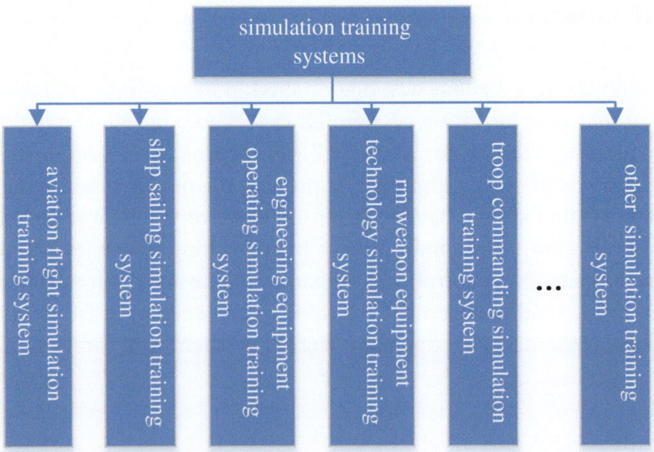

Fig. 80.1 Classification chart for simulation training system

simulation of verisimilar war process, operational environment and weaponry, trainees will practice their commanding, skills, and coordination in an environment very close to the reality, they will be prepared for the real war, shorten the distance between training and actual combat, accelerate the generation of combat effectiveness, increasingly innovate the military strategy training, propel the pace of new military reform, and lay a solid foundation to win the future wars [2].

80.2.2 To Innovate the Methods and Means for Technical and Skill Training, and Improve the Speed and Efficiency of Talent Cultivation

The technical and skill training are mainly for the trained individual or team to improve their job qualification. Use simulation training system to carry out spaceflight, aviation, ship, vehicle driving, and equipment operating and other technical training, allowing the trainees of different jobs to obtain the abilities to operate and control the equipment, the abilities to handle and analyze any faults, the ability to work with others, and the ability to adapt themselves to changing circumstances by setting complicated environments, and to obtain specialized skills in the conditions exactly like the actual combat. They will learn the equipment operating procedures and mechanical trouble shooting steps via the simulation training to solve the problems such as difficult fault setting, highly random and loose connection between training and actual work to greatly improve the training efficiency, shorten the time for talent cultivation, and play the role of booster for specialized talents to take their posts [2].

80.2.3 To Create Verisimilar Training Environment to Improve the Service Efficiency of Time, Space, Manpower and Material Resources

The simulation training system can build a real equipment operation training platform exactly consistent with the real environment via computer technology, network technology, virtual reality technology, and actual-load simulator to carry out full-element training to the trainees. Let us take ship simulation training system, for example, you can randomly set various environments, faults, and other simulating conditions to truly realize the all-weather and full-function simulation of specialized ship training, to strengthen the sense of environment and sense of reality in the training, and to solve the problems that due to equipment, venue, environment, and other objective elements, the specialized ship training is very hard to be implemented and the training level is very difficult to be improved. After the man-in-environment and machine-in-environment full-system and full-process training, the simulation training system carries out complete programming training and complicated environment training within the set time and space, which significantly improves the time and space use ratio, extends and integrates the multi-element training conditions, and effectively reduces the training cost and reduces the equipment loss; and the generated economic benefit and social value are beyond comparison to the traditional training mode.

80.3 Guiding Significance of Man-Machine-Environment System Engineering Theory to Develop Simulation Training System

The man-machine-environment system engineering theory systematically explains the essence and relation of man, machine, and environment and explores the operation law and the scientific method for optimum combination in an in-depth manner. The simulation training system is the operating platform consisting of man, machine, and environment via many technologies and entities; therefore, the man-machine-environment system engineering theory provides scientific theory guidance and technical support to the development of simulation training system. They are mainly reflected in the followings.

80.3.1 The Research Methods of Man-Machine-Environment System Engineering Provide Theoretical Guidance to Develop Simulation Training System

The research methods of man-machine-environment system engineering are based on control theory, model theory and optimization theory to analyze three elements of man, machine, and environment, and after the three steps of plan decision, development and production, and actual use, they can achieve the three goals: secure, efficient, and money-saving. This methodology provides the scientific guidance and fundamental opinions to the development of simulation training system; the simulation training system is used and developed by man, it is the system application platform consisting of machine and environment for the satisfaction of certain needs of man, and machine and environment provides carrier and media for man's training; thus, simulation training system consists of three elements, man, machine, and environment, and it is a complicated correlative system by the information transmission, processing, and control among the three of them; it uses the system engineering method to make the system have the features of "secure, efficient and money-saving," which fully reflects the scientificity, practicability, and guiding nature of man-machine-environment system engineering theory; it is of great guiding significance to apply this theoretical results to develop simulation training system [3].

80.3.2 Man-Machine-Environment Engineering Theories Provide Scientific Method to Develop Simulation Training

The acting object of simulation training system is man, thus leading to the correlations between man and machine, man and environment, and machine and environment; it is the key link for the development of simulation system to build and deal with these relations. The man-machine-environment system engineering provides methods for reference to solve the problems encountered in the development of simulation training system. For example, the reliability research on man focuses to allow man to adapt to machine features via selection and training and optimize the coordination between man and machine. The research on machine characteristics emphasizes that the design of machine shall confirm to man's characteristics and requirements. The environment characteristic research requires that only to take environment as a link of the system, full planning and control to the environment from the overall height of the system can be carried out to make the entire system at its optimum working status. All these system research results provide valuable methods and means to scientifically build the relations between man and weapon, man and training environment, and weapon and environment in the simulation training system [4].

80.4 Several Links Requiring Attentions in Developing the Simulation Training System

The man-machine-environment system engineering provides very instructive and applicable research results to develop the simulation training system, and can provide basic opinions to help the developers to broaden their thoughts and improve their abilities of dialectical thinking and scientific design. Therefore, the developer shall not only use their expertise and special skills in the developing mission, but also carry out the integral consciousness of the system in the full process of the development. They shall pay more attention to the following links.

80.4.1 To Use the Man-Machine-Environment System Principle to Make Clear About the Use Object and Objective and Carry Out the Groundwork: System Requirement Analysis and Design

To develop the simulation training system, first carry out the system requirement analysis, argumentation, and design. The main contents of the system requirement design are: use for whom, solve what problems, achieve what purpose, and adopt what technical path. To make the designed system to correspond to reality and meet user's wish, the developer shall make clear about the correlation of man, equipment, and environment based on the man-machine-environment system engineering theory, and thus to design the system requirements with normalization, guidance, and operability and lay the prerequisite foundation to the completion of the system [5].

80.4.2 To Take Care of the Management Control of System Structure, Man-Machine Interface Design and Process Design According to the Man-Machine-Environment Engineering Characteristics

The important guarantee for whether the developed simulation training system meet user's requirements and to make sure the complete and accurate achievement of the developed products in the requirement framework is to carry out the management control in the system development process. No matter how good the objective for a system, how reasonably the personnel is organized, if there is no strict management control as a guarantee, the objective for the system development will be very hard to achieve. The main task of management control in the system development process is to scientifically analyze the conflicts and problems in the development, and carry out the work like risk control, quality assurance, tracking and regulating, information feedback and deviation correction to ensure the direction and objective of the system development.

80.4.3 To Carry Out System Test, Assessment and Modification, and Realize the Design Functions According to the Man-Machine-Environment Action Principle

The ultimate objective to develop the simulation training system is to serve the users, and the most important thing to meet design requirements and usage requirements is to carry out systematic testing and assessment, timely find problems, and put forward modifying plan based on the man-machine-environment action principles; via the optimization and adjustment to system functions and modules, reach the optimum man-machine function, make the indexes and functions of the system to achieve the design objective, and meet users' requirements.

80.5 Conclusion

Man-machine-environment system theory provides important theoretical guidance and practical method to develop simulation training system; it is of important realistic significance for the development of simulation training system to apply the research results. The scientific researchers shall develop the simulation training system with advanced technology, easiness to use, evident benefit, and convenience for popularization for various industries and contribute to cultivate all types of talents, and improve production efficiency and army combat effectiveness with modest attitude, pragmatic style, and innovative courage and resourcefulness.

References

1. Long S et al (2004) Foundation of theory and application of man-machine-environment system engineering. Science Press
2. Hu X et al (2009) War gaming and simulation principles and system. National University of Defense Technology Press, Changsha
3. Long S et al (1993) Progress in man-machine-environment system engineering, vol 1. Beijing Science and Technology Press, Beijing
4. Long S et al (1999) Progress in man-machine-environment system engineering, vol 4. Ocean Press, Beijing
5. Zhang H (2002) Software engineering. Posts and Telecom Press

Chapter 81
Software Aging Analysis Based on Man–Machine–Environment System Engineering

Baiqiao Huang and Yuanhui Qin

Abstract According to the problem that lacking of systematic analysis of software aging, and the factors people focused on are just empirical. From the man–machine–environment systems engineering perspective, a systematic analysis process for software aging factors is presented here. Through the establishment of the hardware/software environment view and the system data view of the software system, the factors that lead to software aging have been identified from the software/hardware environment and the data. One common characteristic of these factors has been concluded: they are all the function of time. We propose a corresponding software rejuvenation method as a reference for the software design and maintenance stages, which eliminates the influence of software aging factors.

Keywords Software aging · Software rejuvenation · Memory leak · Man-machine-environment system engineering (MMESE)

81.1 Introduction

From the common viewpoint, software is a mathematical and logical product. The algorithms and the logic are eternal with time; thus, software should not be aging from the wear and fatigue that affect a physical body like hardware. However, in the reality of a software application, after a long time running, the fault rate of the software increases, the performance decreases, and the software may even breakdown. A researcher called this the *software aging-related phenomenon* (SAR).

Reference [1] is acknowledged as the first paper to propose the SAR problem. The SAR problem has been paid much attention in later research. Recent research on SAR has focused on memory consumption and the performance degradation [2].

B. Huang (✉) · Y. Qin
System Engineering Research Institute of China State Shipbuilding Corporation, Beijing 100094, China
e-mail: seafury@buaa.edu.cn

© Springer-Verlag Berlin Heidelberg 2015

S. Long and B.S. Dhillon (eds.), *Proceedings of the 15th International Conference on Man–Machine–Environment System Engineering*, Lecture Notes in Electrical Engineering 356, DOI 10.1007/978-3-662-48224-7_81

For example, References [3, 4] studied the mechanisms of software memory leak and software memory fragment. Several researchers have analyzed the SAR problem of special software systems, such as for cloud computation software [5], the Linux operation system [6], and network service software [7]. Several researchers studied the mechanism of the SAR problem; for example, the four reasons causing software aging were identified as resource leak, fragment, data error, and data damage [8]. The ultimate goal of the research on SAR is to mitigate the influence of SAR failure, which is called the *software rejuvenation technique*.

The existing research on SAR does not systemically analyze the mechanism of the SAR problem, which is not suitable for integrated selection of the proper recovery time and methods from the system viewpoint. This paper puts the software within the human–machine–environment system engineering for integrated analysis of the SAR problem, establishes a view of the software/hardware environment and the software system data to analyze the various factors causing the software aging, and proposes the common characteristics of the factors causing the SAR—namely that the factors all are the functions of time. The transformation accumulation of these factors along with the accumulation of time causes the SAR problem.

It is important to note that there is another meaning of software aging. When software requirements have significant changes or hardware is upgraded, the software should be developed again. Then, the old version of software steps off the stage of history, which will be called software aging as well [9]. However, the commonly acknowledged meaning of the SAR is the former.

81.2 MMESE-Based Software System Modelling

81.2.1 MMESE Model of Software System

Man-machine-environment systems engineering (MMESE) is a science that focuses on the relationship between man, machine, and the environment. It deeply studies the optimal combination of man, machine, and environment by using the system science theory and systems engineering approach [10]. A software system is a typical man-machine-environment system. First of all, the software has a specific user. Second, the software has its own operating environment. The software itself cannot run separately: there must be a memory, central processing unit (CPU), network equipment, and other hardware peripherals; also, the software must have its software environment, such as the operating system and database software. Moreover, the software system works in a natural environment. Using the MMESE theory to analyze the factors of software aging after long time running can make the results more systematic.

81.2.2 Analysis of System Theory-Based Software Aging-Related Failure

It makes no sense to talk about software aging if we only focus on the software itself, because software is a mathematical and logical product, and these do not decay with time. Therefore, we must put the software into its man-machine-environment system for an integrated analysis with systems theory.

From the perspective of a system, because the arithmetic and logic of the software cannot be changed with time, the aging factors should be sought from the software/hardware environment and the data from the arithmetic and logic processed. Factors leading to software aging should meet a common characteristic: they are all the functions of time. With the accumulation of passing time, these factors change from quantitative to qualitative, and eventually they lead to the failure of software due to aging. Obviously, the life of the hardware itself is a function of time; the hardware resources will accumulate with time too, while some special data stored in the memory and in the hard disk that are processed by the arithmetic and logic are also the function of time, such as the counter variable, file size, and the database size. These factors result in running the same logic but getting different results. In order to analyze the cause of the software aging phenomenon systematically, we separately constructed the software system model from the software/hardware environment perspective and the data perspective, and then analyzed the factors of software aging base by these models.

81.2.3 Software/Hardware Environment View Analysis

The software/hardware environment view of software system is shown in Fig. 81.1. The hardware environment mainly includes the CPU, memory, storage devices, and network equipment. The software environment mainly includes the operating system.

First of all, hardware has a lifetime because of physical wear; namely, the hardware state is a function of time. People have usually ignored this feature because the hardware lifetime is too long to compare with the software application time. While in the safety critical embedded software system, the hardware state is an important factor that has to be considered. For example, the flash memory's lifetime is measured by read/write times; the state of the flash memory device will become unstable if the read/write times have exceeded the limitation. If the flash memory should be read/written frequently by the software function, with the accumulation of the software running time the device will be unavailable, and the software will activate aging-related failures.

Second, during the time the hardware equipment starts use to wears out, the change of its structure and the reduction of the performance are gradual processes. For some special software systems that are sensitive to the hardware structure, the

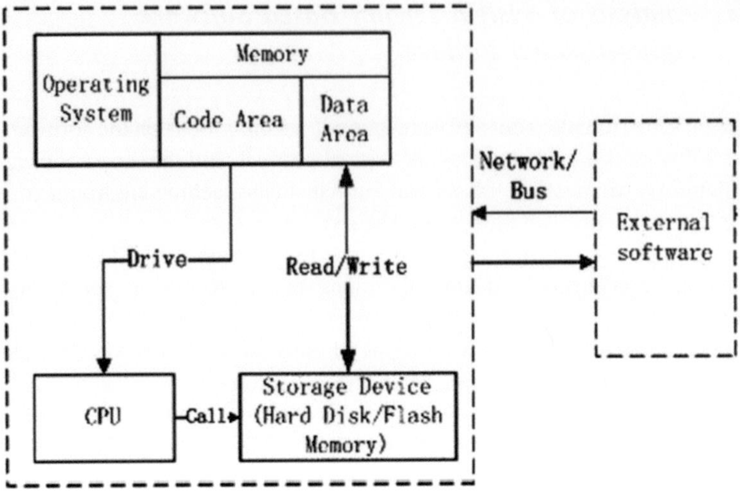

Fig. 81.1 Software/hardware environment view of a software system

effect of this gradual change cannot be ignored. In an optical measurement system, a specific set of control parameters should be configured for the software, and these parameters are adapted to the system's initial structure and state. As time goes on, some offset appears, called system zero offset, and the software will activate aging-related failures because these initial control parameters were not suitable again.

Third, hardware resource consumption is usually a function of time. The resources will run out with the accumulation of application time, such as memory leakage and disk space exhaustion. If there is a memory leak situation in the software, the memory will be allocated but not be released properly. As time goes on, the available memory becomes less and less, and the system performance becomes lower and lower, until the software crashes because of failed memory. The output files of the software and the increased database records should continually occupy the space of the disk; predictably, the disk space will be exhausted. These will all activate aging-related failures.

Fourth, process management and interrupt management, which prevent the application software from running normally, are subject to the operation system. After a long time running, some unpredictable and repeatable problems emerge, such as process deadlock and interrupt priority confusion. These also cause the application software to activate aging-related failures.

Finally, the natural environment where the system is located will affect the hardware device of the software system. A poor working environment will accelerate the aging speed of hardware, and then cause an indirect impact on the software. Therefore, the software system usually has a level of protection for its location in the natural environment. For example, the open platform on board should have its protection level as "IP56".

81.2.4 Data View Analysis of the Software System

The data view of software system, which describes the relationship between different data objects, is shown in Fig. 81.2. Some data objects are also the function of time and make the software activate aging-related failures.

First of all, the value or the accuracy of the variable in the software is sometimes the function of time. Take the counter variable as the first example; with the accumulation of time, the value of the counter variable will go beyond the legal range, or even exceed the maximum value specified by the variable type, and activate the overflow failure. For another, the accumulation of round-off error in a specific algorithm can lead to a huge disaster; this tragedy happened in the Patriot Missile defense system. There is a numerical conversion algorithm in the Patriot Missile defense system, where 0.1 s needs to be changed from decimal to binary. This change results in only 0.000000095 round-off. However, when the software system continues working for 100 h, the accumulated round-off will reach 0.34 s—this deviation is too big to activate the relevant function failure. On 25 February 1991, the Patriot system located at Dhahran, Saudi Arabia, failed to intercept an arriving Scud missile after a runtime of more than 20 h since the last reboot.

Second, the size of the files and the database records are usually a function of time, which will affect the performance of the software. For example, suppose there is a function that needs to iterate a database table. The performance of this function may decrease because the records will become bigger as the application time goes on.

Finally, the CPU clock cycle is also a function of time. Generally, it is considered as a constant value; however, it is really a random value that has a certain expectation. This will become the instability factor for some embedded software that has high real-time requirements.

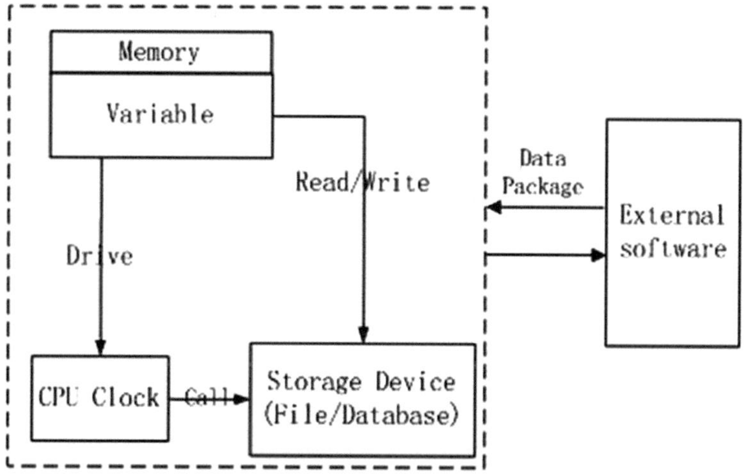

Fig. 81.2 Data view of the software system

81.3 Software Rejuvenation

The final goal of the research on SAR is to mitigate the effect of the SAR failure, which is called the software rejuvenation technique. According to the aging-related factors described above, some rejuvenation methods are recommended.

- *Hardware lifetime factors treatment.* This factor mainly is focused on embedded software. In Windows applications, the hardware states have been managed by the operation system. In embedded software, the build-in test (BIT) program usually has been developed to test the hardware states and the program alarms in time when the equipment fails. A technique called prognostic and health management (PHM) is proposed here. The establishment of a lifetime model to predict the device's downtime can guide repairs or replacement of the hardware promptly.
- *System zero offset factor treatment.* For some software systems that configure a specific set of control parameters, after a period of time, these control parameters should be calibrated again and adapted to suit the system state at that time. This makes the system return to the zero position.
- *Memory leak factors treatment.* The factors of memory leak should be found and removed during software testing. Usually, we can use code static analysis tools such as Klocwork and C ++Test to look for code errors that will lead to memory leak in the unit test, and we can monitor the consumption of the memory when executing the function to judge whether there is a memory leakage in the Computer Software Configuration Item (CSCI) test. If there does exist a memory leak in the software, and after a period of running time there has been a significant decline in performance, we can restart the software regularly to eliminate this kind of aging-related failure.
- *Storage space consumption factors treatment.* The focus of this factor is whether we can forecast the time the storage capacity was exhausted, and what measures we can take to treat this situation. We can use a method of estimation. First, we conduct a typical task scene for a full day, obtain the storage space consumption in 1 day, and then use the disk capacity to divide the day consumption to get the available deadline. Before the deadline, we should backup the files and cleanup the disk.
- *Operation system factors treatment.* Usually, the failures caused by the operation system are unrepeatable. We should restart the operation system to eliminate this kind of aging-related failure.
- *Value cumulative or round-off cumulative factors treatment.* First, we should identify variables that have a value cumulative or round-off cumulative effect and create protection in the code. Also, we can reboot the application software regularly to reset the variables back to their initial values. Unlike the case of the Patriot missile system, if the operator reboots the system after a period of use, the tragedy will not happen.
- *File size and database records factor treatment.* We should assess the impact on the software performance due to file size and the size of the database records

periodically, backup the files or the database, and then cleanup and restore the software to its original performance.

81.4 Conclusions

From the man-machine-environment systems engineering perspective, the factors of software aging have been analyzed systematically. Based on the software/hardware environment view and the data view established in this chapter, the aging-related factors of software were analyzed from the software/hardware environment and the data used in the software. One common feature of these factors has been summered up: they are all the function of time. According to these aging-related factors, some corresponding rejuvenation methods were recommended. These suggestions can mitigate the aging-related failures in the design and maintenance stage of the software.

References

1. Huang Y, Kintala C, Kolettis N, Fulton N (1995) Software rejuvenation: analysis, module and applications. In: Proceedings of twenty-fifth international symposium on fault-tolerant computing, pp 381–390
2. Cotroneo D, Natella R, Pietrantuono R et al (2011) Software aging and rejuvenation: where we are and where we are going. In: Third international workshop on software aging and rejuvenation (WoSAR). IEEE, Hiroshima, pp 1–6
3. Macêdo A, Ferreira TB, Matias Jr R (2010) The mechanics of memory-related software aging. In: Second international conference on software aging and rejuvenation (WoSAR). IEEE, San Jose, pp 1–5
4. Matias Jr R, Costa BE, Macedo A (2012) Monitoring memory-related software aging: an exploratory study. In: IEEE 23rd international symposium on software reliability engineering, software reliability engineering workshops. IEEE, pp 247–252
5. Machida F, Xiang J, Tadano K, Maeno Y (2012) Aging-related bugs in cloud computing software. In: IEEE international conference on software reliability engineering, software reliability engineering workshops. IEEE, Dallas, pp 287–292
6. Cotroneo D, Natella R, Pietrantuono R et al (2010) Software aging analysis of the linux operating system. In: IEEE international conference on software reliability engineering. IEEE, pp 71–80
7. Grottke M, Li L, Trivedi KS (2006) Analysis of software aging in a web server. IEEE Trans Reliab 55(3):411–420
8. Grottke M, Matias Jr R, Trivedi KS (2008) The fundamentals of software aging. In: IEEE international conference on software reliability engineering, software reliability engineering workshops. IEEE, Seattle, pp 1–6
9. Parnas DL (1994) Software aging. In: Proceedings of the 16th international conference on software engineering, Los Alamitos, pp 279–287
10. Long S(1993) Man-machine-environment system engineering theory and its significance in productivity development. In: Progress in man-machine-environment system engineering research, vol 01. Beijing Science and Technology Press, Beijing, pp 2–13

Chapter 82
Application of Human–Machine–Environment System Engineering in Handle Design of Power Hand-Held Tools

Yifeng Du

Abstract The design, selection, evaluation, and use of the power hand-held tool must be combined together with the ergonomics for research. The purpose of this research is to analyze the handle design of the power hand-held tool with the principle of human-machine-environment system engineering technology, starting from the hand anatomy, to optimize the handle design of the power hand-held tool, to study the handle's shape, angle, length, and diameter using the method of data analysis, and to enable the handle design of the power hand-held tool more scientific and rationalized.

Keywords Human-machine environment system engineering · The power hand-held tool · The handle design

82.1 Research Purpose

Human-machine-environment system engineering mainly researches concentrating on human, machine, and environment. Human is the operator and the user of the machine and the final service object of product manufacture. The elements of human's behavior style, psychological reaction, ratio scale, and relationships among human, machine and environment, etc. in process of using the product are the bases for developing the new product. The human-machine-environment system engineering is a subject [1] aiming at improving the production efficiency and ensuring human safety, health, and comfort through the research on the interaction among human, machine and environment; however, the key point of ergonomics study of modeling design is essentially to discuss how the devices directly operated and used by human in the machines and equipments adapting to human use [2].

Y. Du (✉)
Zhuhai College of Jilin University, Zhuhai 519041, China
e-mail: laviedudu@163.com

© Springer-Verlag Berlin Heidelberg 2015
S. Long and B.S. Dhillon (eds.), *Proceedings of the 15th International Conference on Man–Machine–Environment System Engineering*, Lecture Notes in Electrical Engineering 356, DOI 10.1007/978-3-662-48224-7_82

Tools are expansion of human's extremities function, because using tools enabling human increasing actuation range and power and improving work efficiency. For precision operation, many researches must be conducted on the anatomy function of human's hands and tools structure. As the new type tools to improve laborers work efficiency, power hand-held tools have been comprehensively applied in working practice. But, most of the power hand-held tools have not reached optimal form in design, shape, size, etc. Neither accord with human-machine-environment system engineering principle, so it is difficult for human to operate effectively and safely. Power hand-held tools will do harm to human body and influence human health because of vibration, noise, unbalancedness, and impact force, etc. If people use the power hand-held tools and equipment in poor design for long-term and repeatedly in work or daily life, it possibly causes cumulative damage to users and thus causes physical discomfort, injury, and disease, which will not only reduce productivity, but also possibly will cripple labors and increase people's psychological distress and medical burden.

Design, selection, evaluation, and use of the power hand-held tools must be researched in combination with human-machine-environment system engineering. The research purpose is to adopt theory and principle of human-machine-environment, start with human hands anatomy, and analyze handle design of power hand-held tools to enable it not only completing specific function of the tools but also adapting to restrictive conditions and factors of each kind of human index and human body constitution. The research aims at optimizing handle design of power hand-held tools and combining with human-machine-environment system engineering to make the handle design of the power hand-held tools more scientific and reasonable.

82.2 Research Method

The handle design of the power hand-held tools must be of appropriate proportion to the user's body position at the same time when the expected function is realized, in order to decrease the fatigue and maximize work efficiency and fully consider operation safety and comfort [3] when maximizing use rationality. The research starts with human hands structure organization and anatomy analysis and applies human anatomy knowledge to analyze human hands structure and analyze human-machine-environment system engineering with handle design of power hand-held tools of human-machine-environment engineering theory to find optimal scheme for handle design.

82.2.1 Hands Structure Organization and Anatomy Analysis

Ergonomics factors of power hand-held tools mostly depend on physiological basis of human's hands. The hands are complex structures, (Fig. 82.1) composed of

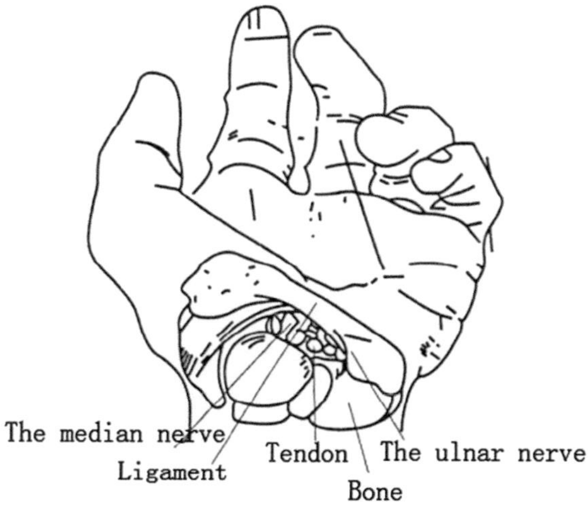

Fig. 82.1 Human hand structure diagram. Reprinted from *Ergonomics.* [256], Copyright 2008, with permission from Beijing Institute of Technology Press

bones, artery, nerves, ligament, tendon, etc. Centered on the wrist joints, hands deflection activities in each direction are also done by pulling muscles. Hands activities are not done only by nearby muscles, but also by affecting multi-bundle muscles and tendons from hands to forearms, upper arms, and elbow joints. Because these muscles are overlapping, if the arm is locked and the wrist is deviated, each muscle bundle mutually disturbs, which will influence these muscles exerting normal functions smoothly.

The wrist is a joint with multi-degree of freedom and many muscles, tendon, blood vessels, and nerves pass by here. The carpal is linked with the radius and ulna on the forearm with the radius linked to the thumb side and the ulna linked to the little finger side. The wrist joint structure and orientation enable it acting on only two sides with the two sides of 90°, respectively. One side enables palm lateral flexion and back flexion and the other side enables ulnar deviation and radial deviation (Fig. 82.2). Therefore, if the wrist joint is in large deviation and deflection

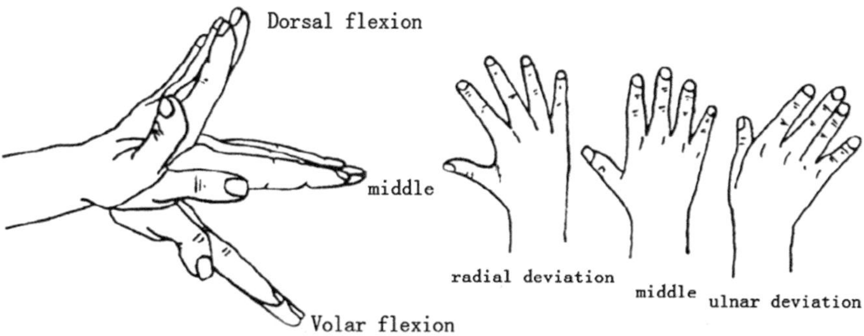

Fig. 82.2 Action state graph of wrist joint. Reprinted from *Ergonomics* [255], Copyright 2008, with permission from Beijing Institute of Technology Press

status, the muscles, tendon, blood vessel, and nerves among will be pressed, and hand and fingers activities will be influenced and thus damage and disease will be caused if it lasts long, such as tenosynovitis and carpal tunnel syndrome [4].

82.2.2 Human-Machine-Environment System Engineering Requirement for Handle Design of Power Hand-Held Tools

To ensure use efficiency and user's health, for designing power hand-held tools, it should follow the below principles of human-machine-environment system engineering: first, the reserved basic function must be effectively realized and the suitable ratio to the operator's body must be require in order to enable the operator exerting maximum efficiency; during design process, it should design in accordance with the operator's power and the work competency, so the differences in sex, training degree, and physical quality should be considered suitably; besides, the work posture required by power hand-held tools should not cause the laborer over fatigue.

82.3 Research Result

By analyzing the relevant data of human-machine-environment system engineering, the elements of handle design of power hand-held tools are obtained; by analyzing the elements of diameter, length, shape, inclination angle, materials and switch trigger, the research result is drawn.

82.3.1 Handle Diameter Design

Because of different arm forces required for operation with different tools, the optimal handle spans are different. According to the suggestion in publication of International Labor Office, *Ergonomics Principle of Hand Tools Design*: the optimal diameter of electric hammer is 2.54–3.81 cm; the optimal diameter of electric saw is about 3.5 6 cm; the optimal diameter of hand-held spray gun is 2.54–4.06 cm. The optimal handle spans vary from handles of different structures and forms. Diameter of 5–6 cm is optimal for pistol type handle, 4.06–6.1 cm for straight tube-shaped handle, and 3.56 cm for carrying type handle. For the gimlet, the longer diameter will increase torque but the shorter diameter will reduce grip strength and thus reduce flexibility and work speed and increase finger tip bones curve, which will cause finger tip fatigue for long-time operation. Suitable diameters are: 3–4 cm for exerting grasp and 8–16 cm for precision grasp.

Table 82.1 Hand size measurement data (unit: cm)

Project	Bare-handed		Wearing inflatable pressure gloves		Non-wearing inflatable pressure gloves	
	Average	Standard deviation	Average	Standard deviation	Average	Standard deviation
Hand length	19.38	1.14	19.84	1.22	19.71	1.35
Hand length (the thumb and index finger fist)	11.89	1.02	13.26	1.19	13.49	1.17
Hand width (the palm forward)	8.84	0.56	9.58	0.58	9.53	0.66
Hand width (hand Bent over)	8.83	0.51	9.28	0.51	9.53	0.66
Hands round	20.98	1.37	22.99	1.45	24.56	1.57
The width of the inside boxing hand (manual hold)	4.88	0.41	4.37	0.41	4.01	0.48
Hand width (handle by hand)	10.19	0.79	11.25	0.64	11.63	0.79

Reprinted from *Research on the Human-machine Design of Electric Manual Drill Design.* [51], Copyright Vol. 34 No. 16 2013-08, with permission from Packaging Engineering

Handle size design is mainly to solve the problem of avoiding forcing partially on palm center. As most users are young and middle-aged males, investigated by *Ergonomics Principle of Hand Tools Design,* see Table 82.1 for the initial data of the sizes when their hands bared, wearing inflatable pressure gloves and wearing non-inflatable pressure. The designer must adopt the shorter diameter for the handle when designing the handle diameter, so that the tools are light to grasp; however, it should not blindly pursue short diameter. If the diameter is too short, hand pressure is large for operating and it is difficult to force, which makes people feel uncomfortable.

82.3.2 Handle Length Design

Because those operators are different in sex, stature, hand length, hand width, and palm length, so the holding strength, finger force peak value and palm force peak value are also different, and the optimal length selection requirements are different. It spends strong force to operate power hand-held tools, which is easy to get fatigue if big hand holds small-span handle, while small hand holds big span handle. To expand range of users, the hand length is generally selected between the fifth percentile (71 cm) for female and the ninety-fifth percentile (97 cm) for male. Overlong handle may cause inflexible operation while over short handle may cause inconvenient grasp or grip strength reduced and palm forced. Therefore, the general length for handle suitable for Chinese is among 100–125 cm.

(1) **Handle Shape and Inclined Angle Design**
In accordance with the relevant principle of human-machine-environment system engineering, when the operator is working, the wrist is straight, the hand joints are relaxed, and the blood vessels and nerves are in natural state. For exerting grasp, the more contact areas between handle and palm are, the less the pressure stress is. Therefore, it is better to choose the handle of circular cross section is better. The most suitable shape should be chosen according to work property. To prevent relative slip between palms, the triangle or rectangle can be adopted to increase stability for placing tools.

It requires a best angle for the pistol type handle and middle type handle. According to ergonomics principle, the handle angle sideline is roughly parallel with the arm joints line when grasping tools, and there should be a best angle between the handle axis and arm joints line. The best angle enables the operator grasping tools saving effort with high precision. In accordance with Contemporary Ergonomics Tools, published by Sweden Atlas Copco, the recommended best angle is generally 70°, see Fig. 82.3.

(2) **Handle Materials Selection**
To transmit the operator's hand force to the tools to maximum limit, prevent slide between hand and tools surface and decrease operator's fatigue, antiskid lines are generally designed on handle surface of hand tools and handle cover also usually adopts plastic materials of ABS, PP, TPR, etc., to increase operation comfort [5]. According to the research studied on the handle materials by the abroad scholars and the actual operation and test adopted with EMG, the handle adopting polyvinyl cover is able to provide suitable and good grip surface. The handle surface is elastic and comfortable for operating.

(3) **Switch Trigger Design**
The size of pistol type switch trigger has relatively large influence on finger force and palm force when the operator holds tools. Different trigger structures should be used in different tools. Design of power hand-held tools trigger is mainly to avoid fingers operating repeatedly.

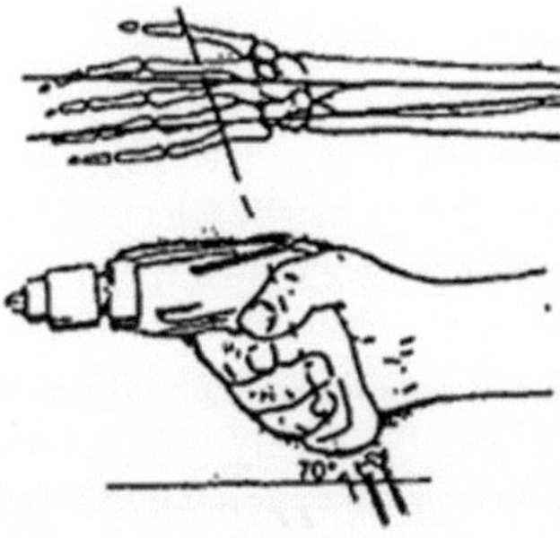

Fig. 82.3 Handle grip angle diagram. Reprinted from *Ergonomics*. [236], Copyright 2006, with permission from Xi'an Electronic and Science University press

Table 82.2 Different switch trigger finger, palm hard situation

Test object	Project name	Ordinary switch		Lengthened switch	
		Finger	Palm	Finger	Palm
Young workers	Peak force/N	130 (71)	153 (85)	119 (66)	142 (83)
	Peak value application of force/% MVC	50(22)	60 (27)	47(21)	56 (26)
	Average of the grip force/N	38(11)	47 (13)	29(9)	35 (12)
	Average of the grip pressure/% MVC	12(5)	16(6)	8(4)	11(5)

Operators must press buttons repeatedly in work but continuous repeat of use will cause fatigue and pain. The size, starting power, position, and shape of the button design decide the fatigue degree of hands and fingers. The size of switch trigger mainly requires the operator's finger force and palm force, especially the finger force. The research on labors operating pistol type handle tools shows that the finger force and palm force for operating lengthened switch trigger (length is 4.8 cm) are less than those for operating ordinary switch trigger (length is 2.1 cm). The research on the finger force and palm force for youth laborers operating pistol type handle tools shows that the average peak value of the finger force for operating lengthened switch trigger is 9 %, less than that for operating ordinary switch trigger, 8 % less than average peak value of the palm force, 65 % finger grip force decreased, and 48 % palm grip force decreased. See Table 82.2.

The size of the button with high using frequency should be suitable, because over-sized button needs more force for operation, which is easy to get fatigue while undersized button is easy to get mis-operation. The button position arrangement and shape design must accord with the hand shape, be switched freely for use. The button should be possibly set on position for convenient operation, because the conversion angle of people's fingers is limited. The most comfortable angle is 20° to the left and 30° to the right. See Fig. 82.4.

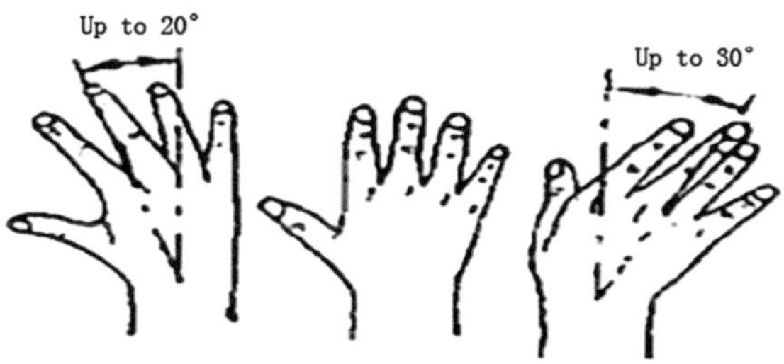

Fig. 82.4 Fingers conversion angle. Reprinted from *Ergonomics*. [255], Copyright 2008, with permission from Beijing Institute of Technology Press

82.4 Research Conclusion

Human-machine-environment system engineering research points out that in each stage of the whole process of product design, human-machine-environment system engineering design must be conducted to ensure full exertion of the product use function. The dominant idea of "human" oriented is embodied in that each design should consider human as the principal line and run human-machine-environment system engineering theory through the whole process of design.

The key point for design is that how does the handle design of power hand-held tools realize ergonomic design of the model in accordance with ergonomics. Here, only the general users are studied. With the development of the times and the development of the power hand-held tools function, the users of the power hand-held tools also change dramatically, from professional personnel initially to home application, model fans even females. For these users, they are left-handers and right-handers, so the characteristics of each group must be considered for design for these different use groups. As the ergonomics values designed are also different, the design idea of "human" oriented must be always insisted in specific design.

At present, the study on handle design is still in initial stage and the influence of handle section and integrated linear change on hand operation comfort and comfort subjective evaluation, etc. worth further study.

References

1. Liu C (2004) Ergonomic applications. Shanghai People's Fine Arts Press, Shanghai
2. Pang Z (2002) Industrial design. Harbin Institute of Technology Press, Harbin
3. Zhang Y (2010) On the relationship between ergonomic pliers handheld tools. Assoc Sci Technol Forum (The Second Half) 2:99–100
4. Ding Y (2011) Human-machine engineering. Beijing Institute of Technology Press, Beijing
5. Li L (2008) Research of the humanized design of hand tools. Dev Innovation Mach Electr Prod

Chapter 83
Situation Analysis of Fire Management Based on Human–Machine–Environment System Engineering Thinking

Xiuwen Zhao

Abstract System thinking emphasizes the essence of safety and advocates people to see the thing as an organic connection and interaction of the development integration, guide people in making a general survey of the whole thing from partial, provide insight into the root cause from the surface, and understand the interaction among different elements from static analysis. On this basis, people are able to clarify the structure relationship hiding behind the things and look for a systematic approach to solve the problem. Viewing from systematic thinking mode and combining with the current situation of fire, the paper analyzes the problem existed in fire management and puts forward the development strategy.

Keywords System thinking · Fire management · Situation analysis

Human have experienced long years on the cognition and management of fire accidents prevention, from fatalism to empiricism to system theory, from unconscious passive receiving to positive response to active prevention, and from the mono-factorial fact-oriented to systematic holistic approach. The fire management work has been constantly improved; however, it is still facing challenges [1]. Table 83.1 lists four indexes statistics of fire from 2001 to 2012 and draws development change curve along the time of each index listed in Table 83.1 on the above basis.

From development change trend of the curve, it can be seen that the overall situation of the current fire safety is improved; however, our country is in special historical period of economy transition and social transformation, and new situation and new problem of fire safety have risen continuously and each kind of fire cause increased; at the same time, there is fluctuation in stable fire situation, although it is not obvious, the increasing signal is obvious, which puts forwards higher requirement for fire safety management. Viewing systematically, Table 83.2 has constructed system elements for fire safety management (Figs. 83.1 and 83.2).

X. Zhao (✉)
Department of Fire Control, The Chinese People's Armed Police Forces Academy, Langfang 065000, China
e-mail: zxw2008happy@163.com

© Springer-Verlag Berlin Heidelberg 2015

S. Long and B.S. Dhillon (eds.), *Proceedings of the 15th International Conference on Man–Machine–Environment System Engineering*, Lecture Notes in Electrical Engineering 356, DOI 10.1007/978-3-662-48224-7_83

Table 83.1 2000–2012 Four indexes statistics of fire

Year	Number of fires	Dead people	Wounding	Direct losses (one hundred million)
2001	216,784	2334	3781	14.0
2002	258,315	2393	3414	15.4
2003	253,932	2482	3087	15.9
2004	252,804	2562	2969	16.7
2005	235,941	2500	2508	13.7
2006	231,881	1720	1565	8.6
2007	163,521	1617	969	11.3
2008	136,835	1521	743	18.2
2009	129,382	1236	651	16.2
2010	132,497	1205	624	19.6
2011	125,417	1108	587	20.6
2012	152,400	1028	575	21.8

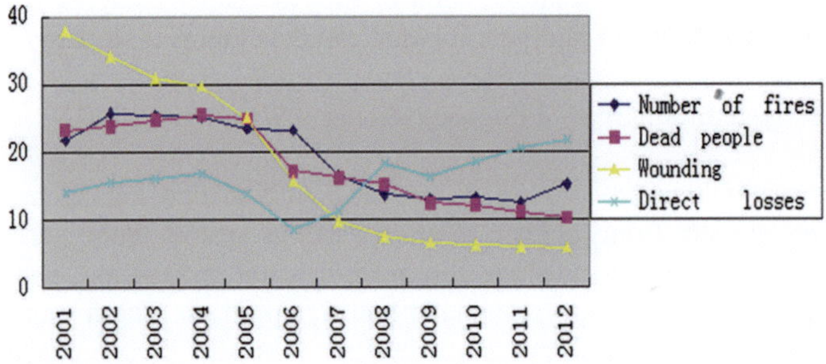

Fig. 83.1 2000–2012 Fire four indexes change curve

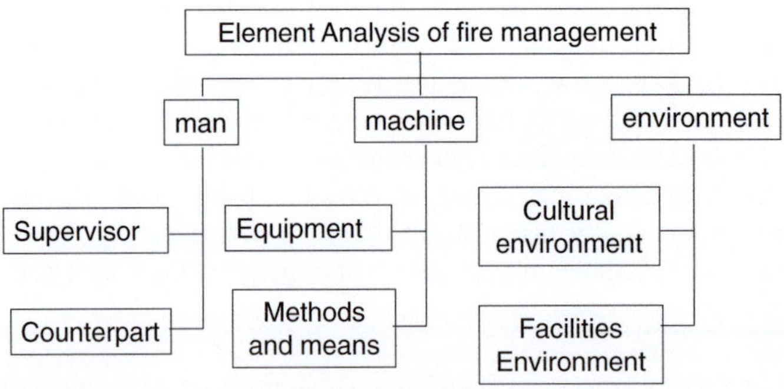

Fig. 83.2 Fire management work elements chart analysis

83.1 Situation Analysis of Human in Fire Control Safety Management

Human is the core element for fire safety management work, the organizer of fire control work, and also the participant of fire control work; the manager of fire control work and also one of the managed objects in fire control work.

83.1.1 Fire Control Safety Situation of the Managed Object

It is not difficult to see that analyzing from fire cause and reason of fire accident; currently our country citizens' quality of fire control safety has large room for improvement.

83.1.1.1 Citizens' Consciousness on Fire Control Safety Is Indifferent

As a probability event, fire has relatively high contingency; however, as unanticipated possibility, it is difficult to draw high attention from social citizens, which causes lacking of initiative and enthusiasm for fire control work, ignorance of overlooking value of fire control management work, inspecting but no correcting for fire risk and frequent accidents after correcting, obvious phenomena of illegal usage of fire and electric power, which have created necessary condition for fire accident.

83.1.1.2 Citizens Lack of Ability in Fire Control Safety

Because of indifferent consciousness on fire control safety and insufficient publicity education and training on social fire control safety, currently citizens lack of each kind of fire control safety ability. If it lacks recognition and correction ability in fire risk in work and life, each kind of risk exists for long time and outbreak of fire probability is increased; when the fire just begins, lack in common sense of extinguishing and ability in putting out fire may cause small fire spreading to serious disaster; once people are trapped in fire, people may lack of basic ability in escaping and evacuating, escape in panic blindly, which will cause mass death and casualty and expand disaster consequence. Besides, the fire accidents in recent years also reveal some staff's poor work capacity of some special working posts, such as operator on duty of fire central control room being indifferent to fire automatic alarm signal or having no idea about how to start automatic fire-extinguishing sprinkler system, etc., which will cause fire spreading on one hand; on the other hand, it also will make function of construction fire control facility failing to be exerted [2].

83.1.2 *Fire Control Safety Situation of Managers*

Managers, especially the supervisors implementing fire control administrative enforcement have mainly revealed the following weak points disharmonious with the fire situation in management process.

83.1.2.1 Lack of Fire Control Supervision and Law Enforcement Strength

It has been one of the difficult problems perplexing fire control supervision management work and its two major reasons are: first, the active system of fire control delimits organizational structure of supervisors by posts, which restricts the quantity of the staff; second, because mass base of social fire control work is poor, mostly depends on the fire administrative monitor, the working strength of fire control supervision management work exceeds far beyond the strength of the present fire control law enforcement.

83.1.2.2 Professional Ability of Fire Control Supervision Enforcers Needs Improvement

Impacted by market economy, part of the fire control managers have active thoughts and unstable sense of responsibility, who are expecting creating better life quality, so they lack of spirit of studying profession diligently, muddle along but do not seek high quality and high efficiency in work. Besides, as fire control supervision managers of administrative enforcement, many comrades have not professional background and have poor fundamentals in science of law and incomplete concept in rule of law and law enforcement tools, which cause certain of hidden danger for fire control supervision management work.

83.1.2.3 Fire Supervision Management Lacks of Artistry

Viewing from the source property, the fire control work belongs to the public service category, namely a kind of work carried out for social public interest, including fire control supervision management, finding out hidden danger, correcting hidden danger and maintaining social fire control safety. Although the mass lack of self-awareness in fire control safety and sometimes assistance from administrative compulsory measures is needed to ensure realization of management effect, but it does not influence the property of the service; therefore, although the management takes law enforcement as precondition, artistry in process should also be noticed. Both of the linguistic organization and measures should be civilized and friendly.

83.2 Situation Analysis of "Machine" in Fire Control Safety Management

Concept of "machine" in system theory refers to machinery equipment; extending to management field, it can be understood as the fire control management method and the measure applied for fire control management instrument and equipment. But, viewing worldwide, currently the instrument and equipment applied for fire control management are relatively fewer, with main reason that most fire accidents inseparable with unsafe acts of people. Therefore, it needs laws, rules and regulations, standard regulation and internal rules of social organizations, cooperating with certain of management method and measure to restrain people's acts and realize the purpose of fire control management. But, in practice, the method and measure of fire control management reveals more and more deficiencies [3].

83.2.1 Administrative Supervision Plays Dominant Role but Less Market Regulation

Because mass' self-consciousness in fire control is insufficient, creating fire control safety environment mainly depends on administrative supervision. But in market economy period, the relevant departments should actively discuss about the market's regulation effect on fire control safety, with the help of measures of fire control safety liability insurance, organization fire control safety grade assessment, etc., to stimulate mass' idea and impetus in fire control safety self-management.

83.2.2 Excessive Management but Less Organization Participation

Currently, fire control safety of the society, organization, and citizen becomes the responsibility of the government and its functional departments, is under all-around supervision, and is corrected for any hidden danger; fast reflect situation when there is fire and rescue all-out; the government should be responsible for it and the person in charge should be investigated and affixed the responsibility when the loss organization is unable to compensate. The above excessive management measure restrains the socialized fire control development pattern of organization taking full responsibility and citizens actively participating and striking the enthusiasm of other social subjects to participate in fire control, which is disadvantageous for the development of fire control safety situation in the long run.

83.2.3 *In Process of Responsibility System Implementation, Ruling by Man Is More Important Than Ruling by Law*

Fire Prevention Law regulates the obligation and illegal liability of each subject of liability. To ensure the responsibility system implementation, strict enforcement of law and legal administrative is necessary. But by making a general survey of law enforcement environment of present fire control supervision, there are many cases departure from responsibility system implementation. For example, some law executors do not respect rules or do not handle procedure in accordance with laws, or orient human feelings other than laws; there is law enforcement intervened by many sides and difficult to realize independent supervision, etc. The entire phenomenon have influenced seriousness of law.

83.3 Situation Analysis of "Environment" in Fire Control Safety Management

83.3.1 *Lack of Social Fire Control Safety Culture Atmosphere*

Viewing from the scope of the whole society, the present fire control safety education mechanism is not sound enough. Most responsibility subjects of fire control safety education have not strong consciousness, lack of rigorous organization and argument on education content design, education form selection, education object popularization, education cycle plan, etc., which have caused invariable education content, monotonous and stereotyped education form, repeated education object, seriously restrained education effect, failure of forming fire control safety culture atmosphere in the whole society or producing subtle influence and infection. It is also the major reason that our citizen's consciousness and quality on fire control safety are improved slowly [4].

83.3.2 *Insufficient Public Fire Control Infrastructure Construction*

Public fire control infrastructure is an important component part of disaster prevention and fight systems, which must insist on the construction speed and degree corresponding with the economy development [5]. But, since public fire control infrastructure construction is a long-term engineering of "invisible" achievement officially, it has not been paid high attention, which has caused the construction degree seriously lagging behind the local economy development level. For example, fire control (water supply) water source lack, fire control communication detention and narrow fire fighting truck access, etc., have reduced efficiency of fire control rescue to a certain degree and also have influenced citizens exerting self-management ability in fire control safety.

83.4 Conclusion

In conclusion, the fire control management work is a systematic engineering and three aspects in man, machine and environment mutually restrict and promote. First, if the awareness and comprehensive quality of the managers and the managed objects of fire control safety are not improved, on one hand, the research enthusiasm and utilization degree of "machine" will be restrained; on the other hand, the construction degree of fire control safety environment will be influenced, which is demonstrated as in-cooperation with fire control safety education and damaging fire control infrastructure, etc; second, for the relevant content of "machine" in fire control management, the contradiction between the managers and the managed objects is difficult to compromise, and the environment is difficult to improve; last, the fire control management environment is the soil for the development of "people" and "machine" in the system, deciding the direction and depth of the two development. Therefore, the fire control management should insist on system thinking, orient three aspects in man, machine, and environment in the system and realize harmonious development and mutually promoting.

References

1. Shan Z (2009) Enterprise security management based on systems thinking. Beijing Petroleum Management Personnel Institute, 14–17 June 2009
2. Smit B, Wandel J (2006) Adaptation, adaptive capacity and vulnerability. Glob Environ Change 16:282–292
3. Wang J (2012) Preliminary measures to strengthen supervision and management functions of the fire. Technol Rich Wizard 20
4. Hu Y (2008) Safety culture construction of building a harmonious society. Chin Collect Econ 5:187–189
5. Shanghai Public Security Bureau (2011). Strengthen public fire protection infrastructure. China Shanghai Government Portal

Chapter 84
Study on Technology System of Human Factors Engineering (HFE)

Baozhi Wang, Guohua Jiang, Zhiqiang Tian, Junrong Li, Jialu Ma, Ning He, Wei Zhao and Wenhao Zhan

Abstract Objective: to establish the technology system of HFE in the manned spaceflight and equipment manufacture of China, evaluate TRL, TAL, and TUL of key technology, and lead the innovation and development of HFE in China. **Method:** by studying the demand of HFE,analyzing the hotspot and trend and learning the successful application of HFE, the systematic and scientific model of HFE technology system was built. **Result (Conclusion):** an explicit definition of HFE technology was given, and a tree-structure model of HFE technology system was presented. The first-order leaf nodes represented three technology directions (research directions) to which seven technology themes (research themes) in the second-order leaf nodes were attached, and the third-order leaf nodes were 31 technology contents (research contents) that belong to the technology themes, and the fourth-order leaf nodes were key technologies for solving the specific problems or research methods in the technology contents. A quantitative evaluation of TRL, TAL, and TUL was given and a general evaluation of HFE technology level was also made.

Keywords Human factors engineering (HFE) · Technology system · Principles and methods of human–computer-interaction · Human capability and reliability · Ergonomic design and evaluation · Digital virtual human · Physical–chemical regenerative life support · Controlled ecological life support · Individual protection

84.1 Introduction

84.1.1 Connotation of HFE

The problem to be solved with Human Factors Engineering, HFE, is "how to design system to reach optimum matching for human-machine-environment for people

B. Wang (✉) · G. Jiang · Z. Tian · J. Li · J. Ma · N. He · W. Zhao · W. Zhan
China Astronaut Research and Training Center, National Key Laboratory of Human Factors Engineering, mailbox:5132, Beijing 100193, China
e-mail: w_bz126@126.com

© Springer-Verlag Berlin Heidelberg 2015

S. Long and B.S. Dhillon (eds.), *Proceedings of the 15th International Conference on Man–Machine–Environment System Engineering*, Lecture Notes in Electrical Engineering 356, DOI 10.1007/978-3-662-48224-7_84

working safely, highly effectively and economically," which applies theories and technologies of fields in system theory, cybernetics, information theory, mechanics, electronics, physical chemistry, psychology, ergonomics, applied mathematics, simulation science and technology, etc., to study human characteristics of physiology, mentality, perception, cognition, etc., expose principles and rules of fields in human–machine interaction, human work competence and reliability, environment control and life support, individual protection, etc. Present human–machine interaction method, space ergonomics design and evaluation method, virtual human modeling and simulation technology, physical–chemical regenerative life support technology controlled the ecological life support technology and individual protection technology and provide scientific theory, method, and technical basis for manned spaceflight and equipment development through human–machine system design regulation formulation, human–machine interface ergonomics design, human life support and guarantee environment building, human–system relationship optimizing, etc., in order to realize optimization of human–machine environment [1, 2].

84.1.2 Present Development of HFE

HFE covers each element of human–machine environment, which evolves to optimized direction of the system at the same time when seeking optimized integrated function of human–machine environment driving HFE side branch technology innovating and developing, namely each side branch technology development not ensuring optimization of only human, machine, or one respect of environment, but also reaching optimization of overall indexes of human–machine environment system. Take America for example, HSEDD (Human Systems Engineering and Development Division) of NASA JSC has developed study and technical development on human system integration (HSI) and spaceflight human factors and habitability factor, etc., from aspects in habitability and human factors and environment factors, etc., and formulated human system integration series standard: NASA-STD3001, MIL-HDBK-46855A (HFE Process and Implementation Procedure) and MIL-HDBK-759C (HFE Design Guide Manual); human system integration research department of ARC has developed research development on human–machine interface, performance and human factors integration, etc. NHHPC has developed research development in advanced human factors and performance, habitability, and environment collaborative innovation.

For side branch technology, virtual human modeling and simulation get more attention and advancement. Institutions of NASA, DOD, and DOE has utilized virtual human dataset (VHK) for research and development, and NASA has implemented DAP (virtual human project). China has developed virtual human kinematics and bio-mechanics, etc., modeling simulation study [3] facing space station mission since 2008.

Physical–chemical regenerative life support of America and Russia has experienced development of space station and then established mature technology system.

Subsequent development trend is to further reduce supply and consumption, increase durability/service life, improve reliability, and reduce complexity. China has developed advanced study on single-key technology and completed ground test of multi-manned and multi-day closed system integration. At present, large spaceflight countries of the world have established each kind of controlled ecological life support system experimental platform and carried out system verification. The subsequent study key point is long-term multi-manned highly closed system integration technology and system design and verification facing lunar mars base. China "863" plan has started controlled ecological life support technology conception demonstration, technology research and system verification and developed multi-manned and multi-day controlled ecological system integration test study [4, 5].

At present, the extravehicular spacesuit is representative of the individual protection technology in complex environments of space orbit extravehicular activity, manned lunar-landing, lunar exploration, etc. America and Russia (the Former Soviet Union) have established relatively complete technology system and developed conception study on advanced spacesuits of Mark-III spacesuit, I-suit spacesuit, biological spacesuit, the chameleon spacesuit and Z-1 spacesuit, etc. Overall, the spacesuit technology develops to the goal of higher ergonomics function, less logistics support, lighter, and less consumption. China is currently carrying out research and development [6, 7] on the second-generation extravehicular activity spacesuit with long service life and high reliability applied for the space station.

To promote development of HFE technology of our country and make up the field difference between our country and foreign countries, through research, HFE key laboratory has presented Technology System of Chinese HFE (see Fig. 84.1).

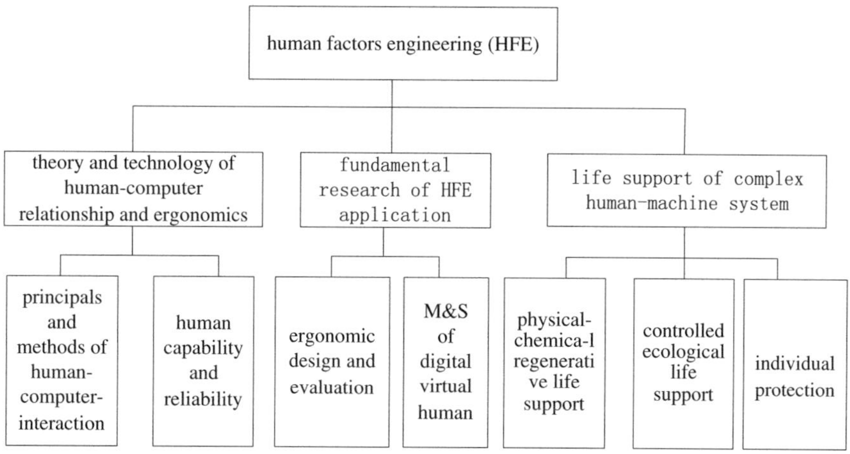

Fig. 84.1 Technology system of HFE

84.2 Method and Result

84.2.1 *Method: Top-Down Hierarchical Modeling Method*

HFE technology system establishment adopts top-down hierarchical modeling method, concentrating on core problem (namely "how to design system to reach optimum matching for human-machine environment for people working safely, highly effectively and economically") to be solved with HFE. On one hand, analyzes and presents human–machine environment factors needing optimum matching, namely human characteristics, machine characteristics, environment characteristics, human–machine relationship, human-environment relationship, machine-environment relationship and human–machine-environment system function (such as security index, efficiency index, and economy index). These characteristics, relationships, and system functions are major objects to be researched fundamentally for HFE, and that principles and rules contained, to be researched and exposed among are the basic theory category of HFE; on the other hand, analyzes and presents engineering technology of the method to ensure people working safely, highly effectively and economically, including design technology, production technology, test technology, evaluation technology, etc. Through the research work of above two respects and combine the present situation of the laboratory, 7 theories and technology research themes are comprehensively confirmed, namely the third layer leaf nodes in Fig. 84.1. Based on that, in accordance with internal connection and technical correction, 7 technical themes are integrated as 3 technical directions, namely human–machine relationship and ergonomics theory and technology, HFE application fundamental research and complex human–machine system life support technology; as 3 first-order leaf nodes of HFE tree-structure model, it makes technology system model more systematic, scientific, and balanced.

84.2.2 *Result: Leaf Nodes of Tree Structure*

84.2.2.1 Human–Machine Interaction Principle and Method

Human–machine interaction principle and method theme includes 3 technical contents of information presentation and perception, manual control characteristics and modeling, and natural interaction technology. Therein to, information presentation and perception includes key technologies of information processing mental model and modeling method, multiple-access information presentation type and work performance and information presentation type and context awareness, etc. Manual control characteristics and modeling include operation complexity grading and behavior characteristics modeling, new type controller operation characteristics research, and complex man-controlled looping characteristics and work performance, etc. Natural interaction technology includes key technologies of brain–machine interface technology theory and method, voice and brain neural signal/mode recognition, human–machine interaction mode and interaction combination, teleoperation natural voice controlling strengthening, human-intelligent robot gesture interaction, etc.

84.2.2.2 Human Capability and Reliability

Human capability and reliability theme includes 2 technical contents of work load and work performance and human reliability. Therein to, work load and work performance includes key technologies of multiple-physiological information convergent technology, physical enhancement and intervention of cognitive competence, rapid detection and early warning of work load, alertness noninvasive/rapid adjustment and perception and cognition decision characteristics modeling and simulation, etc. Human reliability theme includes key technologies of work load and reliability integration analysis, human errors mental mechanism and cognitive neural mechanism and human reliability modeling analysis and evaluation, etc.

84.2.2.3 Ergonomics Design and Evaluation

Ergonomics design and evaluation theme includes 3 technical directions of human–machine interface ergonomic design and evaluation, HFE design and verification of special task, human system integration design and evaluation. Therein to, human–machine interface ergonomic design and evaluation includes human measuring and bio-mechanics parameters test, human measuring and bio-mechanics modeling simulation, quantitative evaluation of human–machine interface availability and information collection and analysis of narrow space habitability, etc. HFE design and verification of special task include key technologies of self-adaption human-machine function division technology, working capability measuring and simulation evaluation of special work and working capability prediction of special task, etc. Human system integration design and evaluation include human-intelligent system combination operation collaborative design and human system integration efficiency evaluation, etc.

84.2.2.4 Virtual Human Modeling and Simulation (M&S)

Virtual human modeling and simulation theme includes 4 technical directions of M&S of geometric, M&S of kinematics, M&S of physiological and M&S of cognition. Therein to, M&S of geometric includes key technologies of M&S of polygon and hook face, M&S of skinned bone, M&S of fluid and M&S of particle, etc. M&S of kinematics includes key technologies of M&S based on parametric key frame, M&S based on kinematics, M&S based on dynamics and M&S based on motion capture method; M&S of physiological includes key technologies of M&S of physiological signal and M&S of biological rhythm, etc. M&S of cognition includes of M&S of key technologies of M&S of multiple-access information acquisition ability, M&S of perception data explanation ability, M&S of perception information memory ability M&S of, mentality and decision, M&S of situation awareness, M&S based on system identification and M&S based on virtual reality, etc.

84.2.2.5 Physical–Chemical Regenerative Life Support

Physical–chemical regenerative life support theme includes of 5 technical technologies of air regeneration, water regeneration, solid waste treatment, integrated environment monitoring, and in situ resource utilization. Therein to, air regeneration includes 4 key technologies of oxygen regeneration, CO_2 removal and utilization, trace harmful gas removal, and integrated air regeneration; water regeneration includes key technologies of condensation water treatment, urine waste treatment, sanitary water treatment, and convergent system water regeneration, etc. Solid waste treatment includes key technologies of solid waste volume reduction, solid waste stabilizing treatment, and solid waste resource utilization, etc. Integrated environment monitoring includes key technologies of atmospheric composition monitoring, water quality monitoring, noise monitoring and fire early warning and monitoring, etc. In situ resource utilization includes key technologies of lunar surface resource utilization, mars surface and air resource utilization, etc.

84.2.2.6 Controlled Ecological Life Support

Controlled ecological life support theme includes 4 technical contents of food production, waste water purification utilization, biomass solid waste treatment and utilization and system integration and building. Therein to, food production includes key technologies of higher plant production, phycophyta production, animal feeding and food processing, etc. Waste water purification utilization includes key technologies of microorganism purification and mineral nutrition, etc. Biomass solid waste treatment and utilization includes key technologies of biomass solid waste stabilization, biomass solid waste treatment and biomass solid waste resource utilization, etc. System integration and building includes system simulation, three generations ecological life support system convergence and system building, etc.

84.2.2.7 Individual Protection

Individual protection theme includes 6 technical contents of overall technology of individual protection, extreme environment protection, ergonomics guarantee technology, portable life support technology, individual communication technology and maneuvering control technology. Therein to, overall technology of individual protection includes key technologies of system design and simulation, integrating design and integration and system test and evaluation, etc. Extreme environment protection includes key technologies of pressure protection, thermal protection, electromagnetic protection, radiation protection, chemical protection, lunar dust protection and dust removal, etc. Ergonomics guarantee technology includes key technologies of highly movable joint technology, highly flexible gloves technology, visual protection and support technology and exoskeleton assistance technology, etc. Portable life support technology includes of 3 key technologies of regenerated air purification, oxygen supply and pressure regulating and non-consumptive thermal control; individual communication technology includes 4 key technologies of multi-sensor fusion and miniaturized technology, portable power technology, control and information display technology and communication technology; maneuvering control technology includes key technologies of independent relief technology and human–machine combination process technology.

Table 84.1 TRL, TAL&TAL

Evaluation index	Quantitating level	Qualitating level
Technology readiness level	TRL1	Clear basic principles
	TRL2	Definite technology concepts and application thoughts
	TRL3	Passed Feasible Argument about the concepts and application thoughts
	TRL4	Passed Experimental verification about the technology proposal and methods
	TRL5	Passed the typical environmental verification about functional modules
	TRL6	Passed the typical environmental verification about demonstrating model machine
	TRL7	Passed the typical environmental verification about productive model machine
	TRL8	Passed the typical environmental verification about demonstrating model machine
	TRL9	Passed the application test about production
Technology advancement level	TALl	General technology
	TAL2	Domestic progressive technology
	TAL3	Domestic advanced technology
	TAL4	International progressive technology
	TAL5	International advanced technology
Technology TAL level	TULl	Required technologies in the future
	TUL2	Required technologies within 15 years
	TUL3	Required technologies within 10 years
	TUL4	Required technologies within 5 years
	TUL5	Urgent technologies at present

Table 84.2 Evaluation of TRL, TAL and TUL of HFE

Topics of technology	TRL	TAL	TUL
Principles and methods of human-computer-interaction	TRL1–TRL4	TAL3–TAL5	TUL3–TUL5
Human capability and reliability	TRL2–TRL4	TAL3–TAL5	TUL2–TUL5
Ergonomic design and evaluation skills	TRL2–TRL5	TAL3–TAL4	TUL3–TUL5
M&S of digital virtual human	TRL1–TRL6	TAL1–TAL5	TUL1–TUL5
Physical–chemical regenerative life support	TRL2–TRL5	TAL3–TAL4	TUL1–TUL5
Controlled ecological life support	TRL1–TRL5	TAL1–TAL2	TUL1–TUL4
Individual protection	TRL2–TRL6	TAL3–TAL4	TUL3–TUL5

84.2.3 Result: Technical Competency Level Evaluation

Technical competency level shall be evaluated quantitatively in respects of TRL (Technology Readiness Level), TAL (Technology Advancement Level) and TUL (Technology Urgency Level), please see Table 84.1.

In accordance with standard of Table 84.1, see Table 84.2 for overall technology level of HFE technology.

References

1. Long S, Huang D et al (2004) Foundation of theory and application of man-machine-environment system engineering. Sci Press 3–12:235–248
2. Jiang G, Wang B et al (2007) The application of virtual reality technology in aerospace and space. National Defence Industry Press, Beijing, pp 13–45
3. Liu Y, Jiang G et al (2004) The application of virtual reality technology in spatial orientation. In: Space Station, 3rd manned spaceflight academy conference
4. Guo S, Dong W, Yang C (2013) Development of space controlled ecological life support technology. Space Med Med Eng 2(3):259–264
5. Hurlbert K, Bagdigian B, Carroll C, Jeevarajan A, Kliss M, Singh B (2010) DRAFT human health, life support and habitation systems, technology area 06, National Aeronautics and Space Administration
6. Li T (2008) Development of "FeiTian" EVA spacesuit. Manned Space Flight 4:8–18
7. Guan C, Li M, Chen J (2010). Advances in the American/Russian lunar extravehicular space suit. Manned Space Flight 2:45–51

Chapter 85
Research Progress on Man–Machine–Environment System Design of Chinese Space Kitchen

Jie Li, Jingtao Zhu, Wei Zhao, Bin Chen, Yanbo Yu and Miao Yuan

Abstract During space flights, repast and preparation will supply chances and sites to assemble and decompress for astronauts. The repast equipment, layout, environment of kitchen, and repast procedure will have effect on astronauts' nutrition and health. In this paper, based on the analysis of man–machine–environment system design of the foreign space kitchen, corresponding Chinese kitchen engineering restricts, the thinking of "Integrated Space Kitchen" is proposed. Integration function of storage, preparation, repast, clean and garbage management of the kitchen is realized through design of region merging and equipments integration. This makes the astronauts repast in a convenient, clean, and beautiful environment. The man–machine–environment system design of kitchen's location, size, layout, equipment, hygiene, environment and repast procedure was also analyzed. Finally, research progresses on layout, putting style to table, minimum repast area, and local smell control in repast area were discussed.

Keywords Space kitchen (SK) · Man–machine–environment system design

85.1 Introduction

With the extension of the manned flight time, repast and preparation before the meal will supply the chances and sites to assemble and decompress for astronauts' intense and busy work. So the repast equipment, repast environment, and corresponding repast process will have direct effect on astronauts' appetite and food-intake, consequently affecting astronauts' nutrition and health. For this reason, it is necessary to establish the space kitchen with Chinese characteristics, have a sufficient consideration of their own characters of Chinese astronauts, and conduct the design for the integration and humanization, which should ensure astronauts to

J. Li (✉) · J. Zhu · W. Zhao · B. Chen · Y. Yu · M. Yuan
China Astronaut Scientific Research Train Center, Beijing 100094, China
e-mail: shibazi_carol@163.com

© Springer-Verlag Berlin Heidelberg 2015

S. Long and B.S. Dhillon (eds.), *Proceedings of the 15th International Conference on Man–Machine–Environment System Engineering*, Lecture Notes in Electrical Engineering 356, DOI 10.1007/978-3-662-48224-7_85

repast in a convenient, clean, and beautiful environment with complete functions, ensure astronauts to be in a convenient, relaxed, and pleasant state during the preparation and repast and contribute to relieve the astronauts' pressure in the face of long-term work and living in orbit with good condition.

NASA has issued a series of standards for man-machine factor studies of the space kitchen, including NASA-STD-3000 Man-System Integration Standards [1] and NASA-STD-3001(the supplement and upgrade of NASA-STD-3000). These standards have discussed the relationship between the kitchen and repast equipment and the astronauts, and the research contents include the location, size, function, equipment, and layout of the kitchen. According to engineering practice of our spacecraft and our astronauts, this article puts forward the guiding ideology and principles of our space kitchen design and states relevant content of man-machine design, which contributes to promote our research and engineering development of man-machine design for our space kitchen.

85.2 Research on Man–Machine–Environment Design of Foreign Space Kitchen

The dining area of the Skylab occupies a large space, which is 2.29 m long, 2.44 m wide and 1.98 m tall with total volume of 11.1 m^3. The dining area is 5.6 m^2 and astronauts can sit besides the collapsible table together with repast and chatting. The heater designs the groove which has the same shape as the heating food, concurrently heats the food for three people and also is used as the dining table or plates. The foot limiter can make astronauts around the table and "sit" down to have dinner. It is equipped with the freezer and the cold closet [2]. See in Fig. 85.1.

During the early stage of space shuttle in America, food preparation was conducted in the pantry in the center of the orbital module. The matched kitchen equipment mainly includes forced convection furnace, water distributor, plate,

Fig. 85.1 Kitchen in Skylab

table-ware, and the vacuum cleaner [2]. The forced convection furnace needs 20–30 min to heat food of four people; the front side of the heater sets the food binding table, which has temporary binding of food; the water distributor is mainly used for food rehydration and drinking water for the astronauts; the tablewares include spoons, forks, knifes, scissors, and chopsticks; the vacuum cleaner completes the cleaning of the cockpit; food waste is in the garbage room. During the later period of the mission, the kitchen equipment is relatively perfect and allows seven astronauts to have dinner at the same time due to advantages such as large space and sufficient power supply. It also equips the dining tables and larders, the temperature of the oven maintains between 71 and 77 °C which can concurrently heat 14 rehydration food packages and drinks, and the temperature of hot water supplied by the water distributor is between 68 and 74 °C [2].

Russian food is mainly hard canned food, so the heating equipment equipped by Russia in the International Space Station (ISS) is the oven which can heat the hard canned food. The oven is integrated with the dining table as a whole and the dining table can partly fold and unfold, which can meet the dining requirement of eight people at the same time; residue collection adopts the concentrating air-exhaust through cockpit bulkhead. The water distributor is set in the cabinet to supply the room temperature water and hot water and the pantry can store 14 days' food (Fig. 85.2). The Chinese American astronaut Edward Tsang Lu thinks that the dining table has a function as a social contact center and charting space in the ISS and what impressed him the most is to mix different foods to match out a new thing. The water fountain is beside the dining table and has the tap to supply warm water and hot water [2].

Through the above analysis, it is found that research on the man–machine–environment design of the space dinner equipment in America and Russia is systematical and deep and it also a processes gradual development and update and upgrade. America and Russia conducted an integrated design of dining equipments and formed a more perfect kitchen system and dining area. The kitchen design has close connection with the operation scale of the spacecraft, food species, and

Fig. 85.2 Russian kitchen in ISS

package as well as astronauts' eating habits. The kitchen mainly supplies such dining equipment to heat food and drinks, store food, provide dining tables and tableware, and waste disposal.

85.3 Design Principles and Constraints

85.3.1 Design Principles

Because of the limitations of space, the concept of space kitchen design normally means that cabinets, kitchen equipment, and dining equipment have a proper layout and clever collocations according to their shapes, sizes, and operating requirements, in order to realize the space integration and function integration of the dining area [3]. In accordance with such factors as astronauts' height, color preference, cultural accomplishment and processing habits, kitchen space structure and lighting, it is combined with principles of ergonomics and human engineering to conduct scientific and reasonable design, integrates functions like the storage, clean, processing, freeze and refrigeration, and dining as a whole and pays attention to the overall style, layout, and functions of the kitchen.

In addition, the space kitchen design should consider relieving labor intensity of the astronauts and convenient usage. The space kitchen should ensure three enough space [4]. First, it should have enough operating space. It should have the turnover place to put tableware and food, which is basic operating space to reduce the astronauts' labor intensity and avoid repeated action of taking. Second, it should have rich storage space. It generally adopts combined cabinets which can utilize the space to store food and goods in a reasonable and combined way. The well-designed combined cabinets will make it convenient to store and take things. Third, it should have sufficient activity space. It is convenient for astronauts to move, bend down, face about, and take. Meanwhile, the layout of the kitchen should be designed according to the operation progress of storage and taking, cleaning and processing, preparation and repast, in order to reduce astronauts' alternating motion and repeated taken actions.

85.3.2 Design Constraints

The main constraints of Chinese space kitchen design include: (a) the scale and configuration of the spacecraft. It should conduct a scientific and reasonable layout to create a comfortable and convenient repast environment and layout plan in a limited space, which can ensure astronauts to have enough operating space and activity space; (b) flight time. The long-term flight time has high requirement for hygiene in the space kitchen, which should provide relevant cleaning methods, equipment, and materials; (c) the number of astronauts. It should fully consider the number of dining people and people who prepare food and then establish enough dining space and preparation space; (d) repast state. During the repast and preparation, astronauts mainly use neutral postures to operate and it should consider parameters such as the activity space of the astronauts and reach range of both hands; (e) astronauts' eating habits. For example, Chinese astronauts like hot food and drinks, so it should conduct the design for equipment to heat food and drinks; (f) space food and package. Different foods and various packaging materials and sizes bring about the corresponding problems which are food preparation and repast preparation, matching between the equipment and food and package as well as the humanization design of the equipment itself; (g) repast progress and time. It should have a reasonable progress planning, including time plan, action plan, and path plan, in order to ensure to finish repast with a little time.

85.4 Design Contents

Based on the analysis of foreign research and reality of spacecraft in our country, it puts forward the design concept to establish Chinese "Integrated Space Kitchen". The integrated design makes repast area and equipment to conduct a scientific and reasonable layout and concentration, in order to realize the integration of kitchen functions (storage, preparation, repast, clean, and garbage disposal). The space kitchen has the following advantages: (a) resource integration. It concentrates and integrates the repast equipment and repast environment to save space, have a reasonable repast progress and effectively shorten the repast time of the whole; (b) it creates a common repast and preparation conditions for astronauts and contributes to increase communications between astronauts and relieve their mental stress; (c) it will improve astronauts' active participation and make it convenient for astronauts to cook food with different tastes and flavors, which can effectively improve their appetites; (d) health and clean make astronauts repast in a clean environment and also make it convenient for cleaning and garbage disposal after repast; (e) multi-functional integration makes the space kitchen as a platform of temporary meeting, universe nutrition guide, and universe popular science, which is worthy of the name "multi-functional hall".

Man–machine–environment of Chinese space kitchen mainly considers seven aspects: location, size, space layout, kitchen equipment, hygiene problems, environmental decoration, and repast progress.

When conducting the design of kitchen location, it should prevent microbial pollution to the kitchen in addition to avoiding smell permeation, so the kitchen must be isolated with garbage area and other hygiene areas and should not be polluted and affected by the garbage and other hygiene areas.

Common repast can promote the communications between astronauts. The basic size of the repast area should ensure that several astronauts can repast together and also consider the astronauts' passage situation during the repast. This index can take an overall consideration combined with the size and layout of the repast equipment.

The kitchen has four function areas which are food storage, food processing, repast, and garbage disposal to highlight the center place and framework function of the dining table and food storage cabinets. When considering the actual working conditions, it plans to adopt single-wall layout or double-corridor layout to try to concentrate the functions of storage, processing, and repast and to ensure enough operating and activity space. In addition, it should take full consideration of the ventilation and heat dissipation of the equipment, power supplies, and other factors, for example, the air inlet and outlet design of food heating devices and its location. The important equipment with frequent operation, for example, food heating devices and water distributor, should be put in a place which is easy for astronauts to reach and ensure to operate and maintain easily. It should consider the classification and sub-package design for all kinds of food and make it convenient to find and take food.

It should provide heating and refrigeration function to food and drinks to enrich the taste of food and improve the acceptability of some food and drinks, so the food heating devices and refrigeration container are the necessary kitchenware in the space kitchen. Currently, many researches are conducting such compound heating technology, temperature control technology, heating uniformity technology, and heat dissipation technology. It should provide the function to cook special tasty food and recipes. For example, the astronaut can use his iPad and other equipment to order and cater.

People have a different requirement of table height when doing different things. The height is between 850 and 900 mm when preparing food, the height used to be 750 mm when dining and higher height is preferred when it is the bar counter. So it is an ideal choice to design the dining table as an "adjustable table" [3].

Food storage cabinets are best designed as push-pull drawers. The pull-out drawer can make inside food completely show in view which is clear at a glance and has good visual sense. It should avoid that gate cabinet needs to be bent down or squat or find food in a cabinet with inadequate light.

The clean and clear capability of the kitchen is a key factor of habitability, including the capabilities to capture and collect food chippings and spill-out liquid, clean tableware, dining table and food cabinet, and dispose garbage. It should dispose more garbage and have a big smell in the long-term flight, which should be in a concealed place. It should set more garbage cans in necessary places to be convenient for garbage collection.

The shining direction, type, and level of lighting should be adjustable when considering the whole layout of the kitchen. Especially during food preparation, it should ensure that the astronaut will find spatters in time and deal with them immediately. When conducting aesthetic design of the environmental decoration around the dining table, it should take full consideration of factors such as visual sense, hearing, smell, taste, and touch, provide the aesthetic design of vertical kitchen environment, bring warm and comfortable feeling to astronauts, and increase the attractiveness of food, in order to increase the appetite of the astronauts [5].

The lights in the kitchen can be divided into two levels: lighting for the whole kitchen and lighting for cleaning, preparing, and operating. The latter generally arranges the part lights and sets convenient switch devices. The main color in the kitchen should be milky white, white, and off-white with a bright, clean, and refreshing feeling matched with some warm and clean cool colors in favor of promoting appetites.

During the whole repast and preparation, the astronauts should finish a series of work and actions such as the choice of recipe and food, food preparation, repast, cleaning after repast, and garbage disposal. The design for sequence, parallel, or serial mode of these actions will directly affect the length of the repast time and the convenience, comfort, and satisfaction of the astronauts when dining.

85.5 Progress and Perspectives

Various layout plans designed by the author have taken comprehensive consideration of such factors as heating and heat dissipation of refrigeration, size matching, operation frequency of astronauts and importance, maintenance, and reachability of the equipment. The research method is to use function diagram and plan of the kitchen and adopt imaginary lines to simulate behavior trace of the astronauts and virtual simulation, in order to research and analysis the rationality and convenience of various layout plans. Next it can use the man-machine design ways and means to establish a physical experiment platform and virtual experiment platform and conduct participation test of subjects and conduct the tests and verification of the rationality for layout and environment design according to such indexes as the whole design of the space kitchen and mental feeling of subjects.

Sitting ways of the astronauts provided by the putting ways, sizes, and shapes of the dining tables are worthy to research. Currently, various putting ways of dining tables such as lengthwise, lateral, and abreast putting ways and their corresponding sitting ways of the astronauts are trilateral sitting-around, bilateral face-face sitting, and unilateral side-by-side sitting. Lengthwise putting dining table is designed as two folding and astronauts can be trilateral sitting-around and bilateral face–face sitting; lateral putting dining table is designed as two folding and the sitting ways of astronauts is trilateral sitting-around; abreast putting way is putting several dining tables side by side and the sitting ways of astronauts is unilateral side-by-side sitting.

The putting ways of the dining tables and sitting ways of the astronauts will affect astronauts' repast mentality and atmosphere and also affect the minimal area requirement of the repast area. Now, comparative study of three ways is being conducted and the minimal requirement of the dining table area is respectively 0.6, 0.8, and 0.6 m^2. When not considering the trafficability, the minimal requirement is, respectively, 2.4, 2.7 and 2.3 m^2 and it is, respectively, 2.9, 3.2 and 3.1 m^2 when considering the trafficability.

Clearing up and cleaning in the kitchen is the key problem of habitability and the long-term manned flight has a high requirement for hygiene of the kitchen. The partial smell in the dining area will have more effect on the appetite of the astronauts. In addition to consider the measures to isolate the dining area and garbage area and other hygiene fields, other technical measures are also considered to increase air current recirculation and ventilation in the dining area.

References

1. NASA -STD-3000 Man-system Integration Standard (1995)
2. Space Food Challenge. http://www.Ag.iastate.edu/center/ftcsc/2008.2.28
3. Ma X (2001) Key issues in the design and planning of central kitchens. Logist Technol 30 (9):88–90
4. Ding Y (2011) Study on integrated design for pipeline and equipment space of kitchen in aggregated housing. Arch J 8:72–75
5. Wang K (2012) On humanized manifestation in kitchen design. Shanxi Arch 38(33):14–15

MIX
Papier aus verantwortungsvollen Quellen
Paper from responsible sources
FSC® C105338

If you have any concerns about our products,
you can contact us on
ProductSafety@springernature.com

In case Publisher is established outside the EU,
the EU authorized representative is:
Springer Nature Customer Service Center GmbH
Europaplatz 3, 69115 Heidelberg, Germany

Printed by Libri Plureos GmbH
in Hamburg, Germany